Urban Energy Systems for Low-Carbon Cities

Urban Energy Systems for Low-Carbon Cities

Edited by

Ursula Eicker

ELSEVIER

ACADEMIC PRESS

An imprint of Elsevier

Academic Press is an imprint of Elsevier
125 London Wall, London EC2Y 5AS, United Kingdom
525 B Street, Suite 1650, San Diego, CA 92101, United States
50 Hampshire Street, 5th Floor, Cambridge, MA 02139, United States
The Boulevard, Langford Lane, Kidlington, Oxford OX5 1GB, United Kingdom

Notices
Knowledge and best practice in this field are constantly changing. As new research and experience broaden our understanding, changes in research methods, professional practices, or medical treatment may become necessary.

Practitioners and researchers must always rely on their own experience and knowledge in evaluating and using any information, methods, compounds, or experiments described herein. In using such information or methods they should be mindful of their own safety and the safety of others, including parties for whom they have a professional responsibility.

To the fullest extent of the law, neither the Publisher nor the authors, contributors, or editors, assume any liability for any injury and/or damage to persons or property as a matter of products liability, negligence or otherwise, or from any use or operation of any methods, products, instructions, or ideas contained in the material herein.

Library of Congress Cataloging-in-Publication Data
A catalog record for this book is available from the Library of Congress

British Library Cataloguing-in-Publication Data
A catalogue record for this book is available from the British Library

ISBN: 978-0-12-811553-4

For information on all Academic Press publications
visit our website at https://www.elsevier.com/books-and-journals

Publisher: Joe Hayton
Acquisition Editor: Lisa Reading
Editorial Project Manager: Mariana L. Kuhl
Production Project Manager: Kamesh Ramajogi
Cover Designer: Matthew Limbert

Typeset by SPi Global, India

Contents

List of contributors **ix**

Introduction: The challenges of the urban energy transition **1**
Ursula Eicker

Part One Methodology, mathematical background, and tools **17**

1 Urban energy system planning: Overview and main challenges **19**
Sébastien Cajot, Nils Schüler
1.1 Introduction 19
1.2 A review of urban planning processes and supporting tools 20
1.3 Main challenges in urban energy planning 29
1.4 Decision support for urban energy system planning 37
1.5 Conclusion 43
References 44
Further reading 49

2 A planning support system using interactive optimization **51**
Nils Schüler, Sébastien Cajot
2.1 Motivation for interactive optimization and overview of URB^{io} 51
2.2 An optimization model for the planning of new neighborhoods 55
2.3 Multi-parametric programming for urban planning 71
Acknowledgments 73
Nomenclature 73
Acronyms 74
Set members 74
Subscripts 74
Sets 75
References 76

3 Building energy demand modeling: from individual buildings to urban scale **79**
Solène Goy, Giovanni Tardioli, Pilar Monsalvete Álvarez de Uribarri, Ana Sancho, Georgios Kazas
3.1 Introduction 79
3.2 Data challenges of large-scale building energy modeling 81

3.3 Clustering methods for representative building identification in urban contexts 81
3.4 Physical dynamic building models 92
3.5 Reduced-order RC models 98
3.6 Electrical appliance modeling 105
3.7 A model based on reference buildings and parametric analysis to generate district energy demand profiles 116
3.8 A data-driven modeling approach for large-scale demand profiling of buildings of the same typology 122
Acknowledgments 131
References 131

4 Load management in buildings 137
Solène Goy, Ana Sancho-Tomás
4.1 Load management 137
4.2 Demand response of electrical appliances using multiagent simulation 144
4.3 Building load management with MPC 167
4.4 Conclusion 173
Acknowledgments 174
References 174

5 Multimodal networks and decentralized renewable generation: Network modeling and energy/exergy performance evaluation 181
Mario Potente Prieto, Nina Solomakhina, Pilar Monsalvete Álvarez de Uribarri
5.1 Urban energy systems for heating and cooling 181
5.2 District heating and cooling networks 195
5.3 Thermo-hydraulic network models and tools 203
5.4 Energy and exergy balances to evaluate performance 206
5.5 System model thermal networks 214
5.6 Multimodal networks 218
5.7 Overview on utility networks models 219
5.8 Modeling of city infrastructure networks 220
5.9 Analyzing the network 225
5.10 Application and use cases 230
References 237
Further reading 239

6 Open data-modeling standards for city-wide energy simulations 241
Giorgio Agugiaro
6.1 Introduction 241
6.2 CityGML 244
6.3 Energy ADE 247
6.4 Utility Networks ADE 249
6.5 Existing tools 251
6.6 Conclusions 253

References 254
Web resources 255
Further reading 255

7 Uncertainty assessment of building performance simulation: An insight into suitability of methods and their applications 257
Parag Wate, Volker Coors, Marco Iglesias, Darren Robinson
7.1 Introduction 257
7.2 Literature review 259
7.3 Methods 261
7.4 Case study designs 266
7.5 Application of the uncertainty analysis framework 268
7.6 Results and discussion 278
7.7 Conclusions and future work 284
Acknowledgments 284
References 284

8 Model integration: A SOA approach 289
Preston Rodrigues
8.1 Introduction 289
8.2 Distributed urban energy infrastructure 291
8.3 Concept and use case 293
8.4 Implementation 294
8.5 Conclusion 297
References 299

Part Two Case Studies and Applications 301

9 The case study cities of Geneva and Vienna 303
Jean-Loup Robineau, Giorgio Agugiaro
9.1 Introduction 303
9.2 Geneva 303
9.3 Vienna 316
9.4 Conclusions 330
References 330
Further reading 331

10 Applying modeling and optimization tools to existing city quarters 333
Mario Potente Prieto, Pilar Monsalvete Álvarez de Uribarri, Giovanni Tardioli
10.1 Energy conversion and district heating systems in Vienna 333
10.2 Actual situation in the urban district in a block of buildings within the urban district of Vienna Meidling: District heating network typologies and reference system 341

10.3 Scenarios for optimizing the energy system in a block of buildings 350
10.4 Results and discussion 354
Annex 375
10.5 Energy performance of decentralized solar thermal feed-in to a Vienna Case Study District Heating Network 375
10.6 Nondeterministic calibration of district scale building energy models 388
References 411
Further reading 413

11 Application of the planning support system URB^{io} 415
Nils Schüler, Sébastien Cajot
11.1 Description of the case studies in Geneva 415
11.2 Common input data for both case studies 420
11.3 Interactive optimization workflow 425
11.4 Demonstration of the developed planning support system URB^{io} 427
Nomenclature 442
Acknowledgments 442
References 442

Conclusions and outlook 447
Index 453

List of contributors

Giorgio Agugiaro Smart and Resilient Cities Unit, Center for Energy, AIT—Austrian Institute of Technology GmbH, Vienna, Austria

Pilar Monsalvete Álvarez de Uribarri Sustainable Energy Research Center (zafh.net), University of Applied Sciences, Stuttgart, Germany

Sébastien Cajot European Institute for Energy Research, Karlsruhe, Germany; Industrial Process and Energy Systems Engineering Group, École Polytechnique Fédérale de Lausanne, Lausanne, Switzerland

Volker Coors Center for Geodesy and Applied Informatics, Stuttgart University of Applied Sciences, Stuttgart, Germany

Solène Goy School of Mechanical and Materials Engineering, University College Dublin, Dublin, Ireland

Marco Iglesias School of Mathematical Sciences, The University of Nottingham, Nottingham, United Kingdom

Mario Potente Prieto Technical University of Vienna (TU WIEN), Wien, Austria

Jean-Loup Robineau Industrial Processes and Energy Systems Engineering Group, Ecole Polytechnique Fédérale de Lausanne, Rue de l'Industrie, Sion, Switzerland

Darren Robinson Sheffield School of Architecture, University of Sheffield, Sheffield, United Kingdom

Preston Rodrigues Hochschule für Technik Stuttgart, Stuttgart University of Applied Sciences Schellingstr, Stuttgart, Germany

Ana Sancho-Tomás Faculty of Engineering, Department of Architecture and Built Environment, University of Nottingham, Nottingham, United Kingdom

Nils Schüler Industrial Process and Energy Systems Engineering Group, École Polytechnique Fédérale de Lausanne, Lausanne, Switzerland

Nina Solomakhina Siemens Munich, München, Germany

Giovanni Tardioli Integrated Environmental Solutions Ltd., R&D; School of Mechanical and Material Engineering, University College Dublin, Dublin, Ireland

Parag Wate Center for Geodesy and Applied Informatics, Stuttgart University of Applied Sciences, Stuttgart, Germany; Department of Architecture and Built Environment, The University of Nottingham, Nottingham, United Kingdom

Introduction: The challenges of the urban energy transition

Ursula Eicker
Stuttgart University of Applied Sciences, Stuttgart, Germany

I.1 Cities, climate change and the energy transition

Accountable for more than 70% of CO_2 emissions, the constantly growing urban metropolises are the key actors of climate change. Three percent of the Earth's land surface is urbanized, and more than half of the world's population lives in urban areas, where 80% of the world's gross domestic product is generated (UN Habitat, 2016). Assuring sufficient food, water, energy, shelter, human health, and safety, while managing a growing population under the challenges of rapid social, economic, and environmental change, will be the hallmark of the 21st century. Given their know-how, wealth, and willingness to act, cities are crucial players to find solutions for a more sustainable future.

Cities face major challenges due to climate change, air pollution, population growth, and other factors. With the worldwide population still growing, many studies still predict that fossil fuel consumption will rise until 2050 despite global agreements on CO_2 mitigation (IEA, EIA, BP, Exxon, MIT, and IEEJ world or international energy outlooks between 2015 and 2017). The average of all six projections shows increases of 23% in world oil consumption, 53% in gas, and 10% in coal over different periods of estimation for 2035, 2040, or 2050. However, to limit global warming to 2°C, fossil-fuel-related emissions would have to start decreasing now, falling by about 25% by 2040 and thereafter to near zero in 2100—a very ambitious target requiring all our efforts in urban areas and beyond.

The main impacts of climate change on cities are infrastructure damages from extreme weather and sea-level rise, effects on health arising from higher temperatures, and water resource availability. They will be briefly discussed before moving to the focus of this book, which deals with mitigation strategies for the urban sector to reduce CO_2 emissions and show pathways to a zero-carbon balance.

According to new data from the Global Rural-Urban Mapping Project (CIESIN, 2011), in coastal environments, 10% of land area is urbanized with a population share of 65%. Worldwide, about 146 million people live only 1 m above sea level and 397 million at 10 m or less (Anthoff et al., 2006). These areas are especially affected by sea-level rises, which are projected between 0.5 and 1.5 m until 2100. Higher rises are possible if, for example, the Greenland or West Antarctica Ice shelf melts: if temperatures around Greenland increase by 3°C compared with preindustrial level, this could

Urban Energy Systems for Low-Carbon Cities. https://doi.org/10.1016/B978-0-12-811553-4.09993-5

cause a 7 m sea-level increase (Mimura, 2013). A 1 m rise would strongly affect many coastal megacities like Rio de Janeiro, New York, Mumbai, Dhaka, Tokyo, Lagos, and Cairo. Resiliency strategies are also needed for natural disasters such as storms, floods, or intense rain, which increased in frequency and intensity with 4000 events between 2003 and 2012 as compared with 80 between 1901 and 1910 (UNDP, 2014).

Another global consequence of climate change is the warming of cities. This was investigated for a range of urbanization scenarios combined with IPCC climate models based on different levels of four greenhouse gas concentration trajectories, the so-called Representative Concentration Pathways (RCPs). The least change in global climate is defined by the RCP2.6 (with 0.9–2.3°C temperature rise and 430–480 ppm CO_2 equivalent); the largest change is outlined in RCP8.5 with over 1000 ppm CO_2 equivalent (with 3.2–5.4°C global surface temperature rise). For the RCP8.5 scenario, monthly mean temperatures from different climate models were overlaid with the predicted urbanized land use in 2100. The results show that 1.5 billion people will live in cities with mean temperatures above 35°C for three consecutive months—compared with 5.6 million today (Marcotullio, 2015). Especially developing-nation local governments will be faced with crisis-level situations. In the face of deteriorating conditions, populations will further move into other urban centers, with ramifying challenges to services and governance. Research and research-based planning are important to anticipate and address these challenges.

While today's cities are taking measures to adapt to the climate-change-related problems such as sea-level rise or urban heat islands, cities also take leadership in climate change mitigation. Mitigation in urban areas means high building energy efficiency standards and high levels of local renewable supply. Such strategies are required to fulfill the ambitious climate protection goals of the Paris 2015 agreement. A major contribution has to come from the existing building stock that represents a major urban energy consumer. In this context, city municipalities, energy suppliers, housing companies, and private owners must be mobilized and bonded around a common low-carbon urban energy strategy. Informed decision-making is an essential prerequisite to find optimum solutions, and the development of tools and methods for such decision support is the main focus of this book.

The Paris agreement is characterized by a bottom-up approach to global cooperation, where each country delivers national inventories of greenhouse gases (NIR) and prepares nationally determined contributions (NDC), which have to be adjusted and strengthened every 5 years (Doelle, 2016). To overcome the very limited participation of countries in the Kyoto protocol (only 33 countries during the first period 2005–12 and major countries did not ratify the protocol afterward), a more flexible approach was chosen in the Paris agreement to allow each country to outline national efforts toward low emissions. This resulted in a high participation of countries, covering 85% of global greenhouse gas emissions. A major problem is that the current pledges for emission reduction will only limit projected global warming to 2.7°C. Instead of binding legal mechanisms for parties to comply with the targets, a managerial approach is taken with a clear articulation of an ambitious collective goal to limit temperature rise to 1.5°C and with transparency and communication mechanisms and flexibility to adjust to changing circumstances (Fig. I.1).

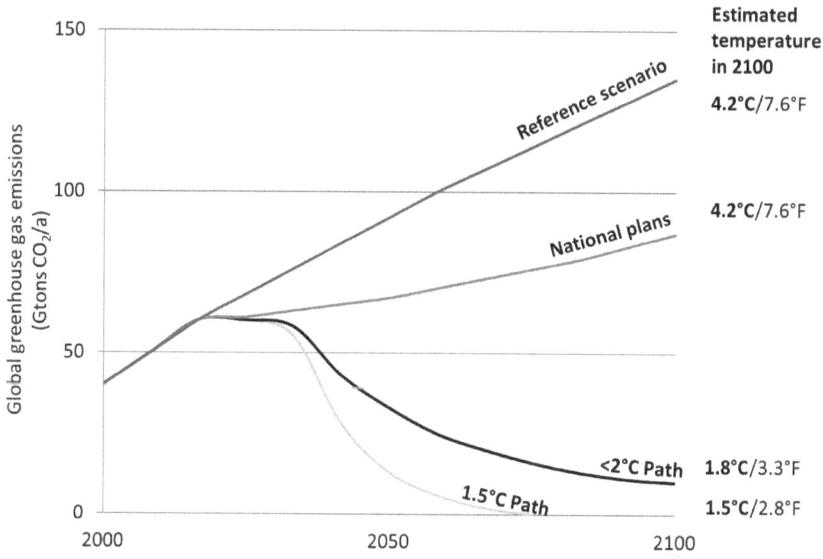

Fig. I.1 Greenhouse gas emissions and estimated temperature increases until the year 2100 for the business as usual reference scenario, the current pledges and national plans, and the more ambitious 1.5 or 2°C paths (climatescoreboard.org).

Parallel to the multilateral international treatises, a whole range of bottom-up voluntary, managerial, and market-based approaches have been adopted on a local, regional, and subnational level. Cities have become increasingly significant actors in international affairs and global governance (Acuto, 2013). They participate in multilevel climate governance networks, the most important among those being the C40 Cities Climate Leadership Group of powerful global megacities (Davidson and Gleeson, 2015); the CCP Cities in the Climate Protection Program of the International Council for Local Environmental Initiatives (ICLEI) with over 1000 cities in 30 countries, representing 15% of global carbon emissions (Hoffmann, 2011); and United Cities and Local Governments (UCLG). These three networks launched the Global Mayors Compact (GMC) at the United Nations Climate Summit 2014, adding to the prominence of urban climate governance and emphasizing the shift from municipal voluntarism to a more strategic urbanism. On a city level, the environmental goals are already often comparable: New York City developed the program "One City: Built to Last" in 2014; the editor's home city of Stuttgart politically adopted its energy concept in January 2016. Both programs have ambitious goals of reducing CO_2 emissions by 80% until 2050 and contain specific measures for energy savings in public buildings, where the cities have direct control of refurbishment actions.

Many cities have adopted sustainable building retrofit programs to strategically address the challenge of refurbishing large numbers of individual buildings. Given the low rates of building refurbishment and the dominance of the fossil fuel sector, it is increasingly obvious that only coordinated strategies in urban areas that interrelate the different sectors with their stakeholders can lead to the desired result of significant carbon reduction. Although many municipalities have decided on ambitious goals to become climate neutral within the next decades, limited experience is available of how demand can be significantly reduced on a city scale and what contributions local renewable energy sources can bring to cities depending on their density. A global framework for coordinated energy research and policy could remove barriers and make sure that energy efficiency and sustainability become key aspects of urban transformation strategies.

A common feature of energy-related decision-making is the dominance of short-term profits without adequately considering life-cycle costs for efficiency measures in the urban building stock and the external costs of today's predominantly fossil fuel supply. Without a more global approach and framework to tackle the energy question, the numerous local initiatives for energy efficiency improvement and renewable integration will not gain the momentum required for climate change mitigation.

I.2 Decision making and urban energy data

The urban energy transition involves vision building and an urban strategy, concrete interventions and physical implementation of measures, and political and legislative processes to facilitate large-scale replication. The transition from the first ideas and visions to a real change easily takes 10 years or more and involves many stakeholders. Visionary and institutional milestones usually precede physical milestones. In urban planning processes, there are multiple domains and actors involved, resulting in a range of parameters other than energy to consider in decision-making, including social and environmental aspects, quality of life, and land use. Informed decision-making is becoming more and more important, as there is a path dependency of the selected technical strategies and the resulting energy infrastructure. However, there are many technical, institutional, and social lock-ins hindering a rapid transition. On an institutional level, visionary leaders and political agents are crucial to promote the transition. On an individual level, decisions on building energy efficiency, for example, a refurbishment strategy, are dominated by expert recommendation such as the architect, followed by criteria such as age and knowledge of the citizen taking the decision.

Informed decision-making requires an understanding of inherent joint features and variability of urban energy systems, their predictability, and the human dimensions. We expect reciprocal benefits when considering local-to-global scales. Thus, a global viewpoint offers context to local-scale phenomena and, in many cases, defines what happens over the smallest of domains (e.g., climate, urban energy, and thermal comfort under global scenarios of climate change). In turn, the local scale offers critical ground-truthing opportunities for global concepts.

An integrative perspective combining qualitative and quantitative research approaches is relevant for an understanding of the energy transition and the dynamics within coupled social and technical systems. There is a need to develop a framework to make case studies comparable, identify indicators, and study feedback between energy systems and their users.

Certification schemes have recently extended to the urban level, mainly to improve awareness and publicity for the developers. Here, simple assessment rules prevail to evaluate a city quarter performance. In existing assessment tools for urban communities such as the Japanese system CASBEE for Urban Development, the UK certification scheme BREEAM Communities, and the US rating scheme LEED for Neighborhood Development, infrastructure aspects dominate the criteria with 24% (BREEAM), 32% (LEED), or 45% (CASBEE) weight, while resources and energy only amount to 14%–18%. This shows the low impact of energy efficiency in the evaluation process.

Decision-making systems support the specification of certain targets, for example, change in transport policies on energy, finances, and pollutants. The follow-up on the implementation of targets is unfortunately often neglected in the process. Monitoring of progress associated with energy auditing concepts on a city scale is necessary to assess the suitability and effectiveness of the decisions in the planning process. As an example in the United Kingdom, the Councils for Climate Protection (CCP) involved 24 municipalities as pilots in the year 2000, which set targets to reduce the emissions on their area by more than 20%. However, monitoring of progress was nearly impossible, and benchmarks were not available, so achieved savings were often negligible compared with the city's total consumption (Fleming and Webber, 2004).

It is thus essential that urban data are made available to plan and monitor the energy transition. Urban data are heterogeneous and often are characterized by data inconsistencies, a lack of quality control, and multijurisdictional coordination challenges (business, NGOs, municipal stakeholders, etc.). Decision-making is difficult as many stakeholders are involved, data are fragmented, and there are no standardized interfaces between different modeling environments. At present, a disconnect exists between the priorities of the multiple stakeholders across multiple scales (governmental, geospatial, and temporal). Combining large data sets from very different sectors and domains remains a major challenge.

Availability of data sets, their interoperability, matching-up analysis tools, and output presentations are ongoing technical challenges. Without common data structures, integration of each new data set is a challenge. Moreover, problems that can be addressed by urban-scale modeling are not yet well articulated and therefore poorly understood by policymakers. Only with strong problem-formulation ("use case") skills, appropriate at the policy level, will urban-scale integrations be continued and utilized.

An important aspect of work with large data sets is visualization, which has spawned an entire subfield. GIS mapping integrates the visual with data. Building-information modeling (BIM) accomplishes this at the building level. Urban mapping tools join these scales with three-dimensional visualization and embedded data for large numbers of buildings. Beyond "tagging" information on maps, data sets feed a wide range of simulation and optimization models and enable model calibration and progress monitoring. Data sets vary from traditional, relatively static, but extremely valuable property tax records to highly dynamic transactions from digital meters found in taxis, subway turnstiles, and electric distribution to buildings. Studies have begun to identify the patterns in such large-scale urban flows, often labeled as "urban metabolism."

Legal requirements, for example, in the EU at the national level or in the United States at the city level, have begun to make building energy use data available that can be readily aggregated and viewed. New York City's energy benchmark database (which is required for properties of 50,000 ft^2 and larger) represents roughly 20,000 buildings with publicly available building characteristics, annual energy data, and underlying monthly data.

For urban energy modeling, a realistic building construction library is essential to address urban districts with several hundreds or thousands of buildings. Such libraries link building typologies, defined by building types and age classes, to representative building physics parameters. These libraries can exist at a national level (e.g., EU Project TABULA or the follow-up EPISCOPE (Balaras et al., 2016)), at a regional level, or for specific city quarters with exemplary monitoring projects (Nouvel et al., 2013). Generally, the more local and specific these building libraries are, the higher the accuracy of the on-site construction characteristics and resulting simulation results.

Available data sets for the building construction, energy systems, and building use enable to refine the urban thermal model and improve the result accuracy. Nevertheless, urban modeling can be started with some minimum building attribute data, namely, the building usage and year of construction, which are necessary to pick up realistic building physics parameters from the building libraries. The refurbishment year or refurbishment state is valuable data that strongly impact the precision of the heating and cooling demand and energy saving results. For the final energy and CO_2 emission calculations, information about building heating and cooling systems, combustible type, and system efficiency are additionally required.

I.3 Urban energy modelling

Urban systems are complex by nature. Their energy modeling involves buildings; energy supply and distribution systems; the urban microclimate; and, last but not least, the user interacting both in operation and planning. Energy and resource flows have to be analyzed not only within the city but also across city boundaries. Changes over time as a result of socioeconomic and technical strategies need to be modeled and monitored to quantify the energy transition and to visualize results in ways that are usable by stakeholders.

The first step of the development of a long-term urban strategy is a precise diagnosis of the actual urban energy consumption, particularly on the consumption of the existing building stock. Energy consumption data today are mostly available on a macroscopic scale such as aggregated national or international data or on a microscopic scale such as individual households but rarely on a city scale. Because metering is still costly at city scale, energy modeling can help to quantify the status quo of actual consumption and the load profiles at higher time resolution. Furthermore, it allows the prediction of energy savings due to energy efficiency upgrades and then the identification of main refurbishment priorities.

Various modeling techniques allow for the calculation of energy demand and supply at city level and will be analyzed in this book. Bottom-up approaches aim at determining the contribution of each end use toward the aggregate energy consumption value and offer a high level of detail and the ability to model technological options. Data-driven statistical and engineering modeling methods are the two main categories of the bottom-up approach. Statistical methods utilize samples of customer energy billing information as data source for energy modeling, analyzing them with parameters such as building age, form, and household type (Howard et al., 2012). They are capable to encompass occupant behavior and to include macroeconomic and socioeconomic effects in the consumption equation. Statistical modeling allows to include many energy end uses (domestic hot water, electric appliances, and cooking). On the other hand, the reliance on historical consumption information limits this method for the modeling of new energy saving options or renewable integration. Engineering methods are based on thermodynamic and heat-transfer calculations (Mata et al., 2013). It is the only modeling technique able to calculate heating and cooling energy consumption without any historical energy consumption data, although these historical values can be used for calibration and for an improved scheduling. These modeling methods require an important amount of geometric and parametric input building data. To tackle this issue, three-dimensional (3-D) city models offer a solution particularly suitable for heating and cooling energy simulations. These spatial semantic data models, which represent the different urban objects of whole cities, have nowadays many applications for environmental, city, and energy planning and have been spreading widely over the last decade. Three-dimensional data management and exchange, visualization features, and the connection to simulation tools are a major focus of the book.

I.4 Building sector energy demand

The urban building sector energy demand has not gone down significantly during the last decades, as building rehabilitation rates are generally very low at about 1% per year.

When comparing building energy use intensity worldwide, the indicators must match to make results comparable. As buildings mostly use several types of energy, from electricity to natural gas, oil, or steam to renewable heat from biomass, consumption has to be expressed in a single common unit to be used for rating and benchmarking of buildings. In the United States, site energy is used to describe heat and electricity consumed in a building, while source energy includes energy conversion and distribution losses from the primary energy of the raw fuel to the site. In Europe, site energy is named final energy, and source energy is primary energy.

For nonresidential buildings, the US Energy Information Agency (EIA) releases regular commercial building energy consumption surveys. The 2015 release covers 6700 commercial buildings with consumption data referring to the year 2012. The total average site energy consumption for all commercial buildings is 252 kWh/m^{2}a. Office buildings are very close to the average nonresidential building consumption of 245 kWh/m^2a. Base load electricity for appliances, lighting, and ventilation dominates the consumption.

To convert this site energy into source or primary energy, the mix of fuel sources needs to be known. For office buildings in the United States, 70% is electricity, 22% natural gas, 6% district heat, and 2% oil. Using the ENERGY STAR source-to-site ratios of 3.14 for electricity, 1.05 for natural gas, 1.2 for district heat or steam, and 1.01 for oil, the total office building source energy use intensity is 618 kWh/m^2a. In New York, the median office building source EUI was 601 kWh/m^2a with a large spread of values (NYC, 2014).

Using the same ENERGY STAR conversion factors for European office buildings, a typical source EUI for an old office building in Germany with a heating demand of 104 kWh/m^2a and an electricity consumption of 54 kWh/m^2a would be around 278 kWh/m^2a or 55% lower. The best German office buildings today built with high insulation standards and efficient lighting and electric appliances have site EUI of 45–60 kWh/m^2a, corresponding to source EUI of about 120 kWh/m^2a. This is five times lower than the average US office building today. This shows the high impact not only of building standards but also of efficient building operation to lower electric baseline consumption.

1.5 Renewables and energy infrastructure

Smart cities with an energy-efficient infrastructure offer good potentials for local renewable integration and demand-side management. There is still an ongoing discussion on how to best match building standards and renewable supply options from an economic, architectural, and sustainable point of view. Sometimes, there are conflicting goals; for example, very high building energy efficiency renders network-based energy distribution more costly.

Many countries have significantly and measurably increased their renewable energy fraction in the electricity but less so in the heat sector. Wind power and photovoltaics now add more capacity worldwide each year (161 GW in 2016) than fossil fuel plants, and PV prices have dropped to as low as 0.03 USD/kWh. A fully decarbonized electricity system in 2050 is feasible and cheaper worldwide with average levelized cost of electricity estimated at €52/MWh, including curtailment, storage, and grid costs and an assumed doubling of worldwide consumption (Ram et al., 2017). Renewable electricity fed into electric grid infrastructure has been common for a few decades, often at the low voltage level but increasingly at medium or even high voltage levels for large renewable power plants.

In the heating and transport sector, the penetration of renewable energy sources has been much slower than in the power sector. Worldwide, the share of renewable heat is still less than 10%, mostly from limited bioenergy sources, and transport renewables are even below 5%. These sectors dominate the world's final energy consumption, with 50% used for heating of buildings and industrial processes and nearly 30% for transport. This leads to an overall share of renewable energy sources of only 20% of the world's final energy consumption, half of which are traditional biomass sources (Renewables 2017 Global Status Report). This dominance of the heating sector emphasizes the need for new building efficiency and heat distribution strategies in urban areas, which is a main focus of the book. Solar thermal systems or PV-driven heat pumps provide decentral heat that can feed into thermal networks and (seasonal) heat storage. New typologies of distribution networks for heating and cooling are now emerging, where a feed-in of decentral prosumers occurs.

A feasibility assessment for distributed generation in urban areas showed typical renewable energy potentials on a municipal level in China of about 20%–30% of the current heating and electricity consumption (Ren et al., 2010). Solar technologies dominate the renewable supply potential within urban areas, whereas wind energy, biomass, and hydropower can provide high renewable fractions outside the cities. In purely residential areas with moderate building height, very high solar photovoltaic fractions of up to 80% were calculated for Southern European locations such as Andalusia in Spain. For the province of Ontario in Canada, an average rooftop photovoltaic area of 70 m^2 per capita was calculated with a potential energy supply of the province of 30% (Wiginton et al., 2010). A case study of commercial and governmental buildings in the city of Chandler in the Metropolitan area of Phoenix, Arizona, considered 30% of the available roof area as suitable for PV production. In total, the PV production on these buildings could cover 10% of the total electricity demand of all consumers in the area (Jo and Otanicar, 2011). The solar thermal potential has been analyzed in a case study of all 8005 municipalities in Spain. The study showed that 70% of warm water requirements could be covered with an average of 17% of the roof area (Izquierdo et al., 2011). If district heating systems are extended significantly, the solar thermal integration could advance at competitive costs, as large installations lower costs and surplus energy can be used or stored more easily. In Denmark, a potential of up to 4 million square meters of large-scale solar thermal heating systems for district heating was identified, which can produce heat at around 20% of the cost of individual solar heating systems. Today, large urban renewable power plants are often still based on biomass due to a good economic return of investment in countries with a green feed-in tariff. Biomass is, however, not a scalable solution for countries with high energy consumption density and limited space.

Recently, very detailed technical potential studies were done using 3-D city geometry information derived from laser scanning data (LIDAR). In the United States, roof geometry data for 128 cities covering 40% of the US population or 23% of all buildings were analyzed to determine the photovoltaic potential. States with the best solar resource such as California could cover 74% of the electricity sold by their utilities in 2013. Florida with a 30% higher household consumption could still cover 47%, and lower-resource states such as Washington could still generate 27% of current electricity. In total, a technical rooftop potential of 1118 GW was calculated with an annual generation of 1432 TWh or 39% of national electricity sales (NREL report, 2016). The highest available potential is available on small buildings with 65% of total energy generation.

Similar detailed investigations were done by the author in Germany using the widely available 3-D CityGML data format for a detailed description of building geometry in urban areas. In low- to medium-density cities in the Southern German county of Ludwigsburg, the photovoltaic technical electricity potential was 76%. The calculation of the economic PV potential using insolation thresholds to select only south-facing roofs and minimum roof areas leads to an average value of 55% of covered electricity demand for the entire region. An economic analysis was also done for this region with 34 communities. To obtain a 2% annual carbon reduction in the region, 3% of all buildings would have to install PV systems annually at a total investment of €64 million or €180 per inhabitant. At annual revenues of €6 million, this measure seems feasible and economically viable with a payback time of 10 years.

In addition to roof surfaces, facades, further sealed surfaces, and surfaces along roads and rails are available and sum up to comparable levels than roof surfaces.

Combined with efficiency measures and intelligent load management, local renewables can thus very significantly contribute to a decarbonized urban energy future.

I.6 Scope of the book

The purpose of the book is to discuss methods and tools for decision support in urban planning, design, and operation of urban energy systems. While focusing on energy, some methods try to go beyond energy by integrating social, economic, or esthetic domains. The target groups are urban and energy planners, engineering consultancies, energy supply companies, scientists, researchers, students, and interested citizens.

In the book, several scales within the city will be addressed, ranging from individual building performance to blocks of buildings and city quarter levels up to the simulation of an entire city. The topics include the following:

- Decision support strategies and optimization techniques for urban planning processes
- Urban energy data models
- Combining 3-D city models with simulation and optimization tools
- Building sector energy demand modeling and scenarios for refurbishment
- Methods to design and operate urban energy supply and distribution systems
- Demand-side management strategies to better match renewable supply and demand of the building sector and increase flexibilities
- Case study applications in the two European Cities Vienna and Geneva

The work of the book has been done in the frame of an initial training network in the Marie Curie framework of the European Union, where 11 PhD students and 3 postdocs worked within a 4-year time frame on urban energy systems.

The two case study cities of Geneva and Vienna are used as examples to put the methods developed into practice and to test the usability for urban planners and city developers. For the two cities, full 3-D models in CityGML format have been set up and used as a basis to connect a range of modeling tools and to integrate monitoring data from the municipal building stock. The CityGML format was enriched by energy relevant data for every individual building on a city quarter and entire city scale. This unique setting allowed to analyze urban refurbishment strategies, demand response in the building sector, heat generation and distribution in large networks, and many other topics.

Whereas the first part of the book presents mainly methodologies and theoretical foundations of urban modeling and decision support, the second part aims to show results of the case study applications.

The book discusses advances on the state of the art of research in a number of key areas:

- Innovative modeling methods of urban energy systems: a real bottom-up modeling approach is used for the simulation of energy consumption at city level, which includes physical and data-driven modeling of every building, energy conversion system, and distribution networks using engineering methods.
- User behavior modeling and model predictive control strategies.
- Clustering techniques to determine the most relevant building typologies and archetypes.
- Innovative multiscale software framework combining statistical and engineering models for improved prediction of energy demand, energy saving, and renewable generation potentials within the city.
- Full 3-D model integration and web service architecture for optimization and decision support on an urban scale.

Part 1 starts with an overview and main challenges of decision support in urban energy systems. In Chapter 1, the background and tools are discussed to optimize urban energy system planning. The analysis shows that the energy transition cannot be solved by energy experts alone. A connection with the multifaceted field of urban planning is necessary for a deeper change of the urban context. Thinking of energy beyond the single building scale within urban planning processes leads to a "wicked problem," where multiple domains, actors, phases, and scales must interact in a highly dynamic context. As a result, appropriate decision support methods must be developed to bridge scientific results with planning practice, by clearly structuring the problems, and communicating results (e.g., spatial data visualization and multi-criterion decision analysis).

Chapter 2 then presents optimization methods for the planning of new city quarters. A method is shown that combines planner vision and intuition with computer capabilities to create a large amount of alternative solution corridors. It consists of two main components: an interface that serves to explore and filter the alternatives and to control the generation of new alternatives by a user and a multiparametric optimization method that allows to incorporate more domains, actors, scales, and phases at an early stage of the planning process.

In Chapter 3, building energy demand modeling is discussed ranging from individual buildings to urban scale. Different options of demand modeling are presented, ranging from white-box full physical models or simplified model using RC models to gray- and black-box approaches. To simplify the urban modeling challenge, clustering methods and archetypes are investigated. Model calibration using metering data is also included.

Chapter 4 presents load management of appliances and heating systems using model predictive control and multiagent stochastic approaches. User behavior models are introduced as a main factor impacting on the building energy use. With renewable energy system requiring more flexibility, demand-side management is crucial for stable system operation. Operational improvement is also possible using model predictive control: energy savings of up to 25% in winter months can be achieved, if the building dynamics are taken into account and the predicted solar and internal gains are fully used to reduce heating energy. A high thermal capacity of the buildings is advantageous for operational control. Building with low resistance-capacitance (RC) products has much less savings (down to 4%), and there is not much point to implement model predictive control.

Chapter 5 then deals with energy supply and distribution. Heating networks and decentralized renewable generation are analyzed including network modeling and energy/exergy performance evaluation. The main questions are which urban energy systems will be used in the future for heating and cooling and whether these systems are central or decentral. Network models are presented, and a workflow of system modeling including control is discussed. Energy and exergy balances are used for indicators.

Chapter 6 gives an overview on 3-D data models, databases, and standardization efforts to harmonize urban data models worldwide. CityGML as a widely used data format forms the basis of the work of all modeling applications in the book, and energy-relevant application domain extensions are presented.

In Chapter 7, the uncertainty of building performance simulation is assessed using a probabilistic approach. The work discusses qualitative measures to account for the uncertainty arising from building modeling and shows its impact on energy demand calculations. The methodology includes a probabilistic simulation practice for the inclusion of building geometry data uncertainties.

In Chapter 8, the software architecture to integrate a wide range of models is analyzed. To allow computing with a wide range of distributed tools, web services are introduced, and the orchestration of such services is described. In addition to this service-oriented architecture approach, cosimulation processes are discussed to combine different software domains and tools in an integrated urban modeling platform.

In the second part of the book, the presented methods and tools are applied to two case study cities.

Chapter 9 introduces the cities Vienna and Geneva and discusses the challenges and research questions for case studies for the cities as a whole and for the individual urban districts.

In Chapter 10, modeling and optimization tools are then applied to existing city quarters. This includes a detailed energy system analysis in Vienna with energy conversion and district heating systems and discusses the integration of renewables. Using detailed data from an urban district in a block of buildings, the models can be validated and scenarios can be calculated. Cluster analysis and data-driven analytics are applied to a district in Geneva.

Chapter 11 applies the interactive optimization for the planning of new city quarters in Geneva. The computational framework for the planning of new neighborhoods employing interactive optimization is demonstrated by a description of an exemplary workflow together with the results discovered during such a workflow. The results are illustrated using parallel coordinate plots and maps. The variety of results confirmed the need of an iterative and interactive process.

The final chapter summarizes the experience from the ITN network and discusses implications and transferability of the work.

In summary, the book analyzes the key challenges of the urban energy transition and the methods to provide solutions. It includes an analysis of the existing building stock and energy efficiency options and the potentials of renewable energy systems in an urban context as a function of urban density. Indicators to evaluate urban energy performance are introduced, and performance evaluation schemes are discussed. As a result, the urban contribution to a zero-carbon future is outlined, and innovative tools to model, monitor, and optimize performance are presented.

Acknowledgements

The book would not have been possible without the support of the European Commission for the Initial Training network CI-NERGY—Smart cities with sustainable energy systems (grant agreement no. 606851, 2013–17). Thanks are due to all the supervisors in the network who jointly selected the PhD and postdoc students and supported them during the course of the work: My colleague Volker Coors from University of Applied Sciences Stuttgart provided much input on the 3-D city modeling issues, Francois Maréchal from EPFL Lausanne was the optimization specialist, Markus Peter and Andreas Koch from EIFER Karlsruhe and Wolfgang Loibl from AIT Vienna brought the urban planning perspective to the project, Ruth Kerrigan from the Integrated Environmental Solutions company transferred the developments to the commercial software domain, Darren Robinson from Nottingham University supported the user behavior model developments, Marco Perino from Politecnico di Turino was our building physics competence, Michael Watzke from Siemens introduced the challenges of multimodal networks, and finally Donal Finn from UCD Dublin was a major player in the supervision and academic quality control and always shared his expertise on building and system modeling.

From the coordinating team in Stuttgart, Dr. Dilay Kesten-Erhart and Lisa Botero very efficiently managed the many workshops held during the 4-year project and held the project together with their enthusiasm and emotional support.

Our two partner cities Vienna and Geneva strongly helped apply the methods to real-world challenges. Thanks are due to numerous partners from different municipal and cantonal departments and the related energy companies.

Last but not least, the greatest thanks go to all the fellows and postdocs of the CI-NERGY training network, who cooperated extremely well and successfully and who produced this book together in parallel to writing their thesis and preparing their future career. I wish them all the best for their future.

Stuttgart, 4th July, Ursula Eicker

References

Acuto, M., 2013. Global Cities, Governance and Diplomacy: The Urban Link (Routledge New Diplomacy Studies). Routledge, Abingdon, Oxon; New York, NY.

Anthoff, D., Nicholls, R.J., Tol, R.S.J., Vafeidis, A.T., 2006. Global and regional exposure to large rises in sea-level: a sensitivity analysis Tyndall Centre for Climate Change. Research Working Paper 96.

Balaras, C.A., Dascalaki, E., Droutsa, K., Kontoyiannidis, S., 2016. Empirical assessment of calculated and actual heating energy use in Hellenic residential buildings. Appl. Energy 164, 115–132.

CIESIN, 2011. Global Rural-Urban Mapping Project, Version 1 (GRUMPv1). Center for International Earth Science Information Network, Columbia University, International Food Policy Research Institute – IFPRI, The World Bank, and Centro Internacional de Agricultura Tropical – CIAT.

Davidson, K., Gleeson, B., 2015. Interrogating urban climate leadership: towards a political ecology of the C40 network. Global Environ. Politics 15 (4), 21–38.

Doelle, M., 2016. The Paris agreement: historic breakthrough or high stakes experiment? Clim. Law 6 (1–2), 1–20.

Fleming, P.D., Webber, P.H., 2004. Local and regional greenhouse gas management. Energy Policy 32, 761–771.

Gagnon, P., Margolis, R., Melius, J., Phillips, C., Elmore, R., 2016. Rooftop Solar Photovoltaic Technical Potential in the United States: A Detailed Assessment National Renewable Energy Laboratory Technical Report NREL/TP-6A20-65298, Contract No. DE-AC36-08GO28308.

Hoffmann, M.J., 2011. Climate Governance at the Crossroads: Experimenting with a Global Response after Kyoto. Oxford University Press, New York.

Howard, B., Parshall, L., Thompson, C., Hammer, S., Dickinson, J., Modi, V., 2012. Spatial distribution of urban building energy consumption by end use. Energy Build. 45, 141–151.

Izquierdo, S., Montanes, S., Dopazo, C., Fueyo, N., 2011. Roof-top solar energy potential under performance-based building energy codes: the case of Spain. Solar Energy 85, 208–213.

Jo, J.H., Otanicar, T.P., 2011. A hierarchical methodology for the mesoscale assessment of building integrated roof solar energy systems. Renew. Energy 36, 2992–3000.

Mata, E., Kalagasidis, A.S., Johnsson, F., 2013. A modelling strategy for energy, carbon, and cost assessments of building stocks. Energy Build. 56, 100–108.

Marcotullio, P.J., 2015. In: Future Urbanization and the Management of Urban Heat Risk. American Geophysical Union, Fall Meeting.

Mimura, N., 2013. Sea-level rise caused by climate change and its implications for society. Proc. Jpn. Acad. Ser. B: Phys. Biol. Sci. 89 (7), 281–301.

Nouvel, R., Schulte, C., Eicker, U., Pietruschka, D., Coors, V. (Eds.), 2013. CityGML-based 3D city model for energy diagnostics and urban energy policy support.13th Conference of International Building Performance Simulation Association, Chambéry, France.

Ram, M., Bogdanov, D., Aghahosseini, A., Oyewo, A.S., Gulagi, A., Chill, M., Fell, H.-J., Breyer, C., 2017. In: Global Energy System based on 100% renewable energy – power sector study. Global Renewable Energy Solutions Showcase event (GRESS), a side event of the COP23, Bonn, November 8, 2017.

Ren, H., Zhou, W., Nakagami, K., Gao, W., Wu, Q., 2010. Feasibility assessment of introducing distributed energy resources in urban areas of China. Appl. Therm. Eng. 30.

Santin, O.G., Itard, L., Visscher, H., 2009. The effect of occupancy and building characteristics on energy use for space and water heating in Dutch residential stock. Energy Build. 41, 1223.

UNDP, 2014. Human Development Report 2014. UNDP, New York.

UN Habitat, 2016. Urbanization and development: emerging futures. World Cities report 2016.

Wiginton, L.K., Nguyen, H.T., Pearce, J.M., 2010. Quantifying rooftop solar photovoltaic potential for regional renewable energy policy. Comput. Environ. Urban Syst. 34, 345–357.

Part One

Methodology, mathematical background, and tools

Urban energy system planning: Overview and main challenges

1

Sébastien Cajot[*,†], Nils Schüler*[†]
[*]European Institute for Energy Research, Karlsruhe, Germany, [†]Industrial Process and Energy Systems Engineering Group, École Polytechnique Fédérale de Lausanne, Lausanne, Switzerland

1.1 Introduction

Since the symbolic tipping point that occurred in 2007, humankind has become an urban species with more than half of its population living in urban areas (UN, 2014). Not surprisingly have cities become a focus in addressing the global issues of climate change and the related energy transition toward low-carbon, renewable, and efficient systems. As a concentration of people, activities, and energy demand, cities can contribute significantly in terms of achieving climate targets. Furthermore, cities are in a convenient position for action, with a high enough overview and authority, while at the same time close enough to interact with local actors and take into account issues that might threaten or hinder project implementation (Jank and Erhorn-Kluttig, 2013; Keirstead and Schulz, 2010). This balance between overview and action therefore sets urban planners, the key urban actors, in a privileged yet challenging situation. Charged with the mission to shape the cities of the future, they are required to consider long-term consequences and coordinate decisions taken in all urban sectors, from urban form to socioeconomic aspects, the environment, and the energy system.

However, energy is hardly their main concern, as urban planning today is typically seen as the field that pulls different actors and sectors together to mediate and arbitrate the stakes and interests at play. Planners are thus concerned, and rightfully so, that energy (and its rather technical considerations) might become an increasing driving force in city planning, threatening the achievement of other important and possibly more qualitative political targets (Masboungi, 2014; Robinson, 2011; Chapter 7). Energy should not impose technically optimal urban forms or infrastructure, but instead, the relationships between energy and other domains should be studied and better understood to support existing synergies with socioeconomic aspects: energy efficiency as a means to alleviate public or private expenses, renewable energy integration to increase local energy security and air quality, efficient district-scale energy systems that provide architects and designers with more—not less—flexibility in choices of material and shapes, and so forth.

While the PhD network CINERGY initially stems from the fields of urban energy system modeling and simulation, its necessary connection with the multifaceted field

Urban Energy Systems for Low-Carbon Cities. https://doi.org/10.1016/B978-0-12-811553-4.00001-9

of urban planning was recognized from the start. A transversal question was thereby as follows: What are the existing challenges in contemporary urban planning, and which obstacles might threaten the proper integration of energy issues in planning processes? From this question follows the realization that the energy problem, despite being one of the most urgent problems of our time (Costello et al., 2009; WEF, 2016), cannot be solved by energy experts and operators alone. Its interdependencies with surrounding fields and actors must be explicitly considered.

This chapter provides some background and basic principles regarding urban planning and its relationship with energy issues. First, a systematic review of urban planning is proposed. Second, a framework for depicting the main challenges in planning energy at the urban scale is presented. Finally, a mapping of the tools that can help overcome the challenges and improve the accounting of energy in urban planning is performed.

1.2 A review of urban planning processes and supporting tools

1.2.1 Methodology of review

Any attempt at generalizing urban planning is not without risks. Indeed, the processes involved and the goals pursued evolve constantly through time and differ strongly by country, culture, or city (UN-Habitat, 2009). Furthermore, *how* the process is to be described can also affect its understanding, whether the focus is set on the chronological sequence of events, the relationships between its actors, or its general context (Masser, 1983). However, even if a universal model of urban planning cannot be achieved, a common one is nevertheless crucial to improve communication and mutual understanding of the topic. This holds especially true given the highly multidisciplinary nature of planning. For example, the notion of "early phases of planning" is very often referred to yet only vaguely and with no specification of the tasks implied or the scale considered. Many studies dealing with urban planning elude this entirely, while some explicitly define their understanding of the process. The results presented hereafter are a synthesis of the process depictions found in literature, evaluating whether they line up or differ and which insights could be gained from such a comparative overview. This task was carried out for 20 scientific- and practice-based references involving explicit urban planning depictions (Table 1.1), answering the following questions: "What are the typical phases in urban planning?," "which tasks are carried out at each phase?," and "which methods and tools exist to support each task?." The published material was collected using the search terms "urban planning process" in the Scopus database and scanning the 263 references by availability, language, and relevance (i.e., containing explicit discussion and/or depiction of urban planning processes). Research projects were found on the European CORDIS database, using the keywords "urban planning" and "energy," and returning 89 projects, three of which contained planning process depictions. Two internal studies from the European Institute for Energy Research were also included in the research project set.

Table 1.1 List of references analyzed in this work, which contain urban and energy planning depictions

Published scientific material (articles, books, and thesis)	Research projects	Practice-based references
- (Banfield and Meyerson, 1955) - (Yeh, 1999) - (Amado et al., 2010) - (Teriman et al., 2010)[a] - (Wallbaum et al., 2011) - (Mirakyan and De Guio, 2013) - (Nault, 2016) - (Masser, 1983) - (deVries et al., 2005) - (Yigitcanlar and Teriman, 2015)[a]	*EU CORDIS database* - CONCERTO (Di Nucci et al., 2009; Di Nucci and Pol, 2010) - ECOCITY (Gaffron et al., 2005, 2008) - (STEP-UP, 2015)[a] *EIFER studies* - MIE2SUP (Cassat et al., 2015)[a] - (Sipowicz et al., 2016)[a]	*EU initiative* - SEAP (Covenant of Mayors, 2010)[a] *National references* - (SIA, 2003, p. 110) - SméO (Roulet and Liman, 2013)

[a]These references present similar planning phases and tasks and as such are reviewed jointly (respectively, Teriman et al., 2010; Yigitcanlar and Teriman, 2015; STEP-UP, 2015; Covenant of Mayors, 2010; Cassat et al., 2015; Sipowicz et al., 2016). However, they are kept distinct in the review as they do propose different support tools and methods.

Finally, three practice-based references were considered: the sustainable energy action plan guidelines from the EU Covenant of Mayors initiative and two Swiss references addressing urban planning.

In the scope of this review, the studied processes focus on the *sequential model* (Masser, 1983). This representation of planning as a sequence of delimited events has the advantages of being widespread (which is suitable for a systematic comparative review) and inherently simple (which allows a straightforward categorization of tasks in each phase). As Masser (1983) notes, "models of this kind provide a useful starting point for a more detailed analysis of tasks within this general framework," which is precisely what is intended here. Other models include the contextual model, which rejects the notion of an ideal, universal model, describing planning instead as deeply influenced by contextual conditions, including social, political, economic, and traditional aspects. Another model, the interaction model, focuses essentially on the stakeholder relationships within the planning process. Planning and implementation process (PIP) diagrams can be cited as a combination of interaction and sequential models (Di Nucci and Pol, 2010). Though such models would undeniably enrich the discussion to come, their systematic review in literature is out of scope, and focus is set on the sequential models.

The various sequences of events observed in the reviewed documents were compiled and synthesized into five generic phases, and the corresponding tasks were classified accordingly. Tools were then mapped to each task. Additional metainformation regarding the process (iterative nature, duration, participation, and thematic focus) and any mention of decision support tools or methods were also collected and discussed.

1.2.2 Results of review

Although some clear patterns emerge in the depiction of urban planning steps, there are some notable disparities across the different sources (Fig. 1.1). Gaffron et al. (2008), Teriman et al. (2010), and Yeh (1999) propose the most fragmented and detailed processes with up to nine steps. For example, they include steps such as plan selection, detailed design, or project delivery. On the other extreme, deVries et al. (2005) and Nault (2016) only use two or three steps to describe the process. Furthermore, some references view urban planning as disconnected from the more practical phases of implementation and operation (Banfield and Meyerson, 1955; deVries et al., 2005; Nault, 2016) and do not mention any steps related to those phases.

Inferring from the observed patterns (Fig. 1.1), all steps were aggregated and classified into five main phases:

(1) Initiative and vision: Identification of needs for action and description of a vision
(2) Strategic planning: Formulation and prioritization of objectives or goals
(3) Design: Elaboration, evaluation, and comparison of solutions
(4) Implementation: Implementation of plan
(5) Operation and monitoring: Operation of infrastructure and services and monitoring

These were chosen as a compromise between a too general subdivision of the process and a too detailed one, both of which would be unpractical for a deeper analysis of tasks and tools. In addition, most references can be closely mapped to these five phases, while three are nearly identical (Cassat et al., 2015; Masser, 1983; Roulet and Liman, 2013).

Additional information regarding the focus, iterative nature, and planning horizon of the processes was also collected (Table 1.2). Nearly all references explicitly recognized the need for some form of iteration in the process, either with stepwise iterations between phases or as an entire repetitive cycle. Even more consensual was the advocacy of public participation in urban planning. Several temporal indications show the long-term investment of planning, up to over a decade for the entire process. Six references were specifically concerned with sustainability aspects and four with energy. Regarding physical scales, planning focus ranged from building to neighborhood and up to regional scales.

Based on the collected information (Fig. 1.1 and Table 1.2), the following definition of urban planning is proposed:

Urban planning is a continuous, iterative, layered, and interconnected process of five main phases that consist in

(i) elaborating a common vision (initiative and vision);
(ii) identifying and arbitrating corresponding objectives (strategic planning);

Fig. 1.1 Compilation of planning steps as depicted by the references on the left and classified according to five generic phases (top row). Key: pale colors indicate secondary steps within a phase; gradients indicate overlapping steps. Note that Sipowicz et al. (2016), STEP-UP (2015), and Yigitcanlar and Teriman (2015) are not depicted (cf. footnote in Table 1.1).

Table 1.2 Additional information regarding the planning processes in the reviewed documents

	Short description	Iterative nature[a]	Public participation	Duration	Special focus
Banfield and Meyerson, 1955	Rational planning model	⇥	–	–	Rational urban
Masser, 1983	Urban and regional plan-generation cycle	↻	Yes	Long time lag	Regional urban
Yeh, 1999	Urban planning process for GIS role mapping	↻	Yes	–	Urban GIS
SIA, 2003	General structure of urban planning project	↔	Yes	–	Urban
deVries et al., 2005	Urban planning and design disciplines	↔	Yes		Design urban
Gaffron et al., 2005	Urban intervention as a cyclical process	↻	Yes	–	Sustainable urban
Di Nucci et al., 2009	Energy-oriented planning	↔	Yes	–	Energy urban neighborhood
Amado et al., 2010	Sequential sustainable urban planning process	⇥	Yes	–	Sustainable urban
Covenant of Mayors, 2010	Sustainable energy action plan	↻	Yes	2 years	Sustainable energy
Teriman et al., 2010	Integrated framework for sustainable urban development	↻	Yes	–	Sustainable urban
Wallbaum et al., 2011	Stages in planning process according to SIA 2004	⇥	Yes	–	Sustainable urban
Mirakyan and De Guio, 2013	General procedure for integrated energy planning in cities/territories	↔	Yes	Long term	Energy integrated long-term model-based
Roulet and Liman, 2013	Quarter-scale planning phases	↔	Yes		Building neighborhood sustainable urban
Cassat et al., 2015	Urban development value chain	↻	Yes	–	Energy urban
Nault, 2016	Urban planning phases	⇥	Yes	9–13 years	Design urban

[a] ⇥, sequential process; ↔, stepwise iterations; ↻, cyclical process.

(iii) translating the vision and objectives in alternative, concrete actions or designs, which are evaluated, compared, and selected (design);
(iv) implementing a plan (or passing it on to the next relevant administrative scale) (implementation);
(v) operating the system of infrastructure and services and monitoring its success until obsolescence or until a new initiative appears (operation and monitoring).

The key concepts used in the definition are illustrated in Fig. 1.2 and discussed below.

Continuous and iterative. As denoted by the examples in Fig. 1.1, urban planning is frequently visualized as a linear sequence of steps. In reality, despite this linear depiction, many references emphasize the importance of *feedback loops* and *iterations* between the different steps (e.g., ensuring that proposed alternatives meet the goals identified earlier or adapting the goals to new information collected during participative activities). Indeed, the disconnect between phases has been noted as a critical obstacle to achieving efficient plans and designs (deVries et al., 2005). Furthermore, urban planning can be viewed and understood as a *continuous*, *cyclical* process. This tendency to view planning as an ongoing activity is supported by the current reality of most planning projects in Europe. The punctual, well-delimited urban expansion and development projects (so-called greenfield projects) of the last century have become the exception rather than the rule. European countries are already over three-quarters urbanized (UN, 2014), and further conditioned by the trend for compact and dense cities, urban planning now consists mainly of a constant monitoring of the urban area and activities, which leads to urban renewal, redevelopment, or infill projects. Such projects possess blurred out physical and temporal boundaries, where precise beginnings and endings are more difficult to define.

Layered and interconnected. These concepts refer mainly to the relationships between the physical and administrative scales at which decisions are taken. The five phases described above are purposely generic and scale independent. To some extent, they are replicable at the three key layers at which urban planning operates the city scale, the intermediate or neighborhood scale, and the building scale. This points to the fact that urban planning is continuous not only temporally, with iterations and permeability between sequential phases and sectors, but also vertically, with reciprocal feedback between upper and lower administrative or geographic layers. Fig. 1.2 illustrates how the implementation phase of the higher scales may consist in the initiation of a lower level (and thus more detailed) planning cycle, where objectives are adapted; more specific and relevant actions or physical plans are elaborated, until their implementation in the form of a construction project; or another planning cycle is attained. Likewise, each completion of a cycle leads to refined information, which in turn feeds back not only to the current scale but also to that above.

Despite the emerging consensus on these properties of urban planning, there remains a certain confusion regarding its exact scope and sphere of action. This confusion is, for example, visible in the ambiguous labeling used in the reviewed material, where generic terms like "planning," "urban planning," or "plan design" are used to describe just a single step belonging to some broader, ill-defined process. We argue

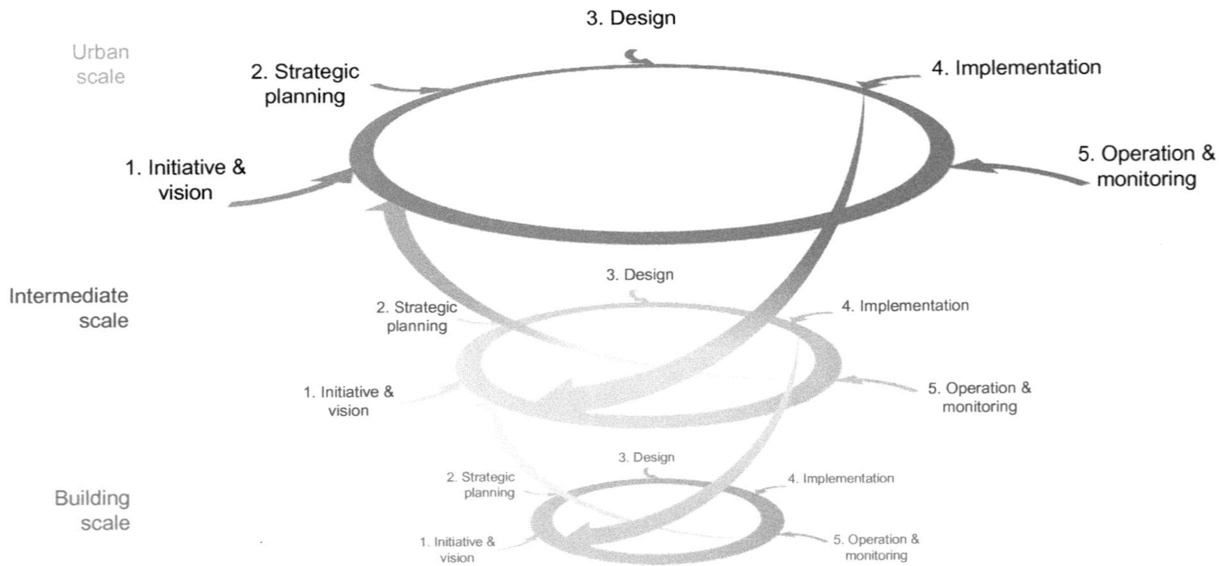

Fig. 1.2 Continuous and interconnected planning cycles.

here that urban planning concerns more than just the *strategic* component of the process or the spatial *design* of the urban area. Given the necessary permeability and interdependencies between each phase, urban planning must encompass the entirety of the process.

Therefore, from the collective identification of intervention needs to the operation of urban infrastructure, urban planners must continuously bring together information and actors from each phase and scale, in order to best anticipate the global and local needs and how best to satisfy them. In this context, their role becomes highly polyvalent and, to some extent, ubiquitous regarding scales and disciplines.

1.2.2.1 Mapping of urban planning tasks and supporting tools

Using the planning phases established in the previous section, it is possible to map in more detail the various tasks that have been mentioned in the reviewed documents (Table 1.3).

1.2.2.2 Urban planning tools and methods

Many tools and methods exist to support urban planners in the tasks described in Table 1.3. The list presented in this chapter does not claim to be exhaustive, but rather shows an overview of tools that are mentioned in the reviewed urban planning documents. This overview is therefore purposely limited to reveal common tools and methods that are currently used or advised in the urban planning field.

Consequently, the same references used for the review of planning process and tasks are used to identify relevant tools and methods (cf. Table 1.1). These are organized into the five following categories:

(i) Problem structuring and project management
(ii) Data collection and visualization
(iii) Public participation
(iv) Alternative generation
(v) Evaluation

The categories are defined loosely, and one method or tool could belong to different categories. For example, some methods in the first and second categories could also be considered helpful for the generation of alternatives (e.g., SWOT analysis and GIS tools) or evaluation (e.g., checklists, certifications, and MCDA). These latter categories are chosen to better emphasize the role of computer tools—particularly optimization and simulation—which shall be discussed in Section 1.4.2.

Fig. 1.3 shows the intensity of use of each category along the planning phases. Problem structuring, project management, and public participation are the most evenly used throughout the process. Data collection and visualization are most used in phases 1 and 3. Alternative generation is predominant in the strategic planning and design phases, while evaluation is found essentially in the design phase. Fig. 1.4 provides the names of tools and methods found at each phase.

Table 1.3 Synthesis of tasks after aggregation across all references (darker shades mean more references mention the task, white indicates 1 reference, and the darkest shade indicates 12 references)

1. Initiative & vision	2. Strategic planning	3. Design	4. Implementation	5. Operation & monitoring
Characterize past and present situation	Determine objectives of intervention	Design different alternatives	Implement plan and construct	Monitor implemented plan against objectives
Establish a shared vision	Prioritize objectives	Evaluate, compare and select alternatives	Engage public for validation, training and organization	Adapt, rectify, maintain infrastructure and site
Involve specific stakeholders and citizens (identify requirements and get feedback)	Elaborate master plan	Communication of results and exchange with public and DM	Create implementation program (budget, timing, resources)	Operate the systems (infrastructure, services..)
Identify needs for intervention	Public participation	Detailed design of single alternative	Decide to carry out plan	Engage, inform stakeholders and train them on new technologies
Store and analyze physical, social and economic data (GIS)	Define measures and policies	Competition/tender	Monitor development to ensure plan compliance	Identify new intervention needs when obsolescence is reached, define new urban context
Identify areas of environmental sensitivity (GIS)	Define spatial aspects	Create scenarios	Coordinate development operations and construction	Reporting and submission of implementation report to political authorities
Contract definition	Establish budget, timing and responsibilities	Establish/revise implementation program (budget, timing, resources)	Legally formalize project	Update plan periodically according to feedback
Cost estimations	Perform initial studies	Dispatch lots and projects to developers	Select appropriate site	
General deadline calendar	Describe barriers preventing goals from being reached.	Determine detailed conceptual model of planning process and formalize in tool	Analyze market trends	
Select neighborhoods	Define eco-district program	Collect, describe and interpret the information	Follow administrative procedures	
Get political commitment for plan	Publish energy concept	Examine strategies by impact assessment, review results and goals	Deliver project	
Adapt city administrative structures for sustainable energy planning	Launch international urban planning competition	Implement the objective hierarchy in a MCDA approach for each scenario		
Initiation		Operationalize the quality targets and the measurable indicators		

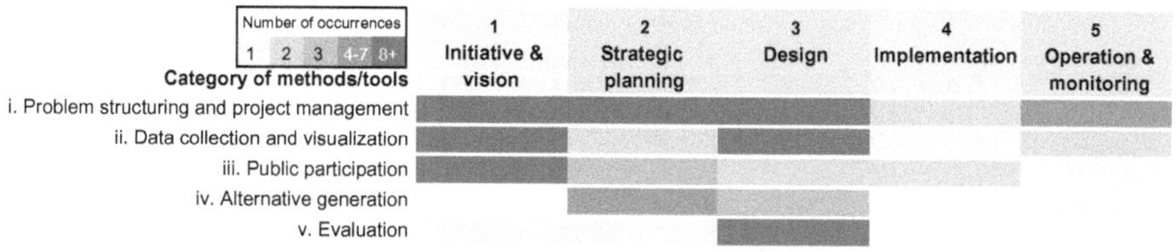

Fig. 1.3 Intensity of use of methods and tools in reviewed documents, by category, throughout the urban planning process.

1.3 Main challenges in urban energy planning

Traditionally, energy planning has been a fairly confined process, focusing in great detail either on the building unit or on large and centralized power and heat plants. Furthermore, in most municipalities, these processes are still found to be decoupled from the more general urban planning process (Petersen, 2018; Schiefelbein et al., 2017).

However, to obtain cost-effective and energy-efficient building stocks, evidence suggests that the building scale may be suboptimal and that innovative measures involving urban planning are needed (Caputo and Pasetti, 2015; Gossop, 2011; Strasser et al., 2017; UN-Habitat, 2009; Zanon and Verones, 2013). Addressing these issues at wider scales than the single building comes however at a price. Despite presenting possibly complex technical problems, energy planning at the building scale is facilitated by the fact that buildings are self-defined and well-understood entities, with relatively fixed life spans, and involved fewer stakeholders with clear power relations (Peter et al., 2009; Strasser, 2015; Zanon and Verones, 2013). As soon as energy issues are considered at city level, these traits disappear, turning a well-defined and fairly delimited problem into an ill-defined, multifaceted, and dynamic one. Despite these difficulties, both the research and planning communities agree today that a new understanding of the role and form of urban planning is required to take on these pressing issues related to energy. Going beyond the traditional tasks of designing the city's spatial aspects and defining strategic targets, urban planning must be carefully and profoundly rethought to take ownership of and appropriately address energy and resource issues. This means that planners are expected to handle simultaneously both qualitative aspects such as aesthetics of urban form or quality of life, along with the more quantitative concerns of energy system design and engineering.

According to Knox (2010), the aims of urban planning have ranged from "mythology and religion to geopolitics, military strategy, national identity, egalitarianism, public health, economic efficiency, profitability, and sustainability." Even though aspects of sustainability and rational energy use can be identified in early concepts of urban planning, it is only in recent times that these issues take on a central role.

As modern urban planning emerged in the mid-1800s, it consisted essentially of a collection of processes to manage the sanitary and social crises that were threatening industrial cities' prosperity (Knox, 2010; Teriman et al., 2010). This top-down, expert-led approach resulted in master plans and land-use regulations (UN-Habitat, 2009),

Fig. 1.4 Detailed heat map of specific methods and tools mentioned in reviewed documents for each planning phase.
Adapted from Cajot, S., Peter, M., Koch, A., 2017c. Embedding Urban Energy Simulation and Optimization in Urban Planning (No. HN-45/17). European Institute for Energy Research, Karlsruhe.

favoring low-density, single-use districts and private cars as a means of transportation. The result was an increase in urbanization and a quick economic recovery that lasted until the end of World War II. After this period of urban expansion, the following decades saw a shift toward urban regeneration. Here, the goal was instead to improve existing urban areas (Duarte and Seigneuret, 2011).

In this evolving context, the oil crisis and related volatility of fossil fuel prices since the 1970s and growing concerns for climate change and the emergence of the concept of sustainability (understood as a balance between social, environmental, and economic interests) in the 1990s (Brundtland et al., 1987; UN-Habitat, 2009) were the key drivers that progressively pushed energy issues into the focus of conventional urban planning approaches. Today's widespread acceptance of sustainable urban planning concepts can be understood as a consequence of the described evolution in urban planning and the associated societal transformation processes. It should be noted as well that in addition to these rather environmental concerns, other global trends as well are contributing to reshape urban planning. Population growth and urbanization rates, as introduced in the opening of the chapter, raise new challenges, for example, in terms of sprawl mitigation, infrastructure development, urban form, and their associated impacts on quality of life (UN, 2014). Additionally, the so-called digital revolution and increasing availability of high-resolution urban data also influence urban planning's focus and operating modes. For example, such data may shift urban planning's current long-term, strategic, and rather static approaches to a more short-term and dynamic understanding of cities (Batty, 2013). This also opens up new possibilities in terms of planning support systems and other computer-based technologies that can be used to inform and rethink planning processes (Derix, 2012; Geertman and Stillwell, 2004; Reinhart and Cerezo Davila, 2016). However, though these trends, alongside the goal of sustainable development, all contribute together to complexify the tasks of urban planners, their detailed analysis is out of the scope of this chapter. Instead, we focus here more specifically on the topic of energy, as this is arguably, with its direct implications on climate change, considered one of, if not the most, pressing issues of our time (Costello et al., 2009; WEF, 2016).

A few words on the current understanding of energy planning can be useful at this point of the discussion. As a composition of a multitude of tasks, the integration of energy planning in the framework of urban planning is by its nature a fragmented and inconsistent matter. A basic problem is the lack of a common agreement on the exact definition of energy planning (Prasad et al., 2014). For example, Thery and Zarate (2009) define energy planning as "determining the optimal mix of energy sources to satisfy a given energy demand." In this sense, the purpose of energy planning is to balance the spatially localized energy supply and demand of a given area (Keirstead et al., 2012). However, this encompasses a variety of processes, energy carriers, and technologies that are rarely managed together, as should be, for example, supply, conversion, storage, and transportation technologies (Løken, 2007). Furthermore, the raising interest for local, decentralized, and renewable energy technologies is currently reshaping our understanding of energy planning in cities (Adil and Ko, 2016). Cities formerly might have been considered only as "centers of passive demand that must be supplied from an exurban source" (Keirstead et al., 2012) but today must play a more active role in organizing their energy systems from within their geographic boundaries. This change of paradigm shifts the responsibility of energy planning from a limited group of specialists, including local and national authorities, energy companies, and operators, to a wider one including as well local producers, energy consumers, transportation companies, technical officers, international institutions, manufacturers of end-use appliances, financial institutions, and environmentalist groups (Coelho et al., 2010). Additionally, unlike urban planning, energy planning is generally not a systematically established institution within administrative departments. Mainly for historical reasons (Merlin and Choay, 1988), the tasks described above are most often dissociated from the planning department, relegated to the management and design of networks and infrastructure by private or public actors (Schiefelbein et al., 2017). Caputo and Pasetti (2015) found that municipal offices often lack knowledge and authority regarding energy planning, even in the technical offices in charge of the built environment. Ideally, however, energy planning—understood as the combination of tasks and actors discussed above—should be better embedded in the spatial and strategic planning processes attributed to urban planners.

1.3.1 A framework for describing complex problems

The discussion in the previous section started to outline the different sources of complexity in including energy issues in urban planning: multiple actors; different scales; long-term implications; and uncertainty in the processes, methods, and basic definitions. The formulation of the problem itself remains somewhat unclear. Scholars since the 1970s have often referred to such ill-defined problems as *wicked problems*, a notion that has been used to discuss social and environmental problems, for example, regarding renewable energies or climate change (Balint et al., 2011; Boxenbaum, 2013; Levin et al., 2007; Vandenbroeck, 2012). Rittel and Webber (1973) argued that trying to formulate a wicked problem was a problem in itself. In the present section, we shall rely on the main characteristics of wicked problems to frame the problem of energy planning in cities in a systematic way.

Wicked problems are essentially characterized by the presence of multiple actors with varying interests and the difficulty to precisely define the problem statement (Boxenbaum, 2013). Additionally, what perhaps distinguishes them most from *tame problems*—their easier-to-solve counterparts—is that they "don't have a right answer" (Camillus, 2008). Balint et al. (2011) elaborate these definitions with nine specific conditions, summarized in Table 1.4. These conditions are grouped here into three categories, whether they pertain to multiplicity and heterogeneity aspects of the problem, to related complexity and uncertainty issues, or to the instability of the context.

Initially, the notion of wicked problems referred explicitly to the intractable nature of economic and social issues in planning. Today, adding an extra layer of energy-related issues certainly amplifies the relevance of the concept. In fact, two central points raised by Levin et al. (2007) describe how the present issue exceeds the "wickedness" of dilemmas faced by urban planners in the 1970s (Rittel and Webber, 1973). First, unlike before, "the central authority needed to address the challenges is weak or nonexistent" (Levin et al., 2007). This was in particular made clear in Section 1.3, since energy planning is still not systematically established in most administrative structures. Second, when dealing with limited energy resources and climate change impacts, "time is running out" (Levin et al., 2007). This means that the problems faced here, if not addressed urgently, can become literally uncontrollable or lead to irreversible consequences.

Table 1.4 Conditions for identifying wicked problems

Category	Condition	Explanation
Multiplicity and heterogeneity	1. The lack of a single problem statement	Difficulty to agree on the exact nature of the problem, as the definition lies in the eye of the beholder
	2. Conflicting values	The existence of conflicting values, driving the different actors, makes it difficult to assess the objective quality of any solution
	3. Conflicting objectives	Objectives and targets set by multiple stakeholders and on multiple scales might not converge when lacking communication or consensus on the values
	4. Multiple tactics to address the problems	The lack of consensus on the best approach to achieve similar results is increased by the lack of clarity on objectives and values
	5. Multiple actors with the power to assert their values	Multiple stakeholders can influence the problem and defend their interests, while the interests of third parties must also be taken into account
Complexity and uncertainty	6. Scientific complexity and uncertainty	Decision-making is hindered by uncertain or incomplete knowledge
	7. Political complexity and uncertainty	Ambiguity among political groups and public opinion leads to a confusion in predominant values
	8. Administrative complexity and uncertainty	Financial limitations and fragmentation of governmental institutions, across multiple levels and sectors, hinder appropriate implementation of solutions
Instability	9. Dynamic context	Evolving environmental, economic, and social frameworks require flexible and adaptive solutions

Adapted from Balint, P.J., Stewart, R.E., Desai, A., Walters, L.C., 2011. Wicked Environmental Problems: Managing Uncertainty and Conflict. Island Press, Washington, DC.

1.3.2 Managing wicked problems

- If wicked problems cannot be solved in the same linear way as a well-defined problem might, they must nevertheless be tackled somehow. The studies reviewed hereafter reveal two opposite views or positions on the matter, namely, simplification or comprehensiveness (Table 1.5). The first view realistically assumes that the only way to overcome a wicked problem is to simplify—or tame—it, by breaking it down into smaller, solvable issues. Rittel and Webber (1973) initially noted that the nature of wicked problems entails such simplifications. Because there is no "stopping rule" to wicked problems, one must at some point

Table 1.5 Summary of reviewed mind-sets and methods recommended for tackling wicked problems

View	Advocated mind-set	Proposed methods	Reference
Realistic/ simplification	Taming the problem	–	(Law, 2014)
	Dialogue between humans and software	Optimization software Pareto analysis	(Du et al., 2006)
	Satisficing decision-making Problem aggregation Expert-based decision-making	–	(Rittel and Webber, 1973)
Ideal/ comprehensiveness	No "taming" or "overstudying" the problem Coherence and sharing of issues Feed-forward orientation	Dialogue mapping	(Conklin, 2006)
	Stakeholder engagement Feed-forward orientation	Framework for responding to wicked issues Pareto analysis	(Camillus, 2008)
	Argumentation Stakeholder engagement Transparency	Issue-based information system	(Noble and Rittel, 1988)
	Knowledge-based approach Interactive group process	Mess mapping Resolution mapping	(Horn and Weber, 2007)
	Precautionary principle Adaptive management Participatory approaches	Learning network process	(Balint et al., 2011)
	Sensitivity to complexity and openness to simplicity Dialogue	Systems thinking Design thinking Soft systems methodology Transition management	(Vandenbroeck, 2012)
	Web-based communication and collaboration	Collaborative innovation networks (COINs)	(Totten, 2012)
	Decision analysis	Multicriteria decision analysis	(Løken, 2007; Pohekar and Ramachandran, 2004; Strantzali and Aravossis, 2016; Wang et al., 2009; Zhou et al., 2006)
	System modeling	Urban energy system modeling (simulation, optimization, LCA, rating systems, etc.)	(Huang et al., 2015; Keirstead et al., 2012; Mendes et al., 2011; Mirakyan and De Guio, 2013; Reinhart and Cerezo Davila, 2016; Sharifi and Murayama, 2014)

> stop working on it for external reasons (e.g., the lack of time, patience, or funding) or if a sufficiently good (or "satisficing") solution is found, even if this means that the problem is left only partly addressed. They also point out other coping mind-sets, such as aggregating problems to higher scales in order to not focus on minor symptoms or entrusting few, selected experts to arbitrate the various contradictions in values. This view is also held by Law (2014), who points out that *simplifications must necessarily happen* to handle "the indefinite extension of value clashes, political controversies, problems embedded in other problems, material heterogeneities, fluidities, and endless and unpredictable feedback loops." He advocates, for example, the "homogenization" of problems (i.e., determining the extent to which they can be reduced to a single facet or metric) and "centering" of observation points (i.e., reducing the plurality of perspectives and influences in favor of centralized command and control). In another study, Du et al. (2006) advocate *dialogue* both between humans and between humans and software as a means for cooperation and increased acceptance of solutions. They suggest that the problem must be sufficiently simplified so that the explanations (e.g., from another human or an optimization algorithm) about the relative quality of an alternative remain understandable for the decision-maker. This implies an iterative process, addressing successively the different viewpoints and issues.

The second view ideally attempts to comprehend simultaneously all facets of the problem and interactions through various analysis methods. Conklin (2006) emphasizes the need for coherence and shared understanding of all issues among stakeholders, to avoid the fragmentation of the problem and failing to address the overall issues at stake. To achieve this coherence, he proposes *dialogue mapping*, a group facilitation method to systematically represent decision-making processes and increase transparency and shared understanding of the issues by all stakeholders. Dialogue mapping is based on the issue-based information system developed by Noble and Rittel (1988), which was developed in response to the understanding that planning processes are in fact nonlinear and may benefit from argumentation, stakeholder engagement, and transparency. Horn and Weber (2007) also present a methodology to represent, understand, and analyze wicked problems. The methodology includes *interactive, knowledge-based group processes, and visual analytics that* enable decision makers to select actions that ameliorate the considered problem. Balint et al. (2011) studied four cases presenting wicked characteristics, including the European cap-and-trade program for greenhouse gases. They discuss in depth the benefits and limitations of three common responses to wicked problems: *the precautionary principle, adaptive management, and participatory approaches.* Additionally, they propose a "*learning network process*," which seeks to identify realistic alternatives or solutions to a wicked problem by discovering the issues, preferences, and values of the involved stakeholders. Vandenbroeck (2012) also emphasizes the importance of acknowledging wicked problems and advocates the application of five methodologies, including *soft systems methodology, transition management, and design thinking.* However, he also points out a possible risk of acknowledging a problem as wicked, as it may "obscure simple and pragmatic ways of making a positive difference." Totten (2012) suggests that the rise of *web-based, self-organizing collaborative platforms* will play a role in solving global wicked problems, on one hand by informing citizens to the risks of such problems to increase cooperation and on the other connecting them to generate new ideas and motivate local action. Camillus (2008) proposes a framework for coping with wicked problems in strategic business decisions. Put forward are the importance of communication and involvement of stakeholders; the identification of common corporate identity and values, focusing on few, prioritized courses of action; and feed-forward mentality to anticipate various contextual changes and possible futures. As additional supporting solutions, to cope with the multiplicity and heterogeneity challenges, *multicriteria decision analysis (MCDA) methods* have been widely used to provide an analytic framework to facilitate discussions and mutual understanding among multiple stakeholders and include a diversity of objectives in the decision process. Løken (2007), Pohekar and Ramachandran (2004), Wang et al. (2009), and Zhou et al. (2006) have reviewed studies applying MCDA specifically to energy planning situations, providing aid in understanding and choosing appropriate decision support methods. *Modeling approaches* are another common way to understand and overcome scientific complexities and uncertainties. Computer models are particularly relevant for wicked problems, for which the implementations of solutions are often associated with high costs (Du et al., 2006) and are typically irreversible over short periods. As Rittel and Webber (1973) note in this line, "planners have no right to be wrong." Consequently, there exist many different methods and tools to predict the consequences of actions on the urban and community scales, before they are implemented. These have been classified in various ways. Keirstead et al. (2012) provide a review of available tools for urban energy system modeling, distinguishing between simulation, optimization, empirical, and econometric methods and identifying the key areas of application for these tools (viz., technology design, building design, urban climate, systems design, and policy assessment). Mirakyan and De Guio (2013) decomposed long-term energy planning in cities and territories into four main phases and reviewed the methods and tools available along these steps. They provide a list of common tools from the fields of simulation and optimization, system dynamics, and life-cycle analysis. On a more local scale, Mendes et al. (2011) surveyed energy models for community-level energy planning and assessed the appropriateness of different bottom-up, simulation, scenario, equilibrium, and optimization tools. Huang et al. (2015) also reviewed *bottom-up computer tools available for community energy planning* and what they call *top-down methods*, which rely on upper policy and global guidelines to guide local planning. These include, for example, rating systems such as BREEAM Communities, LEED-ND, or CASBEE for urban development, tools which have expanded, during the last decade, from the building scale to cover issues at the intermediate community scale (Sharifi and Murayama, 2014). Finally, Reinhart and Cerezo Davila (2016) review urban microsimulation methods at the neighborhood scale. They argue that such bottom-up approaches are necessary to provide detailed information about local integrated energy systems useful to support the planning of neighborhood-sized projects.

Synthesizing the two different views mentioned above, Conklin (2006) highlights the key limitations in both extremes. On one hand, taming a wicked problem may only prove useful in the short term but generally fails to solve the initial wicked problem in the long run or even exacerbates it. On the other hand, overly studying the problem slows down the process, may be extremely costly, and might lead to *analysis paralysis*, in which action is needed for more information, but more information is expected in order to take action, thus freezing the process. A certain balance between the realistic and idealistic views seems desirable and therewith some kind of iteration between a temporary reduction of the problem's scope to allow studying certain precise aspects of it, followed immediately by a recontextualization of the insights into the larger context. What is important is to not get stuck or waste resources in solving only a part of the problem or to be overwhelmed by the size and intractability of the whole.

These few studies provide specific mind-sets and methods that can help better manage wicked problems. In the present work, we propose to use the list of conditions from Table 1.4 as a guide to identify the issues from real-world wicked problems and to systematically map potential responses. Our main hypothesis is that a better understanding of the multifaceted problem should foster a multifaceted response, instead of adopting fragmented or incomplete solutions. This implies not only seeking a diversity of approaches, from the methods and mind-sets reviewed above, but also encouraging new creative responses, tailored to the local specificities found in real planning projects.

As an illustrative example, the case study of Les Cherpines (introduced in Chapter 11 of this book) is used. The various and conflicting interests at stake in this greenfield planning project, pressured by scientific, political, and administrative complexities, as well as an unpredictable, dynamic context, make the progress of the project a highly tedious task for planners. No ideal legal framework, improved planning process, or attempt to thoroughly model the urban system can alone overcome such challenges. Instead, planners in Geneva tackle the *wickedness* with various complementary and often innovative solutions (Table 1.6). For a more detailed discussion on these aspects, the reader is referred to the original paper (Cajot et al., 2017b).

1.4 Decision support for urban energy system planning

So far, this chapter outlined some underlying concepts describing urban planning and current challenges in tackling energy issues at the urban scale. More "solution-oriented," the mission of the CINERGY project was to research how the combined use of data, computers, and human expertise could support decision-making in this challenging context. This last section will thus focus more closely on the topic of decision support. Particular attention is devoted to simulation and optimization approaches, which constitute the bulk of CINERGY contributions. Their main characteristics are discussed, emphasizing their benefits and limitations in urban energy system planning.

1.4.1 Basic concepts of decision support

When taking decisions as complex and high stake as those in the urban context—investing in a district heating network, allocating public funds for the retrofit of a neighborhood, and promoting decentralized renewable energy systems—the common

Table 1.6 Mapping of observed issues in urban energy planning and proposed solutions adopted in Les Cherpines and canton of Geneva

Category	Observed issue	Proposed solution
Multiplicity and heterogeneity	Multiple interdependent public policies to achieve simultaneously	Holistic master plan development on canton, commune, and district scales
		Development of an energy planning instrument (territorial energy concept (TEC)), complementary to urban planning instruments
		Regrouping of the town and country planning services with those of housing and energy policies in a single administrative department
	Conflicting values	Clarification of public goals in master plans, TEC instrument to foster involvement of all stakeholders
		Public consultations and opposition procedures during the planning process
	Multiple interpretations of "ecodistrict" concept	Political "motion" (participative amendment of project proposition)
Complexity and uncertainty	Uncertain effects of key urban parameters (e.g., density, quality of life) and natural resources	Mandate external studies, use of urban energy system simulation and optimization tools
	Difficulty to identify relevant project boundaries	Definition of project scope in terms of energy resources rather than administrative boundaries, inclusion of nearby industrial zone
		Neighborhood-sized urban development projects involving public authorities and citizens ("grand projects")
		TEC instrument to identify relevant scales
	Limited legitimacy of planners to influence the energy system	Application of direct and indirect measures (density targets, preemption right, district heating enforcement, promotion, etc.)
	Financial barriers to densification	Establishment of intercommunal funding scheme
	Data scarcity	Energy law requiring relevant actors to provide the canton with supply and demand data
		Mandate external studies and data collection (e.g., GEothermie 2020)
Instability	Modification of problem's framing conditions	Increased flexibility of localized physical plans
	Long-term project duration	Long-term "grand project" approach, steering by dedicated interdisciplinary working groups, flexible guiding plans
	Evolution of data and knowledge	Development of an online GIS data platform SITG and CityGML model of the city

sense and intuition that we rely on in our day-to-day decisions are often insufficient. Decision analysis, which Keeney (1982) simply describes as "the formalization of common sense," is particularly relevant to demystify the wicked problems outlined in Section 1.3 and to identify the most efficient and/or satisfying options while helping to justify these decisions to the public and other stakeholders.

The methods and tools presented in Section 1.2.2.2 all ultimately aim to help take more informed decisions. Whether related to problem structuring, data collection, public participation, or computer models, they provide insights on the three main components that characterize a decision (Malczewski and Rinner, 2015): the decision-maker (and the values they represent, whether their own or those of others), the possible alternatives from which to choose, and the criteria that allow to evaluate each alternative.

However, even the most advanced supporting methods are no guarantee of finding the best solution, if the governing values are not properly exploited. In a seminal work, Keeney (1992) outlined two common decision approaches, well summarized by the following quote: "Value-focused thinking involves starting at the best and working to make it a reality. Alternative-focused thinking is starting with what is readily available and taking the best of the lot." In alternative-focused thinking, values are used only secondarily to judge an existing set of alternatives. This implies that better alternatives could easily be overlooked. While value-focused thinking could, in principle, considerably improve the quality of complex decisions such as those found in urban energy system planning, it appears that alternative-focused thinking is more likely to be followed *precisely* in such complex and dynamic situations (Cajot et al., 2017a; Parnell et al., 2013).

There is a strong parallel between Keeney's alternative- and value-focused approaches and simulation and optimization approaches discussed throughout this book. Simulation requires the DM to know in advance which alternatives are to be simulated and evaluated. Conversely, optimization starts with the DM's values, expressed as objective functions and constraints, and identifies the alternatives that best satisfy those values. The implications of this parallel—and in particular why both approaches remain equally valid to support urban-scale decisions—are discussed hereafter.

1.4.2 Modeling, simulation and optimization

As simplified representations of real systems, models can help us understand, predict, and eventually shape these systems according to our needs. The use of computer-based, mathematical models allows to do so in a cost-effective way, where it would be too costly or impossible to do in practice, for example, with physical models (Bierlaire, 2015; Pasanisi, 2014; Wang et al., 2012). How the computer model is used—or rather, the purpose for which it is adopted—is what distinguishes simulation and optimization. These distinctions are highlighted in the following sections and summarized in Table 1.7.

Table 1.7 Schematic comparison of simulation and optimization

	Simulation	Optimization
Input	- Input variables - External parameters	- Objective function and direction - Solution space (defined by decision variables and constraints)
Output	- Quantified indicators	- Optimal solution (or solutions)
Approach	- Descriptive - Alternative-focused - Typically determined system of equations (no. of variables = no. of equations)	- Normative - Value-focused - Typically underdetermined system of equations (no. of variables > no. of equations)
Common classifications	- (Swan and Ugursal, 2009) o Top-down o Bottom-up - (Coakley et al., 2014) o Law-driven (forward modeling, deterministic, white box, detailed simulation) o Data-driven (inverse modeling, black box)	- (Maringer, 2005) o Deterministic o Stochastic/heuristic - (Nguyen et al., 2014) o Single objective o Multiobjective

1.4.3 Simulation

Models can be run to mimic the behavior of a system through simulation, by specifying a set of input variables (alternatively control or decision variables) and external parameters (Fig. 1.5) (Bierlaire, 2015). The user assesses the effects on key indicators by answering questions of the type *what if*. For example, if a communal energy company invests in a district heating network, what would be the consequences on greenhouse gas emissions or long-term finances?

Techniques for simulating the built environment can be categorized in different ways. Coakley et al. (2014) distinguish physical law-driven approaches from data-driven approaches. Law-driven approaches are like a white box where physical and environmental processes are explicitly modeled (heat transfer, mass flows, gravity, etc.). These forward models are data intensive but allow to predict behaviors that haven't yet been observed. On the other hand, data-driven approaches are black boxes in terms of physical processes and rely on monitored or historical data to determine the model's coefficients. Easier to implement by relying on statistical methods, the lack of physical meaning makes the results more difficult to interpret. Swan and Ugursal (2009) make a distinction between top-down and bottom-up approaches. Top-down

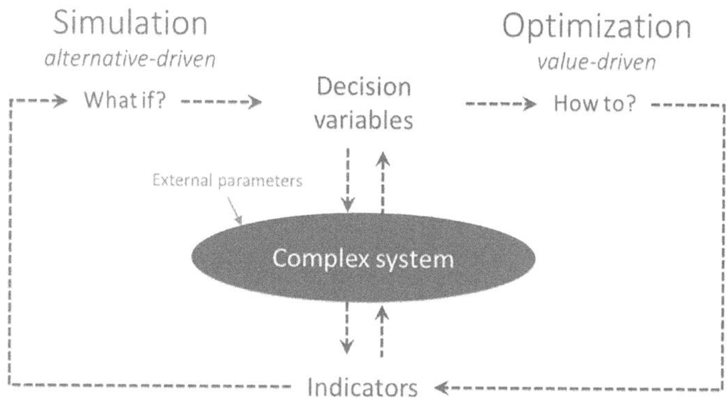

Fig. 1.5 Schematic depiction of simulation and optimization characteristics.
Based on April, J., Glover, F., Kelly, J.P., Laguna, M., 2003. Simulation-based optimization: practical introduction to simulation optimization. In: Proceedings of the 35th Conference on Winter Simulation: Driving Innovation. Presented at the Winter Simulation Conference, Winter Simulation Conference, pp. 71–78; Bierlaire, M., 2015. Simulation and optimization: a short review. Transp. Res. Part C Emerg. Technol. 55, 4–13. doi: 10.1016/j.trc.2015.01.004 (Engineering and Applied Sciences Optimization (OPT-i)—Professor Matthew G. Karlaftis Memorial Issue); Thiriot, S., Solon, A., 2016. From "what if?" to "how to!": evolutionary metaheuristics for decision support. In: ResearchGate. Presented at the SSC2016: Social Simulation Conference 2016, Rome.

approaches rely on aggregations of entire sectors, ignoring individual end uses. Main variables used in these models include global economic attributes (GDP and employment rates) or technological data (survey of owned appliances, building stock age, weather, etc.). Much like black box models, top-down weaknesses are the incapacity to anticipate the effects of new technologies or suggest accurate improvements at the end-user scale. Bottom-up approaches are those that "use input data from a hierarchical level less than that of the sector as a whole" (Swan and Ugursal, 2009). They can use these data either following a black box approach (statistical regression or neural network analysis at the building scale) or a more white box approach (e.g., archetype or microsimulation).

1.4.4 Optimization

Conceptually, optimization can be considered as simulation in reverse (Fig. 1.5): The main inputs here are the preferred direction of the key indicators, defined as objectives and constraints, and the output is the set of decision variables that allow the objectives to be best achieved (Hwang and Masud, 1979). In other words, optimization answers *how to* questions, as in the following: how to design the city's energy system to minimize GHG emissions or public expenses?

Various solving techniques exist to systematically search for the set of decision variables, which optimize the objective function. These techniques can be roughly divided into two categories, whether the objective function is known and explicitly

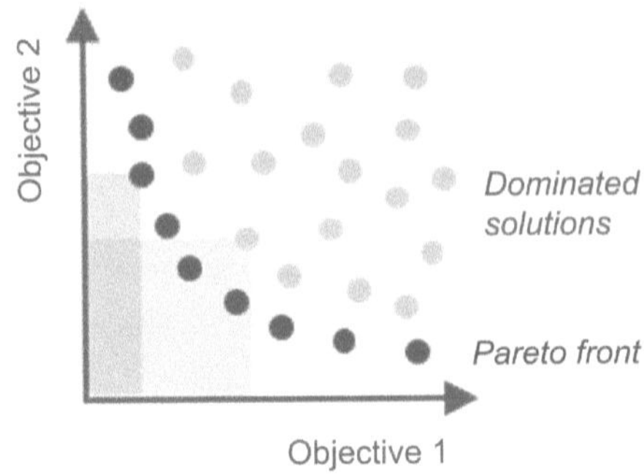

Fig. 1.6 Illustration of a Pareto front and dominated solutions.

solved (mathematical programming) or whether the objective function is too complex to model explicitly and is optimized by means of heuristic or metaheuristic approaches (April et al., 2003; Gosavi, 2015; Maringer, 2005). Another common distinction is made between single- and multiobjective optimization. Although it was observed in a recent review of building design optimization that over half of the studies had optimized a single objective (Evins, 2013), most real-world problems, especially at the urban scale, are multiobjective in nature (Nguyen et al., 2014; Shi et al., 2017). In multiobjective problems, the optimization process itself does not lead to a single best solution, but instead produces a set of multiple, equally interesting solutions. These solutions—referred to as Pareto-optimal or nondominated solutions—cannot be improved in one objective without depreciating at least one other objective (Fig. 1.6). As such, multiobjective optimization does not tell the decision-maker which solution to choose, only, at best, which solutions to avoid (Wierzbicki et al., 2000). It is through the articulation of the decision-maker's preferences that optimization can be used for decision support, that is, to select a preferred option from the nondominated alternatives. How a nondominated set can be generated and how the preferences can be articulated will be illustrated in Chapter 2.

1.4.5 Discussion

By placing objectives before alternatives, optimization is in fact enabling a value-focused approach. This makes it a valuable decision support asset, because it protects the DM from overlooking important alternatives. Simulation however is not without its own benefits. First, simply because the inputs of the simulation are manually selected by the user, it does not mean that they do not reflect carefully identified values, obtained by other means. Second, where optimization techniques tend to harness computer power for the systematic and quick exploration of many alternatives, simulation dedicates this power to more depth in the modeled aspects of carefully selected alternatives, for instance, by considering finer time and space resolutions. This suggests that both approaches might be more suited to different phases of the planning process depicted in Section 1.2. The "value-focused" benefits of

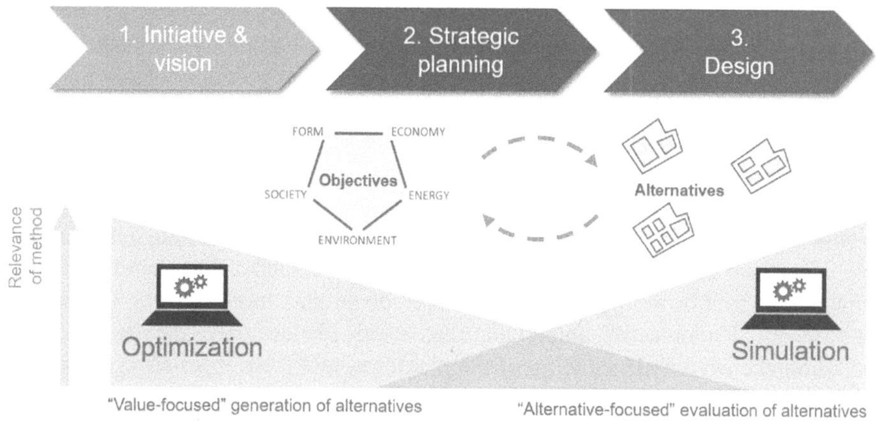

Fig. 1.7 Suggested contributions of optimization and simulation techniques in the first three urban planning phases.

optimization should be used as soon as possible, for example, for generating sketch maps or arbitrating between conflicting political targets in the vision and strategic phases. Indeed, it is important in the early stages to properly scan all feasible options before getting locked into irreversible decisions or decisions that are expensive or time-consuming to change (e.g., investing in central power plants or in distribution network and storage infrastructure). The more precise evaluation provided by simulation can be best harnessed once the systematic screening and design has been performed in a way that satisfies the values at stake. The most promising options identified in the strategic phase can be checked and fine-tuned by relying on dynamic, physical models, able to consider high temporal resolution and precise geometric information, necessary for the urban and architectural design phases. A balanced use of optimization and simulation as suggested here (Fig. 1.7) can thus effectively address what deVries et al. (2005) termed the "planning-design gap," by providing adapted feedback between the strategic and design phases.

1.5 Conclusion

The recent evolutions in the field of urban planning, today recognized as an integrated, collaborative process, make it ideally suited to address energy and sustainability issues. As discussed, however, the price for addressing these issues at wider scales than the single building is the emergence of a wicked problem, preventing the identification of any clear solution.

Several decision support tools and methods were reviewed, and the benefits and limitations of simulation and optimization were discussed in the context of value-focused thinking.

In 1986, Albers (1986) concluded his discussion on urban planning by referring to an "uncomfortable situation," brought up by the new understanding of urban systems as complex and uncertain entities. He noted the unevenness and unpredictability of planning measures in achieving political goals and the inability of mathematical models to fully represent such systems. Thirty years later, despite notable progress in computers and urban models, it seems the situation remains, in (1986) sense, somewhat uncomfortable. There is still a divide between sectors, scales, and people, and the development of appropriate solutions thus entails a close collaboration between researchers and practitioners. This is in line with the *symmetry of ignorance* discussed by Noble and Rittel (1988), which assumes that "knowledge is distributed in unknowable ways" and dismisses the fact that "there could be such a thing as *experts* who know more about how a problem ought to be solved than those directly affected by the problem." Undoubtedly, among all urban actors, planners have the most accurate and comprehensive understanding of the problems, and their experience, creativity, and intuition will remain central in handling the more qualitative and wicked aspects of planning. However, the true contribution of novel quantitative solutions, including urban-scale simulation and optimization tools, has yet to be fully understood and established in planning practices. In this context, the pressing issues related to energy might represent a convenient opportunity to rethink how such tools could be used to inform planning. Perhaps indeed, as Albers (1986) implied, optimizing a system as wicked as a city remains a fallacy. This does not mean, however, that optimization and simulation methods could not be harnessed to provide planners with new quantitative insights regarding the many concurring trade-offs taking place in cities and therewith provide new understandings of the wicked challenges of our time and how to adequately address them.

References

Adil, A.M., Ko, Y., 2016. Socio-technical evolution of decentralized energy systems: a critical review and implications for urban planning and policy. Renew. Sust. Energ. Rev. 57, 1025–1037. https://doi.org/10.1016/j.rser.2015.12.079.

Albers, G., 1986. Changes in German town planning: a review of the last sixty years. Town Plan. Rev. 57, 17–34.

Amado, M.P., Santos, C.V., Moura, E.B., Silva, V.G., 2010. Public participation in sustainable urban planning. Int. J. Hum. Soc. Sci. 5, 102–108.

April, J., Glover, F., Kelly, J.P., Laguna, M., 2003. Simulation-based optimization: practical introduction to simulation optimization. Proceedings of the 35th Conference on Winter Simulation: Driving Innovation. Presented at the Winter Simulation Conference, Winter Simulation Conference, pp. 71–78.

Balint, P.J., Stewart, R.E., Desai, A., Walters, L.C., 2011. Wicked Environmental Problems: Managing Uncertainty and Conflict. Island Press, Washington, DC.

Banfield, E.C., Meyerson, M., 1955. Politics, Planning, and the Public Interest: The Case of Public Housing in Chicago. Free Press New York.

Batty, M., 2013. Big data, smart cities and city planning. Dialogues Hum. Geogr. 3, 274–279. https://doi.org/10.1177/2043820613513390.

Bierlaire, M., 2015. Simulation and optimization: a short review. Transp. Res. Part C Emerg. Technol. 55, 4–13. https://doi.org/10.1016/j.trc.2015.01.004. Engineering and Applied Sciences Optimization (OPT-i)—Professor Matthew G.

Boxenbaum, E., 2013. Nouvelles énergies pour la ville du futur. Transvalor-Presses de Mines.

Brundtland, G., Khalid, M., Agnelli, S., Al-Athel, S., Chidzero, B., Fadika, L., Hauff, V., Lang, I., Shijun, M., de Botero, M.M., 1987. Our Common Future ("Brundtland Report"). United Nations World Commission on Environment and Development.

Cajot, S., Mirakyan, A., Koch, A., Maréchal, F., 2017a. Multicriteria decisions in urban energy system planning: a review. Front. Energy Res. 5. https://doi.org/10.3389/fenrg.2017.00010.

Cajot, S., Peter, M., Bahu, J.-M., Guignet, F., Koch, A., Maréchal, F., 2017b. Obstacles in energy planning at the urban scale. Sustain. Cities Soc. https://doi.org/10.1016/j.scs.2017.02.003.

Camillus, J.C., 2008. Strategy as a wicked problem. Harv. Bus. Rev. 86, 98.

Caputo, P., Pasetti, G., 2015. Overcoming the inertia of building energy retrofit at municipal level: the Italian challenge. Sustain. Cities Soc. 15, 120–134. https://doi.org/10.1016/j.scs.2015.01.001.

Cassat, M., Bahu, J.-M., Bardeau, G., Duforestel, T., Fulda, A.-S., Heyder, M., Leonard, J.C., Mousseau, B., Nogues, P., Payre, C., Sevenet, M., 2015. MIE2SUP—Methods for Integrating Energy in Early Stage Urban Planning (EIFER Report). EIFER, Karlsruhe.

Coakley, D., Raftery, P., Keane, M., 2014. A review of methods to match building energy simulation models to measured data. Renew. Sust. Energ. Rev. 37, 123–141. https://doi.org/10.1016/j.rser.2014.05.007.

Coelho, D., Antunes, C.H., Martins, A.G., 2010. Using SSM for structuring decision support in urban energy planning. Ukio Technol. Ir Ekon. Vystym. 16, 641–653. https://doi.org/10.3846/tede.2010.39.

Conklin, J., 2006. Dialogue Mapping: Building Shared Understanding of Wicked Problems, first ed. Wiley, Chichester, England; Hoboken, NJ.

Costello, A., Abbas, M., Allen, A., Ball, S., Bell, S., Bellamy, R., Friel, S., Groce, N., Johnson, A., Kett, M., Lee, M., Levy, C., Maslin, M., McCoy, D., McGuire, B., Montgomery, H., Napier, D., Pagel, C., Patel, J., de Oliveira, J.A.P., Redclift, N., Rees, H., Rogger, D., Scott, J., Stephenson, J., Twigg, J., Wolff, J., Patterson, C., 2009. Managing the health effects of climate change. Lancet 373, 1693–1733. https://doi.org/10.1016/S0140-6736(09)60935-1.

Covenant of Mayors, 2010. How to Develop a Sustainable Energy Action Plan. Publications Office of the European Union, Belgium.

Derix, C., 2012. Digital masterplanning: computing urban design. Proc. Inst. Civ. Eng. Urban Des. Plan. 165, 203–217. https://doi.org/10.1680/udap.9.00041.

deVries, B., Tabak, V., Achten, H., 2005. Interactive urban design using integrated planning requirements control. Autom. Constr., Education and Research in Computer Aided rchitectural Design in Europe (eCAADe 2003). Digital Des. 14, 207–213. https://doi.org/10.1016/j.autcon.2004.07.006.

Di Nucci, M.-R., Gigler, U., Pol, O., 2009. Integration of demand and supply side activities in CONCERTO communities: a preliminary assessment of planning and implementation mechanisms for sustainable local energy strategies.10th IAEE European Conference-Energy, Policies and Technologies for Sustainable Economies.

Di Nucci, M.R., Pol, O., 2010. Planning and Implementation Process Assessment Report—CONCERTO Report, Concerto Reports, Austrian Institute of Technology, 16 S. Austrian Institute of Technology, Vienna.

Du, G., Richter, M.M., Ruhe, G., 2006. An explanation oriented dialogue approach for solving wicked planning problems. ExaCt. Presented at the Computing and Informatics, Slovakiapp. 223–249.

Duarte, P., Seigneuret, N., 2011. Projet urbain et planification territoriale durables en Europe: négociation et itération. Enjeux Planif. Territ. En Eur. Press. Polytech. Univ. Romandes 77–101.

Evins, R., 2013. A review of computational optimisation methods applied to sustainable building design. Renew. Sust. Energ. Rev. 22, 230–245. https://doi.org/10.1016/j.rser.2013.02.004.

Gaffron, P., Huismans, G., Skala, F., 2008. ECOCITY Book II—How to Make it Happen. Facultas Verlags- und Buchhandels AG, Vienna.

Gaffron, P., Huismans, G., Skala, F., 2005. ECOCITY Book I—A Better Place to Live, Heather Stacey Language Services, Edinburgh ed. Facultas Verlags- und Buchhandels AG, Vienna.

Geertman, S., Stillwell, J., 2004. Planning support systems: an inventory of current practice. Comput. Environ. Urban. Syst. 28, 291–310. https://doi.org/10.1016/S0198-9715(03)00024-3.

Gosavi, A., 2015. Simulation-Based Optimization, Operations Research/Computer Science Interfaces Series. Springer US, Boston, MA. https://doi.org/10.1007/978-1-4899-7491-4.

Gossop, C., 2011. Low carbon cities: an introduction to the special issue.Cities, Low Carbon Cities (45th ISOCARP World Congress, Porto, Portugal, 18–22 October 2009) 28, pp. 495–497. https://doi.org/10.1016/j.cities.2011.09.003.

Horn, R.E., Weber, P.R., 2007. New Tools for Resolving Wicked Problems. MacroVu Inc Strategy Kinet.

Huang, Z., Yu, H., Peng, Z., Zhao, M., 2015. Methods and tools for community energy planning: a review. Renew. Sust. Energ. Rev. 42, 1335–1348. https://doi.org/10.1016/j.rser.2014.11.042

Hwang, C.-L., Masud, A.S., 1979. Multiple Objective Decision Making: Methods and Applications; a State-of-the-Art Survey. Ching-Lai Hwang; Abu Syed Md. Masud.

Jank, R., Erhorn-Kluttig, H., 2013. Annex 51—Case Studies and Guidelines for Energy Efficient Communities: A Guidebook on Successful Urban Energy Planning. Fraunhofer IRB Verlag, Stuttgart.

Keeney, R.L., 1992. Value-Focused Thinking: A Path to Creative Decision Making. Harvard University Press, Cambridge.

Keeney, R.L., 1982. Decision analysis: an overview. Oper. Res. 30, 803–838.

Keirstead, J., Jennings, M., Sivakumar, A., 2012. A review of urban energy system models: approaches, challenges and opportunities. Renew. Sust. Energ. Rev. 16, 3847–3866.

Keirstead, J., Schulz, N.B., 2010. London and beyond: taking a closer look at urban energy policy. Energy Policy 38, 4870–4879. https://doi.org/10.1016/j.enpol.2009.07.025.

Knox, P.L., 2010. Cities and Design. Routledge, London, New York, NY.

Law, J., 2014. Working well With wickedness. In: Textures of the Anthropocene: Grain Vapor Ray. Revolver Publishing, Berlin, pp. 157–176.

Levin, K., Cashore, B., Bernstein, S., Auld, G., 2007. Playing it forward: path dependency, progressive incrementalism, and the "super wicked" problem of global climate change. IOP Conf. Ser. Earth Environ. Sci. 6, 502002. https://doi.org/10.1088/1755-1307/6/50/502002.

Løken, E., 2007. Use of multicriteria decision analysis methods for energy planning problems. Renew. Sust. Energ. Rev. 11, 1584–1595. https://doi.org/10.1016/j.rser.2005.11.005.

Malczewski, J., Rinner, C., 2015. Multicriteria Decision Analysis in Geographic Information Science/by Jacek Malczewski. Claus Rinner.

Maringer, D., 2005. Heuristic Optimization, in: Portfolio Management With Heuristic Optimization. Advances in Computational Management Science, Springer, Boston, MA, pp. 38–76. https://doi.org/10.1007/0-387-25853-1_2

Masboungi, A., 2014. L'énergie au coeur du projet urbain. Éditions du Moniteur.

Masser, I., 1983. The representation of urban planning-processes: an exploratory review. Environ. Plan. B Plan. Des. 10, 47–62. https://doi.org/10.1068/b100047

Mendes, G., Ioakimidis, C., Ferrão, P., 2011. On the planning and analysis of integrated community energy systems: a review and survey of available tools. Renew. Sust. Energ. Rev. 15, 4836–4854. https://doi.org/10.1016/j.rser.2011.07.067.

Merlin, P., Choay, F., 1988. Dictionnaire de l'urbanisme et de l'aménagement. Presses Universitaires de France—PUF, Paris.

Mirakyan, A., De Guio, R., 2013. Integrated energy planning in cities and territories: a review of methods and tools. Renew. Sust. Energ. Rev. 22, 289–297. https://doi.org/10.1016/j.rser.2013.01.033.

Nault, É., 2016. Solar Potential in Early Neighborhood Design. Ph.D. Thesis, EPFL, Lausanne.

Nguyen, A.-T., Reiter, S., Rigo, P., 2014. A review on simulation-based optimization methods applied to building performance analysis. Appl. Energy 113, 1043–1058. https://doi.org/10.1016/j.apenergy.2013.08.061.

Noble, D., Rittel, H.W., 1988. Issue-based information systems for design. ACADIA Conference Proceedings. Presented at the Computing in Design Education, Ann Arbor (Michigan/USA)pp. 275–286.

Parnell, G.S., Hughes, D.W., Burk, R.C., Driscoll, P.J., Kucik, P.D., Morales, B.L., Nunn, L.R., 2013. Invited review—survey of value-focused thinking: applications, research developments and areas for future research. J. Multi-Criteria Decis. Anal. 20, 49–60. https://doi.org/10.1002/mcda.1483.

Pasanisi, A., 2014. Uncertainty Analysis and Decision-Aid: Methodological, Technical and Managerial Contributions to Engineering and R&D Studies (Thesis). Université de Technologie de Compiègne.

Peter, M., Avci, N., Girard, S., Keim, C., 2009. Energy demand in city regions-methods to model dynamics of spatial energy consumption. Presented at the eceee 2009 Summer Study. Act. Innovate. Deliver. Reducing Energy Demand Sustainably, European Council for an Energy-Efficient Economy, Stockholm (Sweden), Sweden, pp. 705–714.

Petersen, J.-P., 2018. The application of municipal renewable energy policies at community level in Denmark: a taxonomy of implementation challenges. Sustain. Cities Soc. 38, 205–218. https://doi.org/10.1016/j.scs.2017.12.029.

Pohekar, S.D., Ramachandran, M., 2004. Application of multi-criteria decision making to sustainable energy planning—a review. Renew. Sust. Energ. Rev. 8, 365–381. https://doi.org/10.1016/j.rser.2003.12.007.

Prasad, R.D., Bansal, R.C., Raturi, A., 2014. Multi-faceted energy planning: a review. Renew. Sust. Energ. Rev. 38, 686–699. https://doi.org/10.1016/j.rser.2014.07.021.

Reinhart, C.F., Cerezo Davila, C., 2016. Urban building energy modeling—a review of a nascent field. Build. Environ. 97, 196–202. https://doi.org/10.1016/j.buildenv.2015.12.001.

Rittel, H.W.J., Webber, M.M., 1973. Dilemmas in a general theory of planning. Policy. Sci. 4, 155–169. https://doi.org/10.1007/BF01405730.

Robinson, D., 2011. Computer Modelling for Sustainable Urban Design: Physical Principles, Methods and Applications. Earthscan, UK.

Roulet, Y., Liman, U., 2013. Introduction à l'utilisation de SméO, Fil rouge pour la construction durable, Mode d'emploi SméO V3.0. Lausanne.

Schiefelbein, J., Slotterback, C.S., Petersen, J.-P., Koch, A., Strasser, H., Am Tinkhof, O.M., Kimman, J., Church, K., Freudenberg, J., 2017. Implementation of energy strategies in communities—results within the context of IEA Annex 63.30th International Conference on Efficiency, Cost, Optimisation, Simulation and Environmental Impact of Energy Systems.

Sharifi, A., Murayama, A., 2014. Neighborhood sustainability assessment in action: cross-evaluation of three assessment systems and their cases from the US, the UK, and Japan. Build. Environ. 72, 243–258. https://doi.org/10.1016/j.buildenv.2013.11.006.

Shi, Z., Fonseca, J.A., Schlueter, A., 2017. A review of simulation-based urban form generation and optimization for energy-driven urban design. Build. Environ. 121, 119–129. https://doi.org/10.1016/j.buildenv.2017.05.006.

SIA, 2003. SIA 110—Règlement concernant les prestations et les honoraires des urbanistes.
Sipowicz, M., Ge, X., Bahu, J.-M., 2016. Portfolio of Solutions for Multi-Energy Systems Optimisation in the Urban Context (No HN-45/16/007). EIFER, Karlsruhe.
STEP-UP, 2015. Developing Enhanced Sustainable Energy Action Plans: A StEP UP Guide for Cities. Riga University, European Union.
Strantzali, E., Aravossis, K., 2016. Decision making in renewable energy investments: a review. Renew. Sust. Energ. Rev. 55, 885–898. https://doi.org/10.1016/j.rser.2015.11.021
Strasser, H., 2015. Implementation of energy strategies in communities—from pilot project in Salzburg, Austria, to urban strategy.ASHRAE Transactions. Presented at the 2015 ASHRAE Winter Conference, Amer. Soc. Heating, Ref. Air Conditioning Engineers, Inc., Chicago, pp. 176–184.
Strasser, H., Kimman, J., Koch, A., Tinkhof, O.M.a., Müller, D., Schiefelbein, J., Slotterback, C., 2017. IEA EBC Annex 63—implementation of energy strategies in communities. Energy Build. https://doi.org/10.1016/j.enbuild.2017.08.051.
Swan, L.G., Ugursal, V.I., 2009. Modeling of end-use energy consumption in the residential sector: a review of modeling techniques. Renew. Sust. Energ. Rev. 13, 1819–1835. https://doi.org/10.1016/j.rser.2008.09.033.
Teriman, S., Yigitcanlar, T., Mayere, S., 2010. Sustainable urban development: an integrated framework for urban planning and development. Rethinking Sustainable Development: Urban Management, Engineering, and Design. IGI Global, Queensland University of Technology, Australia, pp. 1–14.
Thery, R., Zarate, P., 2009. Energy planning: a multi-level and multicriteria decision making structure proposal. Cent. Eur. J. Oper. Res. 17, 265–274. https://doi.org/10.1007/s10100-009-0091-5.
Totten, M.P., 2012. GreenATP: APPortunities to catalyze local to global positive tipping points through collaborative innovation networks. Wiley Interdiscip. Rev. Energy Environ. 1, 98–113. https://doi.org/10.1002/wene.40.
UN, 2014. World Urbanization Prospects: The 2014 Revision, Highlights. 2014, United Nations, Department of Economic and Social Affairs, Population Division.
UN-Habitat, 2009. Global Report on Human Settlements: 2009 Planning Sustainable Cities. Earthscan, London.
Vandenbroeck, P., 2012. Working With Wicked Problems. King Baudouin Foundation, Brussels.
Wallbaum, H., Krank, S., Teloh, R., 2011. Prioritizing sustainability criteria in urban planning processes: methodology application. J. Urban Plan. Dev. 137, 20–28. https://doi.org/10.1061/(ASCE)UP.1943-5444.0000038.
Wang, J.-J., Jing, Y.-Y., Zhang, C.-F., Zhao, J.-H., 2009. Review on multi-criteria decision analysis aid in sustainable energy decision-making. Renew. Sust. Energ. Rev. 13, 2263–2278. https://doi.org/10.1016/j.rser.2009.06.021.
Wang, S., Yan, C., Xiao, F., 2012. Quantitative energy performance assessment methods for existing buildings. Energy Build. 55, 873–888. https://doi.org/10.1016/j.enbuild.2012.08.037.
WEF, 2016. The Global Risks Report 2016, 11th ed. World Economic Forum, Geneva.
Wierzbicki, A., Makowski, M., Wessels, J., 2000. Model-Based Decision Support Methodology with Environmental Applications. Kluwer Academic Dordrecht, The Netherlands.
Yeh, A.G.-O., 1999. Urban planning and GIS. In: Geographical Information Systems. John Wiley & Sons, Inc, New York, NY, USA, pp. 877–888.
Yigitcanlar, T., Teriman, S., 2015. Rethinking sustainable urban development: towards an integrated planning and development process. Int. J. Environ. Sci. Technol. 12, 341–352. https://doi.org/10.1007/s13762-013-0491-x.

Zanon, B., Verones, S., 2013. Climate change, urban energy and planning practices: Italian experiences of innovation in land management tools. Land Use Policy 32, 343–355. https://doi.org/10.1016/j.landusepol.2012.11.009.

Zhou, P., Ang, B.W., Poh, K.L., 2006. Decision analysis in energy and environmental modeling: an update. Energy 31, 2604–2622. https://doi.org/10.1016/j.energy.2005.10.023.

Further reading

Cajot, S., Peter, M., Koch, A., 2017c. Embedding Urban Energy Simulation and Optimization in Urban Planning (No HN-45/17). Karlsruhe, European Institute for Energy Research.

Thiriot, S., Solon, A., 2016. From "what if?" to "how to!": evolutionary metaheuristics for decision support. ResearchGate. Presented at the SSC2016: Social Simulation Conference 2016, Rome.

A planning support system using interactive optimization

2

Nils Schüler, Sébastien Cajot
Industrial Process and Energy Systems Engineering Group, École Polytechnique Fédérale de Lausanne, Lausanne, Switzerland

2.1 Motivation for interactive optimization and overview of URB^{io}

The optimization of real-world problems can be a very computationally intensive and time-consuming task when many variables are involved. This is the case when modeling entire districts or cities, which comprise a myriad of buildings, each with different heights and potential uses, a collection of energy technologies, and respective sizes, all of which can occupy almost any location in a defined space. A consequence and common remedy against these issues is to simplify the problem by considering only a single objective at a time. While this certainly quickens the calculation process, and greatly facilitates the interpretation of results, the consequence of this simplification is a result which does not reflect the real difficulties of urban planning. Indeed, the main challenges faced by planners are the need to balance multiple conflicting interests, often under strict time constraints, and with a strong need for accountability, toward the authorities on one side, and the general public and end-users on the other.

Multiobjective optimization helps understand the problem more comprehensively by revealing tradeoffs between various competing objectives (Schüler et al., 2017). For example, alternatives can be found which maximize the share of renewable energy, minimize investment costs, while simultaneously promoting building layouts which maximize the view on surrounding landmarks. Contrary to single objective problems, however, this implies that not one well-defined solution is found, but a set of equally interesting Pareto-optimal solutions, which cannot be improved in one objective without depreciating the value of another objective. In order to make use of such results, some preferences must be articulated in order to identify a most satisfactory solution. The preference articulation, consisting, for example, in acceptable ranges, tradeoff information or relative importance of objectives, can be done in one of three ways (Hwang and Masud, 1979; Branke, 2008):

(i) A posteriori, once all the Pareto-optimal solutions have been identified. The advantage is that the decision maker has a complete overview of the available options. On the other hand, the calculation can be extremely long if the solution space is vast, the decision maker may not have the time to wait, and the process will certainly compute many wasteful solutions which are of little interest.

Urban Energy Systems for Low-Carbon Cities. https://doi.org/10.1016/B978-0-12-811553-4.00002-0

(ii) A priori, before starting any calculations, this is the most efficient approach, as theoretically only one solution is calculated. However, it is also probably the most difficult from the decision maker's point of view, as it assumes that they are perfectly aware of their preferences and acceptable tradeoffs, and are able to formulate them precisely. In practice, when dealing with complex and interdisciplinary problems, this knowledge is generally unavailable until the solutions are calculated, and therefore the risk of reaching an infeasible or unsatisfactory solution is high.

(iii) Interactively, as the optimization progresses, this is a compromise between the two previous approaches. It allows to efficiently focus the computational effort on the most interesting solutions only, while at the same time allowing the decision maker to rely on a subset of Pareto-optimal solutions to learn and adapt their preferences iteratively. The active involvement of the decision also increases the trust in the identified solution, and its chances of implementation. The downside is that the decision maker's attention is required throughout the process.

In the context of urban planning, the size of the problem makes a posteriori approaches impractical, while the complexity and interdisciplinary nature of decisions make a priori approaches unlikely to work (Balling et al., 1999). Interactive optimization is thus arguably the preferred choice (Cajot et al., 2017), and consequently will be the focus of this chapter. However, several open challenges still need to be addressed if the interactive process is to be usable by urban planners.

The first challenge concerns the ability of the *optimization model* to integrate multiple domains, scales, and actors while finding solutions as fast as possible to reduce latency during the interaction. For this, Mixed-Integer Linear Programming (MILP), an optimization technique, which allows to incorporate both continuous and discrete decisions, is adopted for the modeling of the urban and energy systems. At the price of being restricted to linear formulations, it is known for rapidly converging to globally optimal solutions for a large range of problem instances in terms of varying input parameters. The developed model is described in Section 2.2.

The second challenge concerns the *representation* of the multidimensional solutions. Indeed, the decision maker must be able to intuitively interpret these solutions. While two objectives are easily plotted in a plane to depict the front of Pareto-optimal solutions, for three or more objectives this can become problematic. An increasingly popular approach to visualize multiobjective results is parallel coordinates, which offer an efficient and intuitive way to display multivariate data, revealing correlations between multiple objectives in a two-dimensional plane (Zhe, 2017; Kipouros et al., 2008; Akle et al., 2016). They consist a set of vertical axes (which contain quantitative or semantic information pertaining to the attributes), and horizontally flowing lines, each of which represents a given alternative to be assessed. The user can visually grasp how various criteria interact with each other, and easily evaluate tradeoffs (denoted by a crossing of lines between two axes) or synergies (denoted by relatively parallel lines). Like other visualizations, parallel coordinates also suffer from a lack of readability when displaying many solutions, and the difficulty to view all pairwise relationships in a single chart. However, these limitations are partly diminished by the possibility to dynamically rearrange the axes positions, and to filter the display of lines by brushing (i.e., selecting) the desired range on an axis. The parallel coordinates interface developed in URB^{io} is based on the Data Driven Documents library,[1] and expands on an existing implementation for parallel coordinates.[2]

The third challenge involves generating a representative sample of solutions reflecting the *user's preferences*. A form of optimization that is suited for this purpose is known as multi-parametric optimization where a number of decision variables are decided by the algorithm dependent on parameterized constraints (Gal, 1995; Liu et al., 2011; Pis, 2007). A mathematical formulation of this method is provided in Section 2.3. Here the user must only specify one criterion to be optimized, and indicate preferred ranges of interest on all other criteria. Once the ranges are specified, the parameters characterizing the constraints are automatically established according to either of two sampling methods, depending on the needs: systematic sampling (preferable in case the user wishes to perform a systematic sensitivity analysis on the selected criteria), or a sampling approach based on a Sobol sequence. The Sobol sequence ensures a much quicker and thorough exploration of the solution space when time is limited (Copado-Méndez et al., 2016). This is particularly relevant in the case of interactive optimization, as the user can interrupt the optimization as soon as they are able to refine their preferences, for example, adjusting preferred ranges and narrowing the exploration to more relevant areas.

These different elements come together to form the planning support system URB^{io} (Fig. 2.1). It is an iterative and interactive optimization process, which consists in the following steps:

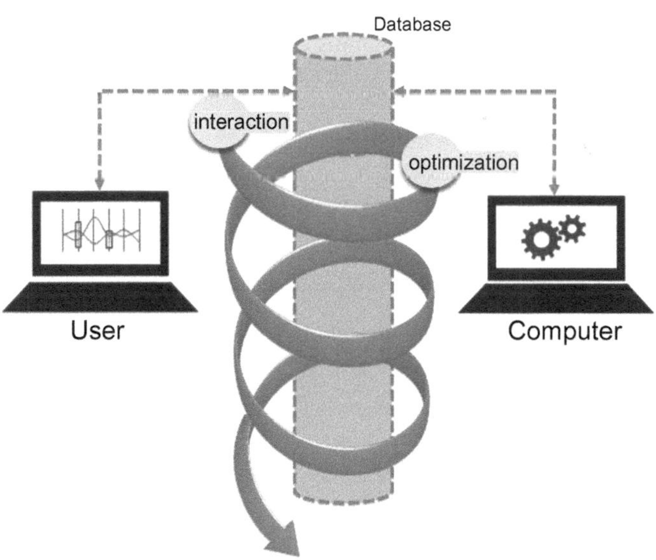

Fig. 2.1 Overview of URB^{io} workflow.

[1] See https://d3js.org/.

[2] See syntagmatic.github.io/parallel-coordinates.

1. initialize the decision process (a few Pareto-optimal solutions are generated for a predetermined objective and constrained set of criteria);
2. present the alternatives to the decision maker for exploration and learning;
3. collect their preferences (in the form of new objective, preferred ranges, and constraints);
4. solve the new optimization problem; and
5. repeat (2) to (4) until a satisfactory solution is found.

Several distinctive aspects of this method can be emphasized. Contrary to some IO methods, URB^{io} does not possess some mathematical stopping criteria, but instead relies on the decision maker's satisfaction (Buchanan, 1997). As such, the "psychological convergence" can be considered more important than a mathematical one (Branke, 2008). In other words, the learning process and clarification of tradeoffs are as important, if not more important, than the actual final solution.

In addition, whereas traditional interactive methods tend to clearly distinguish the learning phase, the preference articulation phase, and the generation of solutions, URB^{io} manages to blend these three phases into an integrated, more immersive experience (Fig. 2.2). Indeed, the exploration of solutions need not be interrupted while the optimization is running: as soon as a solution is found, it is directly included to the chart for the user to interpret. Through various exploratory features described in the following section, the user learns about the relationships between criteria, and how much they are willing to sacrifice in one in order to gain value in others. Each new solution makes more clear the critical contradictions to be resolved, allowing to refine the search toward areas which are found most relevant. As the user narrows the solution space by filtering the various axes, updated and new constraints are automatically regarded by the solver, generating new solutions in those specified areas based on the multi-parametric programming approach. Because all phases merge into a what could be called an "optimization-based thought process," the optimization can

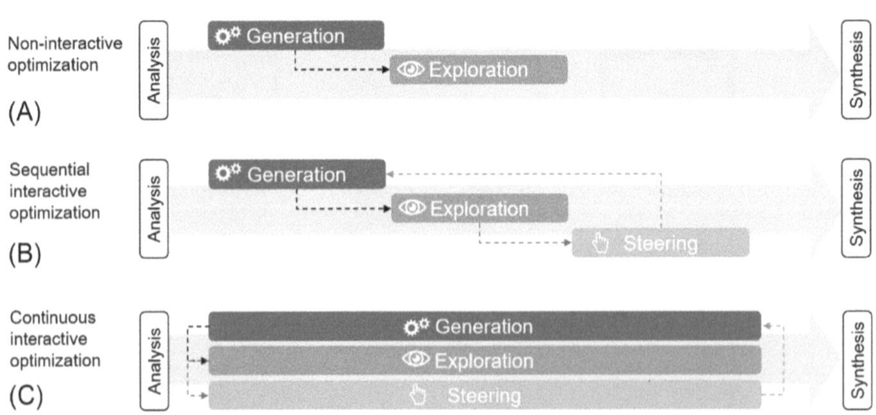

Fig. 2.2 URB^{io}: an integration of the three phases of interactive optimization. Reproduced from Cajot, S., Schüler, N., Peter, M., Koch, A., Maréchal, F., (submitted). Interactive optimization with parallel coordinates: exploring multidimensional spaces for decision support. Frontiers in ICT.

be considered an extension of the user's mind, while at the same time, the user's mind can be considered an extension of the optimization model. Vinge (1993) anticipated the need for less "oracular" and more symmetrical decision support systems, where the program provides the user with as much information as the user provides the program guidance. In other words of his, "computer/human interfaces may become so intimate that users may reasonably be considered superhumanly intelligent." Though it may be a bit early to claim achieving "superhuman intelligence," the close relationship between user and solver enabled by URB^{io} is certainly a step in that direction.

2.2 An optimization model for the planning of new neighborhoods

In this section it is demonstrated how decisions to be taken in an urban planning process can be translated into an MILP model. Although in the European context, urban renewal projects (so-called "brownfield projects") are outnumbering urban expansion projects (so-called "greenfield projects"), the model was developed taking also into account the decisions to be made in the scope of greenfield projects. Aiming for a truly versatile applicability of the model, the main reason for this was the perception that from a mathematical point of view, brownfield problems can be regarded as a subset of greenfield problems where some of the decision variables are fixed prior to the optimization.

The multitude of domains affected by the outcomes of a planning process—and thus to be respected during this process—encompasses the urban form and function as the layout of cities and the distribution of services (Steemers, 2003; Oswald et al., 2003; Schwarz, 2010). A related but yet distinct domain is formed by social aspects such as the spatial variation of density, or the livability, for example, in form of open space (Lehnerer, 2009). Economic aspects comprise a third domain in form of costs and benefits for the individual and for the community (Keirstead and Shah, 2013). To this adds with increasing importance the responsibility of planning cities that are sustainable not only regarding the aforementioned domains but also regarding the demand for and supply of energy and their environmental implications (Moavenzadeh et al., 2002; Steemers, 2003; Oswald et al., 2003).

Furthermore, urban systems inherently span multiple spatial scales ranging from beyond the city borders, concerning, for example, the exchange of material, energy, people, and information (Robinson, 2011), down to the scale of, for example, a single wall, whose insulation impacts energy demand (König et al., 2009; Rode et al., 2014; Kno, 2012; Ratti et al., 2003). The range of considered scales is set to cover the floor scale up to the neighborhood scale. The choice for the smallest scale is mainly due to the planning task of designing both horizontally and vertically mixed-use developments that foster social equality. The largest scale is selected in order to cover most of phenomena that only arise when looking beyond the building scale, while remaining small enough that actual outcomes of the model can have a practical meaning.

This section is structured by ordering the model equations according to the domain they belong to, and the spatial scale they occur on. However, such a separation is of course subjective to some extent, as some aspects of urban systems span several domains and/or scales while other aspects arise only from the holistic view of the separate domains and scales.

2.2.1 Urban form and function

2.2.1.1 Floor scale

2.2.1.1.1 Number of floors and occupancy type

A planning site for a new neighborhood is divided into parcels on which buildings can be placed. These buildings are comprised of floors n_{fl} of potentially different occupancy types *ot*.

$$n_{fl}^{pc,bt,ot} \geq 0 \quad \forall pc \in PC,\ bt \in BT,\ ot \in OT \tag{2.1}$$

The number of floors is a key decision variable of the problem. It is an integer variable, which is indexed over the sets of parcels *PC*, occupancy types *OT*, and building types *BT*. Occupancy types can take values like "residential" (*rsd*), "office" (*off*), or "commercial" (*com*).

2.2.1.2 Building scale

2.2.1.2.1 Building type

Considering different building types allows to account for the differences between mixed-use buildings (*mix*) and single family homes (*sfh*) in, for example, footprint, occupation rates or surface-specific energy demand. A binary decision variable ξ indicates if a specific building type is present at a parcel. Each parcel can be occupied by at most one building type:

$$\sum_{bt} \xi_{blg}^{pc,bt} \leq 1 \quad \forall pc \in PC \tag{2.2}$$

A building's existence is coupled to the number of floors via:

$$\frac{\sum_{ot} n_{fl}^{pc,bt,ot}}{n_{fl,max}} \leq \xi_{blg}^{pc,bt} \leq \sum_{ot} n_{fl}^{pc,bt,ot} \quad \forall pc \in PC,\ bt \in BT \tag{2.3}$$

Special constraints for building types can enforce, for example, only residential floors for SFH.

2.2.1.2.2 Gross floor area

The gross floor area (GFA) per location and occupancy type results from the specific number of floors and the footprint:

$$A_{fl}^{pc,bt,ot} = n_{fl}^{pc,bt,ot} \cdot A_{fp}^{bt} \quad \forall pc \in PC,\ bt \in BT,\ ot \in OT \tag{2.4}$$

2.2.1.2.3 Height

The building height is calculated based on the number and type-dependent height of floors:

$$h_{blg}^{pc} = \sum_{ot} n_{fl}^{pc,bt,ot} \cdot h_{fl}^{ot} \leq h_{max}^{bt} \quad \forall pc \in PC,\ bt \in BT \tag{2.5}$$

The maximum building height h_{max}^{bt} results on the one hand from site-specific legislations. On the other hand, this parameter accounts for the fact that specific building types in very most cases do not exceed a typical height. The resulting difference in absolute building height has an effect again on parameters such as solar potential or view quality. In order to address questions as to where shops should be located within a neighborhood, the building height is calculated based on individual floor heights.

2.2.2 *Social aspects*

2.2.2.1 *Subfloor scale*

The social aspects of cities are naturally arising with the presence of people. Therefore, the number of inhabitants and jobs is estimated for each building on a floor scale based on statistical occupation rates.

2.2.2.1.1 Numbers of dwellings and people

The number of dwellings n_{dw} at each location can be estimated based on the floor area per dwelling A_{dw} with:

$$n_{dw}^{pc} = \sum_{bt \in BT} \frac{A_{fl}^{pc,bt,ot}}{A_{dw}^{bt}} \quad \forall pc \in PC,\ ot \in \{\text{“}res\text{”}\} \tag{2.6}$$

and analogously the number of jobs n_{jobs} based on the floor area per job A_{jobs}.

The number of inhabitants n_{inhab} is then calculated based on average values of inhabitants per dwelling $n_{inhab/dw}$:

$$n_{inhab}^{pc} = \sum_{bt \in BT} \frac{A_{fl}^{pc,bt,ot}}{A_{dw}^{bt}} \cdot n_{inhab/dw}^{bt} \quad \forall pc \in PC, ot \in \{\text{“}res\text{”}\} \tag{2.7}$$

2.2.2.2 Block scale

2.2.2.2.1 Proximity

The proximity of buildings to streets has an effect on access to mobility, noise levels, and visibility, where the evaluation of these different aspects can depend on the building function. While access to mobility is rather positive for all building functions and high noise levels are rather to be evaluated negative, a visibility could be preferable for shops and offices, while unpreferable for residential buildings. This situation can be respected in the model by preventing commercial and office floors being located on parcels, which are not adjacent to major streets.

$$n_{fl}^{pc,bt,ot} = 0 \quad \forall pc \in \{PC : \text{parcel not at major street}\}, bt \in BT, ot \in \{\text{“}com,\text{”}\ \text{“}off\text{”}\} \tag{2.8}$$

2.2.2.2.2 Parks

Several authors covered the optimal allocation of parks on the city scale considering aspects like population density, air and water quality, noise, the urban heat island effect, or land-use patterns (Neema and Ohgai, 2010, 2013; Yuan et al., 2011; Fernández et al., 2015). Assessments on the district to neighborhood scale treat, for example, the accessibility, connectivity, costs, number of beneficiaries, or population density, respectively (Sefair et al., 2012; Yu et al., 2014). This work focuses on the relationships between parks and their surroundings on the building to floor scale. The modeled aspects are inspired by the work of Jacobs (1961), who describes several basic principles characterizing parks with a positive contribution to their environment.

Coherence First, parks should be enclosed and clearly delimited by a certain number of buildings, rather than isolated at the border of the community. Not only does a proper enclosure by buildings visually and spatially define the park as a positive feature, but it also increases its accessibility. This is implemented by providing a constraint on occupied parcels around parks:

$$\xi_{park}^{pc1} \leq \frac{\sum_{pc2,bt} \xi_{blg}^{pc2,bt}}{n_{blg@parks,min}} \quad \forall \left\{ pc1, pc2 \in PC : pc1 \neq pc2, d_{ecl}^{pc1,pc2} \leq d_{blg@parks,max} \right\} \tag{2.9}$$

By adjusting the minimum number of buildings around a park $n_{blg@park,\ min}$ implications of this constraint can be influenced. Higher numbers result to more enclosed parks. Setting the maximum Euclidean distance between the park's parcel and surrounding parcels $d_{blg@parks,\ max}$ defines the perimeter in which surrounding buildings should be considered. Different combinations of both parameters allow to either enforce or prohibit specific layouts such as parks located at corners of blocks.

Diversity Second, and related to the enclosure principle, surrounding buildings should offer a high diversity of uses. This in turns ensures a high diversity of users, leading to a more continuous occupation of the park throughout the day, thus reducing insecurity and vandalism. In the model, this can be enforced by putting thresholds on numbers of floors of a certain occupancy type on parcels around parks:

$$\xi_{park}^{pc1} \leq \frac{\sum_{pc2,bt} n_{fl}^{pc2,bt}}{n_{fl@park,min}^{bt}} \quad \forall \left\{ pc1, pc2 \in PC : pc1 \neq pc2, d_{ecl}^{pc1,pc2} \leq d_{blg@parks,max} \right\} \tag{2.10}$$

Direct sunlight Third, the attractiveness of the park is also influenced by the amount of direct sunlight throughout the year. Complete obstruction of sunlight by tall buildings should be avoided. Here, a constraint for minimum direct solar irradiation per surface ir_{dir} is used:

$$ir_{dir,min} \cdot \xi_{park}^{pc} \leq ir_{dir}^{pc} \quad \forall pc \in PC \tag{2.11}$$

2.2.2.3 Neighborhood scale

2.2.2.3.1 Density

A common planning measure on aggregated scales are densities of different kinds, such as surface densities or population densities. In this work surface densities are employed which can be converted into population densities, once statistical occupation rates are known:

The floor area ratio (FAR) is taken here as the GFA A_{fl} of all buildings divided by the buildable surface of the entire neighborhood, which is defined by the local planning authority:

$$FAR = \frac{\sum_{pc} \sum_{bt} \sum_{ot} A_{fl}^{pc,bt,ot}}{\sum_{pc} A_{par}^{pc}} \tag{2.12}$$

2.2.2.3.2 Surface shares

The shares of surface of the different occupancy types are defined as a measure for mixed development.

$$s^{ot1} = \frac{A_{fl}^{pc,bt,ot1}}{\sum_{ot2 \in OT} A_{fl}^{pc,bt,ot2}} \quad \forall ot1 \in OT \tag{2.13}$$

2.2.2.3.3 Proximity

Next to the proximity parameter on block scale, there are other proximity parameters for which higher scales are relevant. These are proximity to shops, parks, or public transport stops.

The shop share allows to scale the amount of shops with residential density. However, it does not allow to influence the spatial distribution of shops. A more evenly distribution of shops could be desirable since locating shops within certain distances could help to promote soft mobility.

This is implemented by setting a lower bound on the number of commercial floors $n_{fl,com,min}$ within a maximum distance $d_{shops,max}$ around buildings which can have residential floors:

$$\sum_{pc2}\sum_{bt} n_{fl}^{pc2,bt,\text{"com"}} \geq n_{fl,com,min} \sum_{bt} \xi_{blg}^{pc1,bt} \quad \forall \left\{ pc1, pc2 \in PC : d_{mnh}^{pc1,pc2} \leq d_{shop,max} \right\} \tag{2.14}$$

$d_{mnh}^{pc1,pc2}$ is the Manhattan distance as displayed in Fig. 2.3. Specifying more than one commercial floor within the certain distance could help to ensure the provision with shops of different kinds. The combination of shop-to-residence ratio and number of shops within a maximum distance allows to influence quantity and distribution of shops.

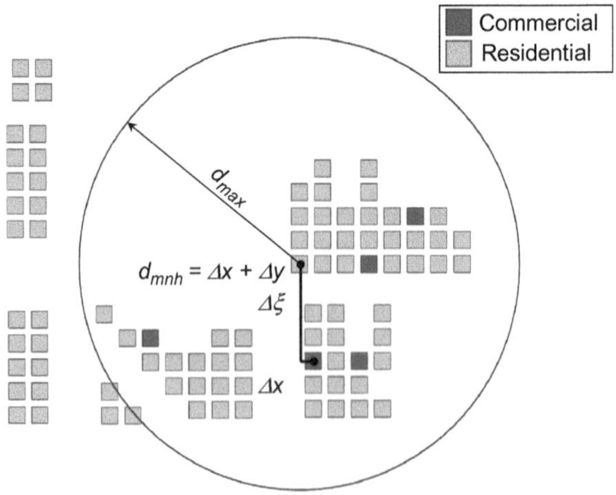

Fig. 2.3 Calculation of Manhattan distance.

Most of the points just stated for shops also hold for parks resulting in a similar constraint:

$$\sum_{pc2} \xi_{park}^{pc2} \geq n_{parks,min} \sum_{bt} \xi_{blg}^{pc1,bt} \quad \forall \left\{pc1, pc2 \in PC : pc1 \neq pc2, d_{mnh}^{pc1,pc2} \leq d_{park,max}\right\} \tag{2.15}$$

It shall be hinted that the proximity of parks is not the same as the coherence of a park as defined before, since a park could be embedded within many buildings while still being not close to all buildings in the neighborhood. Proximity constraints for shops and parks influence implicitly also positively the diversity of functions within a neighborhood.

The frequency and distribution of public transport stops depend on the means of transportation. Bus stops in urban areas are quite frequent and similar constraints as for shops and parks could be implemented to affect their distribution. Other means of transportation, for example, trams or trains, stop less frequently, whereas the definition of "frequently" depends on the considered scale. On a neighborhood scale their might be only one tram stop, while on a city scale only one train stop. In order to influence the placement of such a central mobility hub, the individual distance to buildings weighted by numbers of floors of any occupancy type is accumulated:

$$d_{ts,cum} \geq \sum_{pc} \sum_{bt} \sum_{ot} d_{mnh}^{pc,ts} \left(n_{fl}^{pc,bt,ot} - \frac{\max\limits_{bt \in BT} h_{max}^{bt}}{\min\limits_{ot \in OT} h_{fl}^{ot}} \left(1 - \xi_{ts}^{ts}\right) \right) \tag{2.16}$$

Depending on the question the model shall help to answer, this accumulated distance can then be either constrained or minimized.

For example, the goal of choosing among a predefined set of locations the central transport stop, which minimizes the accumulated distances, can be achieved by constraining the number of transport stops to one:

$$\sum_{ts} \xi_{ts}^{ts} = 1 \tag{2.17}$$

2.2.2.3.4 Landmark view factor

Private view The quality of living space and with it its associated price is influenced by the view it offers on surrounding landmarks. Consequently a performance indicator to assess view quality is needed: The Landmark View Factor (LVF) is defined as the share of floors, either per building or per neighborhood, that have a view on a specified landmark. The number of floors with a view per building can be approximated by calculating the vertical part of a building from which the landmark could be seen and dividing it by the average floor height (see Fig. 2.4A). This number is calculated for each pair of buildings, where one is potentially obstructing the view of the other:

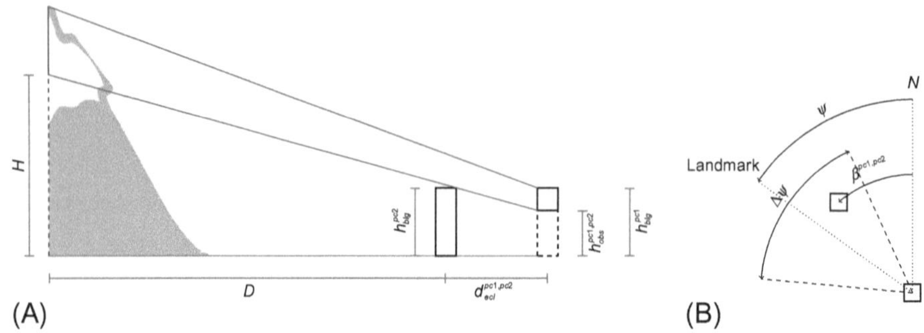

Fig. 2.4 Views of a planning site to illustrate considered heights, distances, and angles for the calculation of the LVF. (A) Side view; (B) top view.

$$n_{fl,view,rel}^{pc1,pc2} = \frac{h_{blg}^{pc1} - h_{obs}^{pc1,pc2}}{\overline{h}_{fl}}$$
$$\forall \left\{ pc1, pc2 \in PC : pc1 \neq pc2, \psi - \frac{\Delta\psi}{2} \leq \beta^{pc1,pc2} \leq \psi + \frac{\Delta\psi}{2} \wedge d_{ecl}^{pc1,pc2} \leq d_{max} \right\} \tag{2.18}$$

h_{blg}^{pc1} is the actual height of the building at location 1, h_{obs}^{pc1} the height of the obstructed part of the building, and $\overline{h}_{fl}$ the average floor height. The height of the obstructed part of the building can be deduced from the height of the landmark H, the distance toward the landmark D and the Euclidean distance between the two considered buildings d_{ecl} using the intercept theorem (Fig. 2.4A):

$$\frac{h_{blg}^{pc2} - h_{obs}^{pc1,pc2}}{H - h_{obs}^{pc1,pc2}} = \frac{d_{ecl}^{pc1,pc2}}{D + d_{ecl}^{pc1,pc2}} \tag{2.19}$$

which becomes:

$$h_{obs}^{pc1,pc2} = \frac{h_{blg}^{pc2} - \dfrac{d_{ecl}^{pc1,pc2} \cdot H}{d_{ecl}^{pc1,pc2} + D}}{1 - \dfrac{d_{ecl}^{pc1,pc2}}{d_{ecl}^{pc1,pc2} + D}} \tag{2.20}$$

which can be substituted in Eq. (2.18). The parcel elevations could be included to make it applicable to uneven planning areas.

Using the Euclidean distance between centers of parcels disregards the horizontal extension of the buildings, which results in an optimistic estimation of the LVF of all buildings and in neglecting the more obstructing effect of buildings with larger footprints. A conservative estimation could be made when using the distances between parcel borders. However, an accurate assessment of the final LVF in an early planning phase is already hindered as the planner might not be able to influence the final buildings' extension and location on each parcel. Thus this indicator should rather serve to detect general trends and trade-offs with other urban aspects.

The number of floors that enjoy a good view is defined for every pair of parcels where a building on one of it could potentially obstruct the view of the other. This is accounted for by regarding only parcels that are both within a certain, predefined margin of angle toward the landmark $\Delta\psi$ (see Fig. 2.4B) and a maximum distance d_{max}. The latter is again deduced from the intercept theorem:

$$d_{max} = \frac{D}{H} \cdot \left(\frac{1}{\max\limits_{bt \in BT} h_{max}^{bt}} - \frac{1}{H} \right)^{-1} \tag{2.21}$$

with the maximum height over all building types $\max_{bt \in BT} h_{max}^{bt}$.

The view range $\Delta\psi$ depends naturally on the type of landmark. In the case of a natural land mark such as a mountain range, most likely only a broader view range would be considered as providing a building with a nice view than it would be the case with, for example, an architectural landmark. The view range has further an effect on the number of generated constraints for the model as a larger view range leads to more constraints.

To finally get the real number of floors with a nice view for each building requires taking the minimum[3] of Eq. (2.18). In the case there is no obstruction by any building, which is especially the case for buildings lying at the neighborhood's border in the direction of the landmark, the LVF equals the number of residential and office floors of the building. Commercial floors are thus not being valorized for a nice view.

$$n_{fl,view}^{pc1} = \min_{pc2} \left\{ n_{fl,view,rel}^{pc1,pc2}; \sum_{bt} \sum_{ot} n_{fl}^{pc1,bt,ot \neq \text{``com''}} \right\} \quad \forall pc1 \in PC \tag{2.22}$$

An LVF per building is then

$$LVF^{pc} = \frac{n_{fl,view}^{pc}}{\sum_{bt} \sum_{ot} n_{fl}^{pc,bt,ot \neq \text{``com''}}} \quad \forall pc \in PC \tag{2.23}$$

In order to compare resulting neighborhood layouts, an overall LVF is calculated. Therefore, the LVFs of all buildings are summed up and normalized by the total number of residential and office floors within the neighborhood:

[3] A linear form of Eq. (2.22) is implemented using inequality constraints.

$$LVF_{tot} = \frac{\sum_{pc} n_{fl,view}^{pc}}{\sum_{pc}\sum_{bt}\sum_{ot} n_{fl}^{pc,bt,ot \neq \text{“com”}}} \tag{2.24}$$

Public view A view from many floors on a landmark is a desirable target, but this view is not enjoyable by everyone, if the buildings are not used for public purposes. The quality of a new neighborhood might, however, also be assessed by the question from how many public spaces a landmark is visible. The above-introduced concept is thus extended by an LVF from open spaces. Therefore, a binary variable is used to indicate if an open space (e.g., a public park) has a view on a landmark and if the open space is used for public purposes (e.g., parks)[4]:

$$\xi_{park,view}^{pc1} = \min_{pc2} \left\{ n_{fl,view,rel}^{pc1,pc2}; \xi_{park}^{pc1} \right\} \quad \forall pc1 \in PC \tag{2.25}$$

The overall LVF of a neighborhood for view from parks is then defined as the share of parks from which the landmark can be seen:

$$LVF_{parks,tot} = \frac{\sum_{pc} \xi_{park,view}^{pc}}{\sum_{pc} \xi_{park}^{pc}} \tag{2.26}$$

2.2.3 Energy aspects

2.2.3.1 Floor scale

2.2.3.1.1 Energy demand

National norms and regulations define energy target values for the construction of new buildings. Surface-specific energy demands e_{blg}, specified by those norms, are used to estimate building demands E_{blg}, in form of heat, electricity, or fuel.

$$E_{blg}^{pc,t} = \sum_{bt}\sum_{ot} sr_{ERA}^{bt} \cdot e_{blg}^{ot,t} \cdot A_{fl}^{pc,bt,ot} \quad \forall pc \in PC, t \in T \tag{2.27}$$

A building type-dependent conversion factor sr_{ERA} allows the estimation of the energetic reference area (ERA) based on the GFA (Eq. 2.4):

[4] A linear form of Eq. (2.25) is implemented using inequality constraints.

2.2.3.2 Building scale

2.2.3.2.1 Energy supply

The resulting energy demand of each building has to be satisfied by a combination of different conversion technologies *CT*, which are installed at the building, and energy form-specific exchange technologies *XT*, which allow to exchange with energy networks *ntw*. Thus local balances for each energy form, location, and time step have to be closed:

$$E_{blg}^{pc,t} + \sum_{ct} E^{lc,ct,t} + E_{xt}^{lc,t} = 0 \quad \forall pc \in PC \subseteq LC,\ lc \in LC,\ t \in T \tag{2.28}$$

The set of locations *LC* comprises building parcels, as well as locations of centralized technologies. The energy networks are modeled by energy balances, which allow to exchange energy between buildings, locations of centralized conversion technologies, and national energy networks *ntw*, *nat*.

$$\sum_{lc} E_{xt}^{lc,t} + E_{ntw,nat}^{t} = 0 \quad \forall t \in T \tag{2.29}$$

The actual energy flows depend on the operation rate *o* and a rate-specific energy flow *r* of each technology. The equation for all energy technologies *ET*, which comprise the subsets of conversion and exchange technologies, is:

$$E^{lc,et,t} = o^{lc,et,t} \cdot r^{lc,et,t} \quad \forall lc \in LC,\ et \in ET,\ t \in T \tag{2.30}$$

The rate-specific energy flow comprises conversion efficiencies (depending on the technology and energy form) or exchange and transport losses. Its sign determines if a flow enters or leaves a balance. The operation rate of each technology is naturally limited by its existence ξ and by its minimum and maximum nominal sizes, σ_{min} and σ_{max}:

$$\xi^{lc,et,t} \cdot \min_{s} \sigma_{min}^{et,s} \leq o^{lc,et,t} \leq \xi^{lc,et,t} \cdot \max_{s} \sigma_{max}^{et,s} \quad \forall lc \in LC,\ et \in ET,\ t \in T \tag{2.31}$$

These minimum and maximum nominal sizes can be further indexed over segments *s* of piecewise linearized cost functions to account for economies of scale (see Section 2.2.5.1).

Photovoltaic For detailed descriptions of each type of conversion technologies the reader is referred to the references stated earlier. In the following only the model for solar conversion technologies is presented for constituting a novel method that integrates cumulative skies into an MILP formulation.

In order to assess the amount of electricity generated by PV panels on the buildings' roofs it is necessary to know the total incident solar irradiation. This depends on the tilt angle and area of the mounted PV panels and on the amount of incident irradiation.

The amount of solar irradiation during a specific time period coming from a specific sky patch can be assessed using cumulative skies. The incident irradiation on the rooftop then depends on obstructions toward this sky vault. In urban surroundings, near-field obstructions depend mainly on the location and height of surrounding buildings. Thus the problem can be narrowed down to the determination of whether a sky patch can be seen from a given parcel. Therefore, it is determined for every parcel and for all azimuth directions of all sky patches, which building in each sky patch direction is obstructing the sky the most by calculating the maximum[5] tangent of elevation angle $\tan \epsilon$[6] over all pairs of building heights (compare Figs. 2.5 and 2.4B):

$$\sum_{el} \tan \epsilon^{pc1,az,el} = \max_{pc2} \frac{h_{blg}^{pc2} - h_{blg}^{pc1}}{d_{ecl}^{pc1,pc2}}$$
$$\forall \left\{ pc1, pc2 \in PC : pc1 \neq pc2, \alpha^{az} - \frac{\Delta\alpha^{az}}{2} \leq \beta^{pc1,pc2} \leq \alpha^{az} + \frac{\Delta\alpha^{az}}{2} \right\} \tag{2.32}$$

where the azimuth angles and the elevation angles of all sky patches form two discrete sets *AZ* and *EL*. Calculating the tangent of the elevation angle instead of the elevation angle itself allows to preserve a linear formulation. Furthermore, there is a tangent variable for each single sky patch in order to calculate the irradiation from the sky dome using a piecewise linear formulation. Therefore for each azimuth direction the highest (partly) obstructed sky patch is determined using binary variables λ:

$$\lambda^{pc,az,el} \cdot \tan \epsilon_{min}^{el} \leq \mathbf{tan}\ \epsilon^{pc1,az,el} \leq \lambda^{pc,az,el} \cdot \tan \epsilon_{max}^{el}$$
$$\forall pc \in PC,\ az \in AZ,\ el \in EL \tag{2.33}$$

$$\sum_{el} \lambda^{pc,az,el} = 1 \quad \forall pc \in PC,\ az \in AZ \tag{2.34}$$

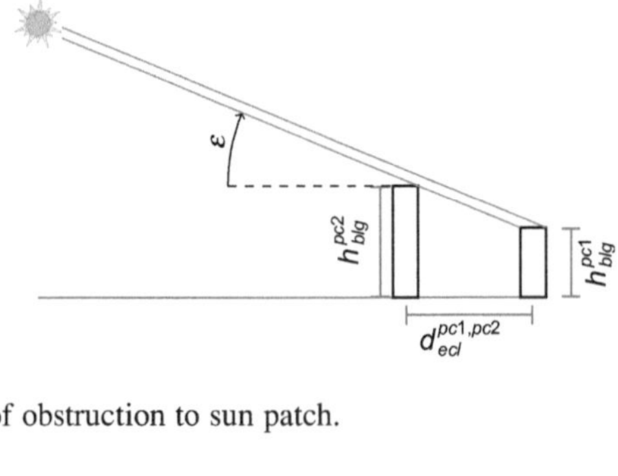

Fig. 2.5 Calculation of obstruction to sun patch.

[5] A linear form of Eq. (2.32) is implemented using inequality constraints. Note that consequently the values of $\tan \epsilon$ are meaningless for problems where the objective function does not drive these values to their bounds.

[6] Assigning a variable to the tangent of the elevation angle rather than to the angle itself allows to maintain a linear formulation.

The surface-specific irradiation ir for each parcel and each time step is then a function of obstructed sky patches:

$$ir_{tot}^{pc,t} = ir_{tot,unobs}^{t} - \sum_{az}\sum_{el}\left(ir_{tot,ic}^{az,el,t} \cdot \lambda^{pc,az,el} - ir_{tot,gr}^{az,el,t} \cdot \mathbf{tan}\ \epsilon^{pc,az,el}\right) \quad \forall pc \in PC,\ t \in T \tag{2.35}$$

with the total irradiation tot from the unobstructed sky vault ir_{unobs} and the intercept ir_{ic} and gradient ir_{gr} of each sky patch. Herein it is assumed that the amount of irradiation not received from a sky patch is linear proportional to its degree of obstruction. For lower or upper bounds on direct sunlight in parks (see the "Direct sunlight" section), the sky domes only comprise the amount of direct irradiation.

Finally the upper bound for the electricity output of a roof mounted PV panel with electric efficiency η_{el} is:

$$E_{pv}^{pc,t} \le \eta_{el,pv} \cdot sr_{pv} \cdot A_{fp}^{bt} \cdot \left(ir_{tot}^{pc,t} + ir_{tot,unobs}(1 - \xi_{blg}^{pc,bt})\right) \quad \forall pc \in PC,\ t \in T \tag{2.36}$$

$$E_{pv}^{pc,t} \le \sum_{bt} \xi_{blg}^{pc,bt} \cdot \eta_{el,pv} \cdot sr_{pv} \cdot A_{fp}^{bt} \cdot ir_{tot,unobs}^{t} \quad \forall pc \in PC,\ t \in T \tag{2.37}$$

where the panel surface depends on the constructed building's footprint and the share of roof surface available for PV panel installation sr_{PV}. Eq. (2.36) assures a panel area chosen according to the building type's footprint while Eq. (2.37) prevents panels on empty parcels.

2.2.3.2.2 Energy networks

An exchange technology extracts from the neighborhood-wide network the amount required for satisfying the local balance plus an amount to account for transfer losses $loss$. Depending on the energy form, this loss can be proportional to both the amount of transferred energy and the network distance between the parcel and the feed-in point of the centralized conversion technologies d_{ntw}. Thus the rate-specific energy flow for an exchange technology is:

$$r^{lc,xt,t} = 1 + r_{loss} \cdot d_{ntw}^{lc} \quad \forall lc \in LC,\ xt \in XT,\ t \in T \tag{2.38}$$

The distance d_{ntw} is the sum of the network distance on neighborhood scale as explained in Section 2.2.3.4.1 and the Manhattan distance between the parcel and the point the block is connected to the network.

2.2.3.3 Block scale

2.2.3.3.1 Energy networks

The network models were developed with the goal of obtaining a good estimation of the network lengths while increasing the problem size as little as possible. A conventional way to include the network layout in MILP formulations is to decide the existence of connections between nodes using binary variables for each potential connection (Söderman and Pettersson, 2006; Weber, 2008; Fazlollahi, 2014). For larger problems this results into a high amount of binaries. In contrary the method proposed here uses binary variables for the decision of whether a block is connected to a network.

Furthermore, the block scale is chosen as smallest scale to determine the length of a network. Thus a building can be only connected if a block is connected:

$$\xi_{hts}^{pc} \leq \xi_{hn}^{bl} \quad \forall bl \in BL, pc \in PC^{bl} \tag{2.39}$$

2.2.3.4 Neighborhood scale

2.2.3.4.1 Energy networks

The neighborhood-scale network model is based on the identification of network branches before the actual optimization. Therefore, the actual planning site is studied and potential network layouts connecting all blocks are designed, where blocks could be connected in different ways. These layouts are then analyzed to identify branches. The main branch is starting at the point where the neighborhood-scale network is connected to a network of an upper scale. Following this branch, at each bifurcation it is checked for each bifurcating pipe if it connects more than one block. If so, an additional branch is defined. Otherwise it is defined as a block connection whose length $l_{ntw,\ block}$ is stored. In the case of several potential block connections of one block, the average is taken. The branches are then reviewed to identify and merge pairs of branches that potentially connect the same blocks. Following this, it is determined for each pair of blocks and branches, how long the branch would at least have to be if the block was connected, resulting in the lengths of the branch segments $l_{ntw,\ seg}$. Each branch length can then be determined by the following equation:

$$l_{ntw,branch}^{br} \geq \xi^{bl} \cdot l_{ntw,seg}^{bl,br} \quad \forall br \in BR, bl \in BL \tag{2.40}$$

The total length of a network is then:

$$l_{ntw} = \sum_{br} l_{ntw,branch}^{br} + \sum_{bl} \xi^{bl} \cdot l_{ntw,block}^{bl} \tag{2.41}$$

Connection rate The connection rate cr is defined as the share of buildings, which is connected to an energy network via an exchange technology:

$$cr_{ntw} = \frac{\sum_{pc \in PC} \sum_{xt \in XT} \xi_{blg}^{pc,xt}}{\sum_{pc \in PC} \sum_{bt \in BT} \xi_{blg}^{pc,bt}} \tag{2.42}$$

It is calculated after the optimization to be able to compare resulting plans.

2.2.4 Environmental aspects

Being an explicit planning target in many countries and cities, including in the chosen case study (see the "Direct sunlight" section), the share of energy originating from renewable sources is chosen as main indicator to measure the sustainability of the energy supply system.

2.2.4.1 Neighborhood scale

2.2.4.1.1 Share of renewable energy sources

This share is defined here as the part of the sum of annual energy required to cover the demand of buildings, network losses *loss*, and potential electricity exports *exp* that are covered by renewable energy sources. Considered energy services are heat required for room heating *rh* and hot water preparation *hw* and electricity required to meet demands due to building services *bs* and room cooling *rc* via compression chillers.

The following energy flows are considered as renewable: Heat from air, soil, and waste heat *wht*, electricity from PV, and partly from the national electricity network *en-nat*.

$$s_{RES} = \frac{E_{air} + E_{soil} + E_{wht} + E_{pv} + s_{RES,en-nat} \cdot E_{imp,net}}{\sum_{pc} \left(E^{pc}_{rh+hw} + E^{pc}_{bs+rc}\right) + E_{hn,loss} + E_{en-loc,loss} + E_{exp,net}} \tag{2.43}$$

The net exported and imported electricity flows are calculated using the binary method for modeling absolute values in MILP:

$$E^{pc}_{exp,net} - E^{pc}_{imp,net} = E^{pc}_{exp} - E^{pc}_{imp} \quad \forall pc \in PC \tag{2.44}$$

$$0 \leq E^{pc}_{exp,net} \leq \xi^{pc}_{exp,net} \cdot E_{exp,max} \quad \forall pc \in PC \tag{2.45}$$

$$0 \leq E^{pc}_{imp,net} \leq \left(1 - \xi^{pc}_{exp,net}\right) \cdot E_{imp,max} \quad \forall pc \in PC \tag{2.46}$$

From Eq. (2.43) a constraint can be derived to ensure a minimum share of RES $s_{RES,min}$ for the energy supply of the neighborhood including its distribution networks:

$$\begin{aligned} &E_{air} + E_{soil} + E_{wht} + E_{pv} + s_{RES,en-nat} \cdot E_{imp,net} \\ &\geq s_{RES,min} \left(\sum_{pc} \left(E^{pc}_{rh+hw} + E^{pc}_{bs+rc}\right) + E_{hn,loss} + E_{en-loc,loss} + E_{exp,net} \right) \end{aligned} \tag{2.47}$$

2.2.5 Economic aspects

Considered cost factors in this work are: Investment and operation costs of the provision of energy and the investment cost of the transportation network. To account for the effect of economies of scales piecewise linearization is employed for the calculation of the investment cost of technologies.

2.2.5.1 Building scale

The operation rates of the technologies for each time step (Eq. 2.31) result from closing the energy balances (Eq. 2.28) (Section 2.2.3.2). The final size of the technology that shall be purchased, results from the operation rates and the existence of technologies:

$$o^{lc,et,t} \leq \sum_{s} \sigma^{lc,et,s} \quad \forall lc \in LC,\ et \in ET,\ t \in T \tag{2.48}$$

$$\sum_{s} \xi^{lc,et,s} \leq \sum_{t} \xi^{lc,et,t} \quad \forall lc \in LC,\ et \in ET \tag{2.49}$$

where the additional index s denotes the segments of the linearized investment cost curve. Finally, the following equations serve to determine the single, active segment of the cost curve corresponding to the chosen unit size.

$$\xi^{lc,et,s} \cdot \sigma_{min}^{et,s} \leq \sigma^{lc,et,s} \leq \xi^{lc,et,s} \cdot \sigma_{max}^{et,s} \quad \forall lc \in LC,\ et \in ET,\ s \in S \tag{2.50}$$

$$\sum_{s} \xi^{lc,et,s} \leq 1 \quad \forall lc \in LC,\ et \in ET \tag{2.51}$$

Investment costs C_{inv} and operation costs C_{op} are then calculated as:

$$C_{inv}^{cb,lc,et} = \sum_{s} \left(\xi^{lc,et,s} \cdot c_{inv,cst}^{cb,et,s} + \sigma^{lc,et,s} \cdot c_{cb,inv,lin}^{et,s} \right) \quad \forall cb \in CB,\ lc \in LC,\ et \in ET \tag{2.52}$$

$$C_{op}^{cb,lc,et} = \sum_{t} o^{lc,et,t} \cdot c_{op,lin}^{cb,et} \cdot \delta^{t} \quad \forall cb \in CB,\ lc \in LC,\ et \in ET \tag{2.53}$$

where the operation costs are dependent on each time step of duration δ. $c_{cb,inv,cst}$ and $c_{cb,inv,lin}$ denote the constant and the linear part of each segment of the linearized investment cost curve. The operation cost factor $c_{cb,op,lin}$ depends on the operation rate of the technology.

In order to compare the different cost, the investment costs are annualized using an annuity factor

$$af^{cb,et} = \frac{i^{cb} \cdot \left(i^{cb} + 1\right)^{\wedge} lt^{et}}{\left(i^{cb} + 1\right)^{\wedge} lt^{et} - 1} \quad \forall cb \in CB,\ et \in ET \tag{2.54}$$

which depends on the actor-specific interest rate i and the equipment-specific lifetime lt.

Costs C and cost factors c are further indexed over cost balances cb to account for costs occurring to different actors. Thus operation costs can be linked to a cost flow into one balance (e.g., the building's tenants) and a cost flow out (i.e., revenues) of a another balance (e.g., the energy supplier).

2.2.5.1.1 Photovoltaic

The PV model presented in Section 2.2.3.2.1 estimates the solar potential based on height differences with surrounding buildings. As these heights are resulting from decision variables, an additional sizing of the panels in terms of area would lead to a nonlinear problem formulation. Thus the assumption is made that if panels are installed, they always cover the entire part of the roof top area that is deemed as suited for PV electricity production. The implication for the cost calculation of panels is that there is no linear cost factor in Eq. (2.52).

2.2.5.2 Neighborhood scale

2.2.5.2.1 Energy networks

The investment costs for energy networks are calculated based on the overall network length l_{ntw} from Eq. (2.41) and an average pipe diameter dm_{ntw} if applicable:

$$C_{inv,ntw} = (c_{inv,l,dm} \cdot dm_{ntw} + c_{inv,l}) \cdot l_{ntw} \tag{2.55}$$

2.2.5.2.2 Transportation network

The costs for the transportation network result from a length-specific cost factor $c_{inv,l}$, a base length for the shortest possible network connection $l_{tn,base}$ and the length of additional sections due to stops placed more centrally in the neighborhood:

$$C_{inv,tn} = \left(l_{tn,base} + \sum_{ts} \xi_{ts}^{ts} \cdot \Delta l_{tn}^{ts} \right) \cdot c_{inv,l} \tag{2.56}$$

2.3 Multi-parametric programming for urban planning

This section shows how the task of urban planning can be stated as a multi-parametric programming problem. Decisions are usually taken based on performance indicators Π, which are themselves a function of the former:

$$\Pi = F(\phi), \quad \phi \in \Phi \tag{2.57}$$

ϕ denotes all variables whose values are to be decided (i.e., decision variables) within their regarding variable space Φ. Based on the recognition that the number of decisions is very large, further that these decisions are made by different actors at different phases of the urban development project, and last that all these decisions influence the final values of the performance indicators, the decisions are roughly separated into decision variables θ, which are typically decided by planners at early phases and larger scales of the urban project, and decision variables x and y, which are typically decided by different actors at later phases and smaller scales. Adopting a common denotation in the field of multilevel optimization (Floudas, 2000; Pis, 2007), these actors will be referred to as followers. In the context of urban planning, this denotation is merely referring to the chronology of decisions and not to the hierarchy of actors.

$$\Pi = f(\theta, x, y), \quad \theta \in \Theta, x \in \mathbb{R}, y \in \mathbb{Z} \tag{2.58}$$

Here x denotes continuous variables and y denotes discrete (i.e., binary or integer) variables. They are decided by an optimization algorithm, which solves the following multi-parametric Mixed-Integer Linear Programming problem:

$$\begin{aligned} \min_{x,y} \quad & f(\theta, x, y) = c_x^T(\theta) \cdot x + c_y^T(\theta) \cdot y \\ \text{subject to} \quad & P_x(\theta) \cdot x + P_y(\theta) \cdot y = p(\theta) \\ & Q_x(\theta) \cdot x + Q_y(\theta) \cdot y \leq q(\theta) \\ & \theta \in \Theta, x \in \mathbb{R}^n_{\geq 0}, y \in \mathbb{Z}^q \end{aligned} \tag{2.59}$$

where optimal values for x and y are determined in dependence on the values of θ. x and y can thus be interpreted as rational reactions of the followers to the decisions of the planners θ. In this work the problem of the followers Eq. (2.59) is solved using a Mixed-Integer Linear Programming algorithm, while the problem of the planners (Eq. 2.58) is explored via systematic sampling of the decision space. This means that the values of decisions of the planners θ are systematically changed and the resulting values for decision variables of the followers x and y as well as the values of the performance indicators Π are captured.

The variables and parameters presented in Section 2.2 take the following roles in formulation (2.59): Core (i.e., not derived by equality constraints) discrete decision variables y for the optimization model are the number of floors of buildings n_{fl} (Eq. 2.1), the existence ξ of buildings (Eqs. 2.2–2.3), parks (Eqs. 2.9–2.11), and energy conversion technologies (Eqs. 2.36–2.37) at specific locations, and the connection of blocks (Eq. 2.39). In addition, auxiliary variables are required for piecewise linearized functions as, for example, λ for the sky dome model (2.33)–(2.35). Core continuous decision variables x are the operation rates of technologies (2.30)–(2.31).

Although the planner could in principle adjust the values of all parameters, the main parameters are those that parameterize constraints by specifying minimum or maximum accepted bounds for performance indicators. A small example: The actual computed value of the share of RES s_{RES} is a performance indicator, while the specification of a minimum threshold on this indicator $s_{RES,min}$ is a decision of the user, and the values for the decision variables of the model are determined by the algorithm in such a way that their combination does not violate the constraints on some performance indicators while optimizing others. In a similar way as presented for the RES share (Eq. 2.47), constraints can be formulated for, for example, economic targets as costs or social targets as FAR (Eq. 2.12), or the LVF (Eq. 2.24). Since the model formulation contains integers, those parametric constraints should often be formulated as inequality constraints and not as equality constraints since the latter would aggravate the convergence of the solver, or even prevent finding feasible solutions. The way a constraint should be formulated, that is, as a lower or upper bound, depends on the problem at hand and the underlying trade-offs.

Acknowledgments

The authors would like to thank the urban planning and energy offices from the canton of Geneva for their constructive inputs in the elaboration of this work.

Nomenclature

α	Azimuth angle of sky patch
β	Angle from north toward direction of building
δ	Duration of time step
$\Delta\psi$	View range around landmark
ϵ	Elevation angle of sky patch
λ	Binary variable for sky patch
ϕ	All decision variables
Π	Performance indicator
ψ	Angle from north toward direction of landmark
σ	Reference size of technology
θ	Decision variables for user, parameters for model
ξ	Existence of …
A	Area
af	Annuity factor
C	Cost
c	Vector containing objective function coefficients
D	Distance to landmark
d	Distance
dm	Diameter
E	Energy flow
e	Specific energy flow (to area or length)
H	Height of landmark
h	Height
hn	Heating network
hts	Heat-transfer station
i	Interest rate
ir	Surface-specific irradiation
l	Length
lt	Lifetime
n	Number of …
o	Operation rate of technology
P	Matrix containing equality constraint coefficients
p	Vector containing right-hand side parameters of equality constraints

Q Matrix containing inequality constraint coefficients
q Vector containing right-hand side parameters of inequality constraints
r Rate-specific energy flow
x Continuous decision variables for computer/optimization
y Continuous decision variables for computer/optimization

Acronyms

ERA Energetic reference area
FAR Floor area ratio
GFA Ground floor area
HTS Heat-transfer station
LVF Landmark view factor
MILP Mixed-Integer Linear Programming
PV Photovoltaic
RES Renewable energy sources

Set members

com Commercial
mix Mixed-used buildings
off Office
rsd Residential
sfh Single family home

Subscripts

air Heat from the air
base Base length of transport network
blg Building
bs Building services
cst Constant part of equation
cum Cumulative
dir Direct
dm Diameter-specific
ecl Euclidean
en-loc Local electricity network
en-nat National electricity network
exp Export
fl Floors
fp Footprint
gr Gradient of piecewise linearized function segment
hw Hot water
ic Intercept of piecewise linearized function segment
imp Import
inv Investment
l Length-specific

lin	Linear part of equation
loss	Network losses
max	Maximum value
min	Minimum value
mnh	Manhattan
net	Net
ntw	Network
obs	Obstructed
op	Operation
par	Parcel
park	Park
rc	Room cooling
rel	Relative
rh	Room heating
seg	Segment
soil	Heat from the ground
tn	Transport network
tot	Total
ts	Transport stop
unobs	Unobstructed
view	View on landmark
wht	Waste heat
x	For continuous variables
xt	Exchange technologies
y	For discrete variables

Sets

Φ	All decision variables
Θ	Decision variables for user, parameters for model
AZ	Azimuth angles of sky patches
BL	Blocks
BR	Network branches
BT	Building types
CB	Cost balances
CT	Conversion technologies
EL	Elevation angles of sky patches
ET	Energy technologies
LC	Locations of buildings or centralized technologies
OT	Occupancy types
PC	Parcels
S	Segments of linearized function
T	Times
TS	Transport stops

Throughout this chapter the following notation is adopted: All sets are fully capitalized, while their according indices are uncapitalized and members of sets are put in single quotes. Furthermore, the convention is made that whenever a sum is made over a variable indexed over some set, the sum is made over the entire set of the given index, if not stated otherwise.

References

Akle, A.A., Minel, S., Yannou, B., 2016. Information visualization for selection in design by shopping. Res. Eng. Des. 1–19. ISSN 0934-9839, 1435-6066. https://doi.org/10.1007/s00163-016-0235-2.

Balling, R.J., Taber, J.T., Brown, M.R., Day, K., 1999. Multiobjective urban planning using genetic algorithm. J. Urban Plann. Dev. 125 (2), 86–99.

Branke, J., 2008. Multiobjective Optimization: Interactive and Evolutionary Approaches, Lecture Notes in Computer Science, State-of-the-Art Survey, 5252. Springer, Berlin, New York. ISBN 978-3-540-88907-6, p. 470.

Buchanan, J.T., 1997. A naïve approach for solving MCDM problems: the GUESS method. J. Oper. Res. Soc. 48 (2), 202–206. ISSN 0160-5682, 1476-9360. https://doi.org/10.1057/palgrave.jors.2600349.

Cajot, S., Schüler, N., Peter, M., Koch, A., Maréchal, F., 2017. Interactive optimization for the planning of urban systems. In: Scartezzini, J.-L. (Ed.), Energy Procedia. CISBAT: Future Buildings & Districts—Energy Efficiency From Nano to Urban Scale, vol. 122. Elsevier, Lausanne, Switzerland, pp. 445–450.

Copado-Méndez, P.J., Pozo, C., Guillén-Gosálbez, G., Jiménez, L., 2016. Enhancing the ϵ-constraint method through the use of objective reduction and random sequences: application to environmental problems. Comput. Chem. Eng. 87, 36–48. ISSN 0098-1354. https://doi.org/10.1016/j.compchemeng.2015.12.016.

Fernández, F.J., Alvarez-Vázquez, L.J., García-Chan, N., Martínez, A., Vázquez-Méndez, M.E., 2015. Optimal location of green zones in metropolitan areas to control the urban heat island. J. Comput. Appl. Math. 289, 412–425. ISSN 03770427. https://doi.org/10.1016/j.cam.2014.10.023.

Fazlollahi, S., 2014. Decomposition optimization strategy for the design and operation of district energy systems. EPFL. Available from: http://infoscience.epfl.ch/record/197534?ln=en.

Floudas, C.A., 2000. Deterministic Global Optimization, Nonconvex Optimization and Its Applications, vol. 37. Springer US, Boston, MA. ISBN 978-1-4419-4820-5 978-1-4757-4949-6. https://doi.org/10.1007/978-1-4757-4949-6.

Gal, T., 1995. Postoptimal Analyses, Parametric Programming, and Related Topics, second ed. De Gruyter, Berlin, Boston. ISBN 978-3-11-087120-3. https://doi.org/10.1515/9783110871203. Available from: https://www.degruyter.com/viewbooktoc/product/143242.

Hwang, C.L., Masud, A.S., 1979. Multiple Objective Decision Making: Methods and Applications; A State-of-the-Art Survey. Lecture Notes in Economics and Mathematical Systems, vol. 164. Springer, Berlin, Heidelberg, New York. ISBN 978-3-540-09111-0.

Jacobs, J., 1961. The Death and Life of Great American Cities. Vintage Books, New York, NY, p. 466 . 978-0-679-74195-4.

Keirstead, J., Shah, N., 2013. The changing role of optimization in urban planning. In: Chinchuluun, A., Pardalos, P.M., Enkhbat, R., Pistikopoulos, E.N. (Eds.), Optimization, Simulation, and Control. Springer Optimization and Its Applications, vol. 76. Springer, New York, NY, pp. 175–193. Available from: http://link.springer.com/10.1007/978-1-4614-5131-0.

Kipouros, T., Mleczko, M., Savill, M., 2008. Use of parallel coordinates for post-analyses of multi-objective aerodynamic design optimisation in turbomachinery. In: 4th AIAA Multidisciplinary Design Optimization Specialists Conference. American Institute of Aeronautics and Astronautics.

Knoll, M., Oertel, B. (Eds.), 2012. Dienstleistungen für die energieeffiziente stadt. Springer, Berlin, Heidelberg. Available from: http://link.springer.com/10.1007/978-3-642-18438-3.
König, H., Kohler, N., Kreissig, J., Lützkendorf, T., 2009. Lebenszyklusanalyse in der Gebäudeplanung. Institut für internationale Architektur-Dokumentation, München. Available from: http://www.muenchner-fachforen.de/jdownloads/2011/Fachforum_Zertifizierung_19_05_11/koenig.pdf.
Lehnerer, A., 2009. Grand Urban Rules. 010 Publishers, Rotterdam, The Netherlands, p. 275. ISBN 978-90-6450-666-6.
Liu, P., Georgiadis, M.C., Pistikopoulos, E.N., 2011. Advances in energy systems engineering. Ind. Eng. Chem. Res. 50 (9), 4915–4926. ISSN 0888-5885, 1520-5045. https://doi.org/10.1021/ie101383h.
Moavenzadeh, F., Hanaki, K., Baccini, P., 2002. Future Cities: Dynamics and Sustainability. Springer Science & Business Media, Berlin, Heidelberg, New York, p. 252. ISBN 978-1-4020-0541-1.
Neema, M.N., Ohgai, A., 2010. Multi-objective location modeling of urban parks and open spaces: continuous optimization. Comput. Environ. Urban. Syst. 34 (5), 359–376. ISSN 01989715. https://doi.org/10.1016/j.compenvurbsys.2010.03.001.
Neema, M.N., Ohgai, A., 2013. Multitype green-space modeling for urban planning using GA and GIS. Environ. Plann. B. Plann. Des. 40 (3), 447–473. ISSN 0265-8135, 1472-3417. https://doi.org/10.1068/b38003.
Oswald, F., Baccini, P., Michaeli, M., 2003. Netzstadt. Springer Science & Business Media, Berlin, Heidelberg, New York, p. 314. ISBN 978-3-7643-6963-7.
Pistikopoulos, E.N., Galindo, A., Dua, V., (Eds.), 2007. Multi-parametric programming: theory, algorithms, and applications. In: Pistikopoulos, E.N., Galindo, A., Dua, V. (Eds.), Process Systems Engineering, vol. 1. Wiley-VCH, Weinheim, p. 309.
Ratti, C., Raydan, D., Steemers, K., 2003. Building form and environmental performance: archetypes, analysis and an arid climate. Energy Build. 35 (1), 49–59. ISSN 0378-7788. https://doi.org/10.1016/S0378 7788(02)00079 8.
Robinson, D., 2011. Computer Modelling for Sustainable Urban Design: Physical Principles, Methods and Applications. Earthscan, London, p. 277. ISBN 978-1-84407-679-6.
Rode, P., Viejo Garcia, P., Bahu, J.M., 2014. Cities and Energy—Urban Morphology and Heat Energy Demand. Available from: https://lsecities.net/publications/reports/cities-and-energy-urban-morphology-and-heat-energy-demand/.
Schüler, N., Cajot, S., Peter, M., Page, J., Maréchal, F., 2017. The optimum is not the goal: capturing the decision space for the planning of new neighborhoods. In: Frontiers in Built Environment 3. ISSN 2297-3362. https://doi.org/10.3389/fbuil.2017.00076.
Schwarz, N., 2010. Urban form revisited—selecting indicators for characterising European cities. Landsc. Urban Plan. 96 (1), 29–47. ISSN 01692046. https://doi.org/10.1016/j.landurbplan.2010.01.007.
Sefair, J.A., Molano, A., Medaglia, A.L., Sarmiento, O.L., 2012. Locating neighborhood parks with a lexicographic multiobjective optimization method. In: Johnson, M.P. (Ed.), Community-Based Operations Research. vol. 167. Springer, New York, NY, pp. 143–171. Available from: http://link.springer.com/10.1007/978-1-4614-0806-2_6.
Söderman, J., Pettersson, F., 2006. Structural and operational optimisation of distributed energy systems. Appl. Therm. Eng. 26 (13), 1400–1408. ISSN 1359-4311. https://doi.org/10.1016/j.applthermaleng.2005.05.034.
Steemers, K., 2003. Energy and the city: density, buildings and transport. Energy Build. 35 (1), 3–14. Available from: http://www.sciencedirect.com/science/article/pii/S0378778802000750.

Vinge, V., 1993. Technological singularity. In: VISION-21 Symposium Sponsored by NASA Lewis Research Center and the Ohio Aerospace Institute, pp. 30–31.

Weber, C., 2008. Multi-objective design and optimization of district energy systems including polygeneration energy conversion technologies. EPFL, Lausanne, Switzerland, 210 pp PhD thesis.

Yu, J., Chen, Y., Wu, J., Liu, R., Xu, H., Yao, D., Fu, J., 2014. Particle swarm optimization based spatial location allocation of urban parks—a case study in Baoshan District, Shanghai, China. In: Third International Conference on Agro-Geoinformatics (Agro-Geoinformatics 2014). IEEE, pp. 1–6. Available from: http://ieeexplore.ieee.org/abstract/document/6910575/.

Yuan, Z., Tiemao, S., Chang, G., 2011. Multi-Objective Optimal Location Planning of Urban Parks. IEEE, pp. 918–921. ISBN 978-1-4577-0320-1. https://doi.org/10.1109/ICECC.2011.6066364.

Zhen, L., Li, M., Cheng, R., Peng, D., Yao, X., 2017. Adjusting parallel coordinates for investigating multi-objective search. In: Simulated Evolution and Learning. Asia-Pacific Conference on Simulated Evolution and Learning, Springer.

Building energy demand modeling: from individual buildings to urban scale

3

3.1 Introduction

Solène Goy, Giovanni Tardioli, Pilar Monsalvete Álvarez de Uribarri

The building sector consumes approximately 35% of the world final energy, which is more than either of the industry or transportation sectors (International Energy Agency, 2013). To address current and future energy challenges, buildings have a crucial role to play; they need to be constructed and operated in a more sustainable way. Modeling building energy consumption is necessary to understand how to mitigate the current impacts, to develop more sustainable policies and technical solutions for the future. More specifically, building energy models (BEMs) have been used, among others, for the prediction of performance indicators including energy consumption or CO_2 emissions.

Modeling buildings is a complex task and requires interdisciplinary expertise. It can involve different spatial scales, time frames, stakeholders, and topics ranging from building physics, control theory to social behavior. In practice, building energy studies have addressed various spatial scales: from a single zone or building to urban districts, cities, or even countries (International Energy Agency, 2014; Mastrucci et al., 2014; Mata and Johnsson, 2014; Petersdorff et al., 2006; Perez et al., 2011; Vivian et al., 2017; Bacher and Madsen, 2011; Heeren et al., 2013; Richardson et al., 2010). Energy predictions have been made over a wide range of timeframes, for example, on a annual, monthly, hourly, or minute basis.

Techniques for modeling of energy use in single buildings are well consolidated, but modeling at large scale, for example at district or city level onwards, is an emerging field where there is still a lack of validated procedures and standardization (Reinhart and Cerezo Davila, 2016). Several challenges may increase the difficulty in modeling a vast number of buildings, such as data uncertainty and data availability, high time requirements to run simulations, modeling of occupant behavior, complexity in simulating different physical phenomena, and providing scalable and replicable solutions. Overcoming all these issues requires a major research effort. Small-scale studies have provided valuable inputs on the building dynamic behavior or the resolution of specific issues; however, urban or country scale work is essential to support decision makers in assessing options and making the best choices for future cities.

At large scale, modeling approaches can first be categorized into top-down or bottom-up, depending on the level of aggregation of the data with respect to the case study (Swan and Ugursal, 2009). Top-down techniques typically rely on aggregated data and do not model each single building nor end-use; they rather consider the building stock as an "energy sink" (Swan and Ugursal, 2009). The bottom-up approach, on

Urban Energy Systems for Low-Carbon Cities. https://doi.org/10.1016/B978-0-12-811553-4.00003-2

the other hand, relies on disaggregated data to model the multiple entities of the case study (e.g., each building or several groups of buildings) before extrapolating to the whole case study (Swan and Ugursal, 2009; Li et al., 2017). In practice, the top-down approach is more of an "macro-economic tradition", while bottom-up has a more "engineering-economic" background (van Ruijven et al., 2010). In the present chapter, the focus is on bottom-up approaches.

Within this category, several building modeling techniques have been proposed. These methods can be divided in physical approaches (White Box/WB), hybrid approaches (Gray Box/GB), and data-driven approaches (Black Box/BB) (Swan and Ugursal, 2009; Foucquier et al., 2013; Zhao and Magoulès, 2012). The classification relies on the way the parameters are identified and how the equations are developed. WB model equations are built using physics-based principles; parameters are identified using construction data or expert knowledge only. BB models, on the other hand, do not require input from building physics as they are entirely built on historical data; their parameters typically do not have physical meaning. BB techniques include, for example, multilinear regression, support vector machine (SVM) or neural network (NN). GB approaches have characteristics from both, WB and BB approaches. For example, they can take the form of simplified physics-based equations with parameters identified using historical data and calibration techniques. The frontier between modeling strategies is not related to the complexity of the model (De Coninck et al., 2014). A reduced order RC model which parameters are identified using knowledge of the building construction materials and geometry would be considered as WB. All three approaches have been used in the literature to model buildings and none is necessarily best overall. Rather, there is a more adequate technique for each case study depending on parameters such as data availability, number of buildings, and the objective of the modeling. For more details about the various techniques, the reader is referred to the reviews of Foucquier et al. (2013), Swan and Ugursal (2009), or Zhao and Magoulès (2012).

Occupant behavior modeling is a crucial part of building energy demand estimation. In real buildings there is a large variability in occupant energy usage behaviors, referring to demand patterns, device interaction or consumption management, which significantly influences electricity and heating demands. Occupant behavior is thus considered one of the reasons for the performance gap between measured and modeled results in building and urban energy modeling. Occupant behavior is naturally ruled by stochastic processes, which should be represented in computing simulations, in order to improve the predictive capability of the simulations.

The present chapter focuses on the modeling of energy consumption and temperature prediction in buildings for urban scale studies. In Section 3.2, the data challenges of large-scale building energy modeling are introduced. In Section 3.3, techniques for large-scale simplification are addressed namely, the selection of representative buildings using clustering techniques. Then, several of the modeling techniques mentioned above will be described, covering a wide range of applications and spatial scales:

- Section 3.4 addresses detailed physical building models.
- Section 3.5 describes reduced-order RC building models.
- Section 3.6 discusses the modeling of electrical appliances.
- Section 3.8 addresses parametric analysis applied to detailed physical models.
- Section 3.9 discusses data-driven models (DDM).

Each method is detailed and the context in which they can be applied is specified.

3.2 Data challenges of large-scale building energy modeling

Pilar Monsalvete Álvarez de Uribarri, Solène Goy

Building energy modeling at urban scale requires a large amount of data covering not only physical parameters (i.e., architecture of the buildings and surroundings) or weather information, but also important aspects such as the number of inhabitants, their behavior, infiltration and ventilation rates, HVAC system configuration, among others. Such data are in most cases very challenging to obtain, if existing at all, and are usually disaggregated into different data sources with many gaps and errors. While some of these databases are open access (e.g., official cadastral maps or census data), in many other cases, access tends to be restricted for privacy or marketing reasons (e.g., energy provider databases). These data availability shortcomings impact the modeling step: acquiring data for large-scale modeling is typically challenging and time consuming. In addition, this problem increases with the scale of the study. Therefore, the appropriate methodology to large-scale building modeling depends not only on the objective of the study, but also strongly on the availability, the degree of detail, and the quality of the required data.

To date, large-scale building modeling typically requires simplifications compared with single building investigations; these simplifications can be made in the model used and/or in the number of buildings considered for modeling. Both options are addressed in the present chapter. Model simplification is, for example, discussed in Section 3.5, where reduced-order models are presented. Reducing the modeling effort to a smaller number of buildings is another option to overcome large-scale challenges. An acknowledged approach relies on the definition of representative buildings; rather than modeling each single building of the dataset, a few buildings, considered representative of a larger ensemble, are modeled. The identification of these typical buildings relies on grouping techniques. An example of application of such technique is detailed in Section 3.3.

3.3 Clustering methods for representative building identification in urban contexts

Giovanni Tardioli

This section presents an adaptable and flexible methodology for representative building identification at an urban scale. The study is based on the application of clustering algorithms for the definition of similar elements in a complex building features space. Several clustering algorithms are investigated. Building geometric features and energy-related variables are considered in the analysis. The quality of the results is assessed by the means of clustering validation indices. The methodology is applied to the residential building stock of the city of Geneva (CH). Results indicate the capacity of clustering algorithms to identify representative buildings. A predictive modeling approach is used to extend the results of clustering to a larger portion of the building stock with high prediction accuracy.

3.3.1 Introduction

Currently, large-scale physics-based BEMs require a characterization process of modeling input variables. Modeling inputs can be gathered from the literature, from experts of the building stock, or from building audits and metered data. However, as highlighted in Section 3.2, building data may be not always available due to privacy issues and economic interests. In addition data gathering is still expensive and time consuming. For this reason, limiting the investigation to representative buildings could alleviate the data-gathering process. Three main definitions and approaches have been recently presented (Ballarini et al., 2014) for national scale archetypes:

- the "real building" approach where representative buildings are defined based on the knowledge of experts of the building stock;
- the "real average building" approach where real representative buildings are identified using statistical data; and
- the "synthetic average building" where representative buildings are a statistical representation of the sample.

Clustering is a possible alternative to traditional statistical approaches. The main difference between statistical approaches and clustering is the possibility to perform element grouping and multidimensional analysis. Further to this, clustering allows the identification of the representative element of a group. Most of clustering studies focus at national scale. Few research has been conducted at smaller scale such as at urban and district level. In addition, few studies provide approaches to help the energy modeler in classifying the building stock and to identify representative buildings appreciating the quality of the partitioning solution. In this context, the current state-of-the-art lacks of methodology, which may reduce subjective decisions in representative buildings definition and identification. This section is structured as follows. Initially, a comprehensive review of clustering techniques applied in different building energy modeling categories is presented; then, the research "gap" and contribution is described; following this, the research methodology is presented and explained in detail; the methodology is then tested on a city case study and finally a description of the main outcomes of this approach is provided. In the final discussion, strengths and limitation of the proposed approach will be discussed and suggestion for future work will be presented.

3.3.2 Literature review

It is possible to summarize the research work related to clustering applied to the building energy modeling sector in different categories.

3.3.2.1 Benchmarking

Several research works adopt clustering to identify groups or categories for benchmarking analysis and baseline case studies for comparison. In this context, representative buildings are the reference energy standard of the group (Gao and Malkawi, 2014).

3.3.2.2 Consumption profiles classification

Clustering is usually employed by utility companies to classify customers based on consumption information derived by smart meter data (Bu et al., 2014). This provides a better understanding of customer portfolio, which enables utilities to perform a better planning of unit commitment and supply requirements.

3.3.2.3 Archetypes identification

Several research studies show that identifying representative buildings by the means of clustering allows classification of group of buildings at regional and national levels (Famuyibo et al., 2012). In addition, in some studies the identification of representative buildings is also linked to the creation of priority intervention lists for applying retrofit cost-optimal solutions or estimations of retrofit potential for the group of buildings (Lara et al., 2014).

3.3.2.4 Predictive modeling framework

Prediction methods, such as regression approaches for the estimation of building consumption, are likely to provide more accurate results in groups of similar buildings. In this context, cluster-wise prediction methods have been applied for the prediction of energy consumption of entire cities (Hsu, 2015).

3.3.2.5 Occupant behavior

Some studies employ clustering for occupant behavior classification and related effects on building consumption. For example, unrelated occupant parameters affecting building energy consumption can be grouped together using a *k*-means clustering-based approach (Guerra Santin, 2011).

3.3.3 Methodology

This section presents an overview of a research methodology related to representative buildings identification in an urban context using clustering techniques. Fig. 3.1 provides a schematic representation of the steps involved in the methodology and the subsequent paragraphs describe the approach.

3.3.3.1 Data gathering and preprocessing

As a first step relevant data available from the building stock are gathered. Building stock data (such as geometry, energy consumption, etc.) are subsequently organized in a database. Thus, data are merged together to structure different building information of different domains. The dataset is classified (filtered) using information of building typology, period of construction, and building final use (e.g., multifamily buildings, single family houses).

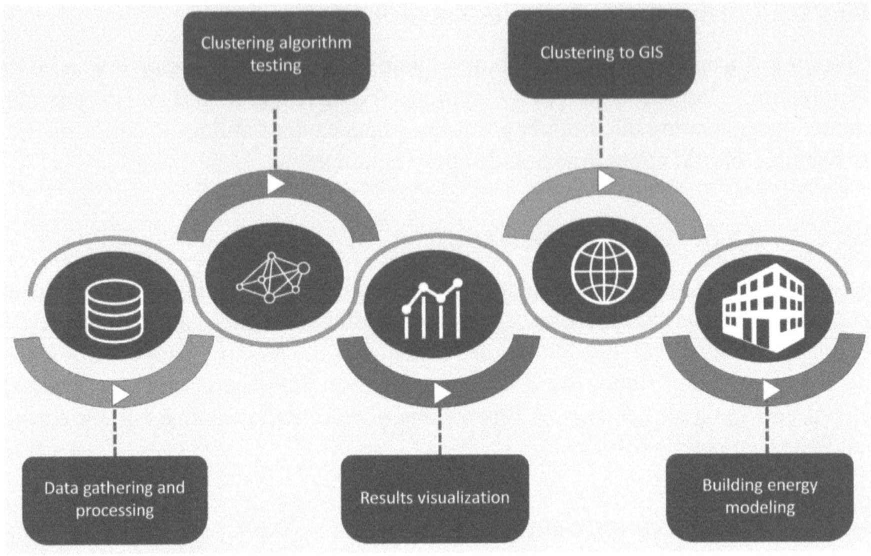

Fig. 3.1 Methodology overview.

3.3.3.2 Algorithm testing

Data are then clustered with four clustering algorithms: *k*-means, partitioning around medoids (PAM), hierarchical agglomerative, and hierarchical divisive. The number of clusters, which is an important modeling parameter, is varied in a large range to appreciate any trends in the results.

3.3.3.3 Validation and results visualization

Results of clustering are then evaluated with reference to validation indices. Different validation indices are used to test the quality of clustering results. Validation indices are characterized by different scales, for this reason to compare together a normalization approach is required. A set of visualization tools are developed to assist the modeler in identifying the best partition and the most adequate algorithm.

3.3.3.4 Clustering results to GIS

Next, data are visualized using a GIS application to locate elements and their position in the urban context. In addition linking data to GIS facilitates the visualization of buildings in a specific cluster.

3.3.3.5 Building energy modeling

The identification of representative buildings enables a wide range of possibilities for the energy modeler who can decide the type and resolution of the information to gather for energy modeling: for example, investigating different data sources, going on site, and conducting surveys.

3.3.3.6 Applied predictive classification

A predictive classification model is developed to expand the results of clustering to a larger part of the building stock of the city case study. The classification approach uses the subset of results generated by clustering for the training of the predictive model.

3.3.3.7 Algorithm description

K-means, PAM, hierarchical divisive, and hierarchical agglomerative algorithms are used in this study. They have been selected on the basis of well-known techniques found in the literature. Using more than one algorithm allows identification of the best partitioning solution across many options.

3.3.3.7.1 *K*-means clustering

K-means clustering is based on a combinatorial optimization procedure where data vectors are relocated between clusters, with minimization of an objective function until an optimal partition is obtained (James et al., 2014):

$$f = \min_{C_1,\ldots,C_k} \left\{ \sum_{k=1}^{K} W(C_K) \right\} \tag{3.1}$$

where $C_1, \ldots, C_k$ denotes sets containing indices of the observations in each cluster and $W(C_K)$ a dissimilarity measure defined as:

$$W(C_k) = \frac{1}{|C_k|} \sum_{i,i' \in C_k} \sum_{j=1}^{p} (x_{ij} - x_{i'j})^2 \tag{3.2}$$

where $|C_k|$ is the number of observations in the kth cluster. Indices i and j refer to elements in cluster k. Considering Eqs. (3.1), (3.2) follows:

$$f = \min_{C_1,\ldots,C_k} \left\{ \sum_{k=1}^{K} \frac{1}{|C_k|} \sum_{i,i' \in C_k} \sum_{j=1}^{p} (x_{ij} - x_{i'j})^2 \right\} \tag{3.3}$$

Each point of the dataset is related to a centroid. The position of the centroids is recalculated and it is updated as the baricenters of the clusters resulting from the previous step. Points are then re-associated to each cluster and cluster memberships are evaluated again. The recursive loop is repeated until the position of the centroids do not move any more.

3.3.3.7.2 Hierarchical clustering

Two types of hierarchical algorithms are considered: agglomerative and divisive (Jain et al., 1999). In the agglomerative method, a distance measure is used to identify similar data points. Initially, the number of clusters is equal to the number of data points.

At each step, observations are compared and aggregated into clusters until all the observations are allocated in different clusters. In the last step of the algorithm, all the observations are grouped in one single cluster.

In the divisive approach, first observations are all in the same cluster. At each step the most different observations are divided and allocated in different clusters until each observation is in a different cluster in the last step. At each iteration clusters are linked together and a common dissimilarity value is decided based on the linkage scheme adopted.

3.3.3.7.3 Partitioning around medoids

The PAM algorithm is based on the search of k representative objects (medoids) among all the elements of a given cluster. The representative objects are the ones with minimal average dissimilarity. Such elements are named medoids. The algorithm is based on the minimization of the objective function (Kaufman and Rousseeuw, 1987):

$$f = \min\left\{\sum_{i=1}^{n}\sum_{j=1}^{n} d(i,j)\right\} \tag{3.4}$$

PAM algorithm can be described in two steps: a build phase where objects are selected as representative of a set (initial medoids), and a swap phase where clustering is improved by switching the representative elements of a cluster. In other words, the algorithm labels each point based on the closest center, then replace each center by the medoid of the points in its cluster.

3.3.4 Case study: Data description

Data of the building stock are gathered from an online repository of the city of Geneva (Le système d'information du territoire à Genève (SITG)). The data offered by the online portal are, for the majority, geo-referenced. Different tables describe the building stock of the city of Geneva. Tables "Cad. batiment.hors.sol" and "Scane.indice. moyennes" are two main datasets used in this research work. These two tables describe several properties of the urban stock at a level of single building. Tables are linked together by a building reference code "EGID," which identify a building in the database. Table "Cad.batiment.hors.sol" provides mostly information on the building typology, period of construction, district location, building final use, and geometric information. Table "Scane.indice.moyennes" provides information regarding the location of the building, its final use, total consumption for heating and domestic hot water, heating reference surface, and geometric information of the building. The first database is characterized by about 88,000 entries, the second by about 11,500. The majority of the information in the database "Cad.batiment. hors.sol" concerns residential buildings (about 45%) followed by a large number of building of different categories classified under "other" (about 33.5%), commercial (about 9.5%), mixed-use (about 7.5%), and public buildings (about 4.5%). The database called "Scane.indice.moyennes" provides information mostly about multifamily buildings (about 6400 entries 56.5%) followed by mixed-use buildings with ground floor occupied by a commercial use (about 2100 entries 18%), mixed-use buildings without floor occupation distinction (about 1900 entries 16.5%), offices (about 285 entries 2.5%), single family houses (about 270 entries 2.3%), and two family houses (about 65 entries 0.6%).

3.3.5 Dataset analysis

Data gathered and analyzed from the two main datasets are preprocessed in order to create a useful structure, which enables the use of clustering. A series of data-cleaning and restructuring procedures are performed to create a complete dataset. In particular, first, a missing value removal is performed for both datasets, then data are merged together using the building ID: "EGID." The final dataset presents 6100 residential buildings information. The resulting dataset is characterized by nine features. These are number of energy meters in the building, heating reference surface, footprint area, perimeter of the building, energy index, height, CO_2 emissions, number of storeys, and annual "average" energy consumption.

3.3.6 Results

3.3.6.1 Validation indices trends

Several validation indices are employed to estimate the quality of clustering results. Validation indices are characterized by different numerical scale. In order to present the results in a meaningful manner, a normalization approach is adopted. Furthermore, indices are re-evaluated to indicate the best results when they are maximized. The sum of the normalized indices is visualized on a graph in Fig. 3.2.

The x-axis is the number of clusters employed in the analysis. The range of variability of the number of cluster is between 2 and 15: 2 is the minimum number of clusters required to run the algorithms, 15 is a relatively high number of clusters, which enable the understanding of trends in the results. In addition, the graph shows the trend of each clustering algorithm in different colors. The mean of the values of the indices at each step is visualized on the graph. The best solution in terms of cluster numbers is the one, which maximizes the trends of the indices. The maximum of the validation indices provides information on the best partition of the dataset. An analysis of these results reveals that generally the best clustering solution is identified for a small number of clusters. This result is expected due to the fact that the pre-classification carried out by filtering categorical variables, already creates groups sharing some characteristics (period of construction, typology, and final use). In addition, trends of validation indices usually decrease increasing the number of clusters across different algorithms. This suggests that creating smaller clusters (increasing the number of clusters in the analysis) does not always guarantee a better solution. The energy modeler can assess the best modeling practice based on the results of the validation indices. If results of two different solutions are comparable, in terms of validation indices, the modeler can decide the number of representative energy

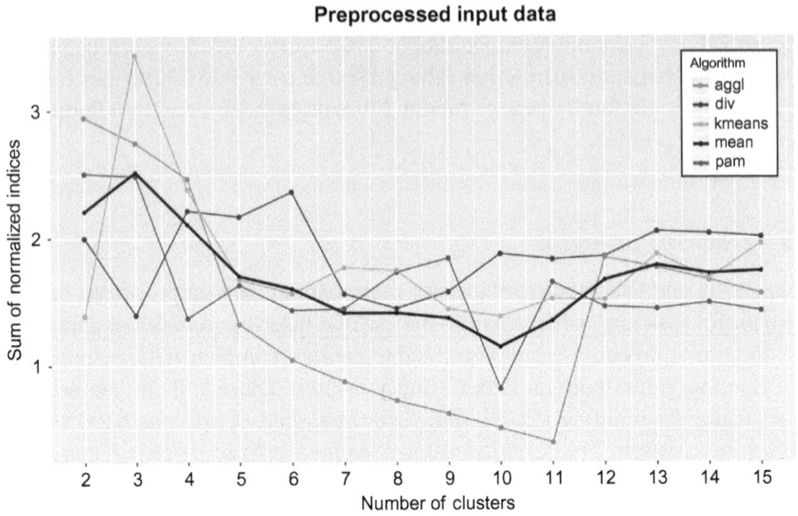

Fig. 3.2 Example of sum of normalized validation indices for different clustering algorithms.

models based on a trade-off between representativeness of the models and efforts and resources available for model creation.

3.3.6.2 Visualization tools

Compactness and separation define the quality of clustering results. A cluster has good compactness when element is close to each other and good separation when clusters do not overlap. To understand these two properties, visualization tools can help the energy modeler. Once a clustering solution is identified in terms of number of clusters and algorithm, feature properties of elements (buildings) in each cluster can be visualized. Fig. 3.3 represents a histogram distribution of building features. On the x-axis, there is the range of variability of a given feature. On the y-axis, the variable "count" indicates how many times data features are in a given range. With reference to the distribution histograms, it is possible to appreciate that selecting the best results indicated by the maximum of the sum of normalized validation indices, clustering provides good results in terms of separation and compactness for the case study under analysis. Features seem to create quite separated groups and a reduced degree of overlapping for the majority of the features analyzed.

Different clustering algorithms allow the identification of the representative element describing the cluster. This element is the one that has the most similar features with every element in the cluster. The identification of a representative element opens a variety of options for the modeler. For example, the modeler may decide to create an energy model of a building using a statistical representation of each feature in each cluster. On the other hand, the identification of a real building allows the decision of the modeling accuracy based on a trade-off between resources, time, and objective of the simulation. Based on the modeling requirements, relevant data are gathered

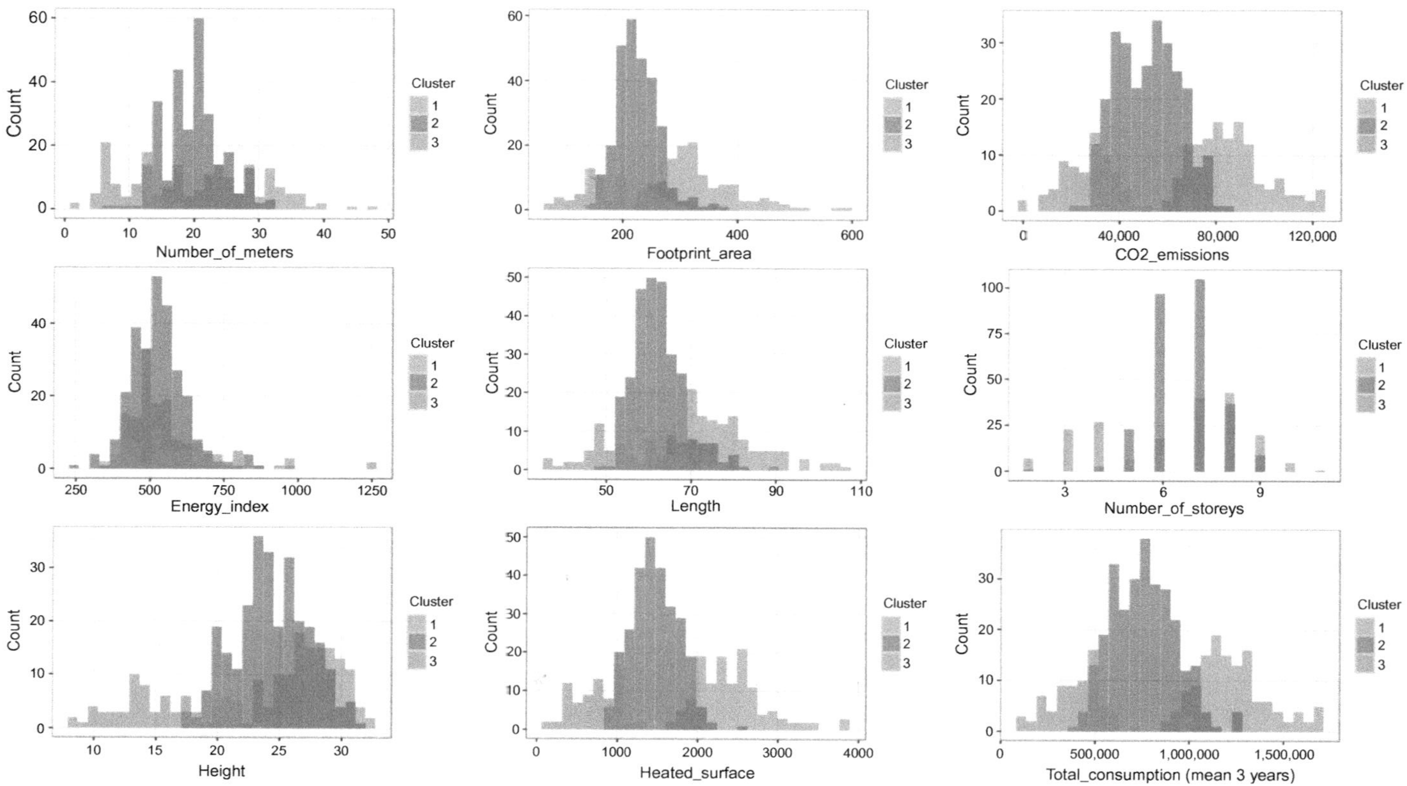

Fig. 3.3 Feature distributions for each identified cluster.

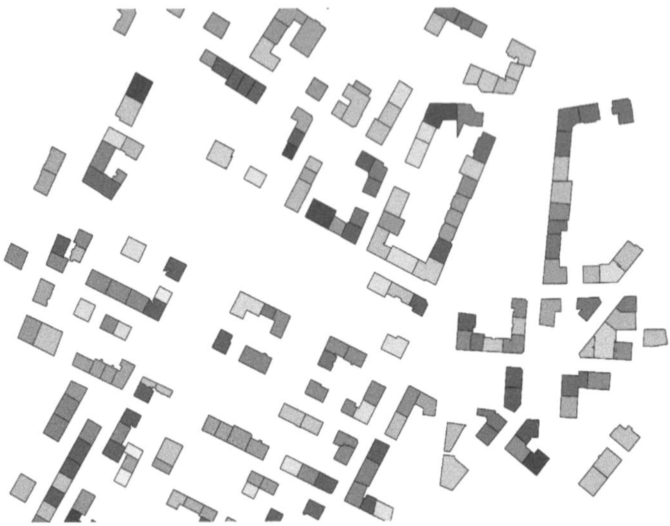

Fig. 3.4 Example of clustering techniques applied on a district of the city of Geneva.

through visual inspection, surveys, or from metering systems if available. Another useful procedure is to link clustering results to GIS visualization maps. Fig. 3.4 shows the results of clustering using geo-referenced data of the city of Geneva. Different colors indicate different class memberships. Linking GIS to clustering results to 3D maps of the city provides additional modeling information for the creation of the energy models such as number of adjacencies, glazing surfaces, urban density, and shading scenarios. In addition creating a link between clustering results and GIS enables the use of automated modeling procedure for models creation.

3.3.7 Extending clustering results to a larger part of the building stock

A predictive classification model is developed starting from a training set of clustering results. The results of this application are shown in Table 3.1. The table shows building typologies, final use, period of construction, number of buildings available in the dataset, number of clusters selected, and prediction accuracy achieved with the predictive classification method. Results show high performances of the predictive model, which is able to achieve an average accuracy across building groups equal to 89%.

3.3.8 Discussion

Clustering, compared with classic statistical method for representative buildings identification, presents interesting features such as the ability to group urban buildings together based on similarities and the possibility to identify representative elements.

Table 3.1 Predictive modeling accuracy and number of new building classifiable

Typologies	Final use	Period of construction	No. of buildings	No. of clusters	Prediction accuracy (%)	No. of new buildings
Residential	Multifamily	1919–45	483	3	90	246
Residential	Multifamily	1946–60	817	3	95	215
Residential	Multifamily	1961–70	1169	2	96	187
Residential	Multifamily	1971–80	921	2	95	207
Residential	Multifamily	1981–85	173	2	92	56
Residential	Multifamily	1986–90	276	3	97	156
Residential	Multifamily	1991–95	437	2	90	202
Residential	Multifamily	1996–2000	545	2	91	266
Residential	Multifamily	2001–2005	331	2	92	264
Residential	Multifamily	Before 1919	634	2	88	643
Total no. of buildings			Average accuracy		Total no. of predictions	
5876			92.6%		2442	

In particular, this last point enables the investigation of several modeling approaches and data-gathering procedure during the creation of urban BEMs. This provides a vast range of possibilities for the modeler, which can decide the level of detail that the energy model of the archetype should reach. The present work addresses the lack of flexible methodology for the use of clustering algorithms for archetypes identification across different building typology providing a framework for testing and investigation of different modeling solutions. Selecting and testing different algorithms, the modeler can appreciate which one perform better on a dataset comparing them on a common scale of normalized validation indices. The selection of the number of clusters which best partitions the data should be done with the aim to have a trade-off between capacity of the archetype to correctly represent the building stock (good value of the validation indices) and simplification of the burden connected with large-scale energy modeling (number of clusters and representative buildings). The application of the methodology on a large and diversified data set shows the flexibility of the method. An innovative method of linking clustering results to predictive classification method allows to overcome problems of completeness of the dataset. With such approach it is possible to extend cluster memberships to a larger part of the city case study.

3.3.9 Conclusions on clustering method

In the present work a total of 23 representative residential buildings are identified by the means of a methodology based on clustering algorithms for the city of Geneva (CH). Validated clustering results and visualization tools enable the identification of best partitioning of the data. The analysis of different building features on common scales improves the quality of clustering in terms of compactness and separation. In addition, with the help of visualization tools it is possible to further extend the capacity of evaluation of clustering results in terms of features considered in the analysis, compactness, and separation of clustering results. Further work may be required to define the applicability of clustering in different contexts and for different city scenarios to create a benchmark of comparison on the applicability of this technique. Exploring a higher number of building features, algorithms and modeling options, may provide a higher insight on the advantages presented in this research work. Furthermore, linking clustering results with an energy modeling approach may provide feedback on features useful to be included for archetyping identification refining the subset of information the energy modeler should analyze to describe the buildings stock.

The rest of the chapter addresses various building modeling approaches for large-scale modeling, ranging from WB to BB. Each method is detailed, their differences highlighted, and the context in which they can be applied is specified.

3.4 Physical dynamic building models

Pilar Monsalvete Álvarez de Uribarri

The dynamic building model aims to calculate the thermal energy demand for air conditioning of one building, based on a physical nodal method. This method is widely used for building simulation. According to Foucquier et al. (2013), the principal assumptions taken by this method are:

- Each building zone is a homogeneous volume characterized by uniform state variables.
- A node represents a room, a wall, a window, or else the exterior of the building.
- The heat-transfer equations are solved for each node of the system.
- The nodal method can be considered as a one-dimensional approach.

Fig. 3.5 illustrates the energy flow paths considered for the building model. The traditional energy flow paths in buildings (Clarke, 2001) also consider furniture and latent flows, which are not included here. However, the ground surface is explicitly studied, which is not done in the traditional model.

Starting from the energy balance in the air point node, the energy flows are broken down step by step to explain the building model.

$$\frac{dU}{dt} = \dot{Q}_{conv} + \dot{Q}_{TB} + \dot{Q}_{vent} + \dot{Q}_{inf} + \dot{Q}_{IGconv} + \dot{Q}_{sys\_conv} + \dot{Q}_{dem} \quad [\mathrm{W}] \tag{3.5}$$

where

$$\frac{dU}{dt} = m \cdot Cp \frac{dT_{room}}{dt} \tag{3.6}$$

$\dot{Q}_{vent}$ and $\dot{Q}_{inf}$ are the mechanical ventilation (external air mass flow voluntary introduced to maintain good air quality) and infiltration (involuntary external air mass flow into the volume) heat inputs, respectively (Fig. 3.6).

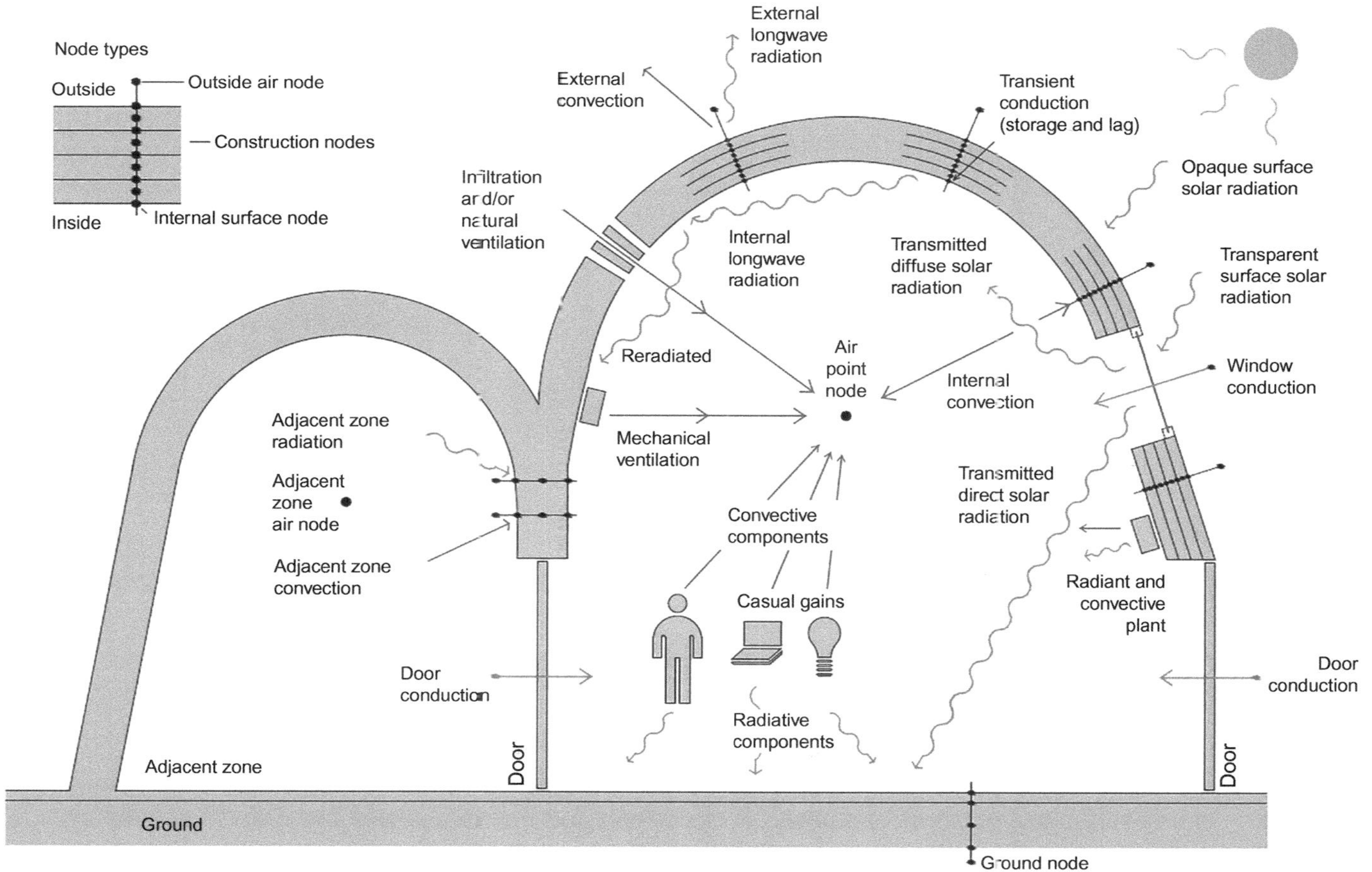

Fig. 3.5 Energy flow paths.
(Adapted by Alfonso García-Leyaristy from Clarke, J.A., 2001. Energy Simulation in Building Design. Routledge, Abingdon. ISBN 978-0-7506-5082-3.)

Fig. 3.6 Ventilation and infiltration.
(Rocío Monsalvete Álvarez-Uríbarri. Arquitectura e Interiorismo (rociomonsalvete.com).)

$$\dot{Q}_{vent} = \dot{m}_{vent} Cp(T_{vent} - T_{room}) \tag{3.7}$$

$$\dot{Q}_{inf} = \dot{m}_{inf} Cp(T_a - T_{room}) \tag{3.8}$$

All parameters of both equations (except T_{room}) are inputs for the model. $\dot{Q}_{IGconv}$ and $\dot{Q}_{sys\_conv}$ (Fig. 3.7), which are the convective heat inputs due to casual gains (or internal gains) and convective plants, are also inputs for the model so, they do not need to be internally calculated.

The term $\dot{Q}_{TB}$ appears due to the one-dimensional heat-transfer model of the nodal method approach. The points (e.g., corners) and linear details (e.g., two walls intersection, window-wall junction) must be treated separately (Fig. 3.8).

$$\dot{Q}_{TB} = H_{TB}(T_a - T_{room}) \tag{3.9}$$

where T_a is the ambient temperature and H_{TB} is the total heat transmittance due to thermal bridges. The last one may come as a fix value for the complete building from the *Construction library* or be calculated during the construction preprocessing as:

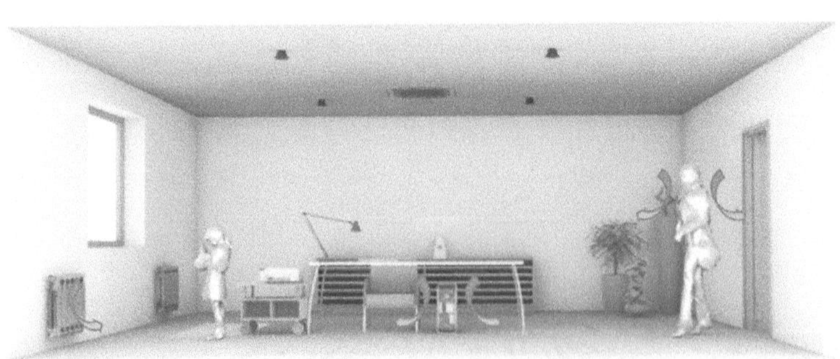

Fig. 3.7 Convective loads.
(Rocío Monsalvete Álvarez-Uríbarri. Arquitectura e Interiorismo (rociomonsalvete.com).)

Fig. 3.8 Thermal bridges.
(Rocío Monsalvete Álvarez-Urîbarri. Arquitectura e Interiorismo (rociomonsalvete.com).)

$$H_{TB} = \sum(\Psi_i \cdot L_i) + \sum(\chi_j \cdot n_j) \tag{3.10}$$

where

- Ψ_i is the linear heat transmittance value of detail "i";
- L_i is the total length of the linear detail "i" in the analysis zone;
- χ_j is the point heat transmittance value of detail "j"; and
- n_j is the number of point thermal bridges of type "j" in the analysis zone.

$\dot{Q}_{conv}$ is a function of T_{room}. Thus, either the demand for a set T_{room} or the temperature in the room in free floating periods ($\dot{Q}_{dem} = 0$) is obtained from Eq. (3.5).

$\dot{Q}_{conv}$ is the sum of all internal convection flows. The internal convection is caused by the difference between the temperatures of the different enclosures (walls, windows, doors, and ground) and the air node. It is obtained by applying the energy balance (Eq. 3.12) for every enclosing surface.

$$\dot{Q}_{conv} = -\sum_i (\dot{q}_{conv\_i} A_{surf\_i}) \quad [W] \tag{3.11}$$

$$\forall i, \quad \frac{du}{dt} = \dot{q}_{conv} + \dot{q}_{cond} + \dot{q}_{rad\_lw} + \dot{q}_{rad\_sw} + \dot{q}_{IGrad} \quad [\mathrm{W/m^2}] \tag{3.12}$$

where

$$\frac{du}{dt} = m \cdot Cp \frac{dT_i}{dt} = \rho \cdot x \cdot Cp \frac{dT_i}{dt} \tag{3.13}$$

$$\dot{q}_{conv} = h_{ci}(T_i - T_{room}) \tag{3.14}$$

x is the virtual thickness of the surface where dT occurs.

$\dot{q}_{IGrad}$ is the radiative component of casual gains and radiant plants, and is an input for the model (Fig. 3.9).

$\dot{q}_{cond}$ is the conduction through room enclosures and is calculated using the one-dimensional heat-transfer equation of the Finite-Difference Approximation Method.

$\dot{q}_{rad\_sw}$ is the incident short wave radiation. This is the fraction of the transmitted solar radiation that enters the room though the transparent enclosures (windows) which is absorbed by the room enclosure. It can be divided between direct ($\dot{q}_{sw\_b}$) and diffuse ($\dot{q}_{sw\_d}$) radiation (see Figs. 3.10 and 3.11).

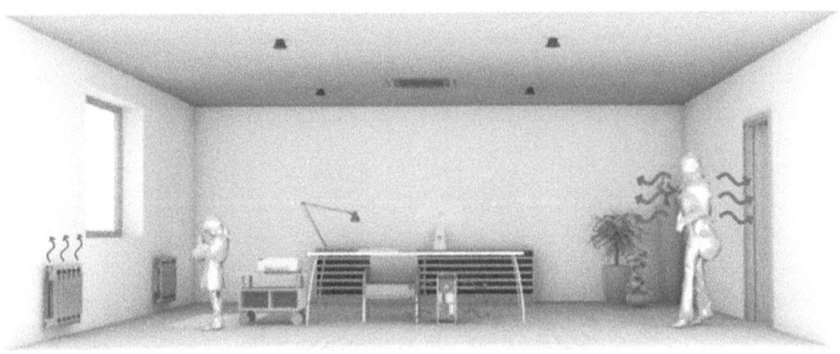

Fig. 3.9 Radiative components of casual gains and radiant plants.
(Rocío Monsalvete Álvarez-Urîbarri. Arquitectura e Interiorismo (rociomonsalvete.com).)

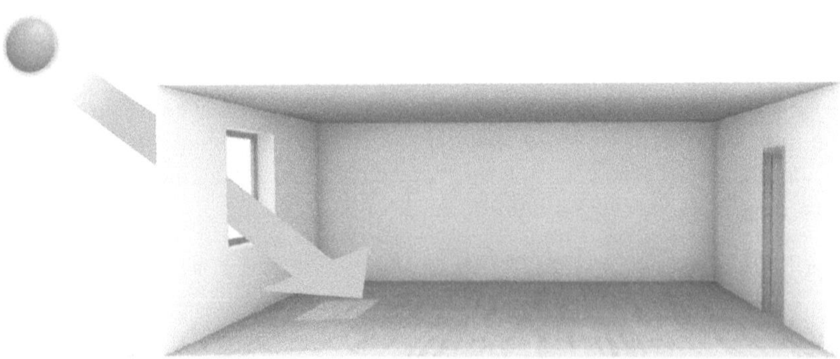

Fig. 3.10 Transmitted direct solar radiation.
(Rocío Monsalvete Álvarez-Urîbarri. Arquitectura e Interiorismo (rociomonsalvete.com).)

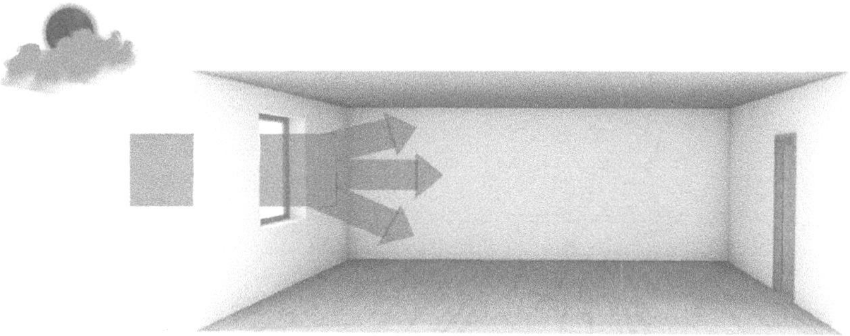

Fig. 3.11 Transmitted diffuse solar radiation.
(Rocío Monsalvete Álvarez-Uríbarri. Arquitectura e Interiorismo (rociomonsalvete.com).)

Fig. 3.10 shows how the direct radiation enters the room through the window and hits a small part of the floor. This solar spot moves along the day with the sun changing the temperature of the surfaces. Following this process has high computational costs, so three approaches are usually used to solve this problem (ATECYR, 2008):

- predefined distribution (60% to the floor and the rest is divided per area between the rest of surfaces);
- approximated distribution (using correlations); and
- backward-ray-tracing method.[1]

This is calculated during the geometrical preprocessing accounting with the walls absorptivity, and as a result, the simulation model receives the direct and diffuse solar fractions (SF_{sw}) on every enclosing surface. Up to now, the method used for this calculation is the *predefined distribution.*

$$\dot{q}_{rad\_sw} = \dot{q}_{sw\_b} + \dot{q}_{sw\_d} \tag{3.15}$$

$$\dot{q}_{sw\_b} = SF_{sw}^{b} \frac{\sum_{wi} g_{wi} I_{bi} A_{wi} f_{wi} (1 - f_{bi})}{A_j} \tag{3.16}$$

$$\dot{q}_{sw\_d} = SF_{sw}^{d} \frac{\sum_{wi} g_{wi} I_{di} A_{wi} f_{wi} (1 - f_{bi})}{A_j} \tag{3.17}$$

I_{bi} and I_{di} are the direct and diffuse solar radiations at the window's tilt, respectively, and include the effect of external shadows and reflections.

A_{wi} is the area of each window, f_{wi} their glass-frame ratio ($f_{wi} = 1$ would mean only glass), g_{wi} their direct-beam transmittance at normal incidence, and f_{bi} the darkness factor due to blind devices ($f_{bi} = 1$ is completely dark).

[1] For a better understanding see Arvo and Chelmsford (1986).

Back to Eq. (3.12), the last term to be defined, $\dot{q}_{rad\_lw}$, refers to the internal long wave radiation. This radiation is calculated as (i refers to the cited surface and j to the rest of them in a given enclosed volume):

$$\dot{q}_{rad\_lw} = E_i - \epsilon_i \frac{\sum_{j=1}^{n} \left(A_j F_{ij} \left(E_j + (1-\epsilon_j)\overline{q_{rad}}\right)\right)}{A_i} \tag{3.18}$$

where

$$E_i = \sigma \epsilon_i T_i^4 \tag{3.19}$$

$$\overline{q_{rad}} = \frac{\sum_{i=1}^{n} E_i A_i}{\sum_{i=1}^{n} \epsilon_i A_i} \tag{3.20}$$

The form factors are also calculated during the geometrical preprocessing.

The special case of low time constant of enclosures is also considered, such as very thin walls or, most commonly, windows (time constant for windows around 1–2 minutes, the environmental conditions can thus be considered constant over the time response (Rubin, 1982)). In that case, the enclosure can be simulated in steady-state conditions and, therefore Eq. (3.12) gets simplified and the conductive heat transfer through the enclosure is calculated with the Fourier equation for steady state:

$$\dot{q}_{cond} = U_{tot}(T_i - T_o) \tag{3.21}$$

$$\frac{du}{dt} = 0 \tag{3.22}$$

where U_{tot} is the overall heat-transfer coefficient of the enclosure, and T_i and T_o are the temperatures of the internal and external surfaces (either outside or an adjacent zone).

Therefore,

$$0 = \dot{q}_{conv} + U_{tot}(T_i - T_o) + \dot{q}_{rad\_lw} + \dot{q}_{rad\_sw} + \dot{q}_{IGrad} \tag{3.23}$$

3.5 Reduced-order RC models

Solène Goy

3.5.1 Theory background

Among the building modeling techniques available, the reduced-order RC approach has been acknowledged as a promising solution for urban-scale case studies (Lauster et al., 2014; Robinson et al., 2009). It relies on a thermoelectric analogy, consisting of representing the building as a combination of thermal resistances (R) and capacitances (C) analogous to an electrical circuit. The R value represents the resistance to the thermal flux, whereas the C value models the capacity to store thermal energy (Bacher and Madsen, 2010). R and C formulas are given in Eqs. (3.24)–(3.26) (Bruzgevičius et al., 2013); conductive ($R_{conduction}$) and convective ($R_{convection}$) resistances are distinguished. For the sake of simplicity, radiative effects will not be discussed here.

$$R_{conduction} = \frac{d}{\lambda} \tag{3.24}$$

$$R_{convection} = \frac{1}{h} \tag{3.25}$$

$$C = \rho^* c_p^* d \tag{3.26}$$

where

- d: thickness of the layer [m]
- λ: thermal conductivity $[\mathrm{W}/(\mathrm{m}\ ^\circ\mathrm{C})]$
- h: convection heat-transfer coefficient $[\mathrm{W}/(\mathrm{m}^2\ ^\circ\mathrm{C})]$
- $R_{conduction}$ and $R_{convection}$: thermal resistances $[(\mathrm{m}^2\ ^\circ\mathrm{C})/\mathrm{W}]$
- ρ: density $[\mathrm{kg}/\mathrm{m}^3]$
- c_p: specific thermal capacity $[\mathrm{J}/(\mathrm{kg}\ ^\circ\mathrm{C})]$
- C: thermal capacitance $[\mathrm{J}/(\mathrm{m}^2\ ^\circ\mathrm{C})]$

In an RC circuit, the thermal resistances and capacitances are connected together. A node represents typically the temperature of a building element (e.g., a wall), a building system (e.g., a heater), the indoor or the outdoor environment. The order of a model is obtained by counting the number of thermal capacitances the model consists of. An example is given in Fig. 3.12 for a first-order model, where T_o and T_i represent the outdoor and indoor temperatures, respectively; this model was considered in

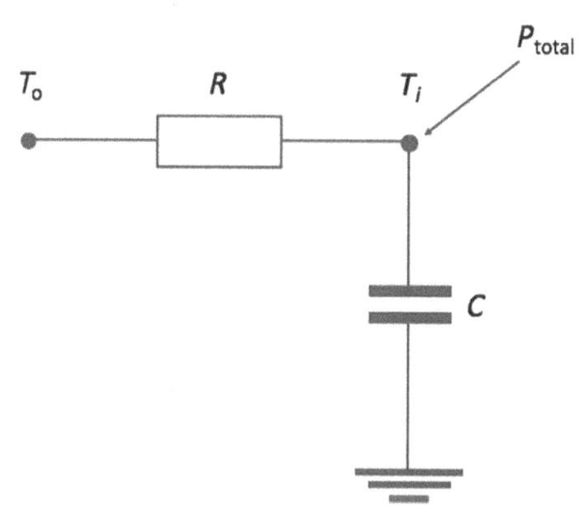

Fig. 3.12 An example of a first-order RC model.

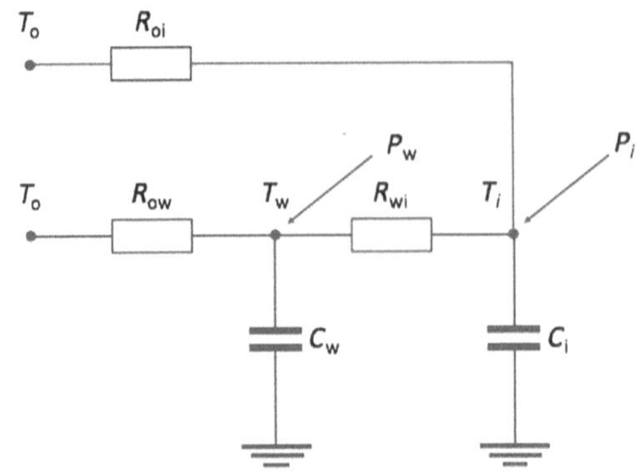

Fig. 3.13 An example of a second-order RC model.

the literature (Ashouri et al., 2015; Fux et al., 2014). An example of another RC circuit is shown in Fig. 3.13, with a second-order model (3R2C), where a node for the building wall is added (T_w) and a general formulation of the thermal inputs is presented. Several second order models have been used in the literature (Berthou et al., 2014; Fux et al., 2014; Harb et al., 2016; Reynders et al., 2014); they vary in terms of number of resistances, definition of the temperature nodes and calculation of the thermal inputs. More complex model examples can be found in the following references (Bacher and Madsen, 2011; Berthou et al., 2014; Fux et al., 2014; De Rosa et al., 2014; Reynders et al., 2014). From an RC circuit, the building model can be formulated by deriving heat-transfer differential equations for each node, as shown in Eqs. (3.27), (3.28) for the first-order model (Ashouri et al., 2015; Fux et al., 2014) and in Eqs. (3.29), (3.30) for the second-order example.

$$C\frac{dT_i}{dt} = \frac{1}{R}(T_o - T_i) + P_{total} \tag{3.27}$$

$$P_{total} = P_{solar} + P_{internal} + P_{hvac} \tag{3.28}$$

- R: thermal resistance between the outdoor and the indoor nodes [°C/W]
- C: total building thermal capacitance [J/°C]
- T_o: outdoor temperature [°C]
- T_i: temperature at node i (indoor) [°C]
- P_{solar}: solar gains [W]
- $P_{internal}$: internal gains [W]
- P_{hvac}: thermal input from the HVAC system at node i (a heat gain is counted positive) [W]

$$C_w\frac{dT_w}{dt} = \frac{1}{R_{ow}}(T_o - T_w) + \frac{1}{R_{wi}}(T_i - T_w) + P_w \tag{3.29}$$

$$C_i\frac{dT_i}{dt} = \frac{1}{R_{oi}}(T_o - T_i) + \frac{1}{R_{wi}}(T_w - T_i) + P_i \tag{3.30}$$

where

- R_{ow}: thermal resistance between node o and w [°C/W]
- R_{oi}: thermal resistance between node o and i [°C/W]
- R_{wi}: thermal resistance between node w and i [°C/W]
- C_i: thermal capacitance of node i (indoor) [J/°C]
- C_w: thermal capacitance of node w (wall) [J/°C]
- T_o: outdoor temperature [°C]
- T_i: temperature at node i (indoor) [°C]
- T_w: temperature at node w (wall) [°C]
- P_w and P_i: sum of the thermal inputs considered (e.g., solar and internal gains, HVAC system input) at nodes w and i, respectively [W]

When developing an RC circuit, the various building layers can be lumped together depending on the model complexity foreseen; lumping decreases the number of R and/or C elements and, as a consequence, the model complexity. The lumping technique applied is similar to an electrical RC circuit and depends on whether the elements are in series or in parallel. Formulas for the combined R value are specified in Eqs. (3.31), (3.32) for a two-layer example characterized by resistances R_1 and R_2 (°C/W). To obtain the combined C value for the same example, the thermal capacitances of layers 1 and 2 are added.

In series:

$$R_{total} = R_1 + R_2 \tag{3.31}$$

In parallel:

$$\frac{1}{R_{total}} = \frac{1}{R_1} + \frac{1}{R_2} \tag{3.32}$$

In the building research literature, an RC model is often written in the state-space form (Hazyuk et al., 2012, 2014; Reynders et al., 2014; Bacher and Madsen, 2010). The state-space form encompasses a wide range of possible formulations. The general formulation of a deterministic Linear Time Invariant (LTI) state-space model is described in Eq. (3.33) (state equation) and Eq. (3.34) (output or measurement equation), where the A, B, C, and D matrices do not vary with time. Stochastic formulations have also been investigated (Bacher and Madsen, 2010). In the previous second order example, the states are the temperatures T_i and T_w, while P_i, P_w, and T_o constitute the input vector. More generally, the variable of interest in the RC models is typically the indoor room temperature or the heating/cooling power.

$$\dot{X}(t) = AX(t) + BU(t) \tag{3.33}$$

$$Y(t) = CX(t) + DU(t) \tag{3.34}$$

where

- $X(t)$: state vector;
- $U(t)$: input vector;
- $Y(t)$: output vector; and
- t: time.

3.5.2 Parameter estimation

Once the model and the equations are known, model parameters need to be quantified. Typically, these latter can be estimated in two ways (De Coninck et al., 2014; Reynders et al., 2014):

- Based on physical building characteristics or expert knowledge. In such case, the model is classified as WB.
- By calibrating the model using historical data. In such case, the model is classified as GB.

The WB approach consists in collecting the building construction information and using it to calculate the R and C values of each model parameter. If no data is available for each single building, construction practices can be identified by architects for example (Kämpf, 2009; Perez, 2014). In the literature, the distinction has been made between the apparent and effective parameters (Antonopoulos and Koronaki, 1998; Reilly and Kinnane, 2017). The calculation of the dynamic thermal characteristics is standardised and has been used in multiple works (Mata and Johnsson, 2014; Building Research Establishment, 2013; International Organization of Standardization, 2008, 2017). At urban scale, validation can be challenging and will depend on the measured data available, if any. The temporal scale of data available for validation will determine at which level the validation can take place. While hourly temperature and energy demand data can be available and used at small scale, such profiles are generally not available at district or city scale, limiting the validation options of urban models. Another factor to account for is the data spatial scale, which also constrains the validation step; if aggregated data are provided, the validation cannot take place at an individual building level. In several cases, models for urban scale application are firstly validated on a small case study, for which detailed measured data is available; then, the model is run on selected representative buildings and validated on the aggregated outcome.

The GB approach is applicable if temperature and/or heating power data are available, typically at an hourly or subhourly level. Data are split into training and validation datasets and used, respectively, to identify the parameters and validate the model. Calibration can also be refined online on a regular basis (e.g., hourly) based on newly measured or synthetic data if available (Ashouri et al., 2015). To calibrate a model, several approaches are applicable. A ready-made package for RC model parameter estimation, called CTSM-R, has been developed by DTU (Denmark) for stochastic state-space modeling (Bacher and Madsen, 2010; CTSM-R Development Team, 2015). The package works within the R framework and can be downloaded online. Another approach consists in writing a specific code for the case study, applying optimization algorithms, which minimize the difference between the measurement (e.g., heating power, indoor temperature) and the model output.

3.5.3 Model selection

RC models are fully flexible in complexity: the *R* and *C* can be added to fit the purpose of the work and the data available. In the literature, models ranging from first to 12th order and more have been developed over varying spatial scales (Ashouri et al., 2015; Berthou et al., 2014; Hazyuk et al., 2012; Lehmann et al., 2013; Fux et al., 2014; Wang and Xu, 2006; Zhou et al., 2008; Vivian et al., 2017; De Rosa et al., 2014). This flexibility is an advantage when it comes to replicability as the RC approach can be used in multiple contexts tuning the model to the case study objective and data available. However, this flexibility involves some investigation to choose the adequate model for the case study considered. Several investigations at building or room level have quantified the impact of an increased model order on the hourly energy needs or the indoor temperature accuracy (Fux et al., 2014; Bacher and Madsen, 2011; Berthou et al., 2014; Reynders et al., 2014). The typical approach is to develop models of increasing complexity and compare the model performance with an index quantifying the error (e.g., RMS error, maximum absolute error). Part of these published works were carried out using real data obtained from highly monitored buildings or test cells; the deployment of such buildings and the collection of data can be challenging, which currently limits the large-scale deployment of such approach. Other works use synthetic data; obtaining synthetic data requires developing a detailed building model but allows to have hourly/subhourly profiles over varying building configurations without the need to install a comprehensive monitoring system.

Small-scale studies have shown varying conclusions regarding the relation between the model complexity and accuracy. This variability can be explained by the multiple changing conditions between case studies (climate, building type, geometry, insulation, control variables, etc). To give a few concrete examples, Fux et al. (2014) report maximum absolute errors of 1.60–0.83°C and root mean square error (RMSE) of 0.47–0.35°C with models ranging from 1R1C to 6R4C for a passive house with a BIPV surface (validation period). Vivian et al. (2017) obtained an average RMSE of 0.57°C for a 5R1C model versus 0.35°C for a 7R2C model (with free floating temperature for an apartment over varying building configurations).

While accuracy is a crucial parameter, moving toward larger-scale case studies implies to account for additional challenges when selecting of the RC model complexity:

- *Computing time*: The higher the model order, the longer the running time tends to be. Computing time can become a constraint when dozens, hundreds, or thousands of buildings are considered. This is specifically the case if the models are to be used for research involving multiple parameter testing, optimization techniques (e.g., Model Predictive Control) and/or fast-decision making (e.g., grid management).
- *Data availability*: As highlighted in Section 3.2, obtaining data for individual buildings at urban scale is challenging. While geometric information is now available in multiple cities as 3D maps shared online, obtaining building-level construction information, consumption, or

occupant behavior profiles is more challenging. The data availability constrains the model development and the validation steps. For a WB approach, an accurate R and C estimation at individual building level is not guaranteed if only generic construction data are available. Also, GB approach cannot be implemented if past temperature or heating profiles are not available. Overall, a lack of input information can lead to parameter identification with multiple sets of solutions for a given accuracy.

The model order selection is thus a compromise between computing time, data availability, and model accuracy and should be analyzed for each case study and application.

3.5.4 On the potential of RC models for large-scale building modeling

RC models offer adequate characteristics for urban scale urban energy modeling; indeed, the previous sections highlighted the flexibility offered by the reduced-order RC approach in terms of:

- technique for parameter estimation: either building information (WB) or historical data can be used (GB) (Section 3.5.2).
- complexity: the number of R and C can be tuned to the case study (Section 3.5.3).

Such flexibility in the identification process and complexity selection is relevant for building modeling at large scale as not all cities offer the same data availability for model parameter identification; thus, RC model can be adapted to various case studies. Moreover, the R and C values have a physical meaning, which is necessary when using the model for specific applications such as studying the potential for retrofit or testing future scenarios.

As mentioned in Section 3.2, to model buildings at urban scale, simplifications are typically necessary compared with the individual building scale: either in the number of buildings considered (archetype approach) or in the model complexity (e.g., steady-state or reduced-order models). These two options, which can be combined, are further described here, within the RC context.

- *Building model development automation*: It involves looping over the available data and calculating directly the model parameters for each building. The full automation is more challenging when using historical data (GB approach) rather than building construction data (WB approach). Computing time can be reduced using parallel capabilities of supercomputers.
- *Reducing the dataset to selected representative buildings*: As shown in Section 3.3, this approach consists of reducing the building stock to a few representative buildings. The modeling effort is concentrated on the typical buildings identified rather than the whole dataset. These representative buildings are identified following a segmentation step (Reinhart and Cerezo Davila, 2016), where the building dataset is divided into groups of distinct characteristics; buildings with similar characteristics will be gathered within the same group. The grouping techniques vary from fully supervised approaches (e.g., age/type categorization) to unsupervised techniques (clustering). The representative buildings are typically chosen as being the centroid or medoid of each group but other options can be

foreseen, including the selection of several buildings per group to better quantify the distribution of the building properties. The final number of buildings to model varies greatly with the case study. Examples are given in Reinhart and Cerezo Davila (2016) review, where the building to archetype ratio varies from 17 to more than 500,000.

The previous paragraphs showed that RC models are an option to consider when modeling buildings at urban scale. The model structure and complexity is to be adapted to the data available and the objective of the study.

3.6 Electrical appliance modeling

Ana Sancho

3.6.1 *Introduction*

In Europe, approximately 20% of energy use in households is due to electrical appliances, and this proportion is higher in better insulated homes. Residential electrical appliance use has direct implications for local low voltage (LV) networks, the loads on them and their integrity; and indirect implications for thermal energy demands, since electrical energy is ultimately dissipated as heat, most of which is emitted within the building envelope. It is therefore important to be able to reliably predict electrical appliance use, in particular the magnitude and temporal variation of the energy use and power demand profiles arising from the aggregation of individual appliances, to support design and regulation of LV networks serving communities of buildings and of building's thermal systems.

But this is a complicated task, for the ownership and use of different types of appliance varies significantly from house to house, and between users. Addressing this diversity requires that we have an appropriate basis for allocating appliances to households depending on their composition and socioeconomic characteristics and for predicting their subsequent use. This in turn implies the use of stochastic simulation and bottom-up approaches that may also facilitate the future testing of Demand Side Management (DSM) strategies.

So far, bottom-up approaches have focused on the modeling of high-load appliances: those who are commonly owned and which contribute significantly to total annual electricity use. Examples include cold (fridge and freezer), wet (washing machine and dishwasher), and cooking appliances. For example, the model of Jaboob (2015) predicts when the appliances are switched on, the duration for which they will remain on and their fluctuating power demands while on. But in our everyday lives we also use myriad low-load appliances. Their individual share of energy use may be small, in some cases even negligible, but it is significant when considering them as a group (or groups).

In this work, we present a successful strategy for modeling low-load appliances. In this we distinguish between four categories of appliance: audio-visual, computing, kitchen, and other appliances, which collectively account for those that are not represented by current device-specific models.

3.6.2 Literature review

As introduced in Section 3.2, *bottom-up* approaches describe the dynamics of a system by explicitly modeling the behavior of the individual parts of that system. For the case of energy use, they consider the individual modeling of every end-use, or aggregates of them, in order to obtain aggregate profiles. These approaches are particularly promising given their potential for (a) improving predictions of energy use of individual buildings or neighborhoods when integrated with building energy simulation, (b) sizing decentralized generation and storage devices, and (c) testing demand side management (DSM) strategies and rules for load management. Moreover, bottom-up approaches have the potential to explicitly include the effects of household composition and individuals' behavioral diversity.

Regarding appliance modeling, bottom-up approaches can be configured at different aggregation levels: from a pure microsimulation where each device is explicitly modeled, to strategies that consider aggregations of device for typologies of them.

Detailed microsimulation approaches are considered in probabilistic empirical models (as defined in Grandjean et al., 2012), which tend to model appliances one by one. Collected data, information on dwelling and household (occupants) characteristics, technical properties of appliances, and aggregated values of energy use are combined in such approaches, and probabilistic methods are applied to generate results with profile diversity. Stokes' model (Stokes, 2005) generates profiles at three aggregation levels: 30-minute resolution average household, 30-minute resolution-specific household (with occupancy considerations), and 1-minute resolution-specific household (including information relating to appliances' cycles). It considers 14 appliances plus miscellaneous, although only 9 different input monitored power cycles are taken into account, resulting in a limitation on the diversity of profiles generated, which in turn leads to poor results estimating the energy demand in the validation of the models. Paatero and Lund (2006) introduce a social random factor (supposed to capture the social variety of the demand) that improves the diversity of patterns obtained; however, only yearly consumption data for the 16 end-uses is used for the generation of the models, together with other aggregate statistics, restricting the resolution to hourly time steps. In general, these approaches do not describe in terms of model parameters the dynamic behavior of appliances, but they generate empirical profiles of power demand as a function of time.

Relatively more aggregated methods are models based on time-use-survey (TUS) datasets. In TUS datasets, the respondents fill in diaries of their activities during the day usually for 1 week periods, such as cooking, sleeping, traveling to work, etc. These data provide a powerful input to bottom-up models, since it encapsulates highly detailed information describing occupants' activities that can be related to the use of appliances. To this end, Capasso et al. (1994) presents a first strategy linking occupants' activities with appliance use, using TUS data. The model produces 15-minute profiles of electricity use, considering aggregations of appliance that correspond to just four type of activities: cooking, housework, leisure, and hygiene; each associated with a blend of large and small appliances, which are allocated by considering the average range of appliances present in the simulated household. The relation between performing an activity and using an appliance is described with a single coefficient α (defined as a human resources). Tanimoto et al. (2008) combines TUS with statistical data of ownership of appliances and its peak and stand-by powers. Thirty-one activities are considered in this case, so that the level of aggregation is low, but this is contrasted by a small dataset size (58 households over 2 days).

In a similar vein, Widén et al. (2009) uses Swedish TUS data to model electricity use by assigning appliances to related activities (9 different categories in this case) and imposing five standard end-use profiles based on the type of their demand profile: demand disconnected from activity, power demand constant during activity, power demand constant after activity (with and without addition of temporal constraint), and fluctuating power demand (only applied to lighting). This approach is further developed in Widén and Wäckelgård (2010), where inhomogeneous Markov chains generate sequences of domestic activities that have an impact on power demand (5 minutes and 1 hour granularity), including dependencies with the number of occupants performing these activities. A yet finer temporal resolution of 1 minute is achieved in the work developed by Richardson et al. (2010). Based on seven different activities, a load curve for the appliances is created using the probability of switching on an appliance when an activity is being performed, and applying a fixed power conversion scheme. Using a calibration procedure based on the total time of use of an appliance, they obtain annual energy predictions. Although this tuning ensures a good overall match in annual energy demand, this does not imply the absence of compensating errors in the modeling of different appliance typologies, or that the dynamic characteristics of appliance use are well represented.

Although activity modeling is a promising method to obtain accurate energy demand profiles, this activity-appliance pairing approach does not facilitate the modeling of the range of appliances, because the activities that are recorded in time use surveys is insufficiently detailed, limiting the applicability of this approach to the modeling of either relatively high-load appliances or aggregations of small and large appliances for which there is weak empirical evidence. There has been no rigorous validation of those bottom-up modeling strategies to date, whether these are based on TUS data or not, to demonstrate their ability to faithfully capture energy use/power demand dynamics. These methods also have no rigorous basis for modeling the dependency of appliance ownership and related use characteristics as a function of household sociodemographic composition.

In partial response to these shortcomings, Jaboob (2015) assigns (exclusively large) appliances to households as a function of their sociodemographic characteristics. The activities of the members of these households are then predicted, from which the conditional likelihood that related appliances will be switch on is modeled, as is the corresponding duration that they will remain on and their time-varying mean power demands while on. Thus, this modeling chain rigorously resolves for dynamic variations in mean power demand, in contrast to static power conversion schemes. Moreover, it presents the possibility of being used together with explicit models of low-load appliances, in order to obtain accurate values of the total electricity use of a house.

To this end and informed by these past endeavors, our task is to develop a parsimonious strategy for the use of relatively low-load appliances, in complement to Jaboob's model of high-load appliances.

3.6.3 Modeling of small appliances

3.6.3.1 Modeling tasks

In this work we are interested in modeling the energy and power demands of low-load appliances to support building, systems, and network design. In order to contribute to accurate predictions of residential energy use, we need to address the diversity in dwelling characteristics and human behaviors. Thus, we identify the following modeling tasks:

1. Low-load appliances are categorized into four groups: audio-visual, computing, small kitchen, and other (miscellaneous housework, garden, and personal care appliances). This classification keeps the possibility of linking the modeling of low-load appliances with occupants' activity modeling, which in turn allows for considering sociodemographic factors. Low-load appliance allocation is performed using a random sampling of the total rated power (sum of all devices in the category) of aggregates of appliances, based on the available data set.
2. Model the characteristic use of these appliances in individual households. To this end, we utilize the fractional energy use $f(t)$: the ratio of the actual to the maximum energy $E_{max\ k,\ i}$ that an appliance i belonging to a category k can use, determined by its rated power. Modeling $f(t)$, we can distinguish between:
   - switching on/off events; and
   - fluctuating demands while the appliances are in use.

   The modeling of fractional power can be applied to modeling single devices ($f_j(t)$, where j is a specific type of appliance) or aggregates of them ($f_k(t)$ where k refers to a category of appliances).

Two considerations need to be taken into account in carrying out these tasks. First, stochastic methods are required, as we are interested in describing the underlying randomness in households' appliance use and investment decisions. These methods rely on the definition of coefficients that represent the system as a probability distribution, which can be dependent on different variables such as time of the day, number of occupants, weather, etc. Second, using the normalized fractional energy of individual appliances $f_j(t)$ instead of absolute energy allows us to evaluate load profiles from different appliances of a similar type, but that do not necessarily have the same magnitude. In this way, appliances can be classified into groups and modeled as a category ($f_k(t)$).

Candidate techniques that have been used to good effect in the modeling of occupants' behaviors include Bernoulli processes (activities; Jaboob, 2015), discrete-time random or Markov processes (presence; Page et al., 2008, blinds; Haldi and Robinson, 2010, windows; Haldi and Robinson, 2009), and continuous-time random processes (blinds; Haldi and Robinson, 2010, windows; Haldi and Robinson, 2009): the latter being a hybrid between discrete and continuous time random process models.

Furthermore, it has been previously shown (Jaboob, 2015; Widén et al., 2009; Richardson et al., 2010) that stochastic methods are successful in describing energy demands and the information listed in Task 2. In the work here explained, a specific statistical approach is presented: survival analysis, as it was found to be the most successful candidate in a comparison of several stochastic models (Sancho-Tomás et al., 2017). The technique considered consists of modeling the duration or survival time that an appliance remains in different energy states.

3.6.3.2 Survival analysis for appliance modeling

Survival analysis models the waiting time until a given event occurs, also referred to as *survival time*. Let T be a nonnegative continuous random variable representing the survival time until an on-appliance is switched off (or an off-appliance is switched on); with probability density function (p.d.f.) $g(t)$. $g(t)$ can follow multiple distributions depending on the problem studied, and commonly $g(t)$ is identified with an exponential decay. In those cases, Weibull distributions can be used to model survival time, and then:

$$g(t) = \begin{cases} \frac{k}{\lambda}\left(\frac{t-\gamma}{\lambda}\right)^{k-1} e^{\left(-\frac{t-\gamma}{\lambda}\right)^k} & t>\gamma \\ 0 t<\gamma \end{cases} \tag{3.35}$$

where $k > 0$, $\lambda > 0$, and $\gamma > 0$ are the shape, scale, and location parameters of the Weibull distribution (Cox and Oakes, 1984). Thus, the cumulative distribution function (c.d.f.) $G(t) = P\{T < t\}$ gives the probability of the event to have occurred by duration t. The *survival function* $S(t) = 1 - G(t) = P\{T \geq t\}$ is then defined as the complement of the c.d.f., and describes the probability to remain in a given state before t:

$$S(t) = e^{\left(\frac{t-\gamma}{\lambda}\right)^k}. \tag{3.36}$$

By inverting Eq. (3.36), it is possible to obtain directly the duration for which an appliance will continue (survive) in a specific energy state s as:

$$t_s = \gamma + \lambda[-\ln(w)]^{1/k}, \tag{3.37}$$

given a number $w \in [0, 1)$ drawn randomly from a uniform distribution.

Fitting survival times to Weibull distributions has been successfully deployed in the past to model the times that appliances are in a particular energy state (on, off, stand-by, or other). An illustrative example of the shape of the survival times in our data is presented in Fig. 3.14, for a group of fractional energy states.

Occurrences of each event and durations are first extracted from the data, and used to fit Weibull distributions, obtaining scale, shape, and location parameters λ, k, and γ. The fitting process makes use of the `scipy` optimization package in Python, and fits the data to a Weibull function using a nonlinear least squares method. Interpolation between points was necessary to fit the curves more effectively, given that the resolution of the original data (10 minutes) was not high enough. Two examples of the fitting process are given in Fig. 3.15.

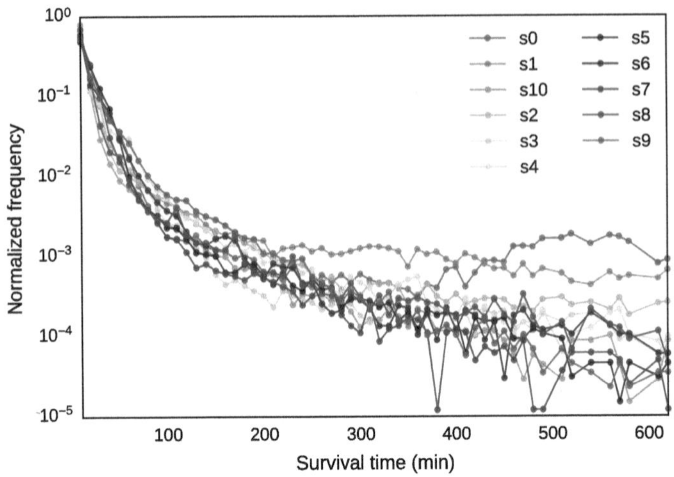

Fig. 3.14 Exponential decay of survival times for 11 different fractional energy states. Frequency is normalized for the distribution.

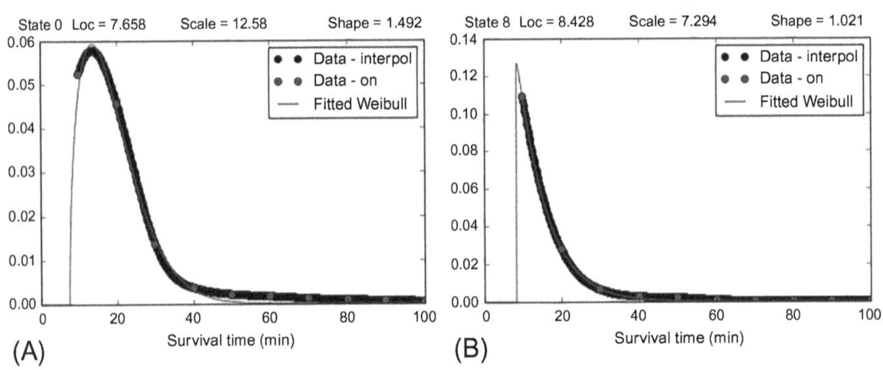

Fig. 3.15 Details of data fitted to Weibull distribution using nonlinear least squares method. (A) State 0; (B) State 8.

These distributions are then used to calculate survival times in a simulation using Eq. (3.37).

3.6.3.3 Monte Carlo simulation of fractional energy states

Monte Carlo methods may be defined as *the representation of a mathematical system by a sampling procedure which satisfies the same probability laws* (Bharucha-Reid, 2012). They provide a method to artificially represent a stochastic process by a sampling procedure, which will be determined by the particular underlying probability distribution of the given process.

For the specific problem posed here, the probability structure is given by either the parameters of the Markov chain or the survival analysis. In both cases, the purposes of the Monte Carlo simulation is to produce a time series of energy states $s_k(t)$ for a given category of appliances. The process to obtain this sequence depends on which of the methods is being used.

At the start of the simulation, state $s(t = 0)$ is assumed. A random number w is drawn from a continuous uniform distribution over the interval [0, 1) and entered in Eq. (3.37) to obtain the survival time $t_{s(t=0)}$ until a change of state occurs, covering a number $n_{s(t=0)}$ time steps. When this time is over, the next state is calculated. In this case, transitions are not modeled. Instead, they were extracted using a sampling following the distribution of hourly probabilities $P_s(t)$ of finding each of the states. Once at the next state $s(t = t_{s(t=0)})$, the process starts again: survival time is computed, until a change of state occurs. Thus, times and transitions between states are successively calculated for the simulation period.

In this work, a multistate survival model has been implemented, with 11 energy states, following an arbitrary division of 10 equidistant fractional energy states, plus the *off* state: s_i for $f(t) = \{0; 0 - 0.1; 0.1 - 0.2; \ldots; 0.9 - 1\}$, respectively. Such a division of $f(t)$ allows us to test the added value of refined characterization of energy states. As it was said above, transitions between the different states are not modeled, but they are calculated using an inverse function method with the hourly likelihood of finding each of the states.

3.6.3.3.1 From energy states to an energy profile

For category k the sequence of energy states $s_k(t)$ now needs to be transformed back into a fractional energy profile $f_{sim\ k}(t)$. Thus, each energy-temporal state is multiplied by its corresponding mean or median fractional energy $\widetilde{F}_k(s)$, depending on the strategy employed, leading to a simulated fractional energy profile

$$f_{sim\ k}(t) = s_k(t) \cdot \widetilde{F}_k(s) \tag{3.38}$$

One final step transforms these profiles from fractional to actual energy values:

$$E_{sim\ k}(t) = f_{sim\ k}(t) \cdot \widetilde{E}_{max\ k} \tag{3.39}$$

where $\widetilde{E}_{max\ k}$ is a statistical measure of the maximum energy (or power $\widetilde{P}$ as required) for all instances of appliance in the category k. At this stage, depending on the measure selected for $\widetilde{E}_{max\ k}$, the approach can be applied to simulate the use of individual devices belonging to a category k, or the use of the category as a whole. When considering the first case, assignment of $\widetilde{E}_{max\ k}$ is performed using the mean value:

$$\bar{E}_{max\ k} = \frac{1}{N_k} \sum_{i=1}^{N_k} E_{max\ k,i} \tag{3.40}$$

where N_k denotes the number of instances i in category k. The second option consists of adding maximum energy values off all appliances in a category for each house, and perform a random sampling of them. In any case, the estimation or selection of $\widetilde{E}_{max\,k}$ (or $\bar{E}_{max\,k}$) becomes critical to calculating accurate aggregate energy profiles. If there is no such data available (as it was in our case), it is essential to produce robust estimation. If the data is available, then the assignment of $\widetilde{E}_{max\,k}$ is trivial. In what follows then, we focus on testing the underlying hypothesis in our modeling strategies rather than in the fidelity of predictions of aggregate energy profiles that require a random assignment process.

3.6.4 Household Electricity Survey data set

The Household Electricity Survey (Department of Energy and Climate Change and Intertek Testing and Certification Ltd, 2010–2011) is an extensive monitoring survey of 250 households in the United Kingdom, carried out during 2010 and 2011. Apart from detailed sociodemographic information, it contains data describing the appliances present in every monitored household and their temporal electrical energy use during 1 or 2 months, with records every 2 minutes. Of the 250 households, 26 were additionally monitored for a whole year, with a 10-minute resolution. Since the 1-month data were not measured during the same month for all households, only the data recorded for the 26 houses during a whole year were utilized in the analysis presented here, in order to avoid possible seasonal effects on the use of appliances.

The relevant low-load appliances found in the dataset were classified into four categories, following the types of activity that relate to their use:

1. audio-visual (excluding TVs that are considered as high-load appliances given their extensive use),
2. computing,
3. small kitchen appliances (excluding cookers, microwaves, and ovens), and
4. other small appliances.

In this section, subscript k denotes a category, and its values correspond to the four categories: $k = \{1\text{: audio-visual; 2: computing; 3: kitchen; 4: other}\}$.

Fig. 3.16 shows the types of device available in the data set and their contribution to annual energy use, with categories depicted in different colors. The height of the bars represents the mean value of annual energy use of the corresponding type of appliance, whereas the width is proportional to the number of instances observed in the 26 households for the given device. Thus, the area of the bar indicates the total energy use of that appliance throughout the stock of houses surveyed.

One shortfall encountered in the data set is that there is no information describing the rated power of the appliances, posing a challenge to the accurate estimation of $E_{max\,k,\,i}$.

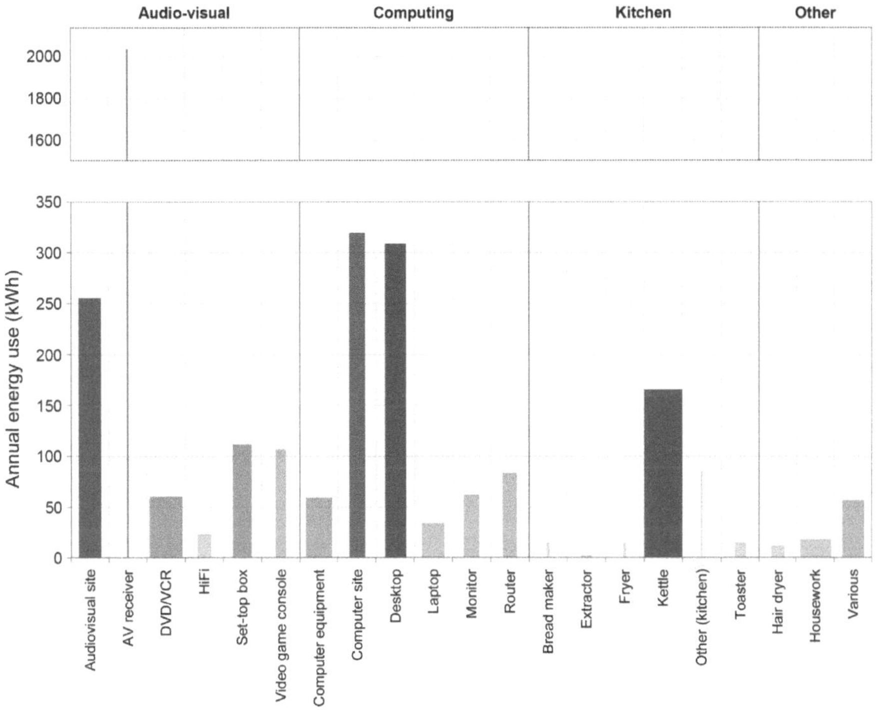

Fig. 3.16 Annual energy use of the types of appliances considered in the modeling, divided in four categories: audio-visual, computing, kitchen, and other. The height of the bars corresponds to the mean annual energy use, while the width is proportional to the number of instances recorded in the 26 houses. Combining this information, *darker bars* identify the dominant types of appliance for the category.

3.6.5 Results

The application of the model is shown in this section in two ways: the first involves a single day simulation for a specific household (labeled in the dataset as "103028"), presented in Fig. 3.17. It contains 13 different low-load appliances, which are described in Table 3.2. As expected, the model predicts usages of different duration and it is able to capture the spikes in the profiles. For the audio-visual and computing categories the simulated profile shows a more dynamic behavior than the observed one, with more predicted on/off switches. The observed data, on the other hand, is showing one or more devices that remain switched on but with low variations during the day. On the contrary, the kitchen and other category predict well the sporadic use of these devices.

The second includes the averaged daily energy usage arising from all the devices in the different categories of appliances over the year, when aggregated to a community of 20 households (Fig. 3.18). This situation is much more representative of the intended usage of the model than for the modeling of individual appliances in a single household. In this case, the total energy use for each category (adding up all the

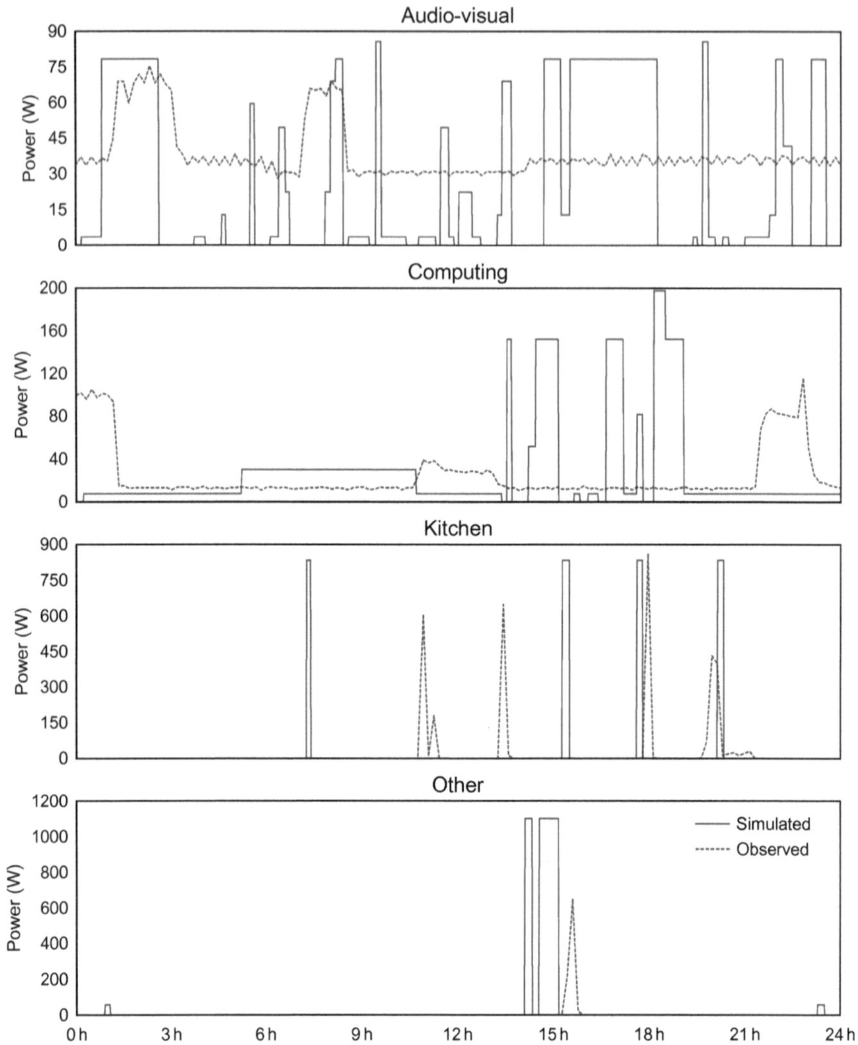

Fig. 3.17 Example of 1-day simulation for aggregates of appliances.

available devices in each household) is averaged in order to create a typical day profile for the community. Then observed and simulated data are compared.

Simulated audio-visual appliances describe temporal variability, although its dependency is not as strong as in reality. Again, the choice of category maximum energy values for appliances with different behaviors impacts the total energy use predicted by the model. This effect is more clear in the high values of energy use at night hours, and lower values than the observed during the evening peak. Something similar occurs for the computing appliances, although in this case, its energy use is consistently overestimated, but still captures variations during the day. The use of kitchen devices is underestimated, and the opposite happens with the other category;

Table 3.2 Available appliances in house "103028" and their rated values, extracted from the data set

	Appliance type	Rated power (W)
Audio-visual	Set-top box	30
	Audio-visual site	43.2
	DVD/VCR	12.6
Kitchen	Bread maker	97.2
	Toaster	720
	Extractor	16.8
Computing	Laptop	48.6
	Computer equipment	3.6
	Desktop	108
	Monitor	30
	Router	7.2
Other	Housework	1248
	Various	54.6

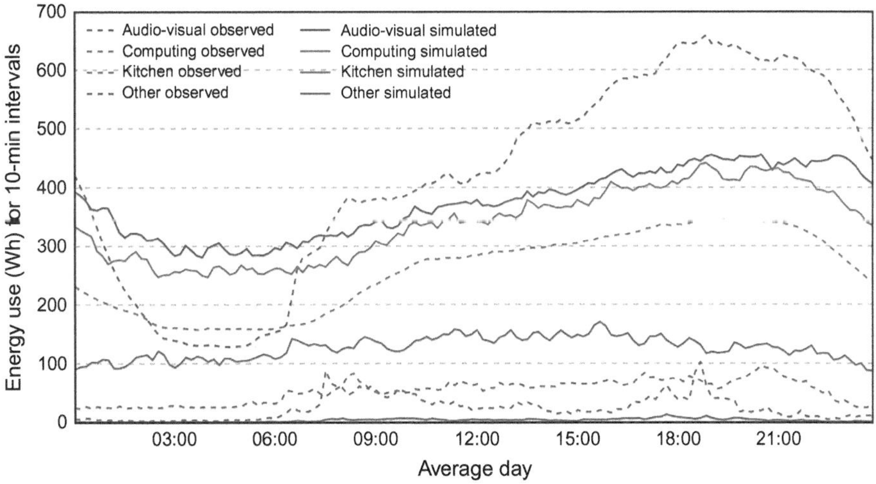

Fig. 3.18 Averaged daily total energy usage from the different categories.

this could be related to the amount and type of devices available for this particular group of households, given that their use is reduced. One way of improving these consistent over/underestimations would be to make a correction after the simulation to the total energy use or to the maximum energy value employed. To conclude, although we are modeling exclusively aggregates of appliance and in a generalized way, realistic magnitudes for the electrical energy use with respect to time can still be obtained.

On the more comprehensive text in Sancho-Tomás et al. (2017) a number of validation measures are explained and applied for the simulation results.

3.6.6 *Conclusion on appliance modeling*

As the integrity of the envelope of both new and existing houses improves, so the proportion of energy that is used by household electrical appliances, which are becoming increasingly ubiquitous, is likely to increase. It is important than that modelers have at their disposal reliable models of appliance energy use, if they are to accurately predict the thermal performance and energy use of future homes. Furthermore, there is increasing interest in the concept of smart grids, to better regulate the distributed supply, storage, and demand of electrical energy. This places increasing focus on the ability to predict the dynamic behavior of household electrical appliances. While good progress has recently been made in the modeling of relatively large appliances: those whose prevalence and cumulative energy use supports the estimation of device-specific models. Poor progress has been made in the modeling of relatively small appliances: those whose cumulative energy use is individually small, but significant when considered as aggregates by typology. To this end we tested a range of strategies for the modeling of small appliance categories (Sancho-Tomás et al., 2017); first predicting discrete states in fractional energy demand, then converting these into absolute energy demand, given an estimate of the corresponding maximum power demand. In this work we presented the application of the most successful strategy: a multistate survival analysis model.

Modeling appliances by their typologies presents many advantages: it provides a straightforward solution for modeling the range of types of appliances, it reduces the amount of input data needed to estimate the model and the risk of overfitting, and it avoids the time-consuming process of modeling appliances individually, simplifying dynamic energy simulation. However, analyzing categories can compromise the fidelity of predictions of aggregate energy use, particularly if modeling small numbers of households. In our case, a successful strategy consisted of allocating energy values with a random assignment process. The survival multistate approach has been effectively deployed to model low-load appliances in four categories: audio-visual, computing, kitchen, and other. The profiles output by the model have satisfactorily compared to those of a community of households.

3.7 A model based on reference buildings and parametric analysis to generate district energy demand profiles

Georgios Kazas

3.7.1 *Introduction*

Physical models of buildings based on a direct and detailed representation of building elements and heat-transfer equations are called engineering or bottom-up physics-based building models and they are widely used (commonly implemented into simulation tool such as EnergyPlus, TRNSYS, ESP-r, IDA-ICE, etc.) to determine heating/cooling loads and assess the energy performance of a single building. Their application at a larger scale can be based on a "brute force" approach—that is, the simulation of each single building in a stock—or on the development of clustering and classification techniques to reduce the number of single buildings to be evaluated in a district. To this regard, two different approaches exist in the literature. The first one, which is called "building archetype," reduces the number of buildings to be simulated by classifying the building stock into clusters of buildings, and has mainly been used in energy decision making and policy tools. The second one, which is called "sample of representative buildings," reduces the number of buildings to be simulated by selecting some real existing buildings as representative of a particular building stock.

The generation of detailed thermal energy demand profiles, at an urban district scale, is nowadays of the foremost importance for a number of reasons (e.g., evaluation of demand side strategies, district heating network design) but at the same time, it is quite burdensome from the computational point of view if addressed with the standard current energy simulation tools. In this chapter, DiDeProM, a model to generate District Detailed thermal energy demand Profiles, at an urban district scale, is presented (Kazas et al., 2017). This model can be classified as a bottom-up engineering model, and it is based on a parametric analysis conducted on some representative buildings of the district. The results from the energy simulations conducted at the first stage of the model, namely a database of normalized thermal energy demand profiles with an hourly time resolution, can later be aggregated to obtain the overall thermal energy demand profile of the complete district.

3.7.2 Methodology

The methodology of DiDeProm follows two main stages: first, an inventory of normalized hourly profiles of sample of representative buildings has to be created (parametric analysis), then at a second stage the average thermal energy demand profile of the district is generated from a statistical aggregation on profiles selected from the thermal energy demand profiles database (aggregation process).

The model structure can be summarized in the following steps. Steps 1–12 refer to the first stage of parametric analysis and creation of the normalized profiles database and steps 13–19 refer to the second stage where the aggregation of the profile of the district takes place.

(1) An insula of buildings (I), which is the case study district, is selected and it is made of a number n of different buildings. Thus $\mathfrak{I} = \{\mathcal{B}_1, \mathcal{B}_2, \mathcal{B}_3, \mathcal{B}_4, \ldots, \mathcal{B}_n\}$.
(2) From Satellite maps, enquiries on GIS, audits, etc., the geometry of the various buildings and their relevant thermophysical properties are inferred.
(3) Based on the data collected at step 2, an analysis is performed to cluster the buildings of the Insula. The clustering process is done according to various parameters (e.g., building age of construction, thermo-physical properties, aspect ratio) in order to identify the minimum number of buildings that can be assumed as representative of the whole insula.

(4) Reference buildings $\mathcal{R}_{B,r}$ are then defined as a result of step 3, choosing some of the real building of the Insula. Each reference building is considered to be representative of the entire cluster of similar buildings.

The number of reference buildings to be identified in an insula/district can vary as a function of the size of the insula, of the urban design, etc., but it has to be kept at a minimum value without losing the variety within the district. To this regard, it can be noted that buildings with the same age, thermo-physical properties, design, etc., but different orientations can be put together in the same cluster since the physical characteristics of the buildings are the same.

(5) Each reference building, $\mathcal{R}_{B,r}$ is modeled by means of detailed white-box models (performing seasonal and/or yearly analysis based on hourly or subhourly time steps).

(6) A number of possible operating scenarios (shading profiles, user behavior, etc.) is set up for covering the most relevant features that influence the thermal energy demand of buildings:

- shading scenarios ($S_{s,k}$)—to take into account different orientations;
- air infiltration scenarios ($S_{v,g}$); and
- internal loads scenarios ($S_{l,m}$).

(7) A database of parametric analysis scenarios, $S_{k,g,m}$, is created whose elements, $S_{k,g,m} = f(S_{s,k}, S_{v,g}, S_{l,m})$, constitute all the possible boundary conditions for the energy simulations.

(8) For each scenario belonging to $S_{k,g,m}$, a simulation of every reference building in $\mathcal{R}_{B,r}$ is performed by means of the detailed multizone model created at step 5.

(9) From simulations conducted at step 8, a database of thermal energy demand profiles, $\vec{\mathcal{P}}_{(\mathcal{R}_{B,r},S_{k,g,m})}$, is created where the number of profiles in the database is equal to the total number of the scenarios, $S_{k,g,m}$, times the number of the representative buildings, $\mathcal{R}_{B,r}$.

(10) The thermal energy demand profiles, $\vec{\mathcal{P}}_{(\mathcal{R}_{B,r},S_{k,g,m})}$, are normalized by the conditioned floor area ($A_{\mathcal{R}_{B,r}}$) of the pertinent reference building, $\vec{p}_{(\mathcal{R}_{B,r},S_{k,g,m})}$, thus obtaining a database of normalized energy demand profiles: $\mathcal{D}$.

(11) Database $\mathcal{D}$ is then grouped in subsets $\mathcal{D}_n$, one for each of the reference building $\mathcal{R}_{B,r}$.

(12) Within each subset $\mathcal{D}_r$, the profiles are then sorted into clusters, $C_{S_{g,m}}$. The variable that is selected for the definition of the clusters is the shading scenario ($S_{s,\ k}$).

(13) The actual building $\mathcal{B}_n$ of the Insula is matched to a reference building, $\mathcal{R}_{B,r}$ and the corresponding subset $\mathcal{D}_r$ is identified,

(14) The solar shading scenario, $S_{s,\ k}$, that is pertinent to the (actual) building of the insula is selected. This identifies the cluster, $C_{S_{g,m}}$, of normalized energy profiles within the subset $\mathcal{D}_r$.

(15) A normalized energy demand profiles $\vec{p}_{(\mathcal{R}_{B,n},S_{S,g,m})}$ is picked-up randomly from the cluster $C_{S_{g,m}}$.

(16) This normalized thermal energy demand profile is multiplied by the conditioned floor area of the actual building to obtain one of the possible hourly time profile of the energy demand of the building: $\vec{\mathcal{P}}_{B_x} = \vec{p}_{(\mathcal{R}_{B,n},S_{S,g,m})} \times A_{B_X,cond.}$

(17) Steps from 1 to 4 are repeated for each building of the insula/district.

(18) The thermal energy demand profiles are summed and the overall thermal energy demand profile of the Insula: is obtained $\vec{\mathcal{P}}_{\mathcal{I}} = \sum_{x=1}^{n} \vec{\mathcal{P}}_{B_x}$.

(19) Steps from 13 to 18 are recursively repeated many times with random selection of normalized profiles and the average thermal energy demand profile with hourly time step of the entire energy demand of the district is assessed as:

$$\vec{\mathcal{P}}_{\mathfrak{I},avg} = \frac{\vec{\mathcal{P}}_{\mathfrak{I},1} + \vec{\mathcal{P}}_{\mathfrak{I},2} + \vec{\mathcal{P}}_{\mathfrak{I},3} + \cdots + \vec{\mathcal{P}}_{\mathfrak{I},j}}{j}$$

$\vec{\mathcal{P}}_{\mathfrak{I},avg}$ can finally be assumed as a representative thermal energy demand profile with hourly time resolution at district scale.

Number j can be between 50 and 100.

In order to represent the complete profile, a postprocessing stage to identify typical monthly days (within the profile) can be performed. A typical monthly day is defined as the day of the month that is closest to the average monthly air temperature $(\overline{T_m})$ and the average monthly rate of solar radiation $(\overline{R_m})$. Once selected for each month, the corresponding energy demand profile (over 24 hours) is extracted from the overall thermal energy demand profile $\vec{\mathcal{P}}_{\mathfrak{I},avg}$. Since the average monthly air temperature $(\overline{T_m})$ and the average monthly rate of solar radiation $(\overline{R_m})$ are not dependent variables, the typical day is assumed as the one that is closest to both parameters (considering air temperature as the primary variable).

3.7.3 Example of application

As an example of application, the DiDeProm model was applied to a rectangular urban district in Turin (Italy) made up of 12 buildings. Since all the buildings of the insula are multifamily residential buildings, with similar construction and architectural features and thermal parameters, only one building was adopted as a reference building $(\mathcal{R}_{\mathcal{B},1})$. The building (five stories for approx. 1000 m^2 of floor area) has a water heating system (radiators) and, as most of the other ones in the insula, it is connected to the district heating system of the city, so it is equipped with a district heating substation (with remote monitoring and metering).

The reference building was simulated into the EnergyPlus engine and the results calibrated against a whole season dataset of subhourly measured data of heating rate demand at the district heating substation provided by the district heating supplier (Monetti et al., 2014, 2015). The results of the calibration showed a good agreement between the model and the measurements. A refinement of the model was necessary in order to correctly estimate the peak morning demand: in the thermal model the peak energy required for the heating of all the water content of the system was added, increasing the goodness of fit between simulated and measured data (Fig. 3.19).

A parametric analysis was then conducted to obtain the database of the normalized heating energy demand profiles of the district. The scenarios were defined considering three main parameters: the relative position of the building (four shading scenarios), the internal loads (such as the occupant's density, lighting, and electrical equipment loads) and the infiltration rates. The combination of the various options of each parameters gave a final number of scenarios equal to 300 (75 for each orientation scenario). From the computational point of view, the parametric analysis was implemented into EnergyPlus through the parametric tool module. Once defined the scenarios settings,

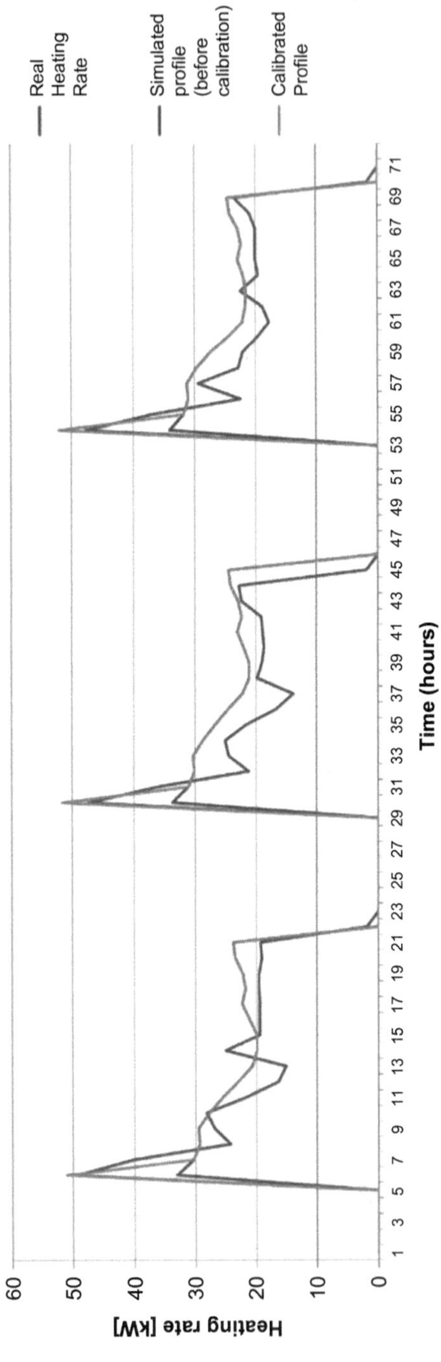

Fig. 3.19 Calibration results for 11, 12, and 13 January.

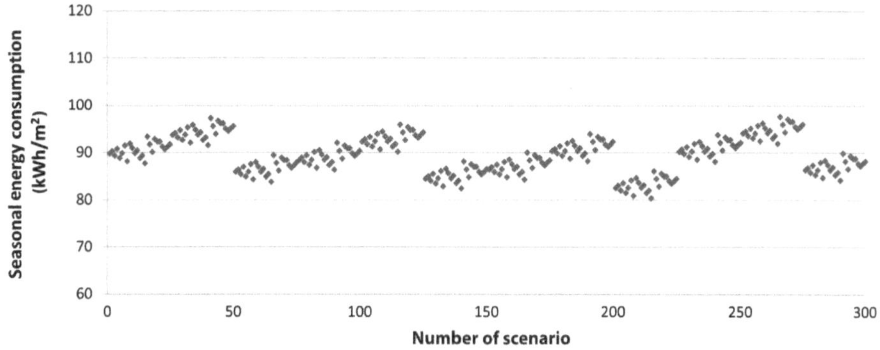

Fig. 3.20 Comparison of seasonal energy consumption of the reference building in the 300 scenarios.

all the parametric simulations were run and the time profiles were normalized by the heated surface. The initial database of the normalized profiles was clustered into four different subsets as a function of the orientation. In Fig. 3.20 the seasonal heating energy consumption of each simulation is reported.

Each real building of the insula was matched to the reference building and the pertinent shading scenario, thus identifying the correct cluster of normalized energy demand profiles. A normalized energy demand profiles was picked-up randomly from each cluster for each building, then all the profiles were summed to obtain the energy demand profile of the whole insula. The same process was repeated 100 times, thus generating 100 heating energy demand profiles of the case study district with an hourly time step resolution. An average heating profile was calculated from these profiles in order to be representative of the block. The choice was to enlarge progressively the sample from $j = 20$ to $j = 100$ to investigate the effect of enlarging the profile sample.

At the end of the calculations, the dispersion between the maximum and minimum seasonal heating demand values at the insula scale (7%) is far less than the same dispersion at the reference building scale (18%). In order to study the reliability of the average heating demand profile, a statistical interpretation of the various profile samples was carried out: the median resulted to be almost stable and 50% of the values of each sample are in almost the same heating energy value. The absolute percentage variation of the average of the samples (i.e., the variation of the average of one sample compared with the average of the previous smaller sample) declined as the sample was enlarged and almost reached a zero value in the end of the process. Thus the generation of the heating demand profiles for the block of 100 buildings can be considered a good and stable sample.

3.7.4 Conclusions

The generation of thermal energy demand profiles with a very detailed time resolution at an urban district scale is of the foremost importance when developing demand side management strategies and integrating active thermal energy storage technologies in buildings. If for the determination of space heating/cooling energy demand of a district of buildings, data-driven and bottom-up statistical models are widely used, for very detailed time resolution energy profiles, bottom-up engineering models can be preferred. However, the "brute force" application of the direct simulation to all buildings of the district can be quite time consuming and does not guarantee reliable results from a statistical point of view, therefore in this section the DiDeProM model was proposed and applied to a case study.

In DiDeProM, an aggregation method, which can cover the uncertainties of a district (parametric analysis), and would be sufficiently precise and general to be suitable for already existing or newly constructed urban districts is developed. It is worth noting that also dynamic occupancy behavior models, developed by others, may be integrated into the model. In case of districts less homogeneous than the one presented as a case study, the parametric analysis should be performed on a larger number of reference buildings from the district. At the same time, if within a district no clusters can be found to classify the buildings, the proposed method can be classified as a "brute force" since each building has to be modeled in detail.

3.8 A data-driven modeling approach for large-scale demand profiling of buildings of the same typology

Giovanni Tardioli

In this section, DDM are coupled with parametric simulations to develop a building modeling framework to simulate demand profiles of a vast number of buildings of the same typology. The proposed methodology is based on the use of a representative building (e.g., archetype), a building energy software for dynamic simulations and a parametric simulator to create large datasets of different scenarios. Three DDM algorithms are used, trained, and tested using the simulation dataset: SVMs, NNs, and generalized linear models (GLMs). Results at an hourly resolution show that DDM approaches can correctly emulate the outputs of the building simulation software. The methodology constitutes a novel example of surrogate modeling at high temporal granularity.

3.8.1 Introduction

The majority of research studies related to DDM analyze single building case studies; in this applications DDM are employed as forecasters of different buildings energy variables. In this context, DDM are employed for predictive control studies (Ferreira et al., 2012) or as emulators for calibration of building modeling parameters, demand side management applications and demand response analysis (Burger and Moura, 2015). Few studies focus on the use of DDM for a large number of buildings and when they do they study aggregated, yearly or monthly building consumption. For this reason, there is a lack of research methodologies based on data-driven methods, to provide estimate demand profiling information at scale. Nevertheless, DDM properties could be exploited to partially alleviate the modeling burdens related with large-scale simulations where simplified methods and speed of calculation are fundamental. Thus, the aim of this section is to describe a scalable methodology regarding the use of DDM as a demand modeling technique for a vast number of buildings of the same typology. The approach is based on the use of a representative BEM and parametric simulations to create a collection of data for the training of DDM. The representative energy model could be, for example, the outcome of clustering analyses conducted on an urban case study (Ghiassi et al., 2015). This section is structured as follows. Initially, a literature review of DDM techniques is presented; then, the research "gap" is identified and described; the research methodology is discussed and explained as well as the contribution and novelty of the present work; after this, the methodology is then tested on a case study and finally a description of the main outcomes of this approach is provided. The strengths and limitations of the presented approach are discussed and scope for further research is presented taking into consideration the employment of the technique at an urban or district level.

3.8.2 Literature review

There are many examples of the use of DDM in the building energy sector. The majority of the studies concerning DDM are related to single building-level applications. An example of the use of DDM for single building application is shown in Fig. 3.21. The application of DDM can be summarized in three steps: data gathering and preprocessing, inputs and models selections, and testing and validation of the results. In the first step, data are gathered from building sensors, from smart metering systems and through building management systems. Then, data are stored in a database and sensitivity analysis is perform to identify the most important parameters related to a dependent variable. A model selection procedure is performed to select the most suitable model for the dataset. Part of the data available is used to train the predictive model and the remaining to test its accuracy. Once trained the predictive model is able to provide future forecasts of the variables under analysis.

DDM are employed mostly as forecasters of building consumption. Jain et al. (2014) developed a sensor-based model using support vector regression methods, for a residential building studying the impact of temporal and spatial granularity of

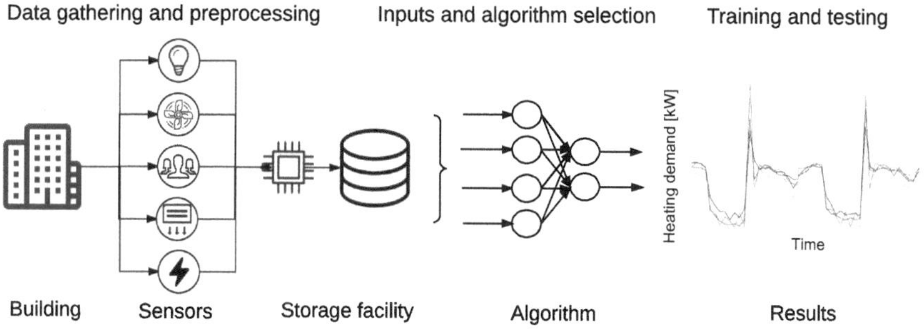

Fig. 3.21 Steps involved in the creation of a DDM for single building-level applications.

input data. Li and Wen (2014) suggested a methodology to estimate building consumption using a frequency domain spectral density analysis for a commercial building. Yang et al. (2014) studied a predictive method to estimate the consumption of a chiller based on building operation data and weather forecasts. Mai et al. (2014) presented an electric load forecasting model at a hourly resolution for a commercial building based on neural network techniques employing as input data outdoor weather information and historical load data. Edwards et al. (2012) compared seven different DDM methods on a residential case study, with input data characterized by 15 minutes resolution. Macas et al. (2016) suggested an artificial neural network (ANN) method for operational cost reduction of an office building based on the estimation of heating energy consumption, internal air temperature, and thermal discomfort. Paudel et al. (2014) showed the potential of a predictive method based on occupancy profiles and operational heating power levels to predict short time horizon heating demand of an institutional building. Fan et al. (2014) created an ensemble model to predict next-day energy consumption and peak power demand for a commercial building. Kapetanakis et al. (2015) presented a DDM to forecast building heating loads of a commercial building using building management systems input data and weather information, presenting an inputs selection technique.

Utility companies employ DDM and smart metering system data at an aggregated level to estimate peak power demand for entire districts. This enables utility companies to forecast supply requirements for trading on energy markets at convenient prices, or to identify stressed points of the networks and to plan ahead network maintenance and upgrades. For groups of buildings the analysis of building energy consumption at an accurate temporal resolution is usually conducted using aggregated data. Data are typically gathered from network substations or from building management systems controlling the entire group of buildings. Some examples in this context are studies conducted on university campuses (Powell et al., 2014). Humeau et al. (2013) analyzed relations of consumption series data by clustering houses according to their consumption profiles and using DDM to estimate aggregate district consumption. Jurado et al. (2015) suggested a methodology for university buildings which combines feature selection based on entropy measures and DDM. Jovanović and živković (2014) suggested a methodology for predicting heating energy consumption of 35 buildings using an ANN. Escrivá-Escrivá et al. (2011) presented an ANN model for short-time horizon predictions of total power consumption divided by end-uses for a large group of university buildings.

At district or city scale, Williams and Gomez (2016) proposed an approach based on DDM to predict monthly consumptions, using regression approaches based on a variety of building features. Nouvel et al. (2015) and Howard et al. (2012) focused on forecasting building consumption for many single buildings at a yearly resolution.

Recently, research studies showed the possibility of coupling parametric simulation and DDM. Catalina et al. (2008), Chou and Bui (2014), and Tsanas and Xifara (2012) presented approaches based on DDM to provide early-stage design consumption estimation for a large number of buildings. Furthermore, these studies underlined the possibility of using DDM not only as forecasters but also as modeling technique.

The literature review process shows that the majority of DDM algorithms are neural networks, SVMs, and multiple linear regressions. Regarding input variables, time variables and weather information are among the most common, along with other data such as building energy consumption. The principal targets of the prediction are electric consumption and thermal requirements for both heating and cooling. Typically, forecasting is performed at an hourly or subhourly basis. The prediction horizon varies based on the nature of the application. In addition, with reference to the literature review, the use of DDM as an accurate forecasting tool is limited to single building studies or aggregated analysis of groups of buildings. DDM have been employed as large-scale modeling techniques for early-stage design demand estimation. At city or district level, DDM are used for single buildings' predictions with an annual or monthly resolution. Thus, it is evident the lack of applications and methods to use DDM as large-scale modeling techniques to provide profiling information at a fine time resolution. Hence, the following sections present a methodology associated with the use of DDM as large-scale modeling technique to estimate building demand information at an hourly resolution.

3.8.3 Methodology

The methodology relies on the use of a representative BEM and parametric simulation to simulate a large number of building scenarios and for the training of DDM. The representative energy model could be an urban representative building identified by the means of statistical analysis or with more advanced methods such as clustering. The simulated large-scale scenario is formed by similar but different buildings of the same typology. This is obtained by permuting building-related parameters such as glazing surfaces, constructions, internal gains, etc.

Fig. 3.22 shows an overview of the research methodology. Initially, the BEM of a residential case study is developed. This will be assumed as baseline model. An input file related to the energy model is provided as input to a parametric simulator. Different energy-related variables are then defined; these are varied using the parametric simulator; to each variation correspond a simulation. For this reason, a considerable synthetic dataset of BEMs is generated. These models are similar but different variations of the baseline model. Outputs from the dataset, including hourly demand information for both electricity and heating, are stored in a database. Next, the DDM is trained with the information from the database and prediction accuracy measures are evaluated. Prediction accuracy measure allows to rank the predictive models based on their performance on the case study.

3.8.4 Building energy modeling

The first step of the methodology requires the creation or the use of a BEM. The case study is a mid-terraced house. The house is located in an eco-village of similar buildings. The mid-terraced house consists a total floor area of 166 m^2 and a volume of approximately 359 m^3. The exterior wall area is of about 122 m^2 and the external openings area is of about 29 m^2. Fig. 3.23 shows a rendering or the case study.

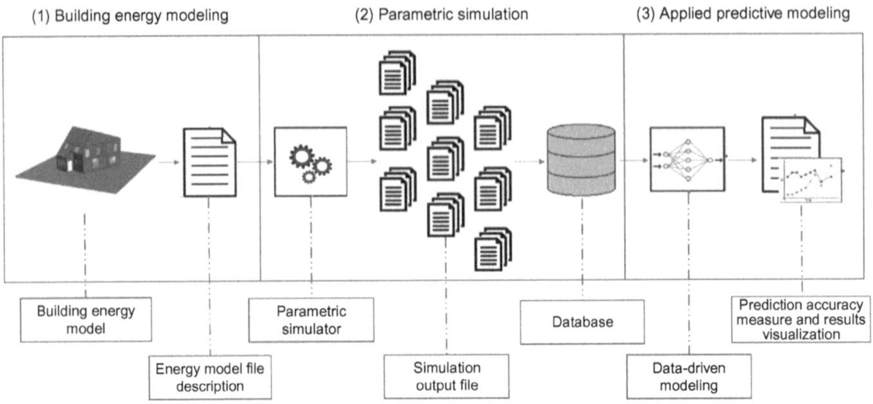

Fig. 3.22 Methodology overview.

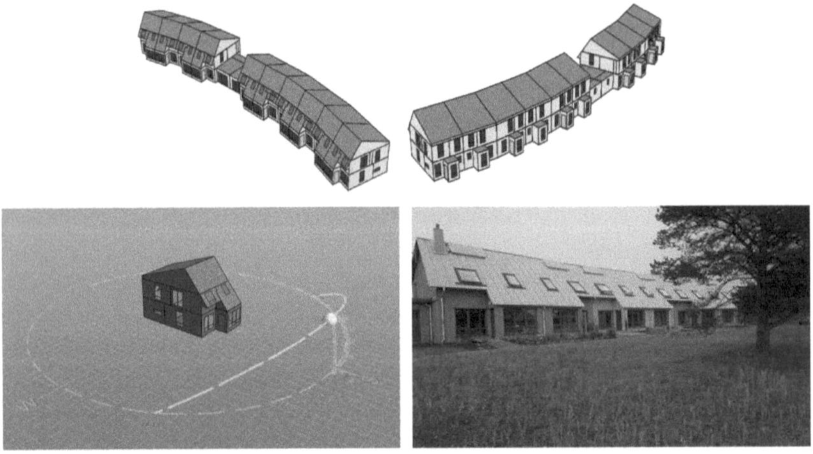

Fig. 3.23 Building case study.

Surfaces are modeled considering roof, ceiling, external walls, internal partitions, ground floor, and doors. Glazing includes roof-light, external, and internal glazed surfaces. The heating system formed by air source heat pump (air to water system) with a seasonal COP of 3.1. The heat pump is connected to a solar heating kit. The solar panel surface is 2.6 m^2. The overall system provides domestic hot water and the thermal requirements for the building. Heating is distributed by an underfloor system. Electric heaters constitute the auxiliary source of heat, directly controlled by the occupants. Room set points are set at 20°C. The heating system is modeled to run continuously during the period of simulation. Cooling systems are not present in the building; if required, cooling is provided by natural ventilation in summer periods. Lighting and equipment gains have been modeled considering measurement from sensors when available to create variation profiles and assigning 2 and 34 W/m^2, respectively.

Occupant presence is defined considering a value of 22 m^2/person and a presence profile for the period of simulation. Furthermore, the energy model is developed taking into account information regarding internal temperature sensors, flow measurement of heat pump, solar systems and underfloor heating, flow and return temperature of heat pump, solar system and underfloor heating, electrical meters data from lighting, ground floor sockets, first floor sockets, heating and main feed, floor plans, sections, and elevation constructions.

3.8.5 Parametric simulation

The energy model is linked to a parametric simulation framework. Parametric simulation is applied to transmittance of external walls, transmittance of glazing surfaces, orientation of the building, infiltration rates, internal gains due to occupancy, lighting and equipment, and glazing surfaces. Internal gains due to occupancy, lighting and equipment are grouped together and considered as a single parameter. Percentages of variation are applied to each of the six selected variables. The total number of simulations is equal to the product of the number of variations considered for each variable. The following equation provides a description of the process:

$$Ns = \prod_{k=1}^{K} n_k \tag{3.41}$$

where Ns is the number of simulations, K is the number of variables, and n is the number of variation of the variable k. If the number of variations is the same for each variable the equation is simplified in:

$$Ns = n^K \tag{3.42}$$

In this study a total number k of six variables are considered with $K = 4$ variations expressed as percentage of the baseline model. Table 3.2 shows the variables and variations considered in this study. Thus, the total amount of simulations Ns is: 4096.

The parametric simulation is performed with the software IES-VE and its parametric simulator (Table 3.3). The period of simulation is two months: November and December 2014. Output is collected with a hourly time step of resolution. The selection of 2 months of simulations is due to limit the very large output dataset obtained from the parametric study. A total of 5,996,544 simulation values regarding average electric demand and heating requirement are stored with weather data. The simulation database will be used in the training of the predictive models.

Fig. 3.24 presents the steps for the creation of the applied predictive modeling. First, data output of the parametric simulation (1) is organized and reshaped in a database (2). A random sample is selected from the dataset for the purpose of model training and model selection. In addition, randomly sampling the dataset may artificially recreate data scarcity and sparsity conditions: a possible scenario in large-scale modeling. The subset is of 40,000 outputs corresponding to 40,000 different hourly

Table 3.3 Parametric simulation: Parameters values and variations

	Baseline	Units	Variations			
U-values (external walls)	0.14	W/m^2 K	0.11	0.17	0.21	0.25
U-values (glazing surfaces)	1.24	W/m^2 K	0.93	1.55	1.85	2.17
Orientation	–	deg	90	180	270	0
Infiltrations	0.15	ach	0.3	0.45	0.6	0.75
Glazing surface	32.3	m^2	25.5	28.7	35	38.2
Internal gains						
Occupancy	22	m^2/person	17.6	26.4	28.6	33
Lighting	2	W/m^2	1.6	2.4	2.6	3
Equipment/miscellaneous	34	W/m^2	27.2	40.8	44.2	51

simulation outcomes (3). Therefore, three predictive models, SVMs, NNs, and GLMs, are used for predicting average hourly electric demand. Inputs to the models are weather variables and values of the parametric simulations. Parameter tuning and the selection of the most accurate predictive model is then performed adopting a re-sampling technique, the *k*-fold cross-validation method. The procedure creates *k* groups of samples of equal size called folders. A model is trained using all the folders except one. Thus, the model is used to predict values of the hold-out folder and to compute performance measures. The procedure is repeated several times, shuffling training and testing folders, to perform an algorithm parameter tuning procedure. In this study a total number of 10 folders are selected to perform parameter tuning and best model selection.

Thus, accuracy measures are evaluated for each model. The best model is the one that has the minimum value of mean absolute percentage error (MAPE) and RMSE (4). In order to test the effective performance of the models on an unseen dataset, another large sample (40,000 points) is selected randomly from the remaining initial data (5). MAPE and RMSE are calculated again on the total unseen dataset (6). Finally, results are visualized in different ways to help the modeler to understand which model performs best (7).

3.8.6 Results

3.8.6.1 Predictive modeling performance

Output values of the predictive models (SVM, NN, GLM) are compared with the parametric simulation results to assess the accuracy of DDM in emulating the building dynamic simulator. Fig. 3.25 represents the comparison between simulated and

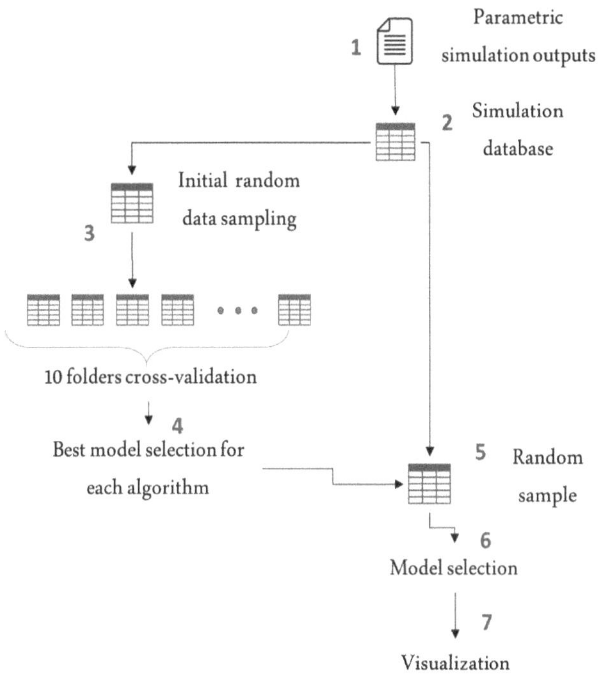

Fig. 3.24 DDM training and validation methodology.

predicted values. Output variable compared in Fig. 3.25 is the electric demand, averaged over 60 minutes. The three predictive models are characterized by different accuracy. Generally, all the algorithms perform well, the results of the comparison follow the 45 degree line attesting that there is a good agreement between simulations and predictions. An analysis of the MAPE and RMSE helps in further ranking the quality of the predictions. The RMSE ranges from 195, 171, and 138 W for GLM, NN, and SVM approaches, respectively. The MAPE values range from 5 for SVM, 7 for NN and to 8.20 for GLM. Based on this results SVM is the one that performs best on the dataset. Another way to assess the accuracy of the predictive models is to graph predicted and simulated data in terms of residuals defined as follows:

$$r = s - p \tag{3.43}$$

where s are outputs of the parametric simulation, p outputs of the predictive models. Fig. 3.26 shows the residuals; if the oscillation of the results is close to 0 it means that simulation and prediction data are similar. Residuals help the modeler to understand if there are particular patterns in the quality of the predictions. For this case study the GLM and the NN algorithms seem to follow a pattern for high and low values of the demand. SVM seems to be the most stable algorithm over the range of variation of the demand.

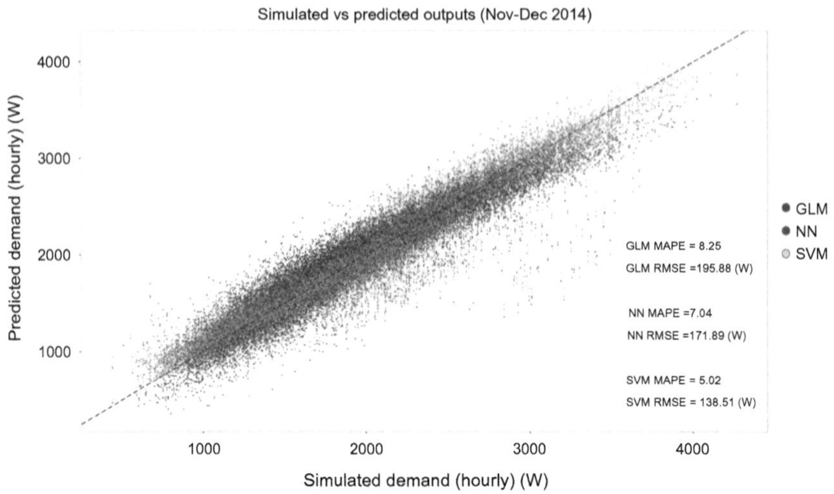

Fig. 3.25 DDM results: predicted versus simulated data.

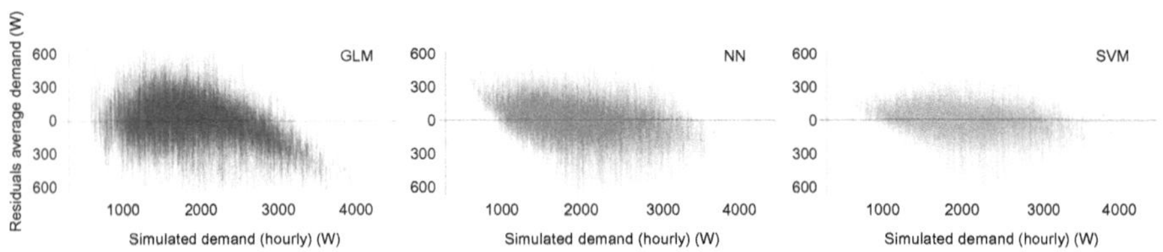

Fig. 3.26 DDM results: residuals.

3.8.7 Conclusion

3.8.7.1 Possible uses of the predictive model

The results of this research work attest that DDM perform well as emulators at an hourly level of resolution of the dynamic simulation software. This could lead to different applications of this methodology in large-scale building energy modeling studies. For example, trained DDM could be employed to characterize and to estimate the demand of a group of buildings of the same typology of the baseline model. The granularity of the outcomes allows to perform studies at different levels of aggregation starting from single buildings to entire groups. After the training process, a DDM could be exploit to analyze different modeling scenarios. For example, it may be employed to assess the effect of changing parameters such as: total transmittances, internal gain magnitudes, glazing surface materials, orientations, changes in building air tightness, or glazing surfaces of the baseline model. Most of the variables, appropriate in retrofit analysis, may be evaluated using a trained DDM.

3.8.7.2 *Advantages and limitations of the proposed approach*

A trained DDM is able to produce fast results. Reducing the time of simulation could be beneficial in large-scale BEMs. However, the proposed methodology presents different limitations due to modeling assumptions that must be taken into account in order to define important improvements. In particular, a series of assumptions must be relaxed. First occupant behavior could be high unpredictable especially in the residential sector and affects the demand pattern of a building. The heterogeneity of constructions, volumes, areas, adjacencies, and shading scenarios could high effect simulation outputs. Different systems configuration efficiencies and operations should be taken into account for heating cooling and electric systems. To improve the reliability of the presented approach all the aforementioned limitations should be considered. Nevertheless, the scalability of the presented approach may play a key role in improving its reliability to characterize realistically an urban environment. The parametric approach is flexible and guarantees the possibility to add numbers of relevant parameters in the analysis and virtually enrich the capability of DDM to represent more realistically a large-scale context.

Acknowledgments

The authors would like to thank Assoc. Prof. D. Finn and Dr. M. De Rosa for their valuable comments on reduced order modelling.

References

Antonopoulos, K.A., Koronaki, E., 1998. Apparent and effective thermal capacitance of buildings. Energy 23 (3), 183–192. https://doi.org/10.1016/S0360-5442(97)00088-1.

Arvo, J., Chelmsford, M.A., 1986. Backward ray tracing. In: Developments in Ray Tracing, Computer Graphics, Proc. of ACM SIGGRAPH 86 Course Notes, pp. 259–263.

Ashouri, A., Fazlollahi, S., Benz, M.J., Maréchal, F., 2015. Particle swarm optimization and Kalman filtering for demand prediction of commercial buildings. In: ECOS 2015, pp. 1–10.

Asociación Técnica Espa nola de Climatización y Refrigeración (ATECYR), 2008. Guía Técnica. Procedimientos y aspectos de la simulación de instalaciones térmicas en edificios.

Bacher, P., Madsen, H., 2010. Procedure for Identifying Models for the Heat Dynamics of Buildings. DTU IMM-Technical Report.

Bacher, P., Madsen, H., 2011. Identifying suitable models for the heat dynamics of buildings. Energy Build. 43 (7), 1511–1522. https://doi.org/10.1016/j.enbuild.2011.02.005.

Ballarini, I., Corgnati, S.P., Corrado, V., 2014. Use of reference buildings to assess the energy saving potentials of the residential building stock: the experience of TABULA project. Energy Policy 68, 273–284. https://doi.org/10.1016/j.enpol.2014.01.027.

Berthou, T., Stabat, P., Salvazet, R., Marchio, D., 2014. Development and validation of a gray box model to predict thermal behavior of occupied office buildings. Energy Build. 74, 91–100. https://doi.org/10.1016/j.enbuild.2014.01.038.

Bharucha-Reid, A.T., 2012. Elements of the Theory of Markov Processes and Their Applications. Courier Corporation.

Bruzgevičius, P., Burlingis, A., Stankevičius, V., Pupeikis, D., Norvaišienė, R., Banionis, K., 2013. Investigation into thermal capacitance of the building envelope. J. Sustain. Archit. Civil Eng. 1. https://doi.org/10.5755/j01.sace.1.2.3543.

Bu, L., Zhao, D., Liu, Y., Guan, Q., 2014. A hierarchical classification algorithm for evaluating energy consumption behaviors. In: 2014 International Joint Conference on Neural Networks (IJCNN)IEEE, pp. 1461–1466.

Building Research Establishment, 2013. The government's standard assessment procedure for energy rating of dwellings—SAP 2012 version 9.92. Building Research Establishment on behalf of Department of Energy & Climate Change, Warford, UK, p. 2013.

Burger, E.M., Moura, S.J., 2015. Gated ensemble learning method for demand-side electricity load forecasting. Energy Build. 109, 23–34. https://doi.org/10.1016/j.enbuild.2015.10.019.

Capasso, A., Grattieri, W., Lamedica, R., Prudenzi, A., 1994. A bottom-up approach to residential load modeling. IEEE Trans. Power Syst. 9 (2), 957–964. https://doi.org/10.1109/59.317650.

Catalina, T., Virgone, J., Blanco, E., 2008. Development and validation of regression models to predict monthly heating demand for residential buildings. Energy Build. 40 (10), 1825–1832. https://doi.org/10.1016/j.enbuild.2008.04.001.

Chou, J.S., Bui, D.K., 2014. Modeling heating and cooling loads by artificial intelligence for energy-efficient building design. Energy Build. 82, 437–446. https://doi.org/10.1016/j.enbuild.2014.07.036.

Clarke, J.A., 2001. Energy Simulation in Building Design. Routledge, Abingdon. ISBN 978-0-7506-5082-3.

Cox, D.R., Oakes, D., 1984. Analysis of Survival Data. vol. 21. CRC Press, Boca Raton, FL.

CTSM-R Development Team, 2015. Continuous Time Stochastic Modeling in R—User's Guide and Reference Manual. http://ctsm.info/pdfs/ctsmr-reference.pdf. Accessed 16 November 2015.

De Coninck, R., Magnusson, F., Akesson, J., Helsen, L., 2014. Grey-box building models for model order reduction and control. In: 10th International Modelica Conference, pp. 657–666 Lund, Sweden.

De Rosa, M., Bianco, V., Scarpa, F., Tagliafico, L.A., 2014. Heating and cooling building energy demand evaluation; a simplified model and a modified degree days approach. Appl. Energy 128, 217–229. https://doi.org/10.1016/j.apenergy.2014.04.067.

Department of Energy and Climate Change, Intertek Testing and Certification Ltd, 2010–2011. Household Electricity Survey. https://doi.org/10.4225/13/511C71F8612C3.

Edwards, R.E., New, J., Parker, L.E., 2012. Predicting future hourly residential electrical consumption: a machine learning case study. Energy Build. 49, 591–603. https://doi.org/10.1016/j.enbuild.2012.03.010.

Escrivá-Escrivá, G., Álvarez-Bel, C., Roldán-Blay, C., Alcázar-Ortega, M., 2011. New artificial neural network prediction method for electrical consumption forecasting based on building end-uses. Energy Build. 43, 3112–3119. https://doi.org/10.1016/j.enbuild.2011.08.008.

Famuyibo, A.A., Duffy, A., Strachan, P., 2012. Developing archetypes for domestic dwellings—an Irish case study. Energy Build. 50, 150–157. https://doi.org/10.1016/j.enbuild.2012.03.033.

Fan, C., Xiao, F., Wang, S., 2014. Development of prediction models for next-day building energy consumption and peak power demand using data mining techniques. Appl. Energy 127, 1–10. https://doi.org/10.1016/j.apenergy.2014.04.016.

Ferreira, P.M., Ruano, A.E., Silva, S., Conceicao, E.Z.E., 2012. Neural networks based predictive control for thermal comfort and energy savings in public buildings. Energy Build. 55, 238–251. https://doi.org/10.1016/j.enbuild.2012.08.002.

Foucquier, A., Robert, S., Suard, F., Stéphan, L., Jay, A., 2013. State of the art in building modelling and energy performances prediction: a review. Renew. Sustain. Energy Rev. 23, 272–288. https://doi.org/10.1016/j.rser.2013.03.004.

Fux, S.F., Ashouri, A., Benz, M.J., Guzzella, L., 2014. EKF based self-adaptive thermal model for a passive house. Energy Build. 68, 811–817. https://doi.org/10.1016/j.enbuild.2012.06.016.

Gao, X., Malkawi, A., 2014. A new methodology for building energy performance benchmarking: an approach based on intelligent clustering algorithm. Energy Build. 84, 607–616. https://doi.org/10.1016/j.enbuild.2014.08.030.

Ghiassi, N., Hammerberg, K., Taheri, M., Pont, U., Owat, S., Mahdavi, A., 2015. An enhanced sampling-based approach to urban energy modelling. In: International Building Physics Simulation Association (IBPSA) Hyderabad, India.

Grandjean, A., Adnot, J., Binet, G., 2012. A review and an analysis of the residential electric load curve models. Renew. Sustain. Energy Rev. 16 (9), 6539–6565. https://doi.org/10.1016/j.rser.2012.08.013.

Guerra Santin, O., 2011. Behavioural patterns and user profiles related to energy consumption for heating. Energy Build. 43 (10), 2662–2672. https://doi.org/10.1016/j.enbuild.2011.06.024.

Haldi, F., Robinson, D., 2009. Interactions with window openings by office occupants. Build. Environ. 44 (12), 2378–2395. https://doi.org/10.1016/j.buildenv.2009.03.025.

Haldi, F., Robinson, D., 2010. Adaptive actions on shading devices in response to local visual stimuli. J. Build. Perform. Simul. 3 (2), 135–153. https://doi.org/10.1080/19401490903580759.

Harb, H., Boyanov, N., Hernandez, L., Streblow, R., Müller, D., 2016. Development and validation of grey-box models for forecasting the thermal response of occupied buildings. Energy Build 117, 199–207. https://doi.org/10.1016/j.enbuild.2016.02.021.4.

Hazyuk, I., Ghiaus, C., Penhouet, D., 2012. Optimal temperature control of intermittently heated buildings using MPC: part I—building modeling. Build. Environ. 51, 379–387. https://doi.org/10.1016/j.buildenv.2011.11.008.

Hazyuk, I., Ghiaus, C., Penhouet, D., 2014. Model predictive control of thermal comfort as a benchmark for controller performance. Autom. Constr. 43, 98–109. https://doi.org/10.1016/j.autcon.2014.03.016.

Heeren, N., Jakob, M., Martius, G., Gross, N., Wallbaum, H., 2013. A component based bottom-up building stock model for comprehensive environmental impact assessment and target control. Renew. Sustain. Energy Rev. 20 (C), 45–56. https://doi.org/10.1016/j.rser.2012.11.064.

Howard, B., Parshall, L., Thompson, J., Hammer, S., Dickinson, J., Modi, V., 2012. Spatial distribution of urban building energy consumption by end use. Energy Build. 45, 141–151. https://doi.org/10.1016/j.enbuild.2011.10.061.

Hsu, D., 2015. Comparison of integrated clustering methods for accurate and stable prediction of building energy consumption data. Appl. Energy 160, 153–163. https://doi.org/10.1016/j.apenergy.2015.08.126.

Humeau, S., Wijaya, T.K., Vasirani, M., Aberer, K., 2013. Electricity load forecasting for residential customers: exploiting aggregation and correlation between households. In: 2013 Sustainable Internet and ICT for Sustainability Conference Proceeding, pp. 1–6.

International Energy Agency, 2013. Transition to Sustainable Buildings—Strategies and Opportunities to 2050. IEA Publications, Paris, France.

International Energy Agency, 2014. World Energy Outlook 2014. IEA Publications, Paris, France. ISBN 978-92-64-20805-6.

International Organization of Standardization, 2008. Energy performance of buildings—calculation of energy use for space heating and cooling (ISO 13790:2008).

International Organization of Standardization, 2017. Thermal performance of building components—dynamic thermal characteristics—calculation methods (ISO 13786:2017).

Jaboob, S., 2015. On the ownership and use of residential electrical appliances (Ph.D. thesis), University of Nottingham.

Jain, A.K., Murty, M.N., Flynn, P.J., 1999. Data clustering: a review. ACM Comput. Surv. 31 (3), 264–323. https://doi.org/10.1145/331499.331504.

Jain, R.K., Smith, K.M., Culligan, P.J., Taylor, J.E., 2014. Forecasting energy consumption of multi-family residential buildings using support vector regression: investigating the impact of temporal and spatial monitoring granularity on performance accuracy. Appl. Energy 123, 168–178. https://doi.org/10.1016/j.apenergy.2014.02.057.

James, G., Witten, D., Hastie, T., Tibshirani, R., 2014. An Introduction to Statistical Learning: With Applications in R. Springer Publishing Company, Incorporated, New York, NY. ISBN 9781461471370.

Jovanović, R.ž., živković, B.D., 2014. Prediction of heating energy consumption in university buildings based on simplified artificial neural networks. In: 18th International Research/ Expert Conference "Trends in the Development of Machinery and Associated Technology" TMT, vol. 18. pp. 155–158.

Jurado, S., Nebot, À., Mugica, F., Avellana, N., 2015. Hybrid methodologies for electricity load forecasting: entropy-based feature selection with machine learning and soft computing techniques. Energy 86, 276–291. https://doi.org/10.1016/j.energy.2015.04.039.

Kämpf, J.H., 2009. On the modelling and optimisation of urban energy fluxes (Ph.D. thesis), EPFL, Lausanne, Switzerland.

Kapetanakis, D.S., Mangina, E., Ridouane, E.H., Kouramas, K., Finn, D.P., 2015. Comparison of predictive models for forecasting building heating loads. In: 14th Conference of International Building Performance Simulation Association, Hyderabad, India, pp. 1087–1094.

Kaufman, L., Rousseeuw, P., 1987. Clustering by means of medoids. North-Holland, Amsterdam.

Kazas, G., Fabrizio, E., Perino, M., 2017. Energy demand profile generation with detailed time resolution at an urban district scale: a reference building approach and case study. Appl. Energy 193, 243–262. https://doi.org/10.1016/j.apenergy.2017.01.095.

Lara, R.A., Cappelletti, F., Gasparella, A., 2014. Selection of representative buildings through preliminary cluster analysis. In: International High Performance Buildings Conference West Lafayette, USA.

Lauster, M., Teichmann, J., Fuchs, M., Streblow, R., Mueller, D., 2014. Low order thermal network models for dynamic simulations of buildings on city district scale. Build. Environ. 73, 223–231. https://doi.org/10.1016/j.buildenv.2013.12.016.

Lehmann, B., Gyalistras, D., Gwerder, M., Wirth, K., Carl, S., 2013. Intermediate complexity model for model predictive control of integrated room automation. Energy Build. 58, 250–262. https://doi.org/10.1016/j.enbuild.2012.12.007.

Li, X., Wen, J., 2014. Building energy consumption on-line forecasting using physics based system identification. Energy Build. 82, 1–12. https://doi.org/10.1016/j.enbuild.2014.07.021.

Li, W., Zhou, Y., Cetin, K., Eom, J., Wang, Y., Chen, G., Zhang, X., 2017. Modeling urban building energy use: a review of modeling approaches and procedures. Energy 141, 2445–2457. https://doi.org/10.1016/j.energy.2017.11.071.

Macas, M., Moretti, F., Fonti, A., Giantomassi, A., Comodi, G., Annunziato, M., Pizzuti, S., Capra, A., 2016. The role of data sample size and dimensionality in neural network based forecasting of building heating related variables. Energy Build. 111, 299–310. https://doi.org/10.1016/j.enbuild.2015.11.056.

Mai, W., Chung, C.Y., Wu, T., Huang, H., 2014. Electric load forecasting for large office building based on radial basis function neural network. In: 2014 IEEE PES General Meeting—Conference & ExpositionIEEE, pp. 1–5.

Mastrucci, A., Baume, O., Stazi, F., Leopold, U., 2014. Estimating energy savings for the residential building stock of an entire city: a GIS-based statistical downscaling approach applied to Rotterdam. Energy Build. 75, 358–367. https://doi.org/10.1016/j.enbuild.2014.02.032.

Mata, S.K.A., Johnsson, F., 2014. Building-stock aggregation through archetype buildings: France, Germany, Spain and the UK. Build. Environ. 81, 270–282. https://doi.org/10.1016/j.buildenv.2014.06.013.

Monetti, V., Fabrizio, E., Filippi, M., 2014. Influence of different temperature control patterns through TRV on district heating loads. In: Li, A., Zhu, Y., Li, Y. (Eds.), In: Proc. 8th Int. Symp. Heating, Vent. Air Cond, pp. 251–258.

Monetti, V., Fabrizio, E., Filippi, M., 2015. Impact of low investment strategies for space heating control: application of thermostatic radiators valves to an old residential building. Energy Build 95, 202–210. https://doi.org/10.1016/j.enbuild.2015.01.001.

Nouvel, R., Mastrucci, A., Leopold, U., Baume, O., Coors, V., Eicker, U., 2015. Combining GIS-based statistical and engineering urban heat consumption models: towards a new framework for multi-scale policy support. Energy Build. 107, 204–212. https://doi.org/10.1016/j.enbuild.2015.08.021.

Paatero, J., Lund, P., 2006. A model for generating household electricity load profiles. Int. J. Energy Res. 30 (5), 273–290. https://doi.org/10.1002/er.1136.

Page, J., Robinson, D., Morel, N., Scartezzini, J.L., 2008. A generalised stochastic model for the simulation of occupant presence. Energy Build. 40 (2), 83–98. https://doi.org/10.1016/j.enbuild.2007.01.018.

Paudel, S., Elmtiri, M., Kling, W.L., Corre, O.L., Lacarrière, B., 2014. Pseudo dynamic transitional modeling of building heating energy demand using artificial neural network. Energy Build. 70, 81–93. https://doi.org/10.1016/j.enbuild.2013.11.051.

Perez, D., 2014. A framework to model and simulate the disaggregated energy flows supplying buildings in urban areas (Ph.D. thesis), EPFL, Lausanne, Switzerland.

Perez, D., Kämpf, J.H., Wilke, U., Papadopoulou, M., Robinson, D., 2011. CITYSIM simulation: the case study of Alt-Wiedikon, a neighbourhood of Zürich City. In: Proceedings of CISBAT 2011—CleanTech for Sustainable Buildings, pp. 937–940.

Petersdorff, C., Boermans, T., Harnisch, J., 2006. Mitigation of CO2 emissions from the EU-15 building stock. Beyond the EU directive on the energy performance of buildings. Env. Sci. Poll. Res. Int. 13 (5), 350–358. https://doi.org/10.1065/espr2005.12.

Powell, K.M., Sriprasad, A., Cole, W.J., Edgar, T.F., 2014. Heating, cooling, and electrical load forecasting for a large-scale district energy system. Energy 74, 877–885. https://doi.org/10.1016/j.energy.2014.07.064.

Reilly, A., Kinnane, O., 2017. The impact of thermal mass on building energy consumption. Appl. Energy 198, 108–121. https://doi.org/10.1016/j.apenergy.2017.04.024.

Reinhart, C.F., Cerezo Davila, C., 2016. Urban building energy modeling—a review of a nascent field. Build. Environ. 97, 196–202. https://doi.org/10.1016/j.buildenv.2015.12.001.

Reynders, G., Diriken, J., Saelens, D., 2014. Quality of grey-box models and identified parameters as function of the accuracy of input and observation signals. Energy Build. 82, 263–274. https://doi.org/10.1016/j.enbuild.2014.07.025.

Richardson, I., Thomson, M., Infield, D., Clifford, C., 2010. Domestic electricity use: a high resolution energy demand model. Energy Build. 42 (10), 1878–1887. https://doi.org/10.1016/j.enbuild.2010.05.023.

Robinson, D., Haldi, F., Leroux, P., Perez, D., Rasheed, A., Wilke, U., 2009. CITYSIM: comprehensive micro-simulation of resource flows for sustainable urban planning. In: 11th International IBPSA Conference, pp. 1083–1090.

Rubin, M., 1982. Calculating heat transfer through windows. Int. J. Energy Res. 6 (4), 341–349. https://doi.org/10.1002/er.4440060405.

Sancho-Tomás, A., Sumner, M., Robinson, D., 2017. A generalised model of electrical energy demand from small household appliances. Energy Build. 135, 350–366. https://doi.org/10.1016/j.enbuild.2016.10.044.

Stokes, M., 2005. Removing barriers to embedded generation: a fine-grained load model to support low voltage network performance analysis (Ph.D. thesis), De Montfort University.

Swan, L.G., Ugursal, V.I., 2009. Modeling of end-use energy consumption in the residential sector: a review of modeling techniques. Renew. Sustain. Energy Rev. 13 (8), 1819–1835. https://doi.org/10.1016/j.rser.2008.09.033.

Tanimoto, J., Hagishima, A., Sagara, H., 2008. Validation of probabilistic methodology for generating actual inhabitants' behavior schedules for accurate prediction of maximum energy requirements. Energy Build. 40 (3), 316–322. https://doi.org/10.1016/j.enbuild.2007.02.032.

Tsanas, A., Xifara, A., 2012. Accurate quantitative estimation of energy performance of residential buildings using statistical machine learning tools. Energy Build. 49, 560–567. https://doi.org/10.1016/j.enbuild.2012.03.003.

van Ruijven, B., de Vries, B., van Vuuren, D.P., van der Sluijs, J.P., 2010. A global model for residential energy use: uncertainty in calibration to regional data. Energy 35 (1), 269–282. https://doi.org/10.1016/j.energy.2009.09.019.

Vivian, J., Zarrella, A., Emmi, G., De Carli, M., 2017. An evaluation of the suitability of lumped-capacitance models in calculating energy needs and thermal behaviour of buildings. Energy Build. 150, 447–465. https://doi.org/10.1016/j.enbuild.2017.06.021.

Wang, S., Xu, X., 2006. Simplified building model for transient thermal performance estimation using GA-based parameter identification. Int. J. Therm. Sci. 45 (4), 419–432. https://doi.org/10.1016/j.ijthermalsci.2005.06.009.

Widén, J., Wäckelgård, E., 2010. A high-resolution stochastic model of domestic activity patterns and electricity demand. Appl. Energy 87 (6), 1880–1892. https://doi.org/10.1016/j.apenergy.2009.11.006.

Widén, J., Lundh, M., Vassileva, I., Dahlquist, E., Ellegård, K., Wäckelgård, E., 2009. Constructing load profiles for household electricity and hot water from time-use data-modelling approach and validation. Energy Build. 41 (7-1), 753–768. https://doi.org/10.1016/j.enbuild.2009.02.013.

Williams, K.T., Gomez, J.D., 2016. Predicting future monthly residential energy consumption using building characteristics and climate data: a statistical learning approach. Energy Build. 128, 1–11. https://doi.org/10.1016/j.enbuild.2016.06.076.

Yang, C., Létourneau, S., Guo, H., 2014. Developing data-driven models to predict BEMS energy consumption for demand response systems. Lect. Notes Comput. Sci. 8481 (Part 1), 188–197.

Zhao, H.X., Magoulès, F., 2012. A review on the prediction of building energy consumption. Renew. Sustain. Energy Rev. 16, 3586–3592. https://doi.org/10.1016/j.rser.2012.02.049.

Zhou, Q., Wang, S., Xu, X., Xiao, F., 2008. A grey-box model of next-day building thermal load prediction for energy-efficient control. Int. J. Energy Res. 32 (15), 1418–1431. https://doi.org/10.1002/er.1458.

Load management in buildings

4

Solène Goy[a,*], *Ana Sancho-Tomás*[a,†]
*School of Mechanical and Materials Engineering, University College Dublin, Dublin, Ireland, †Faculty of Engineering, Department of Architecture and Built Environment, University of Nottingham, Nottingham, United Kingdom

4.1 Load management

4.1.1 Introduction

The electrical power system at the distribution level is evolving toward decentralized monitoring, supply, and control. Small-scale renewable energy generation and storage devices are increasingly incorporated into the network, which in turn needs an intelligent management, information, and communication system to fulfill local power demands. Such an evolved grid is part of a broader concept, the Smart Grid. There is no unified definition of Smart Grid in the literature. Common concepts among the multiple definitions cover: bidirectional communication between supply and demand, use of monitoring technology, asset utilization and optimization, integration of renewable power sources, and intelligent management of all the actors involved in the transmission and distribution systems. As an example, the EU definition characterizes a smart grid as "an electricity network that can intelligently integrate the actions of all users connected to it—generators, consumers and those that do both—in order to efficiently deliver sustainable, economic and secure electricity supplies" (Gelazanskas and Gamage, 2014; European Commission, 2006).

The development of a smart grid is bringing new challenges at technical, regulatory, and policy levels. The deployment of these smart grids, together with the electricity market evolution and the increasing share of renewable energy, requires a smarter control of the demand side consumption (Reynolds et al., 2017; Biegel et al., 2014a,b). A considerable shift is thus happening, where the frontiers of the network are being redefined. When the aim was to supply a given load (demand), a smarter approach consists of considering the load as an active part of the system. Thus, the system goal would not solely consist of meeting the demand, but of further improving/optimizing the way the service is delivered. The users with their loads and decentralized energy systems become an active part of the network thanks to bidirectional communication (Soares et al., 2017); such users are called "prosumers" or "flexible consumers" (Biegel et al., 2014b). Prosumers can provide flexibility to the system, by agreeing to shift/reduce their consumption in exchange for a reward from the network operator or the aggregator.

[a]Both authors contributed equally to this chapter.

Urban Energy Systems for Low-Carbon Cities. https://doi.org/10.1016/B978-0-12-811553-4.00004-4

Buildings and occupants have a crucial role to play in this smart grid deployment. Indeed, buildings consume approximately one-third of the world final energy and emit a similar share of the total CO_2 emissions (International Energy Agency, 2013). Moreover, people spend close to 90% of their time indoors, so occupant comfort should be accounted for and occupant involvement is a sine qua noncondition to the successful deployment of flexibility providers in the residential and commercial sectors.

Overall, smart grids allow an improved management of the loads connected to it, giving the opportunity to address concerns of different stakeholders.

4.1.2 Definitions

Demand Side Management (DSM) and Demand Response (DR) encompass actions that can be deployed in the smart grid context.

There are several definitions for DSM (Behrangrad, 2015; Gellings, 1985; Stifter et al., 2016). In simple words, DSM refers to actions aiming at modifying the demand profiles of users. DSM encompasses DR and energy efficiency initiatives (Behrangrad, 2015). DR can be defined as "changes in electric usage by end-use customers from their normal consumption patterns in response to changes in the price of electricity over time, or to incentive payments designed to induce lower electricity use at times of high wholesale market prices or when system reliability is jeopardized" (Behrangrad, 2015; US Department of Energy, 2006). In practice, the modification can be load curtailment or demand shifting over time, for example (Siano, 2014). Overall, DR actions allow the provision of flexibility for a concerned stakeholder such as the grid operator. Energy efficiency measures are typically effective on a longer time frame than DR, that is, whenever the concerned load is activated (York and Kushler, 2005; Siano, 2014). An example is the replacement of the installed appliances or lights with devices that are more energy efficient (Siano, 2014). To achieve their respective goals, DSM and DR actions can be applied to various types of loads, such as:

- residential electrical appliances
- electric vehicles
- heating and cooling systems
- industrial loads

The present chapter mainly focuses on demand optimization and flexibility within the context of the electrical grid. In the literature, DSM is often defined and applied to the electrical context. However, other networks, such as district heating, can also benefit from DSM actions.

4.1.3 DR program classification

There are different ways to characterize existing DSM programs. In the following section, the focus is put on DR programs, discussing several of their characteristics.

4.1.3.1 Dispatchable vs. nondispatchable programs

A broad classification of DR programs distinguishes between event-based (dispatchable) and nonevent-based (nondispatchable) programs.

Dispatchable DR concerns responses to emergency reliability events and/or peak load reduction events. They are usually triggered by the appropriate system operator (after previous agreement with end-users). In most of the programs in this category, the decision to activate a certain DR service relies on contingency events, and users are obligated to respond. In some other cases (such as demand-bidding), the users follow dispatch instructions from a third party in response to some form of pricing signal. Alternatively, nondispatchable methods allow the end-user to choose to activate DR, following some form of time-sensitive pricing.

Solutions for both dispatchable or nondispatchable DR can be classified into price-based and incentive-based schemes (Deng et al., 2015; Vardakas et al., 2015). Price-based programs concern the modification of the electricity profile in response to price variations. Different tariff structures and regulation contribute to load shifting in peak times. Examples of those are time of use (TOU) tariffs, critical peak price, and real time pricing. Incentive-based programs, on the other hand, are related to fixed or time-varying financial incentive plans. They have been commonly implemented at industrial end-user level for a long time now; however, they are less common in domestic trials. Some examples include direct load control (DLC), interruptible/curtailable services, demand bidding, or emergency programs.

4.1.3.2 Optimized vs. nonoptimized programs

DR actions can be categorized according to the technique used to define them: a differentiation can be made on whether or not optimization methods are used. In nonoptimized cases, some predefined scenarios are applied and there is no objective being optimized. For example, this approach has been applied theoretically when addressing preestimation and/or very large spatial scales and a wide range of loads.

More commonly, the problem of identifying optimal DR strategies is formulated as an optimization problem, in which an objective function $F(x)$ must be optimized under certain constraints. The formulation of the objective function can take different forms, such as optimizing the daily cost of the energy bill or the fraction of renewable energy use on a given period of time (Deng et al., 2015). Depending on the case, the optimization will involve maximizing or minimizing $F(x)$. For complex cases, in which x represents a set of parameters and functions, this optimization problem can be difficult to solve analytically. In this scenario, there are various algorithmic formalisms that enable the parameter space to be explored in the search for a global optimum. The mathematical formulation for the optimization algorithms will depend on the type of objective function $F(x)$, the parameters and functions considered in x, and the specific DR mechanisms considered (such as load shifting or thermostat regulation). Consequently, there are various techniques available for DR optimization. They range from traditional calculus methods to heuristic optimization (Deng et al., 2015). Some specific and widespread optimization techniques for DR are listed in Table 4.1, with considerations and examples for each one. Each of the methods presented in the table

Table 4.1 Examples of optimization problems and methods in DR

Method	Considerations	Examples
Linear programming	Including integer/mixed linear/ nonlinear programming. Challenging for complex scenarios	Faria et al. (2010)
Convex optimization	Basic approach to DR. Challenging for complex scenarios	Mohsenian-Rad and Leon-Garcia (2010)
Game theory	Useful to study interactive decision-making processes, with the potential to address interactions between different actors in the network. Restricted to rather simple constrained problems	Chen et al. (2010)
Particle swarm optimization	Bioinspired heuristic approach, to tackle challenging optimization problems. Other options consider neural networks, genetic algorithms	Saber and Venayagamoorthy (2009)
Dynamic programming	Basic approach to deal with time-varying parameters, such as power generation or price	Zheng et al. (2014)
Stochastic programming	Addresses time-varying parameters whose probability distributions are known. It is a special type of dynamic programming	Clement et al. (2009)
Markov decision problem	Specially addresses time-varying parameters. Usually solved with a combination of dynamic programming and reinforcement learning	Shi and Wong (2011), Dusparic et al. (2015), and Wen et al. (2015)

is linked to a specific problem, and formulated under certain considerations. It is not the purpose of this table to rank the different techniques, but to provide an overview of those available, and how they are adapted to a given DR problem (Zheng et al., 2014).

4.1.3.3 Centralized vs. decentralized programs

Another categorization is made between centralized or decentralized DR (Deng et al., 2015; Vardakas et al., 2015), and how they relate to which component(s) in the system is (are) allowed to react to stimuli.

In the former, decisions on load rescheduling or power dispatch are made by a central controller, informed by the operation of each individual element in the network. Optimization methods used in these cases range from traditional calculus methods (Mohsenian-Rad and Leon-Garcia, 2010) to heuristic optimization (Gudi et al., 2010). However, these approaches become challenging when large and complex networks are considered. In the case of linear programming or convex optimization problems, they attempt to provide a definite optimal solution; when a large number of customers or devices are considered for DR, such problems become increasingly complex and heuristic or other methods capable of finding near-optimal solutions are necessary. When applied in real life situations, centralized systems sometimes may raise consumer privacy protection concerns, as they require a central authority to collect data and information for decision making.

On the other hand, decentralized DR schemes incorporate the ability of distributed decision-making, assuming a certain degree of intelligence of the devices involved (such as smart meters and appliances, power electronics, etc.), which ensures direct communication between the elements in the network. End-users can directly access indicators of the state of the grid, or of other points of the network, and react based on them. Certain software architectures, such as object-oriented and agent-based programming, are naturally configured to deal with decentralized algorithms. For such architectures, some optimization algorithms are more suitable; for instance, game theory methods (adequate for relatively constrained problems, not highly complex typologies), or Markov decision problems are easily adaptable to distributed approaches.

4.1.3.4 Grid vs. building programs

Finally, DR programs and studies can be differentiated based on the application they address.

Some applications are "grid-oriented" and address issues a network operator or an aggregator would be interested in. Grid-oriented analysis tend to be about large-scale applications and examples include requirements for market participation, potential of DR for a more efficient electrical system or potential of DR for future scenarios with higher share of renewable energy (Biegel et al., 2014b; Bergaentzlé et al., 2014; Critz et al., 2013). These studies provide concrete quantification of the gains that could be obtained at the network level and contribute to clarifying the conditions for a larger deployment of DR.

Other applications are more focused on the building physics and aim at quantifying the interactions between the building characteristics and the potential for storage that can be used for DR activities (Reynders et al., 2013; Wolisz et al., 2013). These studies account for the building dynamics and provide a better understanding of the characteristics required for "DR-friendly" buildings.

Finally, both building and grid considerations have been addressed jointly in various ways. For example, using models of representative buildings and quantifying the DR potential at national scale considering current electricity prices (Nyholm et al., 2016). Research has also investigated bringing together detailed building and grid modeling within a single framework, to better study the supply/demand interactions (Patteeuw et al., 2015).

4.1.4 DSM/DR deployment

As described earlier, the deployment of DSM has the potential to address some of the world energy challenges. More specifically, DSM impacts the power system operation, the trading of energy, and indirectly the environmental conditions we live in: as a result, it concerns multiple stakeholders. Considering the electricity market structure, six DR parties can be identified: system operator (SO), generation stakeholder, transmission/distribution stakeholder, retailing stakeholder, load stakeholder, and the demand resource provider (DRP, e.g., an aggregator, ESCO) (Behrangrad, 2015). In addition to these core stakeholders, other parties are involved in DSM deployment such as universities, research centers, consultancies and public institutions for example (Gangale et al., 2017).

DSM actions have been implemented for/by several of these stakeholders. In Europe, DSM was particularly promoted in the Energy Efficiency Directive of 2012 (Bertoldi et al., 2016; The European Parliament and the Council of the European Union, 2012). Research has a crucial role to play in the deployment of DSM programs. Several research or demonstration projects have been implemented allowing to collect data and gain insight into the challenges and possible solutions. In 2017, an analysis of more than 900 European smart grid projects showed that DSM was among the three most addressed topics, representing 25% of the smart grid project investment. More specifically, distribution system operators, ICT companies, and universities represented the top three investors in these DSM projects (Gangale et al., 2017). In parallel to the active research field and the multiple demonstration projects, DSM initiatives are already available for large to small customers. Industries have been providing flexibility to the electric SO through various price-based or incentive-based schemes. Moreover, in multiple countries residential and commercial consumers can choose peak/off-peak tariffs, such as day/night, which indirectly support the reduction of peak demand. A final example is the recent deployment of smart thermostats in the residential sector for heating/cooling systems. These control devices are typically based on machine learning algorithms and account for the occupancy profiles, resulting in a more efficient heating/cooling of the building.

While DSM initiatives have been implemented, for some of them since decades, there are still challenges to overcome before a widespread deployment. In Europe, the status of DR implementation varies across the Member States, from countries where multiple regulatory and technical challenges are yet to be overcome, to states where DR and independent aggregation are possible even though some challenges remain (Bertoldi et al., 2016). First of all, advanced metering systems (AMS) need to be widely deployed. These represent "an integrated system of smart meters, communications networks, and data management systems that enables two-way communication between utilities and customers" (US Department of Energy—Office of Electricity Delivery and Energy Reliability, 2016). AMS typically consist of (US Department of Energy—Office of Electricity Delivery and Energy Reliability, 2016):

- a smart meter on the customer side to measure energy consumption at hourly or subhourly frequencies. These advanced meters are typically capable of other actions such as switching on/off devices or signaling outage;
- a data management system to store and analyze the collected data; and
- communication lines to exchange data between the customer, the other stakeholders, and the data management systems.

AMS are coupled to systems on the customer side to control the loads and/or provide real-time information about their consumption and the energy prices. Examples of systems include DLC devices, programmable communicating thermostat, and in-home display (US Department of Energy—Office of Electricity Delivery and Energy Reliability, 2016). Within this context, advanced load control techniques are essential and need to be deployed together with the AMS; these techniques would allow to manage the building loads depending on the energy price, the network status, and the future demand among others. They consist, for example, of predictive and optimized approaches such as the ones described in Sections 4.2 and 4.3. Countries around the world have been engaging in such AMS deployment (US Department of Energy—Office of Electricity Delivery and Energy Reliability, 2016). Due to the quantity and sensitivity of the data that will be exchanged, the deployment of AMS cannot go without addressing cybersecurity and customer data privacy (US Department of Energy—Office of Electricity Delivery and Energy Reliability, 2016).

In terms of electricity market, redesign is necessary. A first obstacle lies in the current constraints set for participation in DR, which are rather restrictive for the new DR sources. These constraints include, for example, minimum bid size, advance warning, or event duration (O'Connell et al., 2014; Nolan and O'Malley, 2015; Bertoldi et al., 2016). Another challenge is to clarify or fully establish, depending on the country, the role, and responsibilities of each DR stakeholder, including the definition of new parties (e.g., aggregator) and the technical requirements for DR delivery (Bertoldi et al., 2016). Finally, defining business models for DSM is not an easy task. While various schemes have been proposed (Behrangrad, 2015), difficulties lie among others, in the quantification of the value and the definition of a baseline. Ultimately, the DR business model should not destabilize the market, while fairly incentivizing and retributing the stakeholders and ensuring long-term investment in the grid. The lack of data and lesson learnt make the DR business model definition more complex (O'Connell et al., 2014; Nolan and O'Malley, 2015).

Eventually, the DSM challenges from the occupants/buildings and energy markets point of views are interlinked. Indeed, to ensure a successful deployment at large scale, the application of DSM measures requires close interaction and involvement of stakeholders. At domestic level, previous studies (Beckmann, 2013) have consistently shown the relevance of behavioral interventions in the overall effectiveness of DSM programs. On one side, there is a need to inform better building occupants about the needs and benefits of DSM. On the other hand, customer behavior and motivations need to be better understood, to develop more adequate programs (Nolan and O'Malley, 2015). Overall, the deployment of DSM actions and opportunities has started, but challenges remain to be overcome before a complete integration in the energy market of all DSM stakeholders.

4.1.5 Summary

The section presented background information on the concepts of DSM and DR. These are broad concepts that can be applied to diverse loads, customers, contexts and to serve various objectives. Both are tools to support the deployment of a more efficient and sustainable energy system.

In the rest of the chapter, two techniques are presented to explore further DSM strategies within the context of urban energy management. First, in Section 4.2, DR strategies for residential electrical appliances are investigated, using a Multi-Agent Stochastic Simulation framework, and the effects on the local energy self-consumption of a small community of buildings are analyzed. Then, in Section 4.3, Model Predictive Control (MPC) is addressed showing how it can contribute to the improvement of thermal load management in buildings and the provision of flexibility.

4.2 Demand response of electrical appliances using multiagent simulation

4.2.1 *Introduction*

There exist a multitude of different mathematical approaches to define the DR problem, and maybe more ways to computationally implement the optimal operation of devices and resources. Having reviewed the pros and cons of candidate strategies, we have opted for a multiagent simulation (MAS) approach, for the following reasons. First, the landscape of microgrids and community energy concepts is intrinsically represented as a system of multiple actors (*agents*), which are technologically ready to employ varying degrees of intelligence to pursue an objective (*make individual decisions*), and are enabled to communicate with other actors in the network (*agent interaction*). Second, the architecture of No-MASS (Nottingham Multi-Agent Stochastic Simulation) lends itself to the ready extension from modeling occupants as agents to devices as agents, with potential interactions between the two typologies. In this section, we present ideas for testing DR, such as load shifting and battery operation, in an MAS framework. Our general methodology, including Q-learning for system optimization, is explained here.

We are interested in supporting the design of successful DR programs using computing simulation to model and optimize different strategies and scenarios. To carry out such task, we have identified a set of requirements that we acknowledge a DR simulation platform should satisfy; they are articulated in Section 4.2.2. A literature review and major developments in the area are described in Section 4.2.3. Section 4.2.4 introduces the software tool developed and used: No-MASS. The methodology for DR simulation is detailed in Section 4.2.5, with an application study in Section 4.2.6. Finally a conclusion is given in Section 4.2.7.

The work presented in this section has been extracted from published work (Sancho-Tomás et al., 2017a,b) and work to be published (Sancho-Tomás et al., 2018).

4.2.2 *Formulation of DR software requirements*

In our research, we have identified the need of simulation software that is capable of modeling and optimizing DR strategies. To this end, we can articulate some key requirements R that a DR simulation platform should aim to satisfy. It should be capable of:

(1) Simulating demands for four appliance archetypes:
   (a) switched on, regulated by and switched off by the user (e.g., cooker);
   (b) switched on by the user and off when a program is complete (e.g., washing machine);
   (c) switched on and off according to some program or schedule (e.g., hot water system); and
   (d) continuous cycling (e.g., refrigerator).
   The user-interaction should be stochastic.

(2) Drawing power to satisfy demands from:
   (a) local generation capacity;
   (b) local storage devices; and
   (c) the local microgrid and/or the national grid;
   similarly of diverting locally generated power to either local demand and storage devices, or to the local/national grid.

(3) Deciding, to satisfy some objective function, where power should be drawn from/diverted to; rescheduling demands (or the provision of energy-related services), given some predefined constraints related to service delivery (e.g., a washing machine may be activated after having been enabled, but must complete the wash by a predefined time).

(4) Presenting information to the user and emulating the users' decision making rationale regarding the rescheduling of user-controlled devices.

(5) Accounting for diversity in the extent to which users' are willing to relinquish control and to actively engage in behavioral change.

(6) Facilitating the above for communities consisting of buildings with numerous demand devices and potentially numerous supply and storage devices, which can communicate within and between buildings to achieve individual homeowners' requirements; potentially also those of the local low voltage network to which they are connected (e.g., in terms of network stability and safety).

4.2.3 Literature review

In the introduction of this chapter, methods for optimization of DR strategies were presented, classifying DR programs between centralized and decentralized ones. In this section, we first focus on the particular case of decentralized multiagent systems. Second, we revise those software that aim at becoming DR simulators. Finally, we propose our modeling framework.

4.2.3.1 DR and multiagent simulation

Multiagent systems are inherently a successful way of designing distributed DR programs, with the option of employing agent learning for optimization. MAS systems have been proposed in different power engineering applications (McArthur et al., 2007), from marketplace simulations to operation and control methods of the power system. They have been exploited in different ways such as monitoring and diagnostics, distributed control (Vasirani et al., 2013), fault protection, and modeling and simulation. Particular interest is posed in multiagent systems being used for DR (Iacovella et al., 2016; Wijaya et al., 2013; Gudi et al., 2010). Being inherently decentralized, they present a highly flexible and extensible modeling approach for simulating Device-to-Device (D2D) communication between supply, storage, and demand devices: systems are defined in terms of agents, which fulfill individual and collective goals, being able to compete, collaborate, negotiate, and learn behaviors.

Multiagent systems can be found extensively in the literature to tackle DR modeling (Zheng et al., 2014; Shi and Wong, 2011; Dimeas and Hatziargyriou, 2010; Dusparic et al., 2015). The implementation of machine learning techniques allows for agent learning, meaning that different agents can learn to optimize their behaviors. Reinforcement learning (RL) is a popular choice for agent learning, and in particular *Q*-learning has been proved effective for DR applications, from explicit Vehicle-to-Grid control (Shi and Wong, 2011), to microgrid coordination (Dimeas and Hatziargyriou, 2010). A useful mathematical description of the *Q*-learning algorithm in a device-centric approach is presented in Zheng et al. (2014), and applied to a single household. Finally, the *Q*-learning algorithm is taken a step forward in Dusparic et al. (2015), adding another "learning layer" to design a "distributed-W-learning" algorithm, in which agents learn how their local actions affect their neighbors' performance. Although the formulation is promising for microgrid studies, it has only been put into practice for the case of maximizing energy use of electric vehicles from wind-generated electricity.

4.2.3.2 Application into DR simulation software tools

The previous sections reveal that there are multiple modeling strategies and algorithms available for DR simulation. Although many interesting candidate models and ideas are found in those studies, usually they are focused on developing certain aspects of DR, or testing specific algorithms. They do not have the potential to address all the requirements articulated in Section 4.2.2. However, we are interested in developing flexible, scalable, and extensible software for DR simulation. In that sense, this section provides an insight on those studies aiming to be DR simulation tools, which in our view should target to address the range of these requirements in a comprehensive way.

DRSim (Wijaya et al., 2013) is a simulator for DR, which mainly focuses on R1, R4, and R5 requirements. They use an agent-oriented approach, defining house, human, and appliance agents, which makes DRSim a highly modular software, capable of incorporating new models easily, and to confer different behaviors to its agents. Demand in appliance agents (R1) is inferred using conditional probabilities, combining information of the correlated distribution of type of occupant and house, rooms in the house, set of appliances in each room, activity taken place, and the use of an appliance during that activity. More interestingly, user behavior is incorporated into human agents, defining a responsivity parameter (how agent reacts on receiving a signal) to perform a DR action. Both responsivity and action are modeled using three parameters: price, perception, and communication sensitivities. This way, users' decision making (R4) and its engagement (R5) are sophistically parameterized. Two types of DR mechanisms are allowed to the human agents: reduce load and shift load; but "the specific algorithm (R3) used to create a revised schedule is left unspecified as many candidates can be used." On the downside, there is no attempt on this study to consider supply agents (R2) or to involve network operation in any way (R6).

Another DR simulator called DemSi (Faria et al., 2010), "aims at providing a flexible tool to analyze DR actions and schemes" by developing "a software application that can be used by DNO's [Distributed Network Operator] and consumers to optimize their resource management." DemSi focuses on distributed generation issues (R2), covering a broad range of renewable energy technologies (photovoltaic, wind, co-generation, fuel cells, small hydro and biomass), and its effects on the electrical network and energy market elasticity (R6). To do so, they use PSCAD[1] software for network simulation. Their objective is to simulate a variety of DR methods that minimize the costs of a generation shortage situation: minimizing the total cost that the DNO and the suppliers have to pay for nonsupplied loads, which is implemented as a mixed-integer linear model in (commercial) software MATLAB and GAMS[2] (R3). The case of an isolated microgrid with only available electricity from DG gives some hints on how to simulate DR for network faults. It presents some interesting ideas for DR from a DNO perspective; however, energy use is considered with typical profiles, making this simulator incapable (and inextensible) of dealing with R1, R4, and R5 in its present form.

SMASH (SiMulated Adaptable Smart Home) (Dittawit and Aagesen, 2014) is claimed to be a demand side focused simulation platform, "specifically built to provide insights on the effect of different approaches to consumers, in terms of discomfort and decreased electricity costs." It mainly addresses R4, proposing an interesting DR program: consumer-centric load control. It is an incentive-based program similar to DLC, but with maintained freedom and comfort to consumers. The authors formulate a power reduction request sent by the DSO, and the home management system performs a load control action based on policy reasoning and extended finite-state machines.[3] The user will receive a reward based on the energy reduction, and a penalty fee in case of rejection. The objective is to minimize times of reduced comfort through the modification of space heating (R3). However, the other requirements are poorly fulfilled: only space heating and electric water heating (shower) are included (R1); it does not consider local generation or storage so far (R2); the interaction between end-users and the home management system is unclear (R5) and there is not intention for a multibuilding simulation.

4.2.3.3 Proposed modeling framework

Although good progress has been made in the development of DR software, with interesting initiatives available in the literature, there remains no comprehensive framework capable of addressing R1 to R6 requirements.

The remainder of this section describes the extension to an existing Multi-Agent Stochastic Simulation platform (No-MASS), with the aim of comprehensively address the shortfall in DR simulation capability. We present a tool capable of dealing

[1] See https:/hvdc.ca/pscad/.

[2] See https:/www.gams.com/.

[3] Decisions are made based on a set of possible states and a set of transitions or outcomes from each state.

Fig. 4.1 Device-to-device and person-to-person interaction.

(completely or partially) with R1, R2, R3, and R6, and with potential to incorporate R4 and R5 in the future.

No-MASS was initially developed to model the presence, activities and related behaviors of synthetic occupants (agents) of buildings that are co-simulated with EnergyPlus using the Functional Mockup Interface standard (Chapman, 2016). In this way it is straightforward to model occupants, interactions between them, (person-to-person: P2P) and their impacts on the energy performance and indoor comfort of the buildings they occupy, in contrast with most DR tools, which were designed to address D2D interactions (see Fig. 4.1).

Interestingly, the architecture of No-MASS is readily extensible to also consider D2D communications. Thus creating a platform that has the capability to simulate:

- occupant-agents' behaviors and interactions between them (P2P, the prior No-MASS);
- device-agents' behaviors and interactions between them (D2D); and
- interactions between occupants and devices (D2P and P2D).

In this work we focus on developing the second one of these capabilities. Next, the fundamentals of No-MASS are described in more detail, its workflow and the models that have been incorporated. The algorithms implemented to support load rescheduling and battery operation in a multiagent representation are then explained. The following section illustrates the application of this new prototypical platform to estimate the effects of appliance load shifting and electrical storage with the objective of maximizing the use of locally generated renewable energy of a group of buildings.

4.2.4 No-MASS background

As noted earlier, No-MASS (Nottingham Multi-Agent Stochastic Simulation) is a platform that was originally developed to model the presence, activities and related behaviors of occupants of buildings, or groups of buildings, and the energy and indoor comfort consequences of these behaviors. To this end No-MASS employs four complementary modeling strategies:

1. *Data-driven stochastic models* predicting occupants' presence (Page et al., 2008) and associated metabolic heat gains, their activities while present (Jaboob, 2015) from which location is also inferred, and their use of lights (Reinhart, 2004), windows (Haldi and Robinson, 2009), and shading devices (Haldi and Robinson, 2010).
2. The processing of votes for multiply-occupied spaces to emulate agents' *negotiation mechanisms* (e.g., agents' influence-weighted votes to open [1] or close [0] a window; the outcome receiving the greatest number of votes, weighted by influence, being effected).

3. Use of a *Belief-Desire-Intention* (BDI) framework to emulate agents' behaviors that are simple in character (e.g., closing curtains when it is dark, the closing of a window while bathing) but for which data is scarce.
4. *Agent learning* mechanisms to emulate agents' behaviors that are complex in character (e.g., choice of heating set-points) and for which data is also scarce.

The architecture of No-MASS (and its coupling with EnergyPlus), the data-driven (Strategy 1) models, and their application to the simulation of both domestic and nondomestic buildings are described in Chapman et al. (2018). The extension of No-MASS to model agents' negotiations and the data scarce modeling of both simple and more complex behaviors (Strategies 2–4) are explained in Chapman et al. (Data Scarce Behavioural Modelling and the Representation of Social Interactions, Unpublished manuscript).

To facilitate the extension of No-MASS to handle DR simulation and optimization (and indeed the more complete modeling of the impacts of occupants' behaviors), our Strategy 1 models have been complemented with models of occupants' ownership and use of large (Jaboob, 2015) and small (Sancho-Tomás et al., 2017a) electrical appliances. These models have themselves been complemented with models and data of electrical storage (an electric battery), supply or conversion (a photovoltaic panel) devices, and of a low-voltage network model.

4.2.4.1 No-MASS workflow

Fig. 4.2 illustrates the newly extended No-MASS architecture. First (upper left), a population generator creates a household of a size and demographic composition that is suited to the house being simulated. Next the conventional No-MASS modeling tasks are executed.

Household member agents are assigned archetypal properties (room associated with activities, clothing, and activity characteristics, which randomly assign unique behavior probabilistic models). Activities are then computed for the length of the simulation in a preprocess (for activities are not dependent upon environmental inputs); likewise electrical appliances are assigned to the household.

A loop then commences in which indoor/outdoor environment parameters are parsed from EnergyPlus to No-MASS for the present time step. From the preprocessed activities, agents' location, activity, and clothing level are set, from which metabolic heat gains are calculated.

A series of models predicting interactions with windows, shading devices, heating systems, and lights are then called. For heating interactions, agent learning is employed to determine the transient setpoints that minimize a cost function that combines heating costs and discomfort costs. In the case of multiply-occupied spaces, social interactions are considered at this stage through the vote processor to determine the negotiated outcome. Finally, BDI rules are used for straightforward interactions, for which data is scarce.

The workflow proceeds with the prediction of demands for small and large appliances, supply from energy conversion systems (only PV being enabled at this stage) and storage (only from batteries at present). Appliances are also shifted at this stage,

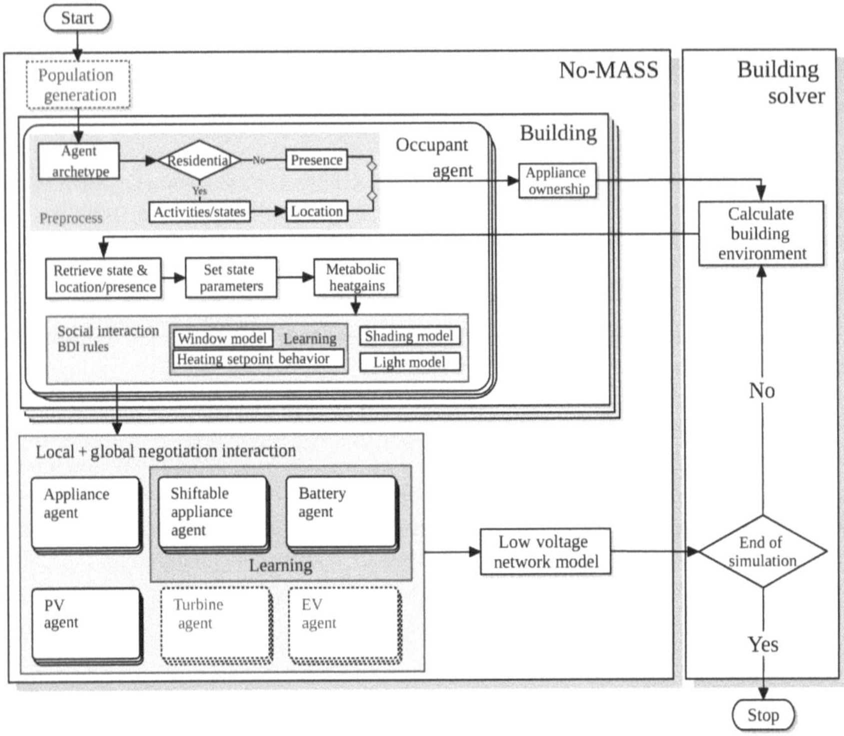

Fig. 4.2 No-MASS flowchart.

again using agent learning, to maximize a cost reward function (of which more later). An LV network model has been integrated with No-MASS. The user can choose to enable it for simulation. The calculation then proceeds to the next time-step, exiting this loop at the end of the simulation period. The corresponding physical models (for D2D modeling) are described next.

4.2.4.2 No-MASS for DR

The extension of No-MASS will, once complete, address the range of requirements outlined earlier. In its present form No-MASS/DR addresses requirements R1 (demand models), R2 (various power flows), and R3 (DR optimization), further explained in this section. The strategies for achieving this are as follows:

- Develop mechanisms for D2D communications between energy conversion, storage, and demand devices represented using the models described below.
- Implement strategies for load-shifting and optimal charge/discharge of a battery using agent learning algorithms. This will reflect whether or not MAS and machine learning are effective approaches to test different DR strategies on different sociodemographic groups.[4]

[4]Sociodemographic considerations are not included in the work reported.

- Extract or quantify differences when simulating residential electrical self-consumption/ autonomy (or other DR objectives) for different sociodemographic groups.

Thus, No-MASS could potentially (given available data) be used to evaluate scenarios involving human interaction and behavior change (R4 and R5) and to support tariff design, using for instance BDI rules and/or agent-learning. For instance, we could assess to what extent active engagement of users increases electrical energy autonomy, or to what extent price signal impacts on behavior change.

The following sections explore in more detail the methods and algorithms used.

4.2.5 Methodology

There are a broad range of available technologies for Distributed Energy Resources in residential areas. In regard to supply technologies, there is solar PV, micro-wind turbines, CHP plants (that can be disaggregated into fuel cells and fuel combustion), small-hydro power, or biomass plants. For small-scale electrical storage, there are different battery technologies available (chemical), supercapacitors (electrical), and flywheels (mechanical); other electrical storage options are available for larger systems, such as water pumping or compressed air storage. Additionally, electric vehicles can be considered as storage systems.

Within this landscape, for our proof-of-concept, we are selecting solar PV for power generation and electric batteries for storage systems.

4.2.5.1 Electricity demand models (R1)

The demand forecast models developed for No-MASS are based on stochastic methods. Devices have been classified in large appliances (high-load and commonly owned) and small appliances (range of low-load appliances), and follow two different modeling methodologies.

1. *Large appliances* include the cooker, TV, microwave, washing machine, and dishwasher. They have been modeled as a three-step process (Jaboob, 2015). First, the probability of switching on is predicted using a time-dependent Bernoulli process. Second, the duration for which devices will remain on is predicted using survival analysis. Finally, transitions between categories of fractional power demand (the fraction of maximum possible) are predicted as a Markov process, at 10-minute resolution.
2. *Small appliances* are modeled as aggregations of appliances, following four categories: small appliances in the kitchen, audio-visual appliances, computing appliances, and other appliances (Sancho-Tomás et al., 2017a). A multistate survival model is used.
3. Electrical demand for *heating* can be obtained thanks to the coupling of No-MASS with Energy Plus. Water heating is not considered for these simulations.

4.2.5.1.1 Supply and power flow (R2)

1. *Photovoltaic data.* For the current purpose of demonstrating the proof of concept of our proposed modeling approach, we are using measured performance data that characterize PV systems' output in-lieu of a predictive model. In the future, a suite of predictive energy conversion system models will be integrated.

2. *Electric battery storage system*. Electric storage technologies provide valuable services in most energy systems. Small-scale systems are becoming (nearly) cost-competitive in certain situations, such as remote or off-grid scenarios (International Energy Agency, 2014). Common applications for the use of small-scale electric storage are arbitrage (response times >1 h), demand shifting and peak reduction, variable supply resource integration (response times <15 min), frequency regulation (response times ~1 min), or voltage support (response times: 0.001–1 s). In No-MASS/DR, an approximated model has been implemented for the charge and discharge of the battery, which considers the State Of Charge (SOC) of the battery independent of its open circuit voltage. Conversion efficiency is fixed to $\eta = 0.98$, leading to constant conversion losses of 2% with respect to SOC.
3. *Low voltage network model*. A forward/backward sweep solver has been used for power-flow analysis. It has been implemented as a recursive algorithm. In the case of a distribution network, recursion constitutes a method to efficiently solve branched networks. During the forward sweep, the currents and voltages at each node depend on the currents and voltages at the child nodes connected to them (one if linear network, several if branched); whereas during the backward sweep, currents and voltages depend on the parent node.

4.2.5.2 *Software orchestration with agent representation*

4.2.5.2.1 Agent representation

In the initial methodology of No-MASS, household occupants were identified as software agents. In extending No-MASS to No-MASS/DR to handle D2D interactions, electrical appliances and energy systems are represented as an agent subclass. Each agent has an ID, a peak power (either for demand or supply) and a priority of service. The priority of service is used for demand agents, and it allows to rank all energy requests based on their priority, and it is activated when restrictions in supply are considered. This would be useful, for example, for simulations of an islanded network, to account for situations when not all services can be delivered. Also, the operation starting time and profile can be calculated (type of prediction depends on the type of device), given that information of the associated model is entered in the agent definition.

During each time step, each device communicates the amount of energy (power integrated over the duration of the time step) to be requested or delivered. First, energy from our PV panel is allocated to appliances based on their priority. Any shortage is provided (perfectly) by the upscale grid. When an electrical storage device is unavailable, any surplus is exported to the grid. When it is available, the battery can be charged using excess PV energy, and can be later discharged to run electric appliances when price conditions favor this strategy.

4.2.5.2.2 Type of agents

Every actor in No-MASS/DR is configured as an agent. Thus, there are seven types of regular agents: LargeAppliance, SmallAppliance, PV, Battery, Grid, CSV, and Functional Mock-up Interface (FMI) agents. FMI agents are those linked to another software through FMI for co-simulation; in the case of No-MASS this software is EnergyPlus. CSV agents offer the possibility to enter .csv data files as input, instead of specific models, giving flexibility to add demand or supply devices for which models are not available yet.

Additionally, two of these agents have been conferred with learning capabilities: LargeApplianceLearning and BatteryLearning. The specific details of the mathematical formulation for their operation are presented in the following section.

4.2.5.3 Agent learning for DR optimization (R3)

In its present form, No-MASS/DR combines the following components (in relation to Fig. 4.3):

(1) Price-based strategies are considered: a cost signal based on indicative prices for electricity import is used as a *driver* stimulus to modify demand patterns. The specific tariff structure can be defined and modified for each simulation scenario. The current demand/supply status influence as well the response.
(2) Two *DR mechanisms* are implemented: load shifting (through appliance reschedule) and optimal electrical battery operation. An RL algorithm is incorporated into learning agents: (a) LargeApplianceLearning or *shiftable* appliance agents that can be regulated autonomously: washing machines and dishwashers, and (b) BatteryLearning agents that learn optimized discharge operation.
(3) The selected *DR mathematical formulation* consists of a Q-learning algorithm for the optimization of the objective function.
(4) In this work, we have focused on the improvement of renewable energy self-consumption using cost minimization as our *objective function*. This cost is defined inside the reward function for each case, and consists of a combination of electricity tariff price, available renewable energy, and intensity of the energy demand. We aim at finding a formulation for the reward function such that minimization of the objective measure leads to an indirect increase of self-consumption.

All these four elements are explained in more detail in the following sections.

4.2.5.3.1 Q-learning algorithm

There are multiple algorithms available for RL. Q-learning algorithms (Watkins, 1989) are one of the most widely used RL methods, with its major advantage being their great simplicity. They require a minimal amount of computation and, on its basic formulation, they can be expressed by single equations (Sutton and Barto, 1998), and easily implemented in computer programs.

Q-learning algorithms (Watkins, 1989) allow agents to learn a response from a reward, to an action. This allows agents to develop an understanding of their preferences over time. An agent learns the best action in a given state by trying every action in a state and updating the expected reward with the actual reward for that action. This is particularly useful, compared to other machine learning methods, for modeling appliance shifting, as the appliances can test different strategies for maximizing their reward where there is a clear link between an action and a driving stimulus. For example, does a chosen action (turn on at a later time) reduce peak power demand over a time-period. This is a quick and effective methodology that would be difficult to model through rules due to the complexities involved, especially when considering multiple shifting appliances. Each appliance would need its own set of rules to ensure

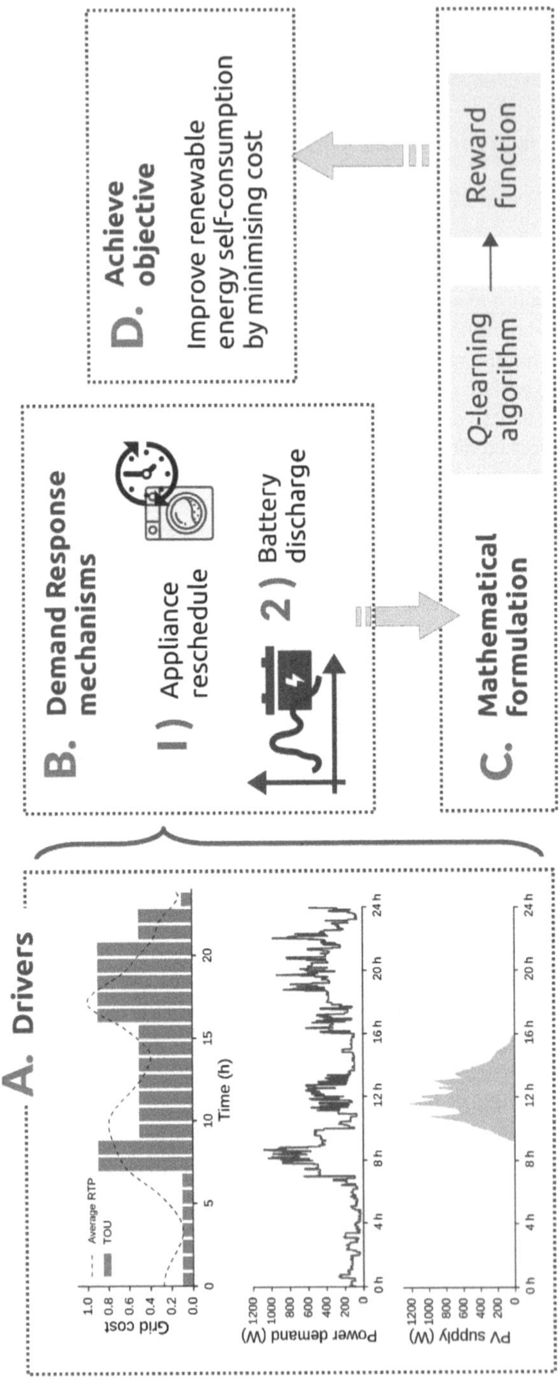

Fig. 4.3 DR methodology.

they would not turn on at the same time, whereas Q-learning allows them to learn their own preferences, considering other appliance demands.

In Q-learning algorithms, an agent chooses an action at a given state based on a Q-quantity, which is a weighted reward based on the expected highest long-term reward (Watkins, 1989). The Q-quantity is defined for each state-action combination, creating a Q-table. The values in the Q-table are updated each time an agent selects an action. Let s_t express the agent's state at time step t, and a_t a chosen action. Using this information, the Q-value for the corresponding combination of (s_t, a_t) is updated:

$$Q_t(s_t,a_t) = Q_t(s_t,a_t) + \alpha[\mathcal{R}_t + \gamma^* \max(Q(s_{t+1},a)) - Q_t(s_t,a_t)], \tag{4.1}$$

where $\mathcal{R}_t$ is the reward observed for the current state and the action taken. For each combination of (s_t, a_t), $\mathcal{R}_t$ can be a single value or it can be a function depending also on other variables. $\alpha \in (0, 1]$ is the learning rate and $\gamma \in (0, 1]$ is the discount factor. The term $\max(Q(s_{t+1},a))$ is an estimate of the optimal future value; thus, the discount factor specifies how soon the agent cares about the reward: near terms goals when $\gamma \sim 0$ (myopic agent), otherwise long-term rewards when $\gamma \sim 1$. In summary, at time-step t, when the agent at state s_t is taking action a_t, the Q-value is updated using: its former value $Q_t(s_t,a_t)$, the reward $\mathcal{R}_t$, and an estimate of the optimal future value $\max(Q(s_{t+1},a))$. The effectiveness of the learning process is highly dependent on the selected parameters α, γ, and ϵ.

Once the Q-table is updated, the next state is set as the current state, and the agent will select a new action. The selection of an action is not completely deterministic (using the Q-table), but uses an epsilon-greedy approach: the best action is selected with $1 - \epsilon$ probability, and a random action is selected otherwise. For instance, if $\epsilon = 0.1$, the (currently) best action will be adopted 90% of the time. This randomness is introduced so that the agent explores more in order to discover the best action over the whole period of time.

4.2.5.3.2 Reward function

The reward function $\mathcal{R}_t$ can consider different variables of interest, such as cost, power demands, or voltage stability, allowing to use the same methodology to explore a range of objectives. An advantage of Q-learning over other black-box machine learning methods such as neural networks is that the definition of the reward function allows to tune the algorithm in a explicit way, using knowledge of the system. Convergence of the learning process is achieved when the change in values of the Q-table no longer affects the results. The reward function is different for each learning agent type. Thus, it needs to be formulated for each case study.

4.2.5.4 DR mechanisms

4.2.5.4.1 Load shifting optimization

For each time step, the switch-on model is run. When a switch-on event is predicted, the profile of use and the new starting time are calculated. The temporal window within which appliances may be shifted has a maximum extent of 24 h, but this may be reduced to meet predefined delivery time constraints.

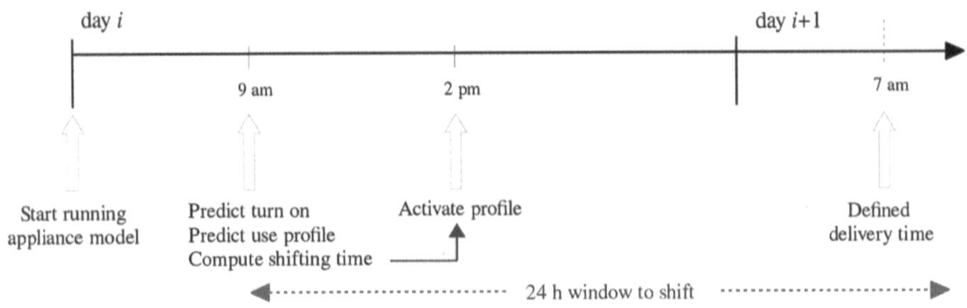

Fig. 4.4 Appliance reschedule diagram. In this example, state is $s_t = 9$ (h) and action $a_t = 14$ (h).

Q-learning for load shifting For a LargeApplianceLearning agent, we map the state (hour of the day) to an action (future hour of day to initiate the appliance). This creates a mapping of 24–24 h, making the Q-table space of 576 combinations in size. An example of a profile shift might involve calculating whether at state s_t the appliance is required to be turned on (e.g., the dishwasher is loaded and ready). If so the appliance demand profile should be calculated using the large appliance model. However, this should not yet be initiated. Instead the Q-table should be used to retrieve a time (action) at which the appliance program should be initiated (see Fig. 4.4).

A graphical example of a 24 × 24 Q-table for appliance reschedule is presented in Fig. 4.5, in which each pixel is false colored according to the Q-value of an action from its state s_j to s_{j+1}. Shifting to peak times (07.00–09.00 and 16.00–21.00 hours) has a smaller reward (darker pixels) than shifting to sunshine hours, or off-peak (23.00–07.00 hours) times. Scope of reduce dimensionality should be explored, either visually or using cluster analysis, as it could allow a faster learning period.

Reward function for load shifting The reward function $\mathcal{R}$ for appliance reschedule consists of two components:

1. The inverse of a cost signal (effectively then an income signal) is based on indicative prices for energy imports. A TOU tariff can be defined (specified for each case study). Values of the tariffs have been normalized [0, 1], since only relative differences are useful for the learning algorithm.
2. Negative reward (punishment) when the service is not satisfied on time (the washing cycle is not complete within the delivery-time-constrained window).

4.2.5.4.2 Battery discharge operation

Our battery has been implemented to charge whenever there is a surplus of PV power available (supply > demand), before it is exported to the grid (while storage capacity is surpassed). For the discharge process, Q-learning is used.

Q-learning for battery discharge For a BatteryLearning agent, we map the state (hour of the day) to an action (either discharge [1] or not discharge [0]). It leads to a mapping of 24 h to 2 actions, and a Q-table space of 48 combinations in size.

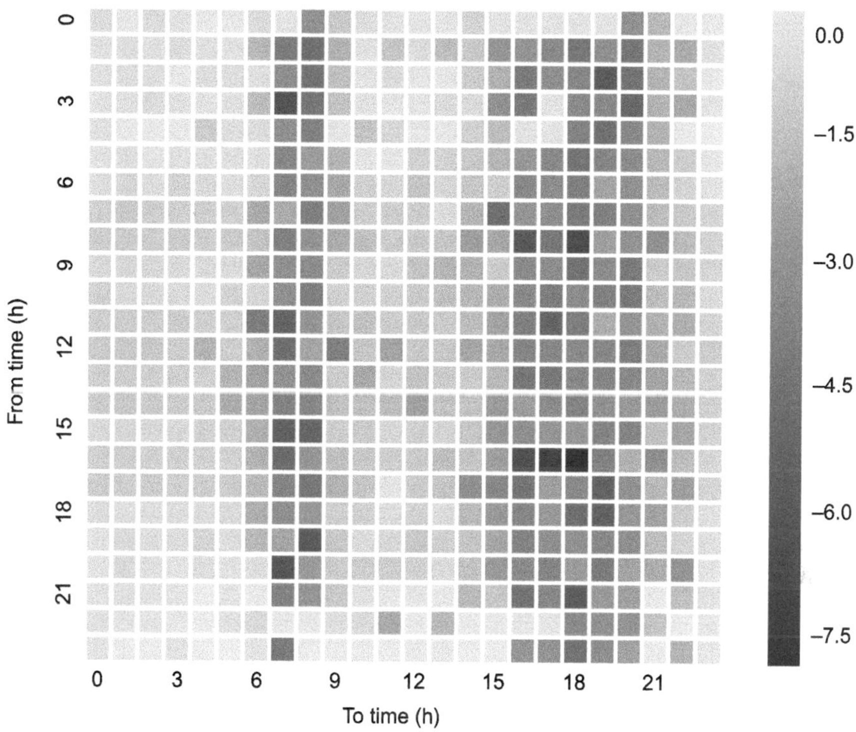

Fig. 4.5 24 × 24 Q-table for appliance reschedule. The units of the legend correspond to reward values, derived from the corresponding reward function.

A graphical example of such 24 × 2 Q-table is presented in Fig. 4.6, in which each pixel is false colored according to the Q-value of the action $a^{(1)}$ of discharging and $a^{(0)}$ for not discharging, for each hour of the day.

Reward function for battery discharge The reward function $\mathcal{R}$ for the discharge process is dictated by either one or both of these two considerations:

1. *Alleviate energy intense periods.* The battery has been configured to relieve peaks of high demand that require electricity imports. The battery learns when the highest hourly peak demand $P_{grid\_import\_MAX}$ is and when imports from the grid exceed some threshold δ (e.g., 70% if $\delta = 0.7$) of this demand:

$$P_{grid\_import} > \delta P_{grid\_import\_MAX}. \tag{4.2}$$

Otherwise, it learns to turn itself off in that hour, storing the remaining energy for peak periods.

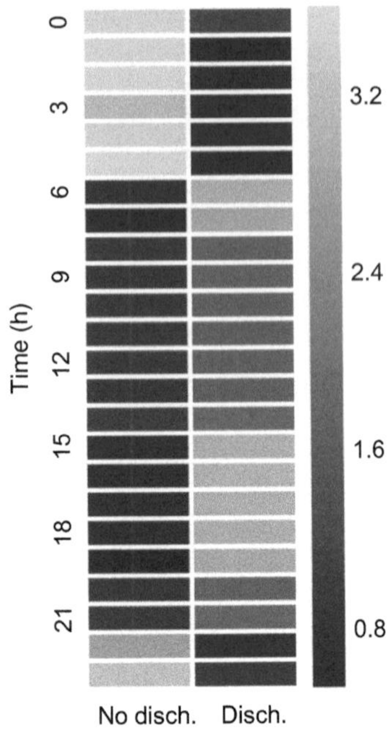

Fig. 4.6 24 × 2 Q-table for battery discharge. The units of the legend correspond to reward values, derived from the corresponding reward function.

2. *Minimize high-cost grid imports.* The battery can also recognize favorable price conditions. It will learn when to discharge based on the price signal, discharging in high-price periods and not discharging during cheaper periods. This mechanism is commonly known as arbitrage.

The activation of such considerations is defined in the reward function for each case, which needs to be tuned accordingly. The specificities of the reward functions used for this work are described in the corresponding case studies.

4.2.5.4.3 Self-consumption maximization

Self-consumption (SC), also referred to as a load matching index (Salom et al., 2011; Luthander et al., 2015), can be defined as the ratio of energy use from on-site generation to the total energy used. It is inversely related to the amount of PV (or other sources) power exported to the grid.

Two main arguments support the efforts in improving on-site power consumption. First, SC increases the economic value of distributed generation (Widén, 2014). In a context of constantly increasing retail electricity prices, and declining investment costs for renewable technology, improving the rates of use of self-generated electricity contributes directly to achieve grid parity.[5] For the case of PV, which has the highest share in DG, grid parity has already been achieved in different European countries, and this is expected to become more widespread in the coming years (Dehler et al., 2015). Second, SC can lower the stress on the electricity distribution grid, caused by high levels of penetration of DG. Integration of intermittent power sources is intrinsically challenging, due to the disparity between power generation and power demands. Moreover, it can cause power faults such as voltage rise due to the peaks in generation during low-demand periods. Increased rates of SC can help to mitigate such effects, as well as reducing transmission losses.

Although maximizing self-consumption is our goal, this is not formally optimized by our Q-learning algorithm, as it is not explicitly expressed in our reward function. But this is indirectly achieved through the tariff signal, through which we pay a low price while using off-peak centrally generated energy and close to zero for locally generated energy. In this way, smart appliances learn to turn on during low-cost periods.

4.2.6 Improving self-consumption of a community of buildings

Self-consumption can be further achieved in a community energy framework. Community energy projects "cover aspects of collective action to reduce, purchase, manage and generate energy."[6] Distributed generated electricity, from collective or individually owned resources, can be used for demands across the community, instead of being fed directly into the grid, enhancing the use of on-site generation. This idea can be extended also to storage devices. In a community energy model, storage systems could be charged or discharged with generation surplus or demands from multiple households.

In this section, we test our tool for self-consumption maximization of a local neighborhood (not necessarily detached from the city) that can operate as a community. In this, we assume that occupants are willing to adopt new technology and/or to modify their energy-using practices to improve upon their energy performance (albeit expressed through financial rewards).

4.2.6.1 Scenario description

A small residential neighborhood of six households is considered for this case study. Four of them have installed 3.8 kW-peak PV panels on their rooftops, and two have installed 1.9 kW-peak panels, following a smaller investment. As a community, they are willing to adopt smart appliances (washing machine and dishwasher) and individual electrical storage in their homes. No-MASS/DR enables to simulate the effects of the introduction of these technologies for this neighborhood, including an estimation of the improvements in self-consumption, both individually and collectively.

Price-incentive-based DR control strategies to: (a) reschedule autonomously controlled washing machine and dishwasher appliances; (b) discharge the battery. A TOU

[5] For PV, grid parity is defined as "the coming of age moment, when electricity from PV will be cost competitive with that from conventional generation sources, without subsidies so that deployment will take off driven by economic fundamentals" (Intelligent Energy Europe—European Commission, 2018).

[6] See https:/www.gov.uk/guidance/community-energy.

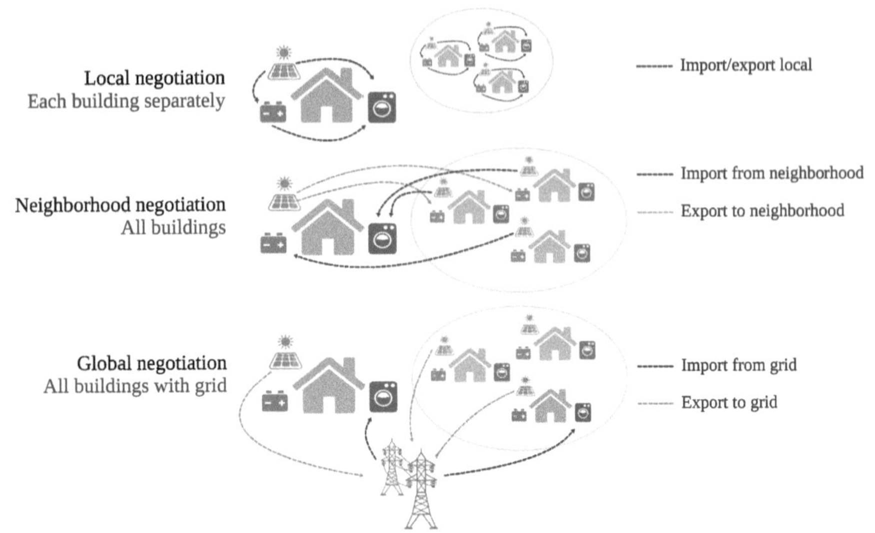

Fig. 4.7 Multilevel coordination.

tariff is tested, with three different pricing periods: on-peak (between 07.00–9.00 hours and between 16.00–21.00 hours), off-peak (at night between 23.00–07.00 hours), and mid-peak for the rest of the day. Values of the tariffs have been normalized [0, 1], since only relative differences are useful for the learning algorithm. For both DR mechanisms, device operation can be tuned using this tariff signal as a driver, which fosters the switch on of autonomous appliances while sunshine is likely, and the discharging of the battery while there is no PV generation. This operation is aimed at maximizing self-consumption over time, while reducing on-peak demand.

4.2.6.2 Implementation: Multilevel coordination

When generalizing No-MASS/DR from a single building to a multibuilding simulation, new software challenges arose. There were limits in computational speed, data storage, and data exchange, which had to be overcome, using data compression methods, or by incorporating parts of the postprocess analysis into the simulation, so that less data had to be written. In that sense, although the multibuilding simulation did not involve major conceptual changes, the software architecture had to be revised and enhanced.

The current operation of No-MASS/DR for multibuilding simulation requires three-level coordination. The agent interaction and negotiation occurs then in three steps (see Fig. 4.7):

1. *Local level.* Each house individually attempts to match power demand and supply.
   - **(a)** Appliance agents request power for their operation.
   - **(b)** If solar power is available:

- PV agents supply power to the appliances (following their priority of use, defined in the configuration file), until all the demands are fulfilled.
- The excess of PV is used to charge the battery.
- After charging the battery, any surplus of PV is made available for the neighborhood level.

(c) If solar power is not available, or it is not enough to fulfill appliance power demands, the battery discharge algorithm is run, and a decision to discharge or not discharge is taken.

- If discharging, appliance demands are satisfied by the battery.
- If the battery supply is not enough to fulfill the appliance power demands, information on the power shortage is communicated to the neighborhood level.

2. *Neighborhood level.* At this level, information is available for each house concerning: any appliance demand shortage, any PV surplus, and any battery which is uncharged. A second process of power matching occurs.

(a) Demand shortages are satisfied by solar energy from the neighborhood.

(b) PV power surplus is used to charge the uncharged batteries (which were not used for discharging during the same time step).

3. *Global level.* Unsatisfied demands are supplied by grid power imports. Excess solar power is exported to the grid.

4.2.6.3 Reward functions

As in other machine learning algorithms, a trial and error process is necessary to identify suitable algorithm parameters. In RL, this is the case for the definition of the reward function. However, in contrast with other black-box machine learning algorithms (such as neural networks), Q-learning allows the modeler to incorporate "expert knowledge" about the operation of the system, through the explicit formulation of the reward function. Thus, different formulations lead to more or less efficient learning. Two ways of evaluating this are: to look at the relative reward values, which should increase with time, as the agents learn the more profitable actions; or to look at the expected behavior of the system (for instance, checking when the appliance switch on events occur).

4.2.6.3.1 Appliance reschedule

The reward function for load shifting was explained above; being a combination of the inverse cost signal (from the TOU) and a punishment when the service is not satisfied within the defined delivery time t_D, which is specified for each appliance in the simulation configuration file. The value for the delay punishment $p \geq 0$ is also input through the configuration file. For this simulation, we have set $p = 1$.

$$\mathcal{R}_{a(t)} = \begin{cases} -C_{cycle}, & t < t_D \\ -(C_{cycle} + p), & t \geq t_D, \end{cases} \tag{4.3}$$

where C_{cycle} refers to the cost of running once the appliance a starting at time t, and depends on the electricity use and price.

4.2.6.3.2 Battery discharge

Battery discharge considers the two mechanisms defined above: a combination of alleviation of demand-intense periods and an arbitrage strategy. Thus, the reward values depend on the variable χ (defined as the ratio between the hourly imported power and the maximum hourly imported power registered throughout the day) and the time-varying cost of grid electricity C_{grid}. Analogous to the case of a single-building, the reward function when discharging is:

$$\mathcal{R}_{a^{(1)}} = \begin{cases} 2\chi + C_{grid}, & \chi \geq \delta_1 \\ \chi + 2C_{grid}, & \delta_1 > \chi \geq \delta_2 \\ 3\,C_{grid}, & \chi < \delta_2 \end{cases} \tag{4.4}$$

and when not discharging:

$$\mathcal{R}_{a^{(0)}} = \begin{cases} 1/(3\chi), & \chi \geq \delta_1 \\ 1/(\chi + 2\,C_{grid}), & \delta_1 > \chi \geq \delta_2 \\ 1/(3C_{grid}), & \chi < \delta_2, \end{cases} \tag{4.5}$$

where $\delta_1 = 0.8$ and $\delta_2 = 0.7$. In the discharging process specified in Eq. (4.4), bias toward C_{grid} increases as χ decreases: when there is a peak in demand (high χ), the cost of the grid C_{grid} is less important, therefore the battery sees a greater reward the greater the peak if it chooses to discharge, independently on the cost of the grid imports C_{grid}, learning to alleviate peaks. On the other hand, when power demand is not as intense (low χ), the battery obtains greater reward for discharging the higher the grid cost C_{grid}. These processes are similar for the nondischarging reward, in Eq. (4.5), but in this case, rewards are inversely proportional to C_{grid} and χ values.

As mentioned earlier, *Q*-learning is a flexible approach to model adaptable behaviors. In this case, a variety of options for Eqs. (4.4), (4.5) were implemented and tested, with different weights for χ and C_{grid}, leading to different learning behaviors for the discharge process. Eqs. (4.4), (4.5) were found to function satisfactorily for the battery.

4.2.6.4 Results

In this section we present the simulation results for the six-building neighborhood detailed above, considering three operation scenarios: the base case, the introduction of smart appliances in all houses, and the additional introduction of electric storage, evaluated for a winter and a summer week. For each case a set of 100 week simulations[7] is considered, and the averaged results are presented here.

Fig. 4.8 presents the weekly energy imports/exports from/to the grid. It shows results for winter (on the left) and summer (on the right), and for the three scenarios: base case on top, load shifting in the middle and load shifting plus battery in the

[7] The number of simulations considered takes into account when new simulations do not vary average values.

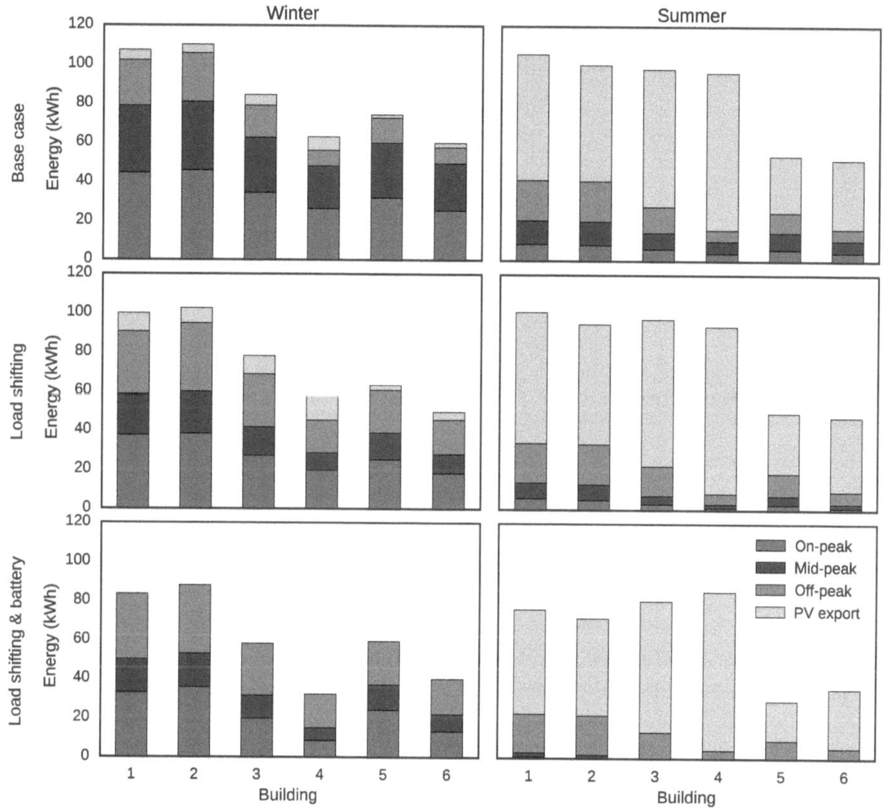

Fig. 4.8 Imports and exports from the grid for winter (on the *left*) and summer (on the *right*), and for the three scenarios: base case on top, load shifting in the middle, and load shifting plus battery in the bottom. Buildings 5 and 6 are those having 1.9 kWp solar panels.

bottom. The results are disaggregated for each building, with averaged values presented in Table 4.2.

For the sake of comparison, the PV profile for the summer case corresponds to a bright and clear day, when solar energy is abundant. This is reflected in Fig. 4.8, with large differences between the PV exports in summer and winter.

In winter, the top left graph shows that for the base case, grid imports occur at peak and mid-peak times. The introduction of the load-shifting algorithm has two effects. First, energy use is shifted to off-peak times. Second, PV exports increase a 77% when appliance rescheduling takes place. This is due to the shifting process: smart appliances learn that it can be more beneficial to operate during cheap night time hours than during sunshine hours, as limited winter solar generation may not be enough to power the whole cycle, leading to expensive on-peak grid imports. When batteries are introduced in the community, practically no solar power is exported to the grid. In addition, total imports reduce to almost 24%, with its main use being off-peak. The improvements in on-peak to off-peak shifting are maintained. This represents a

Table 4.2 Average weekly import/export

	On-peak	Mid-peak	Off-peak	PV export
		Winter (kWh)		
Base case	34.6	28.6	15.6	4.30
Load shifting	27.6	14.9	25.1	7.60
Load shifting and battery	22.3	12.5	25.4	0.013
		Summer (kWh)		
Base case	6.13	9.11	12.7	56.4
Load shifting	3.14	4.79	13.1	59.6
Load shifting and battery	0.31	0.689	11.8	50.2

success of the algorithm, as it achieves our two goals: alleviating energy intense periods and minimizing high-cost grid imports.

In summer, the effect of the large amounts of PV available is clear, leading to a situation where grid imports are a third of the winter value. Also, total demands are lower, due to heating being switched off. Another consequence is that shifting appliances learn to switch to sunshine hours (slight decrease in PV exports) as well as to off-peak night periods (increase in off-peak imports). On average, on-peak and mid-peak imports are reduced over by 48% with load shifting alone. Again, a significant impact is achieved with the introduction of storage systems: local generation exports are cut by a 11%, and imports are shifted to off-grid periods even more, with a reduction over 93% for on- and mid-peak periods.

Results for summer and winter show that the reward functions used work well in shifting demand from expensive grid imports toward locally generated energy. This has been achieved by imposing a grid import tariff, which proves to be successful for our purposes. A powerful potential of the Q-learning technique used is that we could try a different price structure, and the system would adapt to make the necessary changes to achieve a relatively optimized operation.

Fig. 4.9 shows the boxplots for the distribution of self-consumption for each simulation of each building. In this case, we differentiate SC at two levels: local and neighborhood. They relate to the energy use arising from local resources (their own PV/battery system) and from their neighbors' resources (PV/battery systems owned by neighbors), representing individual and collective self-consumption. Mean values for the different groups are given in Table 4.3.

Winter values for individual SC in the base case average to 13.6%. This figure increases to 17.1% for the neighborhood case. For the reasons explained above, there are not major changes in local and neighborhood SC when the load shifting algorithm is applied: local SC is slightly larger and collective SC is slightly smaller, although buildings 5 and 6 are benefited from their neighbors' resources. One way to improve SC more dramatically with load shifting would be to consider a more expensive off-peak tariff, so that appliances learn to shift to sunshine hours, even if grid imports are necessary. As expected, the introduction of batteries has a significant impact, doubling

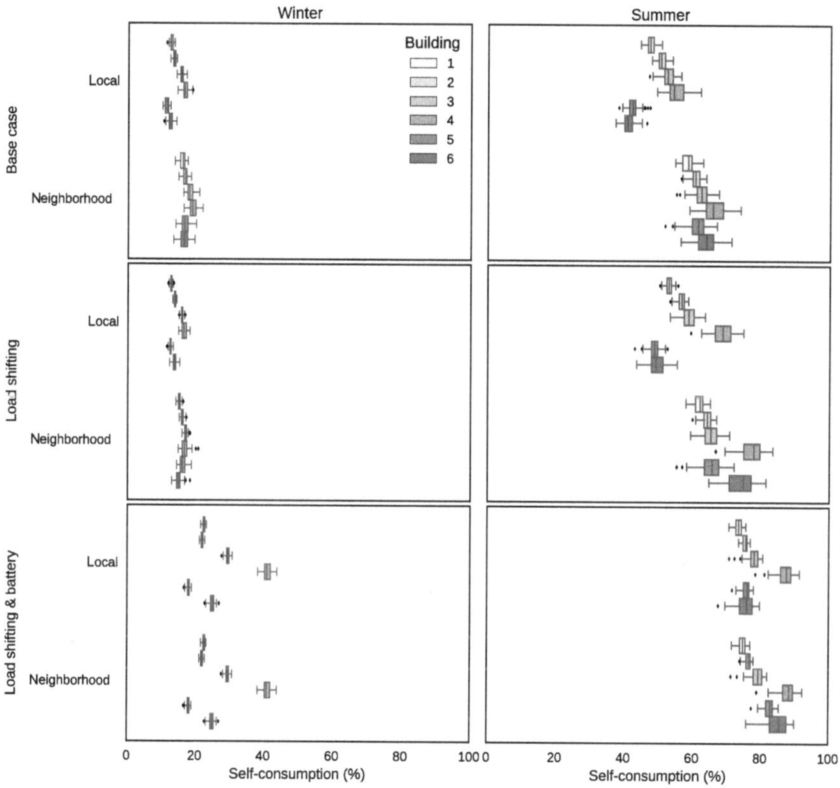

Fig. 4.9 Self-consumption index for the three scenarios, in winter and summer.

the value of local SC. On the other hand, individual batteries available for all buildings neutralizes the positive effect of collective use, as the excess of solar energy is individually stored for later use instead of shared. Allowing the batteries to discharge to other houses' appliances could be useful for improving neighborhood SC, given an appropriate business model.

In summer, SC increases in all cases. When the load shifting algorithm is in use, local SC increases from 48.0% to 55.7%, and neighborhood SC from 61.9% to 67.8%. Maximum levels of SC are achieved with the storage systems in place during summer, reaching values of over 75%. With respect to the base case, this represents a 61% and a 31% increase in individual and collective SC, respectively.

In both summer and winter, households achieve larger values in SC when allowed to share their resources (except in winter battery case, that same results are obtained). This suggests the need for a new way of looking at tariffs. Different pricing systems could be put into practice, which regard not only the interaction with the electricity grid of each individual customer, but also the interaction between customers, creating for example some sort of incentive structure to share power resources between members of the same neighborhood. For example, in our simulation, households 5 and 6 made a smaller

Table 4.3 Average self-consumption (%)

		Winter	Summer
Base case	Local	13.6	48.0
	Neigh.	17.1	61.9
Load shifting	Local	14.1	55.7
	Neigh.	16.0	67.8
Load shifting and battery	Local	26.4	77.5
	Neigh.	26.4	81.0

investment in their PV system, reducing their local energy autonomy. However, the base case and the load shifting scenarios in Fig. 4.9 show how strongly they benefit from participating in the community, with comparable collective SC values.

4.2.7 Conclusion

In this section, we have demonstrated the integration of price-based DR strategies to optimize for load-shifting and the charging and discharging of a battery, by incorporating learning abilities into our device agents. Q-learning has been proven to be a successful candidate for that task. Different reward functions have been explained and put into practice. Single and multibuilding case studies have been presented; the latter requiring an extended negotiation logic.

More specifically, the learning algorithms have effectively: shifted demand from on- and mid-peak periods to off-peak periods, particularly when combining shiftable demand devices with battery storage; increased remarkably the energy autonomy percentage (the fraction of energy demand that is satisfied by on-site generation). The multibuilding approach demonstrates the positive effects in SC when communities of buildings are able to share their resources.

Regarding the Q-learning approach, it is important to note that the current implementation is unable to respond to irregular events (e.g., weather) or to act well with extreme values caused by particular situations. Being able to handle such circumstances would be possible if specific "extreme" state-action pairs are defined.

Although we believe that we have successfully demonstrated this proof-of-concept; there is considerable scope for improvement to this framework. In particular with respect to (and still only considering requirements R1–R3):

1. The battery model should be improved to model state-of-charge-dependent losses.
2. In this work, battery and heating demand have been combined to reduce on-peak grid imports. In a more realistic situation, the heating system can be considered as an additional learning device in its own right, which contributes to the global goals.
3. The negotiation for the multibuilding system should be upgraded to consider battery discharge to appliances in neighboring households. Battery agents would need to take into consideration not only the operation status of their own local system (household), but also the

global system (other households). Different reward structures would be necessary. Another upgrade could be to allow the batteries to charge from the grid. This could be achieved by providing the battery agents with the ability to learn charging strategies.

4. For islanded scenarios in which power cannot be drawn upstream of the mains network, a formal bidding mechanism needs to be integrated, favoring devices of relatively high priority when reserves are limited (e.g., emergency lighting, or freezers that have undergone a long delay since the compressor was last enabled).
5. Alternative tariff structures should be explored that better favor self-consumption.

Looking further to the future (R4–R6):

6. No-MASS/DR should be generalized to solve for multiple buildings interconnected via our LV network model. That way, the above tariffs might also consider local network integrity, when the LV model is utilized as intended.
7. Empirical evidence from DR field trials should be used to predict the extent to which users are willing to: (a) devolve control to autonomous devices, and (b) adjust their behaviors in response to information (e.g., tariff and/or CO_2 emissions) feedback.

Finally, field trials to test the validity of this new more general multiagent stochastic simulation framework would be of considerable value.

4.3 Building load management with MPC

4.3.1 Introduction

Thermal loads have been the focus of previous studies using smarter control techniques for DSM applications. The aims were typically to reduce the energy consumption or to investigate the load shifting possibilities relying on the storage capabilities of the building and its systems. Thermal energy storage can be split into active and passive systems. Storage water tanks and building thermal mass are two examples of active and passive systems, respectively. In particular, the building mass can be seen as a free storage opportunity, available in every building, for preheating or precooling (Zong et al., 2017). The magnitude of this thermal capacity is unequal across buildings of a city or country. To fully harness the potential of such storage systems, advanced control techniques are required.

In the following sections, the background about building load control is first given (Section 4.3.2). Then, focus is put on an advanced technique called Model Predictive Control (MPC); the formulation of the technique is detailed (Section 4.3.3), before addressing its advantages and challenges (Section 4.3.4). In the last section, an example of MPC application with TOU tariffs is addressed (Section 4.3.5).

4.3.2 Building load control

Building thermal load control is a vast field characterized by a wide range of control techniques. Multiple classifications exist, depending on whether the differentiation is made on the type of control approach (classical, hard, soft, hybrid, etc.), the characteristics of the algorithms (optimal, adaptive, predictive, etc.) or the hierarchical level at which they are implemented (local, supervisory). Several comprehensive reviews are available in the literature (Afram and Janabi-Sharifi, 2014; Dounis and Caraiscos, 2009; Mirinejad et al., 2012; Shaikh et al., 2014; Yu et al., 2015; Wang and Ma, 2008). A simple classification distinguishes conventional from advanced control (Shaikh et al., 2014; Mirinejad et al., 2012). Conventional control encompasses among others ON/OFF or P-PI-PID controllers. These techniques have been, respectively, criticized for jeopardizing occupant comfort with overshoots/undershoots, or for their potential tuning/instability challenges. Moreover, in the context of the evolution toward smarter energy networks (Section 4.1), buildings are to move toward more energy efficiency and flexibility. Conventional control techniques are not adequate with this regard, as they fail at comprehensively accounting for the complexity of thermal energy use in a building, which is influenced by multiple parameters, time-varying signals, and requires finding trade-off solutions. For example, conventional control is not ideal to address load flexibility, which requires making a trade-off between energy use and comfort, considering the building characteristics, the future outdoor weather, the presence of occupants and potential time-varying incentive signals. Advanced control encompasses machine learning-based methods, MPC, or agent-based control (Shaikh et al., 2014). These techniques are characterized by their adaptive, predictive, and/or optimal features among others.

MPC has gathered the interest of the building research community. MPC development commenced in the 1970s/1980s in the petrochemical sector. Since then, it has been implemented in various fields from chemical to food or aerospace industries (Qin and Badgwell, 2003). However, it is still not widespread in the building field and it has been mostly limited to theoretical studies or small demonstration projects. Interest is growing though, with an increasing number of papers on the topic. The MPC technique is further detailed in the next section within the context of thermal load control.

4.3.3 MPC formulation

MPC is a flexible technique that can address a variety of thermal load management cases. MPC relies on optimization and a model of the system controlled. The approach is schematized in Fig. 4.10 (adapted from Cigler, 2013; Široký et al., 2011; Gyalistras and Gwerder, 2010). In this figure, the energy price is not a mandatory input, but allows investigation of cost-based optimization problems. An MPC problem more specifically includes:

- a model of the system;
- predictions of the disturbances;
- an objective function; and
- a set of constraints.

At each time step t, an objective is optimized over a prediction horizon (hp) and only the first control input is applied to the system. The new system state is then measured or estimated and fed back as a starting point for the next time step. The whole process is repeated at $t + 1$, starting with the optimization over the prediction horizon

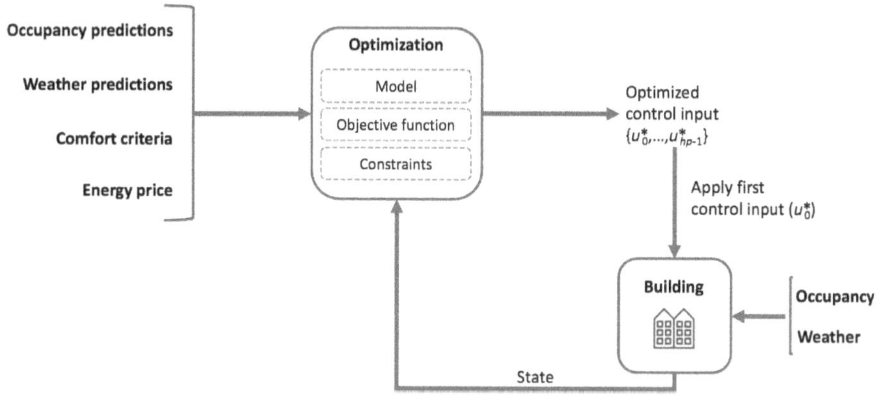

Fig. 4.10 MPC scheme for building load control.

(Afram and Janabi-Sharifi, 2014; Široký et al., 2011). Given these characteristics, MPC is also known as "receding horizon control" (Killian and Kozek, 2016). This control technique is predictive, corrective, and relies on optimization. A typical MPC formulation is given in Eqs. (4.6)–(4.9), specifying the objective function, the constraints, the current state, and the system model (Oldewurtel et al., 2012).

$$J(x_0) = \min_{u_0 \ldots u_{hp-1}} \sum_{t=0}^{t=hp-1} l_t(x_t, u_t), \tag{4.6}$$

subject to

$$(x_t, u_t) \in X_t \times U_t, \tag{4.7}$$

$$x_0 = x, \tag{4.8}$$

$$x_{t+1} = f(x_t, u_t), \tag{4.9}$$

where

- hp: prediction horizon;
- x_t: states at time step t;
- u_t: inputs at time step t; and
- X_t and U_t: constraint sets at time step t for the states and inputs, respectively.

The MPC scheme relies on a building model in the optimization step, which can take various forms, from data driven to physics based (Qin and Badgwell, 2003); examples include reduced order RC models, ARX, neural networks, or models developed with simulation software (Ma et al., 2012a,b; Costanzo et al., 2013; Collazos et al., 2009; Coffey et al., 2010; Zong et al., 2017; Killian and Kozek, 2016; Oldewurtel et al., 2012; Sturzenegger et al., 2016). The use of a middleware is necessary to couple simulation softwares to a computing environment. For example, the building physics can be modeled in Energy Plus, the control algorithm can be implemented in Matlab, and cosimulation can be run with an adequate interfacing tool (Wetter, 2011; Bernal et al., 2012). In the building literature, linear state space formulations tend to be used quite often (Gorecki et al., 2015). As MPC relies on optimization techniques, the question of solvability is a concern with the more complex formulations, which may require the use of linearization techniques. For theoretical/simulation-based case studies, a second model is used to mimic the building response.

The objective function formulation varies from case study to case study. Typical formulations include linear cost, quadratic cost, or less frequently probabilistic cost (Oldewurtel et al., 2012; Gyalistras and Gwerder, 2010). Through the objective function, MPC can deal with the antagonistic objectives of building energy management, which typically oppose energy demand/cost to occupant comfort. In practice, the objective function can be a weighted sum of both terms. The tuneable weight allows to give more or less importance to comfort versus cost or energy use. Other configurations for the objective function can be considered, including a single objective or more than two subobjectives; for example, the joint minimization of heating demand, comfort violation, and peak power. The different stakeholders mentioned in Section 4.1.4 can be addressed by adapting the objective function, to optimize the control variable(s) that address(es) their concern. Building occupants may be interested in minimizing energy cost, while municipalities might want to encourage improved energy efficiency. Finally, a network operator may want to focus on smoothing peak demand, which can be addressed with a cost-based objective function and a well-designed TOU tariff.

Constraints are used to account for the problem restrictions such as the thermal system limits or the occupant comfort. Constraints can be differentiated by their mathematical formulation and be classified as linear, convex quadratic, nonlinear, etc. (Gyalistras and Gwerder, 2010). They can be enforced in different ways: as hard constraints, soft constraints, or setpoint constraints (Qin and Badgwell, 2003). Hard constraints cannot be violated, which can lead to infeasibility issues. Relaxing constraints leads to soft or setpoint constraints, where overshoot and/or undershoot are allowed but penalized in the objective function. In practice, temperature constraints are often dealt this way, adding a term that penalizes discomfort in the objective function.

MPC can be formulated and run considering various perspectives. The term "performance bound" (PB) is used when the system behavior and predictions are assumed perfect (Oldewurtel et al., 2012). In practice, weather predictions are replaced by weather realizations and the receding horizon is not implemented, instead optimization is carried over the whole period of interest (Gyalistras and Gwerder, 2010). PB is an ideal case used as a theoretical benchmark. In the literature, MPC is often formulated in a deterministic way (deterministic MPC). However, accounting for prediction uncertainties contributes to computing a more adequate optimal control input. There are different ways to consider uncertainties, for example, with Robust or Stochastic MPC. Examples for building control applications are addressed in the following references (Oldewurtel et al., 2012; Long et al., 2014; Thieblemont et al., 2017; Gyalistras and Gwerder, 2010).

4.3.4 MPC advantages and challenges

MPC has been investigated in various building types: houses and apartment blocks, universities, offices, etc. Improved performances compared with simpler control techniques have been acknowledged in the literature over varying case studies (Afram and Janabi-Sharifi, 2014; Gyalistras and Gwerder, 2010; Sturzenegger et al., 2016; Hazyuk et al., 2014; Ma et al., 2012b; Shaikh et al., 2014).

Savings are very dependent on the case study. Based on the literature, impacting parameters include, among others, the active/passive storage system capacity, the climate, the incentive signal, the occupancy profiles, and internal gains (Thieblemont et al., 2017). For cases leading to energy/cost savings and/or comfort improvement, higher performances of MPC can be explained by (Coffey et al., 2010; Killian and Kozek, 2016):

- the use of optimization techniques, which screen the solution space and find the best (or near-best) solution for a given objective, while accounting for the case study constraints. Moreover, optimization allows accounting for conflicting objectives by providing a framework for tuneable trade-off.
- the use of a model of the building and predictions of future weather and occupancy profiles, which allow to take into account the building inertia and the future gains to adapt the heating schedule accordingly.
- the feedback loop that allows correcting the trajectory accounting for nonperfect predictions and modeling approximations.

MPC performance is not systematically significantly better than a simpler control technique. Authors have discussed a level of problem complexity above which MPC would be worth deploying (Coffey et al., 2010; Thieblemont et al., 2017). Moreover, research highlighted the importance of having reliable model, weather, and occupancy predictions, as it impacts the MPC performance (Coffey et al., 2010; Ma et al., 2012b).

While multiple papers have addressed MPC, this control technique is not widely deployed in buildings yet but is "on the cusp of commercial availability" (Killian and Kozek, 2016). A challenge to MPC deployment lies in the engineering efforts for the building modeling and the design of the MPC scheme (Thieblemont et al., 2017; Killian and Kozek, 2016; Sturzenegger et al., 2016; Cigler et al., 2013). Multiple modeling techniques have been investigated, but none is best overall. Moreover, each building is unique in terms of thermal characteristics, occupant behavior, and data availability, which makes it more difficult to apply systematic approaches. "Plug-and-play software" for control purposes that would facilitate the modeling task are lacking (Killian and Kozek, 2016), thus the modeling step is time-consuming. Similar remarks apply to the MPC scheme design which parameters have to be tuned for each building. Work has been initiated to address this challenge with toolboxes such as BRCM (Building Resistance-Capacitance Modeling) for RC model generation or OpenBuild, which provides a framework for design and validation of building control schemes (Sturzenegger et al., 2014; Gorecki et al., 2015). Moreover, the MPC controller design can be a challenge in larger buildings in terms of solvability. The issue could be improved with distributed configurations (Cigler et al., 2013).

Finally, MPC deployment will require the training of experts in the field for the design, implementation, and maintenance of such systems (Sturzenegger et al., 2016; Killian and Kozek, 2016); lessons learnt from MPC implementation in industry can be used with this regard.

While an overall background on MPC formulation, advantages, and challenges was given in the previous two sections, MPC is now discussed with a more concrete example considering flexibility with TOU tariff.

4.3.5 Example of MPC application

An application of MPC in buildings is the provision of flexibility in response to a TOU tariff for electricity. The present section describes a common formulation and addresses parameter tuning considering a space heating case.

In a cost-based MPC framework, TOU tariffs can lead to a modification in the building energy use profiles by encouraging consumption during specific periods. Building is preheated when the electricity tariff is lower to later reduce the demand when prices are higher. Advanced techniques such as MPC can be used to automatically adapt the building consumption to this price signal, while accounting for occupant comfort and other potential constraints.

Multiple electricity tariff schemes can be envisaged. Tariff profiles can vary on an hourly basis (e.g., real time pricing) or can be defined over periods of several hours (e.g., peak/off-peak pricing). Reduced prices can be enforced during period of lower demand to shift loads and reduce peaks. Alternatively, lower tariffs can be envisaged during periods of renewable energy production to support the integration of intermittent energies. For solar energy, for example, a lower tariff during periods of higher irradiance could encourage electricity consumption with heat pump systems during the periods of higher PV production.

The MPC objective function is often a weighted sum of two or more terms, one of which uses the energy tariff. A general formulation is given in Eq. (4.10) considering a case with two subobjectives: cost and discomfort.

$$\min \sum_{t=0}^{t=hp-1} [OF_1(t) + w^* OF_2(t)], \tag{4.10}$$

where

- $OF_1(t)$: energy consumption cost at time step t;
- $OF_2(t)$: discomfort at time step t;
- w: weight; and
- hp: prediction horizon.

The energy consumption cost (OF_1) is simply calculated considering the heating consumption and the energy cost at each time step. Discomfort (OF_2) is typically quantified considering temperatures, but other comfort metrics can be considered. The discomfort can be assessed calculating the difference between the indoor temperature and the setpoint. In such cases, the difference is to be squared to account for positive and negative deviations (Avci et al., 2013); the resulting OF_2 term will be quadratic. Another option consists in quantifying the discomfort as the comfort range violations (undershoot and overshoot); in such cases, the comfort range is formulated as soft constraints and two additional control variables, representing the comfort range violations, are minimized in the objective.

The control outcomes are influenced by the multiple parameters that can be tuned, among others: the MPC parameters (objective function weight, prediction horizon, etc.), the comfort range, the building characteristics, the tariff profile, etc. The prediction horizon is often set to 24 h in the literature, but values ranging from 4 to 72 h have been observed in the building energy field (Thieblemont et al., 2017; Killian and Kozek, 2016). This parameter is fixed depending on the building and HVAC system characteristics and the confidence in the predictions (Thieblemont et al., 2017; Oldewurtel et al., 2012). In practice, multiple horizons can be tested and the impact on the resulting performance can be quantified (Zong et al., 2017).

Regarding the objective function weight (w in Eq. 4.10), a higher value leads to scenarios with better comfort to the detriment of the overall energy cost and the opposite for lower w values. An extreme case is a weight value of 0, where comfort is not accounted for and cost is minimized leading to profiles with zero heating demand if comfort is not enforced as hard constraints. In practice, the weight is to be adapted depending on the occupant tolerance to discomfort.

In a context of minimization of the comfort range overshoots/undershoots, a larger range will give more freedom to the MPC algorithm to adapt the heating; in other words, there is more potential for preheating. In a context with minimization of setpoint deviations, these setpoints can be provided as fixed inputs for all buildings of a same type; however, they can also be adapted to the occupants tolerance to discomfort (Avci et al., 2013).

The previous paragraphs discussed the impact of selected parameters influencing an MPC scheme for space heating application with a TOU tariff. Finding the correct parameters of the MPC scheme typically requires test runs for each building.

4.4 Conclusion

The present section addressed MPC, an advanced technique relying on optimization, that has been used for DSM applications. MPC accounts for the system behavior and the future disturbances to optimize an objective over a rolling horizon; a feedback loop allows correction of the trajectory at each time step. Focusing on thermal loads, MPC has been applied to multiple theoretical cases in the literature and to a lesser extend to demonstration projects. Challenges to a large-scale deployment include the modeling efforts and the need for more experts in the field.

MPC has the potential to address some of the concerns of multiple stakeholders, thanks to the flexibility it offers in the formulation of the objective function, constraints, model, and incentive signal. The potential energy/cost savings or flexibility provision is, however, case study-dependent, thus preassessment tools are needed for a time-efficient estimation of the gains before investigating further if considered viable. An identification of the building characteristics and operational parameters leading to higher gains would prove useful for the design and operation of future buildings.

Acknowledgments

The authors would like to thank Assoc. Prof. D. Finn and Dr. P. Spodniak for their valuable comments on MPC and demand response deployment, respectively.

References

Afram, A., Janabi-Sharifi, F., 2014. Theory and applications of HVAC control systems—a review of model predictive control (MPC). Build. Environ. 72, 343–355. https://doi.org/10.1016/j.buildenv.2013.11.016.

Avci, M., Erkoc, M., Rahmani, A., Asfour, S., 2013. Model predictive HVAC load control in buildings using real-time electricity pricing. Energy Build. 60, 199–209. https://doi.org/10.1016/j.enbuild.2013.01.008.

Beckmann, M., 2013. The Challenge of Shifting Peak Electricity Demand. Technical Report. Centre for Carbon Measurement.

Behrangrad, M., 2015. A review of demand side management business models in the electricity market. Renew. Sust. Energ. Rev. 47, 270–283. ISSN 1364-0321. https://doi.org/10.1016/j.rser.2015.03.033.

Bergaentzlé, C., Clastres, C., Khalfallah, H., 2014. Demand-side management and European environmental and energy goals: an optimal complementary approach. Energy Policy 67, 858–869. https://doi.org/10.1016/j.enpol.2013.12.008.

Bernal, W., Behl, M., Nghiem, T., Mangharam, R., 2012. MLE+: a tool for integrated design and deployment of energy efficient building controls. In: 4th ACM Workshop on Embedded Sensing Systems for Energy-Efficiency in Buildings (BuildSys '12), pp. 1–10.

Bertoldi, P., Zancanella, P., Boza-Kiss, B., 2016. Demand response status in EU member states. EUR 27998 EN. https://doi.org/10.2790/962868.

Biegel, B., Hansen, L.H., Stoustrup, J., Andersen, P., Harbo, S., 2014a. Value of flexible consumption in the electricity markets. Energy 66, 354–362. https://doi.org/10.1016/j.energy.2013.12.041.

Biegel, B., Westenholz, M., Hansen, L.H., Stoustrup, J., Andersen, P., Harbo, S., 2014b. Integration of flexible consumers in the ancillary service markets. Energy 67, 479–489. https://doi.org/10.1016/j.energy.2014.01.073.

Chapman, J., 2016. Multi-Agent Stochastic Simulation of Occupants in Buildings (Ph.D. thesis). University of Nottingham.

Chapman, J., Siebers, P., Robinson, D., 2018. On the multi agent stochastic simulation of occupants in buildings. J. Build. Perform. Simul. https://doi.org/10.1080/19401493.2017.1417483.

Chen, L., Li, N., Low, S.H., Doyle, J.C., 2010. Two market models for demand response in power networks. In: IEEE International Conference on Smart Grid Communications, pp. 397–402.

Cigler, J., 2013. Model Predictive Control for Buildings (Ph.D. thesis). Technical University in Prague, Prague.

Cigler, J., Gyalistras, D., Siroky, J., Tiet, V.N., Ferkl, L., 2013. Beyond theory: the challenge of implementing model predictive control in buildings. In: 11th REVHA World Congress Clima, pp. 1–10.

Clement, K., Haesen, E., Driesen, J., 2009. Coordinated charging of multiple plug-in hybrid electric vehicles in residential distribution grids. In: Power Systems Conference and Exposition, IEEE/PES, pp. 1–7.

Coffey, B., Haghighat, F., Morofsky, E., Kutrowski, E., 2010. A software framework for model predictive control with GenOpt. Energy Build. (42), 1084–1092. https://doi.org/10.1016/j.enbuild.2010.01.022.

Collazos, A., Maréchal, F., Gähler, C., 2009. Predictive optimal management method for the control of polygeneration systems. Comput. Chem. Eng. 33 (10), 1584–1592. https://doi.org/10.1016/j.compchemeng.2009.05.009.

Costanzo, G.T., Gehrke, O., Bondy, D.E.M., Sossan, F., Bindner, H., Parvizi, J., Madsen, H., 2013. A coordination scheme for distributed model predictive control: integration of flexible DERs. In: Innovative Smart Grid Technologies Europe (ISGT EUROPE), 2013 4th IEEE/PES, pp. 1–5.

Critz, D.K., Busche, S., Connors, S., 2013. Power systems balancing with high penetration renewables: the potential of demand response in Hawaii. Energy Convers. Manag. (76), 609–619. https://doi.org/10.1016/j.enconman.2013.07.056.

Dehler, J., Keles, D., Telsnig, T., Fleischer, B., Baumann, M., Fraboulet, D., Faure, A., Fichtner, W., Deane, P., Stelzer, V., 2015. Self-Consumption of Electricity From Renewable Sources. INSIGHT-E—An energy think tank informing the European Commission.

Deng, R., Yang, Z., Chow, M.Y., Chen, J., 2015. A survey on demand response in smart grids: mathematical models and approaches. IEEE Trans. Ind. Inf. 11 (3), 570–582. ISSN 1551-3203. https://doi.org/10.1109/TII.2015.2414719.

Dimeas, A.L., Hatziargyriou, N.D., 2010. Multi-agent reinforcement learning for microgrids. In: IEEE PES General Meeting, Minneapolis, MN, pp. 1–8. https://doi.org/10.1109/PES.2010.5589633.

Dittawit, K., Aagesen, F.A., 2014. Demand side focused simulation platform for the evaluation of demand side management approaches. In: IEEE/PES Innovative Smart Grid Technologies Conference Europe (ISGT-Europe), pp. 1–6.

Dounis, A.I., Caraiscos, C., 2009. Advanced control systems engineering for energy and comfort management in a building environment—a review. Renew. Sust. Energ. Rev. (13), 1246–1261. https://doi.org/10.1016/j.rser.2008.09.015.

Dusparic, I., Taylor, A., Marinescu, A., Cahill, V., Clarke, S., 2015. Maximizing renewable energy use with decentralized residential demand response. In: IEEE First International Smart Cities Conference (ISC2), pp. 1–6.

European Commission, 2006. European smart grids technology platform: vision andstrategy for Europes electricity networks of the future. Office for Official Publications of the European Communities, Luxembourg, EUR 22040.

Faria, P., Vale, Z.A., Ferreira, J., 2010. DemSi—a demand response simulator in the context of intensive use of distributed generation. In: 2010 IEEE International Conference on Systems Man and Cybernetics (SMC), pp. 2025–2032.

Gangale, F., Vasiljevska, J., Covrig, F., Mengolini, A., Fulli, G., 2017. Smart grid projects outlook 2017: Facts, figures and trends in Europe. EUR 28614 EN, https://doi.org/10.2760/701587.

Gelazanskas, L., Gamage, K.A.A., 2014. Demand side management in smart grid: a review and proposals for future direction. Sust. Cities Soc. 11, 22–30. ISSN 2210-6707. https://doi.org/10.1016/j.scs.2013.11.001.

Gellings, C.W., 1985. The concept of demand-side management for electric utilities. In: Proc. IEEE 73 (10), 1468–1470. https://doi.org/10.1109/PROC.1985.13318.

Gorecki, T.T., Qureshi, F.A., Jones, C.N., 2015. OpenBuild: an integrated simulation environment for building control. In: 2015 IEEE Conference on Control Applications (CCA), pp. 1522–1527.

Gudi, N., Wang, L., Devabhaktuni, V., Depuru, S., 2010. Demand response simulation implementing heuristic optimization for home energy management. In: North American Power Symposium (NAPS), 2010, pp. 1–6.

Gyalistras, D., Gwerder, M. (Eds.), 2010. Use of weather and occupancy forecasts for optimal building climate control (opticontrol): two years progress report. Terrestrial Systems Ecology ETH Zurich, Switzerland and Building Technologies Division, Siemens Switzerland Ltd, Zug, Switzerland.

Haldi, F., Robinson, D., 2009. Interactions with window openings by office occupants. Build. Environ. 44 (12), 2378–2395. https://doi.org/10.1016/j.buildenv.2009.03.025.

Haldi, F., Robinson, D., 2010. Adaptive actions on shading devices in response to local visual stimuli. J. Build. Perform. Simul. 3 (2), 135–153. https://doi.org/10.1080/19401490903580759.

Hazyuk, I., Ghiaus, C., Penhouet, D., 2014. Model predictive control of thermal comfort as a benchmark for controller performance. Autom. Constr. 43, 98–109. https://doi.org/10.1016/j.autcon.2014.03.016.

Iacovella, S., Vingerhoets, P., Deconinck, G., Honeth, N., Nordstrom, L., 2016. Multi-agent platform for grid and communication impact analysis of rapidly deployed demand response algorithms. In: 2016 IEEE International Energy Conference (ENERGYCON), pp. 1–6.

International Energy Agency, 2014. Technology roadmap: Energy storage. IEA. Available from: https://www.iea.org/publications/freepublications/publication/technology-roadmap-energy-storage-.html.

Intelligent Energy Europe—European Commission, 2018. Definition of grid-parity for photovoltaics and development of measures to accompany PV applications to the grid parity and beyond. Available from: https://energy/intelligent/projects/en/projects/pv-parity.

International Energy Agency, 2013. Transition to Sustainable Buildings—Strategies and Opportunities to 2050. International Energy Agency, https://doi.org/10.1787/9789264202955-en.

Jaboob, S., 2015. Stochastic modelling of occupants' activities and related behaviours (Ph.D. thesis). University of Nottingham.

Killian, M., Kozek, M., 2016. Ten questions concerning model predictive control for energy efficient buildings. Build. Environ. 105, 403–412. ISSN 0360-1323. https://doi.org/10.1016/j.buildenv.2016.05.034.

Long, Y., Liu, S., Xie, L., Johansson, K.H., 2014. A scenario-based distributed stochastic MPC for building temperature regulation. In: 2014 IEEE International Conference on Automation Science and Engineering (CASE), pp. 1091–1096.

Luthander, R., Widén, J., Nilsson, D., Palm, J., 2015. Photovoltaic self-consumption in buildings: a review. Appl. Energy 142, 80–94. https://doi.org/10.1016/j.apenergy.2014.12.028.

Ma, J., Qin, J., Salsbury, T., Xu, P., 2012a. Demand reduction in building energy systems based on economic model predictive control. Chem. Eng. Sci. 67 (1), 92–100. https://doi.org/10.1016/j.ces.2011.07.052.

Ma, Y., Kelman, A., Daly, A., Borrelli, F., 2012b. Predictive control for energy efficient buildings with thermal storage: modeling, simulation and experiments. IEEE Control Syst. 32 (1), 44–64. https://doi.org/10.1109/MCS.2011.2172532.

McArthur, S.D.J., Davidson, E.M., Catterson, V.M., Dimeas, A.L., Hatziargyriou, N.D., Ponci, F., Funabashi, T., 2007. Multi-agent systems for power engineering applications—part I: concepts, approaches, and technical challenges. IEEE Trans. Power Syst. 22 (4), 1743–1752. https://doi.org/10.1109/TPWRS.2007.908471.

Mirinejad, H., Welch, K.C., Spicer, L., 2012. A review of intelligent control techniques in HVAC systems. In: IEEE Energytech, pp. 1–5.

Mohsenian-Rad, A.H., Leon-Garcia, A., 2010. Optimal residential load control with price prediction in real-time electricity pricing environments. IEEE Trans. Smart Grid 1 (2), 120–133. https://doi.org/10.1109/TSG.2010.2055903.

Nolan, S., O'Malley, M., 2015. Challenges and barriers to demand response deployment and evaluation. Appl. Energy 152, 1–10. https://doi.org/10.1016/j.apenergy.2015.04.083.

Nyholm, E., Puranik, S., Mata, E., Odenberger, M., Johnsson, F., 2016. Demand response potential of electrical space heating in Swedish single-family dwellings. Build. Environ. 96, 270–282. https://doi.org/10.1016/j.buildenv.2015.11.019.

O'Connell, N., Pinson, P., Madsen, H., O'Malley, M., 2014. Benefits and challenges of electrical demand response: a critical review. Renew. Sust. Energ. Rev. 39, 686–699. https://doi.org/10.1016/j.rser.2014.07.098.

Oldewurtel, F., Parisio, A., Jones, C.N., Gyalistras, D., Gwerder, M., Stauch, V., Lehmann, B., Morari, M., 2012. Use of model predictive control and weather forecasts for energy efficient building climate control. Energy Build. 45, 15–27. https://doi.org/10.1016/j.enbuild.2011.09.022.

Page, J., Robinson, D., Morel, N., Scartezzini, J.L., 2008. A generalised stochastic model for the simulation of occupant presence. Energy Build. 40 (2), 83–98. https://doi.org/10.1016/j.enbuild.2007.01.018.

Patteeuw, D., Bruninx, K., Arteconi, A., Delarue, E., D'Haeseleer, W., Helsen, L., 2015. Integrated modeling of active demand response with electric heating systems coupled to thermal energy storage systems. Appl. Energy 151, 306–319. https://doi.org/10.1016/j.apenergy.2015.04.014.

Qin, S.J., Badgwell, T.A., 2003. A survey of industrial model predictive control technology. Control. Eng. Pract. 11 (7), 733–764. https://doi.org/10.1016/S0967-0661(02)00186-7.

Reinhart, C.F., 2004. Lightswitch-2002: a model for manual and automated control of electric lighting and blinds. Sol. Energy 77 (1), 15–28. https://doi.org/10.1016/j.solener. 2004.04.003.

Reynders, G., Nuytten, T., Saelens, D., 2013. Potential of structural thermal mass for demand-side management in dwellings. Build. Environ. 64, 187–199. https://doi.org/10.1016/j.buildenv.2013.03.010.

Reynolds, J., Rezgui, Y., Hippolyte, J.L., 2017. Upscaling energy control from building to districts: current limitations and future perspectives. Sust. Cities Soc. 35, 816–829. https://doi.org/10.1016/j.scs.2017.05.012.

Saber, A.Y., Venayagamoorthy, G.K., 2009. Unit commitment with vehicle-to-grid using particle swarm optimization. In: PowerTech, IEEE Bucharest, pp. 1–8.

Salom, J., Widén, J., Candanedo, J., Sartori, I., Voss, K., Marszal, A., 2011. Understanding net zero energy buildings: evaluation of load matching and grid interaction indicators. In: Proceedings of Building Simulation, vol. 6, pp. 2514–2521.

Sancho-Tomás, A., Sumner, M., Robinson, D., 2017a. A generalised model of electrical energy demand from small household appliances. Energy Build. 135, 350–366. https://doi.org/10.1016/j.enbuild.2016.10.044.

Sancho-Tomás, A., Chapman, J., Sumner, M., Robinson, D., 2017b. Extending no-mass: multi-agent stochastic simulation for demand response of residential appliances. In: Proceedings of the 15th IBPSA Conference, pp. 189–198.

Sancho-Tomás, A., Sumner, M., Robinson, D., 2018. Generalisation of multi-agent stochastic simulation architecture to support Demand Response. In preparation.

Shaikh, P.H., Nor, N.B.M., Nallagownden, P., Elamvazuthi, I., Ibrahim, T., 2014. A review on optimized control systems for building energy and comfort management of smart sustainable buildings. Renew. Sust. Energ. Rev. 34, 409–429. https://doi.org/10.1016/j.rser.2014.03.027.

Shi, W., Wong, V.W.S., 2011. Real-time vehicle-to-grid control algorithm under price uncertainty. In: IEEE International Conference on Smart Grid Communications, pp. 261–266.

Siano, P., 2014. Demand response and smart grids—a survey. Renew. Sust. Energ. Rev. 30, 461–478.

Soares, N., Bastos, J., Pereira, L.D., Soares, A., Amaral, A.R., Asadi, E., Rodrigues, E., Lamas, F.B., Monteiro, H., Lopes, M.A.R., Gaspar, A.R., 2017. A review on current advances in the energy and environmental performance of buildings towards a more sustainable built environment. Renew. Sust. Energ. Rev. 77, 845–860. https://doi.org/10.1016/j.rser.2017.04.027.

Stifter, M., Kamphuis, R., Galus, M., Renting, M., Rijneveld, A., Targosz, R., Widergren, S., Nordstrom, L., Brodén, D., Esterl, T., Kaser, S., Koponen, P., Galsworthy, S., Friedl, W., Doolla, S., 2016. IEA DSM task 17—demand flexibility in households and buildings, IEA DSM, http://www.ieadsm.org/wp/files/IEA-DSM-Task-17-Subtask-13-conclusions-recommendations-2016-09-27.pdf.

Sturzenegger, D., Gyalistras, D., Semeraro, V., Morari, M., Smith, R.S., 2014. BRCM Matlab toolbox: model generation for model predictive building control. In: 2014 American Control Conference, pp. 1063–1069.

Sturzenegger, D., Gyalistras, D., Morari, M., Smith, R.S., 2016. Model predictive climate control of a Swiss office building: implementation, results, and cost-benefit analysis. IEEE Trans. Control Syst. Technol. 24 (1), 1–12. ISSN 1063-6536. https://doi.org/10.1109/TCST.2015.2415411.

Sutton, R.S., Barto, A.G., 1998. Reinforcement Learning: An Introduction, vol. 1, 1. MIT Press, Cambridge, MA.

The European Parliament and the Council of the European Union, 2012. Directive 2012/27/EU of the European Parliament and of the Council of 25 October 2012 on Energy Efficiency, Amending Directive 2009/125/EC and 2010/30/EU and Repealing Directives 2004/8/EC and 2006/32/EC. The European Parliament and the Council of the European Union, Brussels, Belgium.

Thieblemont, H., Haghighat, F., Ooka, R., Moreau, A., 2017. Predictive control strategies based on weather forecast in buildings with energy storage system: a review of the state-of-the art. Energy Build. 153, 485–500. https://doi.org/10.1016/j.enbuild.2017.08.010.

US Department of Energy, 2006. Benefits of demand response in electricity markets and recommendations for achieving them: a report to the United States congress pursuant to section 1252 of the Energy Policy Act of 2005 (February 2006). https://www.energy.gov/oe/downloads/benefits-demand-response-electricity-markets-and-recommendations-achieving-them-report.

US Department of Energy–Office of Electricity Delivery and Energy Reliability, 2016. Advanced metering infrastructure and customer systems—results from the smart grid investment grant program. US Department of Energy, Office of Electricity Delivery and Energy Reliability.

Vardakas, J.S., Zorba, N., Verikoukis, C.V., 2015. A survey on demand response programs in smart grids: pricing methods and optimization algorithms. IEEE Commun. Surv. Tutorials 17 (1), 152–178. https://doi.org/10.1109/COMST.2014.2341586.

Vasirani, M., Kota, R., Cavalcante, R.L.G., Ossowski, S., Jennings, N.R., 2013. An agent-based approach to virtual power plants of wind power generators and electric vehicles. IEEE Trans. Smart Grid 4 (3), 1314–1322. https://doi.org/10.1109/TSG.2013.2259270.

Široký, J., Oldewurtel, F., Cigler, J., Prívara, S., 2011. Experimental analysis of model predictive control for an energy efficient building heating system. Appl. Energy 88 (9), 3079–3087. https://doi.org/10.1016/j.apenergy.2011.03.009.

Wang, S., Ma, Z., 2008. Supervisory and optimal control of building HVAC systems: a review. HVAC R Res. 14 (1), 3–32. ISSN 1078-9669. https://doi.org/10.1080/10789669.2008.10390991.

Watkins, C.J.C.H., 1989. Learning From Delayed Rewards (Ph.D. thesis). University of Cambridge.

Wen, Z., O'Neill, D., Maei, H., 2015. Optimal demand response using device-based reinforcement learning. IEEE Trans. Smart Grid 6 (5), 2312–2324. https://doi.org/10.1109/TSG.2015.2396993.

Wetter, M., 2011. Co-simulation of building energy and control systems with the building controls virtual test bed. J. Build. Perform. Simul. 4, 185–203. https://doi.org/10.1080/19401493.2010.518631.

Widén, J., 2014. Improved photovoltaic self-consumption with appliance scheduling in 200 single-family buildings. Appl. Energy 126, 199–212. https://doi.org/10.1016/j.apenergy.2014.04.008.

Wijaya, T.K., Banerjee, D., Ganu, T., Chakraborty, D., Battacharya, S., Papaioannou, T., Seetharam, D.P., Aberer, K., 2013. DRSim: a cyber physical simulator for demand response systems. In: IEEE International Conference on Smart Grid Communications, pp. 217–222.

Wolisz, H., Harb, H., Matthes, P., Streblow, R., Mueller, D., 2013. Dynamic simulation of thermal capacity and charging/discharging performance for sensible heat storage in building thermal mass. In: BS2013, pp. 1–8.

York, D., Kushler, M., 2005. Exploring the relationhsip between demand response and energy efficiency: A review of experience and discussion of key issues. American Council for an Energy Efficient Economy, Report No. U052.

Yu, Z.J., Huang, G., Haghighat, F., Li, H., Zhang, G., 2015. Control strategies for integration of thermal energy storage into buildings: state-of-the-art review. Energy Build. 106, 203–215. https://doi.org/10.1016/j.enbuild.2015.05.038.

Zheng, M., Meinrenken, C.J., Lackner, K.S., 2014. Agent-based model for electricity consumption and storage to evaluate economic viability of tariff arbitrage for residential sector demand response. Appl. Energy 126, 297–306. https://doi.org/10.1016/j.enbuild.2015.05.038.

Zong, Y., Böning, G.M., Santos, R.M., You, S., Hu, J., Han, X., 2017. Challenges of implementing economic model predictive control strategy for buildings interacting with smart energy systems. Appl. Therm. Eng. 114, 1476–1486. https://doi.org/10.1016/j.applthermaleng.2016.11.141.

Multimodal networks and decentralized renewable generation: Network modeling and energy/exergy performance evaluation

5

Mario Potente Prieto, Nina Solomakhina†, Pilar Monsalvete Álvarez de Uribarri‡*
*Technical University of Vienna (TU WIEN), Vienna, Austria, †Siemens Munich, München, Germany, ‡Sustainable Energy Research Center (zafh.net), University of Applied Sciences, Stuttgart, Germany

5.1 Urban energy systems for heating and cooling

Urban energy systems are defined as all those technologies and facilities intended for the energy supply to the habitants of an urban area, whether at neighborhood, district, or whole city scale.

Urban energy systems are usually classified by different methods, depending on for what kind of purpose they are used. For example, simple electric heaters could be used just on an individual apartment level. Gas boilers are used at building floor or building level, supplying hot water to radiators. Combined heat and power (CHP) facilities are basically heat and electricity generators. Water is the heat-transfer fluid circulating through the pipeline from power plant to consumers. Those systems are known as district heating networks and will be the focus of this chapter and Chapter 10, where a real case study of a thermal network located in one of the districts of the city of Vienna will be described and developed.

Urban energy systems thus serve the following purposes:

- Satisfying electricity demand
- Covering heat demand during cold seasons and cooling demand during warm seasons
- Providing domestic hot water

Both this chapter and Chapter 10 will be focused on thermal energy, notwithstanding that the interaction between appliances and electric supply is also covered in Chapter 3 and Sections 4.1 and 4.2. In practice, each scenario is a thermodynamic study of the district heating network in the defined urban area, analyzing behavior and efficiency.

Urban Energy Systems for Low-Carbon Cities. https://doi.org/10.1016/B978-0-12-811553-4.00005-6

5.1.1 Central and decentralized energy systems

The difference is based on the proximity of the energy generator source with regard to the consumption points. In a centralized system, the heat is produced in a specific location, and from there, it is distributed along a wide pipe network that covers the entire urban area, reaching the users. This is the most common variant and followed by most of the actual power grids (see Fig. 5.1, centralized systems).

In the case of decentralized systems, the heat is generated close to the consumption points, so that large pipelines are not required, reducing considerably the fixed construction costs and the subsequent associated heat losses (see Fig. 5.1, decentralized systems).

Decentralized systems are the most suitable for less populated urban areas that are far away from the main city, commonly known as "bedroom cities." In these cases, large distances make not worth the extension of the thermal network. However, it is easier for the electric network to be extended to reach these areas, giving rise to the creation of an individual and localized thermal network, for example, powered by heat pumps (see Fig. 5.1, urban area 2).

There is also the case of maximum decentralization in low population density areas, in which consumers are not only far away from the city but also widely dispersed throughout a large part of the territory. This is the case of small towns and rural areas. Each consumer procures individually his own energy supply, either from fossil or renewable fuels (see Fig. 5.1, urban area 3, urban area 4, and farm).

Another possibility within the decentralized strategy is the existence of several networks serving several zones within the same urban area (see Fig. 5.1, urban area 4); in this case, a part of the city receives electricity from a main CHP plant, receiving in turn the heat coming from a thermal solar plant, while the other part uses the thermal and electric network coming connected to a biomass CHP plant. This decentralized model is also very common in large cities, such as New York (Howard et al., 2014).

Finally, there are hybrids combining both strategies. In this book, we analyze a Swiss case study of Les Cherpines as a newly created residential neighborhood, in which two main possibilities are defined; on the one hand, in a centralized strategy,

Central and decentralized energy systems

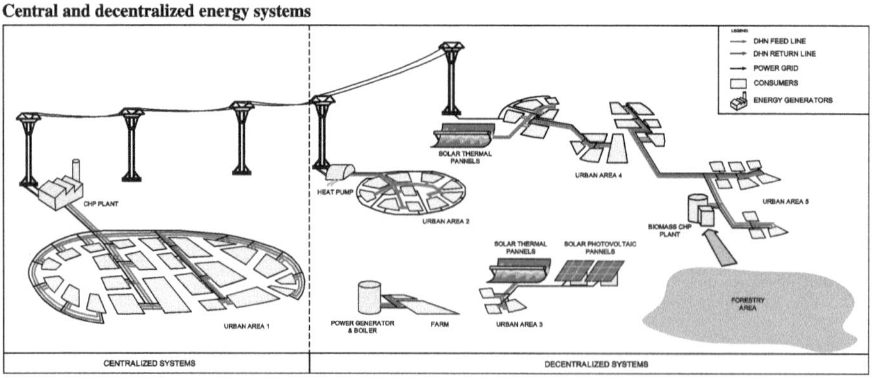

Fig. 5.1 Explanatory diagram distinguishing between centralized and decentralized systems.

the district will be connected to the general district heating network of Geneva, and on the other hand, it has its own network utilizing residual heat coming from the industrial area located nearby. The task of urban planners is to study the different possibilities and decide the best option for each case.

5.1.2 Main urban energy systems

A brief description about the main energy systems considered in this work is explained below:

5.1.2.1 Cogeneration

Cogeneration is the process in which the internal energy of a given fuel is simulta neously transformed into useful thermal energy and power.

In the conventional generation of electricity through the operation of thermal machines, a part of the consumed heat is unrecoverable released to the external environment, obeying the second principle of thermodynamics (Kittel and Kroemer, 1980). In thermodynamic cycles, which are optimized for the maximal production of power, the necessary released heat flux to the environment due to the second law of thermodynamics is released at the lowest possible temperature, limited by the temperature of the environment.

In the CHP concept, the heat is released at a higher temperature, so that it can be used for heating. In this case, less power can be produced, but the heat is not waste heat anymore. Cogeneration represents an evolution compared with classic thermal plants, which were based solely on power generation.

Cogeneration has a wide application in the chemical, paper or food, mining, or ceramics industry and serves in general any heating or cooling consumer.

Advantages of cogeneration are as follows:

- Saving of primary energy: fuel is converted in a more efficient way.
- Reducing fuel consumption and generating power. Aids and subsidized rates are also a possibility in several countries.
- Environmentally friendly: reducing CO_2, NOx, and SO_2 emissions: reducing fuel consumption and allowing the use of other less pollutant sources such as natural gas: and even introducing renewable energies such as biomass.
- Greater independence and safety in the power supply.
- Distributed electricity generation:
  On-site production = investment savings in transportation and distribution lines.
  Decrease of losses in transport and distribution lines.

Disadvantages of cogeneration are as follows:

- High investment and maintenance needed: expensive infrastructures
- Increase of local pollution
- Cogeneration electricity pricing complexity

Cogeneration facilities can be used in two ways:

- As a source of electric and/or thermal energy for internal use of the plant
- As a source of electricity generation that is directly sent to the network for sale

The basic technology of cogeneration includes the following basic elements:

- Thermal engine
- Electric generator
- Heat recovery
- Control system

5.1.3 Types of cogeneration

Steam power plants are primarily designed to change chemical energy into electric energy. Due to the second law of thermodynamics, this conversion is limited. During the conversion process, heat is generated that has to be transferred to the environment.

The basic thermodynamic cycle of a steam power plant is the Rankine cycle, in which the thermal fluid is water circulating through the circuit, changing its phase as it passes through the different stages. The high-pressure steam is generated from water, generally using fossil fuels, such as natural gas, fuel oil, or even gasified coal. Renewable energy variants are, for example, biomass plants.

The superheated steam generated at high pressure in the boiler (Fig. 5.3, point 1) is adiabatically expanded in the turbine by changing its pressure from P_{BOILER} to the $P_{CONDENSER}$, generating power. The result of this process is a saturated steam at the outlet of the turbine (Fig. 5.3, point 2) with all its latent heat available to be transferred at the condenser, obtaining saturated liquid water (Fig. 5.3, point 3). Then, the liquid water is compressed by a hydraulic pump, implying power consumption. The fluid pressure is increased to the P_{BOILER} value (Fig. 5.3, point 4). In total, there will be electric power generation in the turbine (Fig. 5.3, steps 1–2) and heat generation in the condenser (Fig. 5.3, steps 2–3).

Some general features of steam power plants are the following:

- The fuel does not need to be of high quality. It can be of a fossil source (e.g., coal) or of a renewable source (e.g., biomass).
- The heat release (due to the second law of thermodynamics) happens at constant temperature. That can be at the temperature of the environment (to produce a maximum amount of electricity) or at a higher temperature for running processes, for example, drying or heating. In this case, less electric energy can be produced.
- Typical steam inlet values: P: 40–250 bar and T: 400–500°C.
- More frequent in conventional generators.
- High powers are possible (>10 MW).
- Large facilities—Investments raised but with low ratio €/kW.
- Long lifetime (>20 years).
- High degree of reliability of electric supply.

Classic thermal power plants are optimized to produce as much electric energy as possible. This requires the lowest possible heat release temperature in the thermodynamic cycle. Most efficient (at the lowest temperature) is cooling with water from a well, in larger units from a river. If not enough water is available (the temperature of the cooling water would rise too high), cooling towers are used. Cooling towers work the better, the more water can be evaporated. If absolutely no water for cooling is available, dry recooling units are in operation.

While the condensing temperature in the condenser is a few degrees above the river temperature in the case of river cooling, in dry recooling units, the condensing temperature is a several degrees above the air temperature in the environment of the power plant, which causes significant less thermal efficiencies.

If thermal energy (heat) is needed in the nearer surroundings of the power plant as well, the condensation can be done deliberately at higher temperatures. In these cases, the production of electricity is reduced, but the heat is not wasted anymore; it can be used for thermal processes or as source for a district heating network.

Operating in this way, the cooling tower is removed and substituted by the heat consumers in the city (Fig. 5.4, point 5).

This operation follows the evolution of classic thermal plants (Fig. 5.2) leading to second-generation district heating networks.

5.1.4 Gas turbines and its use for district heating networks

Similar to steam power plants, gas turbine power plants are designed to change chemical energy into electric energy as well. The conversion is limited too, which means thermal energy (heat) has to be transferred to the environment.

The basic thermodynamic cycle of a gas turbine power station is the Brayton cycle, using a gas as the working fluid. The advantage of this cycle is that the mean temperature of energy feed to the working fluid is much higher compared with the Rankine cycle, and therefore, this cycle has the potential for higher thermal efficiencies. On the

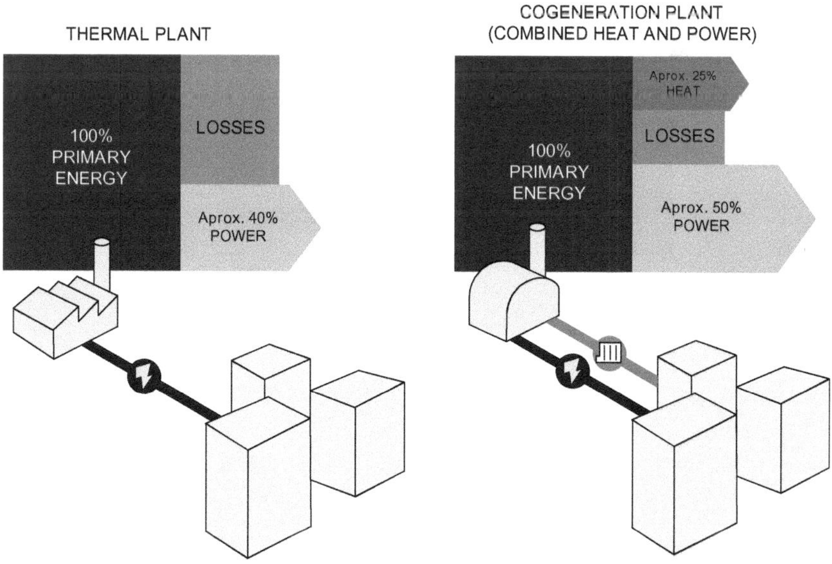

Fig. 5.2 Evolution from thermal to cogeneration technology, based on Höckner (2014).

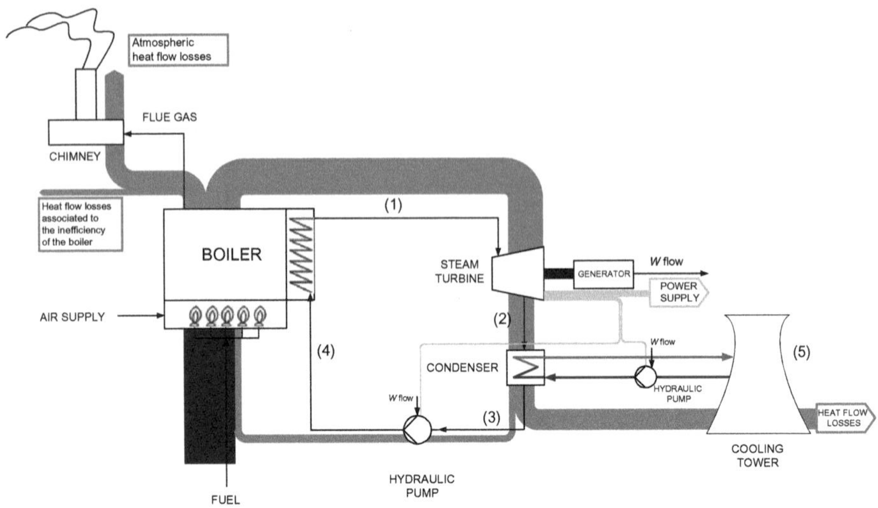

Fig. 5.3 Steam power plant with classic Rankine cycle.

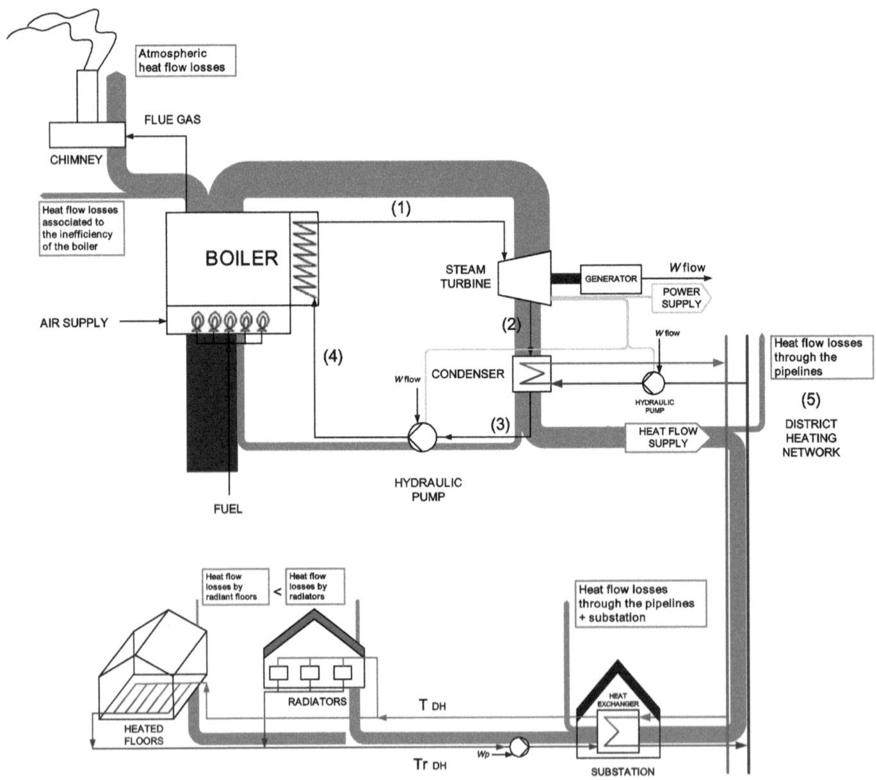

Fig. 5.4 Steam turbine power plant adapted for district heating.

other side, heat release of the cycle is at higher temperatures compared with the Rankine cycle, which would reduce the high thermal efficiency if the high-temperature exhaust gas would not be used.

In combined cycles, the exhaust gases of a gas turbine unit are the energy source for a steam power plant. These combined cycles can reach thermal efficiencies for the production of electric energy up to 60%.

If thermal energy (heat) is needed in the nearer surroundings of the gas turbine power unit as well, the high-temperature exhaust gases of the gas turbine can be used for thermal processes or as source for a district heating network, too. One disadvantage of the gas turbine cycle is the necessity of high-quality fuel. These plants can just operate with gas or oil. The fuel can be of fossil nature but regenerative (biogas or solar produced hydrogen) as well.

As shown in Fig. 5.5, the open Brayton cycle is formed by a gas turbine whose shaft is solidly attached to a compressor. The turbine generates mechanical work for two uses: one part serves for the movement of the compressor, and the other is converted into power.

The main features of gas turbine units are as follows:

- The most used in medium to large cogeneration
- Fuel: mainly natural gas or similar
- Possibility of direct use of the flue gas in a heat recovery boiler
- Possibility of postcombustion before using the flue gas in the recovery boiler
- Average power: 50–250 MW
- Possibility of postcombustion: noise and emission problems (NOx, etc.)
- Bad behavior in partial load or stop conditions

The flow of fresh air at ambient temperature (Fig. 5.5, point 1) is adiabatically compressed in the compressor, giving the required combustion pressure to the air and increasing its temperature, until it reaches the combustion chamber (Fig. 5.5, point 2).

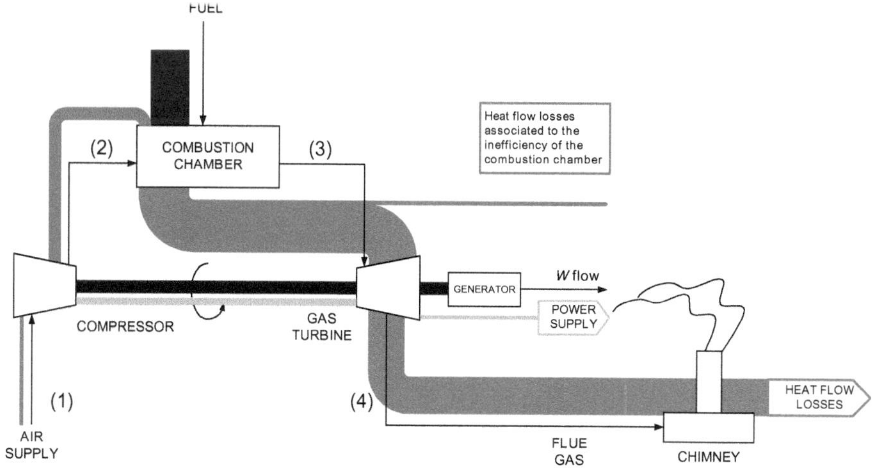

Fig. 5.5 Open gas turbine unit based on the Brayton cycle.

In this enclosure, the fuel is burned at constant pressure producing high-temperature flue gases (Fig. 5.5, point 3). Those are conducted to the turbine, where they are adiabatically expanded, generating power. Since it is an open cycle, the flue gases are evacuated at constant pressure from the turbine and are directly expelled to the atmosphere without any recirculation.

5.1.5 Combined heat and power cycle and its use for district heating networks

Two thermodynamic cycles are joined together in such a way that the heat generated by one of the cycles is partially or totally used as heat that is absorbed by the other cycle. The most common way is coupling a Brayton cycle with a Rankine cycle, as shown in Fig. 5.6.

Both power cycles are coupled in such a way that the heat absorption occurs at high temperature in the gas turbine, and the heat transfer occurs at low temperature in the steam cycle. Therefore, the electric efficiency is higher than having a single cycle in operation (may exceed 55%, considering that a gas turbine rarely exceeds 40%, and the most common values are around 35%).

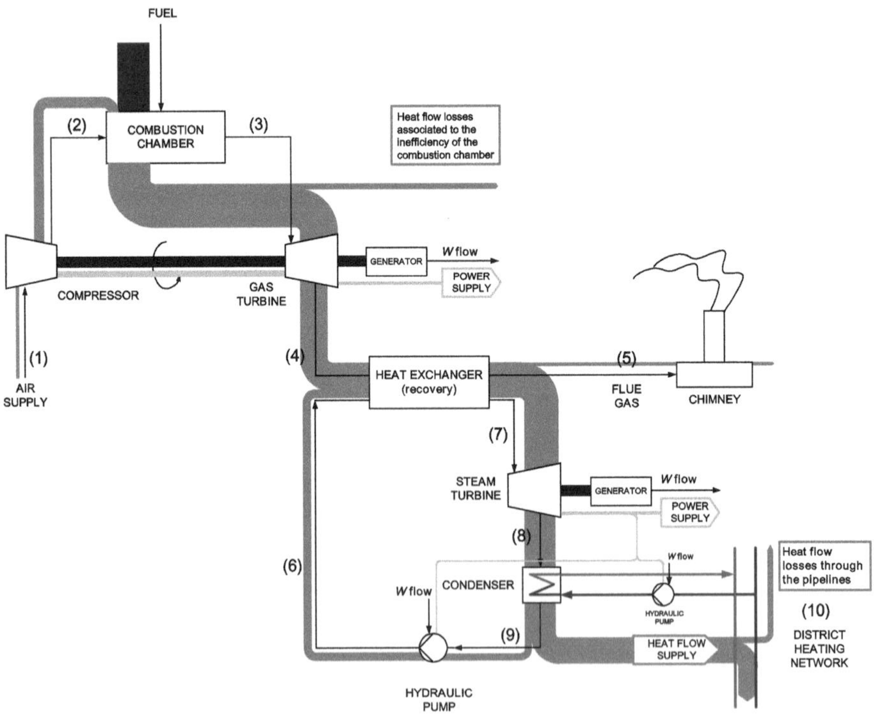

Fig. 5.6 Typical combined heat and power cycle.

The crucial component is the recovery boiler, in which both cycles are connected. In this device, hot exhaust gases and high-pressure water of the Rankine cycle are exchanging heat (Fig. 5.6, points 4–5), that is to say, the heat from the exhaust gases is totally utilized, decreasing their temperature to the lowest possible value. The cooled gases are discharged to the atmosphere through the chimney (Fig. 5.6, point 5).

5.1.6 Internal combustion engine or cogeneration engine

In this kind of system, the thermal engine behaves like any other standard engine receiving fuel, producing mechanical energy in the transmission and hence electric energy in the alternator (Fig. 5.7, point 3). It is a highly versatile device and relatively easy to install. It is able to use different high quality of fuels, both at gaseous and liquid phase. Low-quality fuels, such as coal, waste, or biomass, cannot be used. The heat is mainly extracted from the flue gases at two different points: inside the engine (Fig. 5.7, point 1) and in the exhaust gas outlet points (Fig. 5.7, point 2).

These are small, functional, and compact systems that can be used in areas with low electricity demand, and even as a complement to certain locations where a rapid conversion from fuel into power is needed, such as hospitals, water treatment plants, and biogas plants in agricultural and livestock areas.

5.1.7 Heat pumps

The heat pump is one of the practical applications of the Carnot cycle, whereby it is possible to extract heat from the ambient (low-temperature heat source) and supply it to the environment that is to be heated (heat sink). This heat transfer is done in the opposite direction to that that occurs spontaneously in nature. According to the second principle of thermodynamics, this can be only possible through work consumption (electric power).

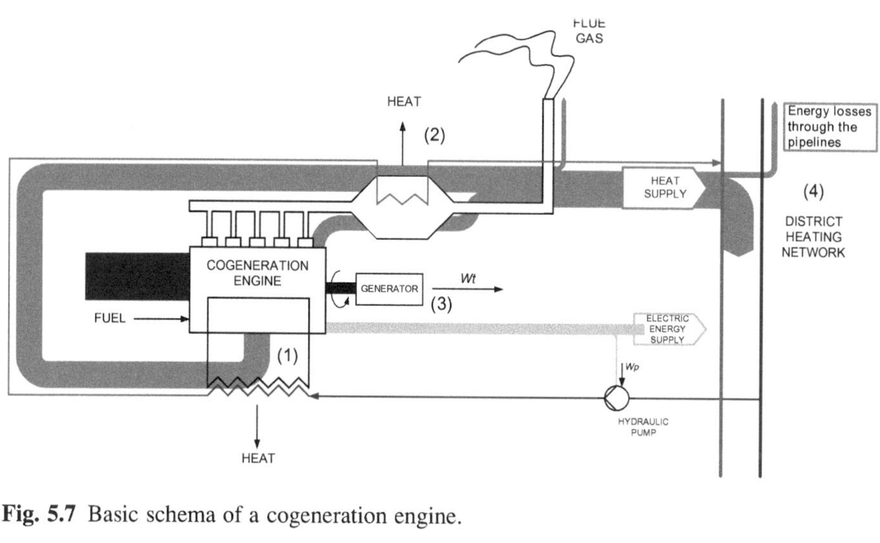

Fig. 5.7 Basic schema of a cogeneration engine.

This heat transfer is carried out by means of a refrigerant flow undergoing pressure, temperature, and phase transformations (liquid and gas) and that absorbs heat from the source and transfers it to the heat sink. The first principle of thermodynamics is therefore fulfilled:

$$\dot{Q}_H = \dot{W} + \dot{Q}_C \tag{5.1}$$

Consequently, the use of a heat pump will always be more advantageous than using simple heating using Joule-effect electric resistances, since this kind of systems produces heat from the work carried out by the electric current. An interesting application is the reversible heat pump, consisting of the Carnot cycle adapted to the needs of each season. By a four-way valve, it is possible to change the direction of the refrigerant flow, converting the heat pump into a cooling unit (air conditioning) in the summer months (Fig. 5.8).

The direct and best-known application of this type of systems is the domestic air conditioning-heating appliances, in which the heat extracted from the heat source is the indoor air of the house to be cooled, while in the winter months, the device switches to heat pump mode, so that the heat source is ambient air, from which the heat is extracted. The efficiency of the heat pumps is described by the COP (coefficient of performance), which measures the performance of the Carnot machine in both modes, either as a refrigerating machine or as a heat pump. In cooling mode, it is calculated as the quotient of the heat extracted from the source and the electric work that must be provided.

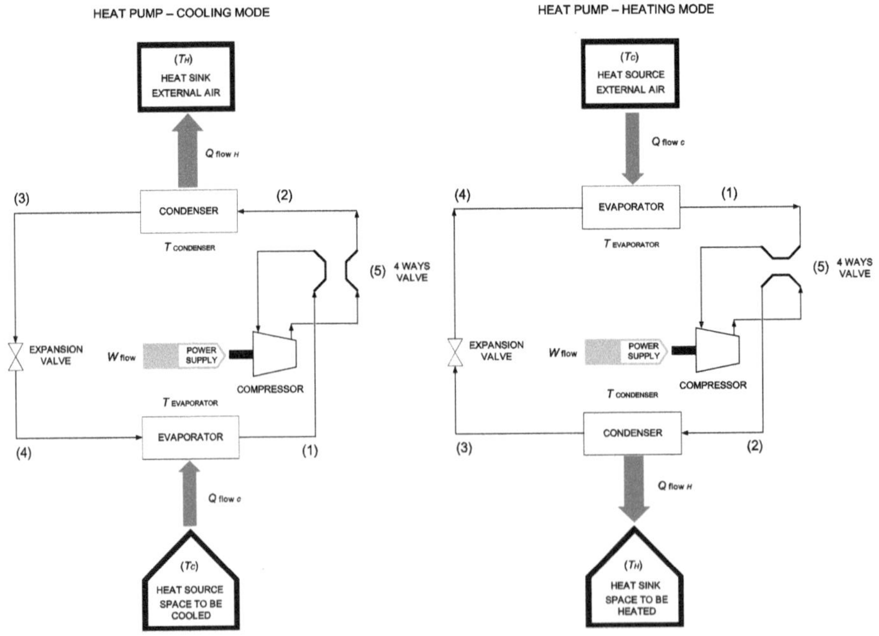

Fig. 5.8 Heat pump operation and cooling-heating modes.

If considered as an ideal process, the equations are the following:

$$COP_{\text{CARNOT COOLING MACHINE}} = \frac{Q_C}{W} = \frac{T_C}{T_H - T_C} \tag{5.2}$$

Therefore, in order to increase the efficiency of a refrigerating machine, it is important that the difference between the two temperature levels ($T_H - T_C$) is as low as possible. This means that the efficiency of the refrigerating machine decreases when low cooling temperatures are required (T_C) or when the heat sink temperatures such as ambient air are high (T_H).

$$COP_{\text{CARNOT HEAT PUMP}} = \frac{Q_H}{W} = \frac{T_H}{T_H - T_C} \tag{5.3}$$

Also, for the heat pump, the efficiency increases when the temperature difference between the heat source and sink (here the useful heat) is low, that is, when the ambient temperature is relatively high. Considering the irreversibilities, the performance of a real heat pump is obtained from the following expression:

$$COP_{\text{HEAT PUMP}} = COP_{\text{CARNOT HEAT PUMP}} \times \gamma \tag{5.4}$$

γ is the coefficient associated with the irreversibilities of the heat pump, which are typically around 0.55 (Cheng and Cheng, 1995; Chen, 1994).

The use of heat pumps in combination with district heating networks involves the use of accessible heat sources such as water from rivers, lakes, aquifers, or the ambient air itself. Heat pumps are important urban energy systems, well suited for rural areas and low population densities (see Fig. 5.1), where the installation of larger thermal power plants is not easily possible. The fact of combining heat pumps with CHP plants through district heating networks entails the introduction of a new degree of freedom that basically allows changing the heat to electricity ratio generated in the system. This is the path toward low-temperature thermal networks.

Hence, the heat needs of the users can be satisfied by two different ways:

1. Heat from thermal power plants
2. Electricity from various sources, not only from cogeneration but also from other renewable energy sources such as solar photovoltaic panels coupled with heat pumps (Fig. 5.9).

In essence, this means diversifying the energy sources that feed the district heating network, which will lead to some advantages and disadvantages that will be analyzed throughout Section 10.1.4.

5.1.8 Deep geothermal energy

Geothermal energy is a type of renewable energy based on obtaining natural heat present inside the earth, transmitted by conduction and/or convection through the various rock and natural groundwater reservoirs called aquifers. The temperature increase with depth is about 30–32 K/km (Agemar et al., 2014).

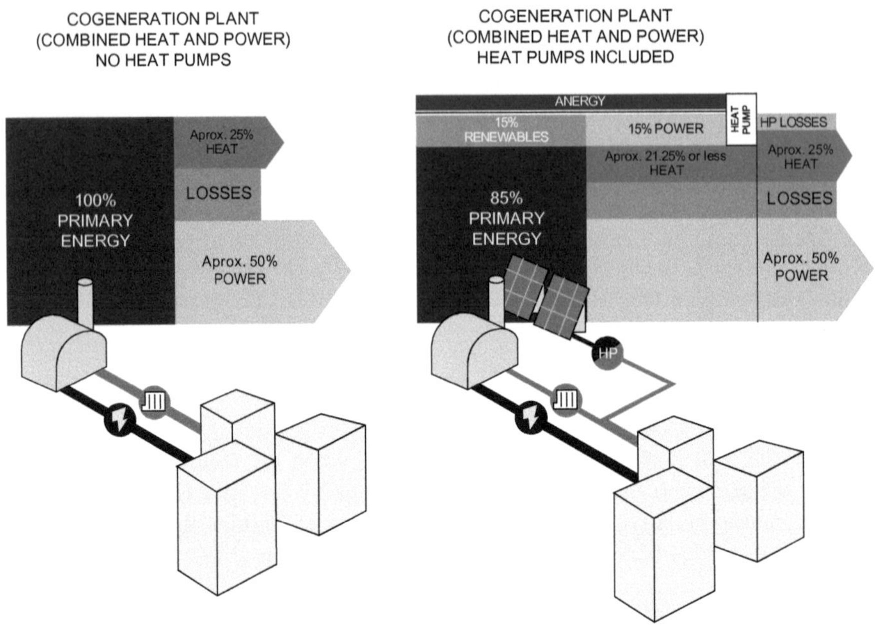

Fig. 5.9 Heat pumps connected to CHP plants and district heating systems (based on Höckner, 2014).

There are mainly three types of geothermal deposits:

1. Aquifers: there are mainly water-bearing sediments of at least 20 m thickness, considered as geothermal reservoirs. To be usable, they must be located deeper than 400 m at temperatures above 20°C. In addition, aquifers must be refillable, which means that the total amount of water taken from them must never be greater than the returned one. In another case, the reservoir could be drained. In order to utilize geothermal water in combination with district heating, it is important to dig two or more geothermal wells. At least one of them must directly extract the hot water; used returning water is reinjected, recharging the aquifer (Fig. 5.10). The water flow rate is between 35 and 150 L/s. If geothermal heat is used for CHP, temperatures of at least 110°C must be reached, in order to ensure economic profitability (Agemar et al., 2014).
2. Dry deposits: also called enhanced geothermal systems. No aquifers are necessary for those kinds of installations, but high-pressure water is directly injected from the surface to the deep hot rock through one or several vertical pipes. After the water is heated, it is extracted by hydraulic pumps (Fig. 5.11).
3. Geysers: it is a special type of thermal springs that emit from time to time a column of high-pressure hot water and steam to the atmosphere. It happens because of the sudden evaporation of accumulated water inside the cavities of the earth due to contact with volcanic magma. It is a natural and rare phenomenon, occurs in very localized areas around the planet, not widespread and quite intermittent, so it is very difficult to use it as a stable source of energy.

Both Figs. 5.10 and 5.11 show the various possibilities offered by geothermal energy, from CHP installations, through direct heat exchange or by the use of a heat pump.

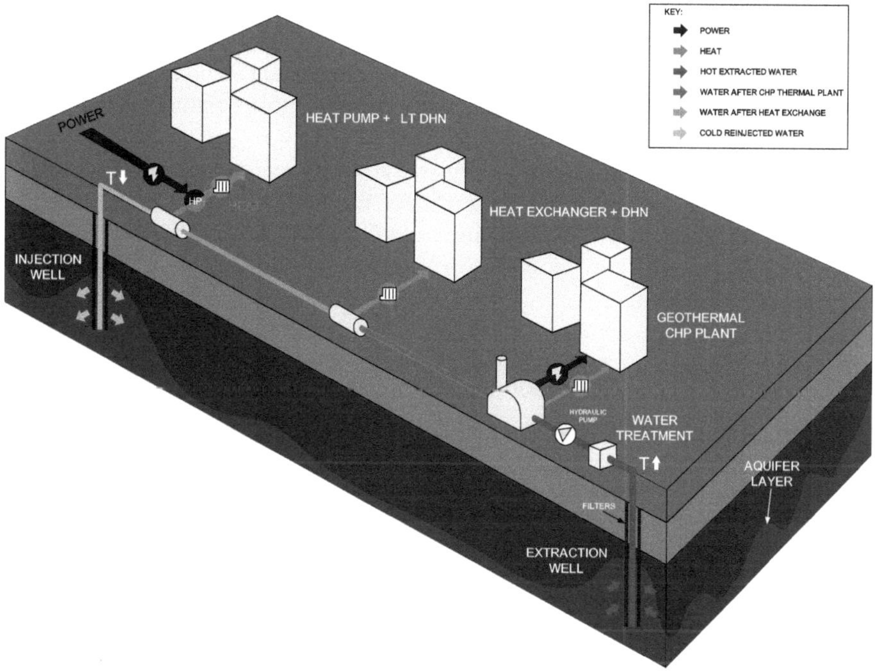

Fig. 5.10 Operation facility based on aquifers as geothermal reservoirs.

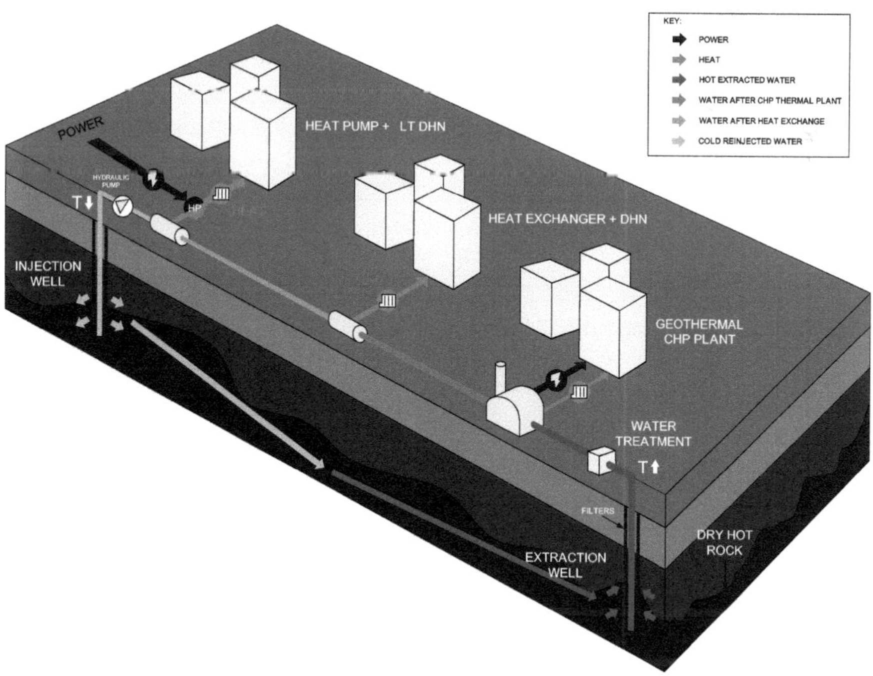

Fig. 5.11 Operation facility based on the hot rock layer as geothermal reservoir.

Depending on the temperature of the water extracted from the underground, the appropriate option is chosen.

In all these cases, the depth of excavation is a fundamental economic and operational factor, ranging from 3 to 12 km, although the most common values differ between 4 and 5 km.

Among the most important current projects are the direct-use facilities applied to district heating located in the Bavarian part of the Molasse Basin (Agemar et al., 2014); the Geysers geothermal fields in North Carolina (the United States) (Khan, 2010); and the exploitations located in Iceland (Geothermal Plant of Nesjavellin), Italy, and New Zealand.

5.1.9 Solar thermal energy

Solar thermal energy is the utilization of energy obtained from shortwave solar irradiance, transferred to a heat carrier medium. It also offers large-scale possibilities, such as high-temperature hot water for steam production and power generation. Interesting applications for urban energy planning include installations at building level or at a possible connection with district heating networks.

There are three temperature levels through which this type of energy can be used; for each of them, a different technology is put into operation:

1. Low-temperature collectors with temperature levels around 65°C. Here, all individual hot water systems for buildings and houses are included.
2. Medium-temperature collectors: they concentrate solar radiation supplying useful heat at higher temperature, usually between 100°C and 300°C.
3. High-temperature collectors: temperatures are higher than 500°C, and steam production is possible for power generation.

Solar thermal energy is versatile using water as a heat-transfer fluid or producing steam and also by the ease of storage and use of hot water and its possible connection with thermal networks. These hypotheses will be developed and studied in Sections 10.1.5 and 10.1.6.

As shown in Fig. 5.12, the components of this type of system are the following:

(1) The solar collector is connected to the heat exchange system through the primary circuit, through which the heat carrier liquid is circulating.
(2) Heat exchange system: performs the transfer of heat from heat-transfer fluid of the primary circuit to the secondary system, which can be potable water or the heating circuit fluid. In general, a plate heat exchanger is used because it is a compact, clean, and easy-maintenance device.
(3) Storage system.
(4) Backup system: complements the solar contribution by supplying the additional energy needed to cover the expected consumption. Possibilities are simple boilers, heat pumps, or connections to district heating networks.
(5) Consumption system: consisting of a set of equipments and components through which domestic hot water or heating demand is satisfied, including pipelines, faucets, and valves.
(6) Inlet of fresh water.

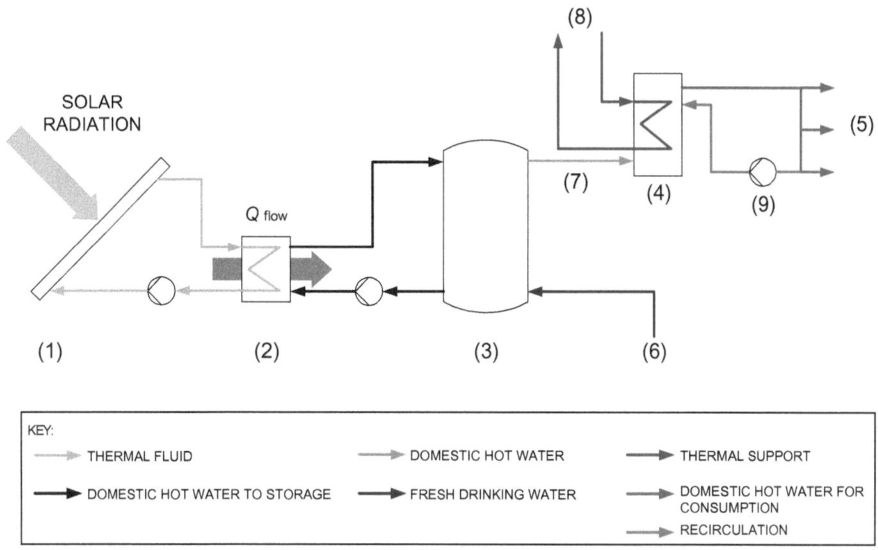

Fig. 5.12 Basic diagram of a simple solar thermal system for domestic hot water (based on Energy foundation (FENERCOM) (2016)).

(7) Domestic hot water for consumption.
(8) Possible water or heat inlet coming from lateral systems.
(9) Recirculation system.

5.2 District heating and cooling networks

As already indicated above, a district heating network is defined as a type of thermal network whose purpose is to send heat to the consumers for heating and domestic hot water preparation. There are different types of consumers, such as residential buildings; public buildings; office blocks; individual homes; and buildings with special requirements such as hospitals, nursing homes, and industries. Buildings that fulfill similar characteristics can be grouped forming archetypes. For a classification based on the energy demand and considering other topographic, technological, and economic factors, the appropriate design of the thermal network is carried out.

The basic idea of district heating is the establishment of a stable, manageable, and adjustable thermodynamic system able to work as a sink and at the same time as a heat source. The network performs as a sink if it receives heat from the thermal plants and functions as a source when the heat is transferred to the consumption points. It is also possible that consumers are heat producers and therefore energy suppliers to the network, in which case they are called "prosumers," increasing the versatility and complexity of the networks.

Traditionally in Western Europe, individual heating systems are used, in which fossil fuels such as natural gas or oil are combusted in boilers, raising the temperature of radiators' water. These types of individualized systems, besides of needing large gas supply networks throughout the city, are not very efficient from the exergy point of view and entail high fuel consumption. In addition, from the municipal/regional point of view, urban energy planning becomes very difficult if each consumer chooses their own source of energy or installation.

The main advantages of networks in comparison with individual systems are the following:

1. Greater energy efficiency: When thermodynamic cycles are operating in large facilities such as thermal plants or combined cycles (see Section 5.1), they use fuel more efficiently.
2. Safety: Hot water networks in a liquid or vapor phase, whose pressure and temperature values are constantly measured and controlled, are safer than natural gas or other potentially flammable fuels.
3. Versatility: A thermal network can provide or take up heat and connect various systems such as heat exchangers, heat pumps, or thermal storage systems

Disadvantages of district heating networks are the following:

1. Significant heat losses
2. Highly unexplored integration potential of different available energy sources (e.g. renewables and waste heat) into the network
3. High installation costs

Engineers, designers, and urban planners were encouraged by these initial deficiencies to improve the existing facilities (insulation, hydraulic systems, etc.) and to work on new urban energy systems (heat pumps and thermal storage) that can be incorporated into these networks.

There are five stages in the evolution of district heating networks.

The first generation of district heating networks proliferated between 1880 and 1930 in the United States and Europe. These systems were based on high-temperature pressurized steam as a heat-transfer fluid, which offers certain advantages from the hydraulic point of view. The steam is generated in thermal power stations distributed along the city and circulates through ducts reinforced with concrete to withstand possible sudden pressure changes, delivering the latent heat by condensation in the radiators. The city of New York is the most representative case of this type.

The most important problem related to this type of networks is the very corrosive condensate that destroys the pipes producing wears, breakages, and leakages in various segments of the return lines. The second problem is the energy inefficiency of this type of systems when using high temperatures (150–200°C) generating high heat losses. In addition, some cases of accidents and explosions occurred, leading to the gradual replacement of this type of systems (Belson and DePalma, 2007).

The second generation was built during the 20th century, between 1930 and 1970, with the Soviet Union and the Eastern European countries as the main actors. In this case, the heat-transfer fluid is pressurized liquid water, with feed temperatures around 100°C, that is, without allowing steam formation.

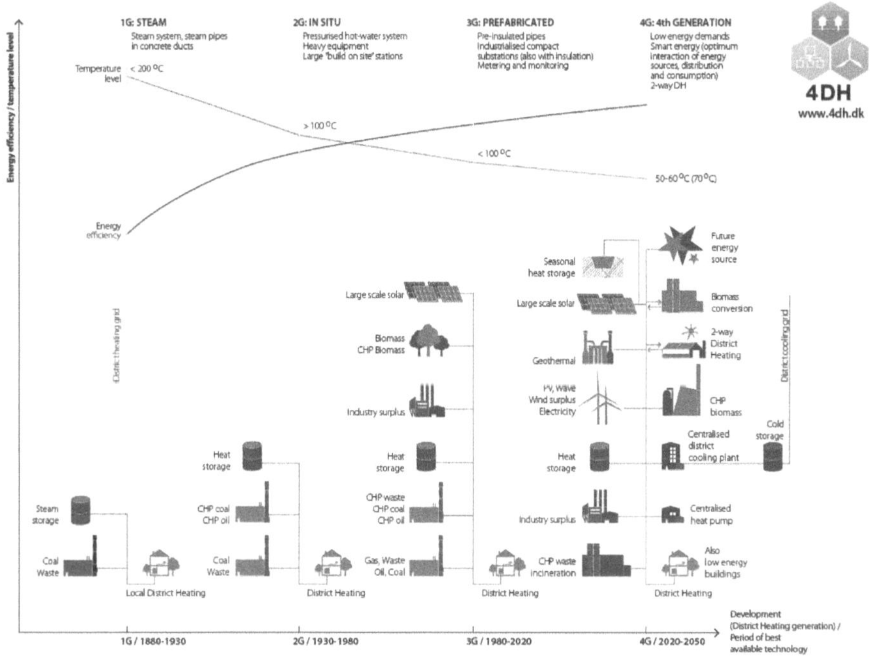

Fig. 5.13 Summary diagram containing the evolution of district heating networks, extracted from Lund et al. (2014).

The water is pumped through insulated pipes and concrete ducts installed under the pavement of the streets. Shell and tube heat exchangers are used, suitable for large volumes and high temperatures, such as in district substations or groups of buildings. Simple control loops were used for this system, without taking into account the variable energetic demand that changes depending on the season of the year (Fig. 5.13).

The third generation of networks emerged in the decade of the 1960–70s in Western and Central Europe with multiple district heating projects. Scandinavian countries carried out and influenced decisively the development of this concept. The main feature in this third generation is the introduction of prefabricated construction elements used in a modular way. Following this procedure, the installation and expansion of networks to new customers became relatively fast and easy, reducing considerably fixed assets and installation costs. There is also a significant progress in heat exchange technology, introducing plate changers, which are more compact, easy to maintain, and suitable for moderate temperatures and smaller flow rates. Another feature was the introduction of new complementary energy sources that feed the network, such as solid urban waste, biogas, biomass, solar thermal, or geothermal energy. This is the status of district heating in most countries, including Asia (mainly China, Japan, and South Korea), Eastern Europe, Central Europe, and North American countries.

From then on, the further evolution of district heating networks is based on the following:

1. Reduction of DHN water temperature, trying to decrease heat losses as much as possible
2. Reduction of fixed costs and easier maintenance of facilities and installations
3. Incorporation of new alternative energy sources, substituting fossil fuels as much as possible
4. Establishment of new energy politics, gathering all administrative proceedings and technical actions, using networks as the main and central element to work on.

The fourth generation of district heating networks is related to the smart grid concept and summarized in the following scheme (Fig. 5.14):

Following a chronological order, the different phases within the next district heating network evolution are described:

- Institutional framework:

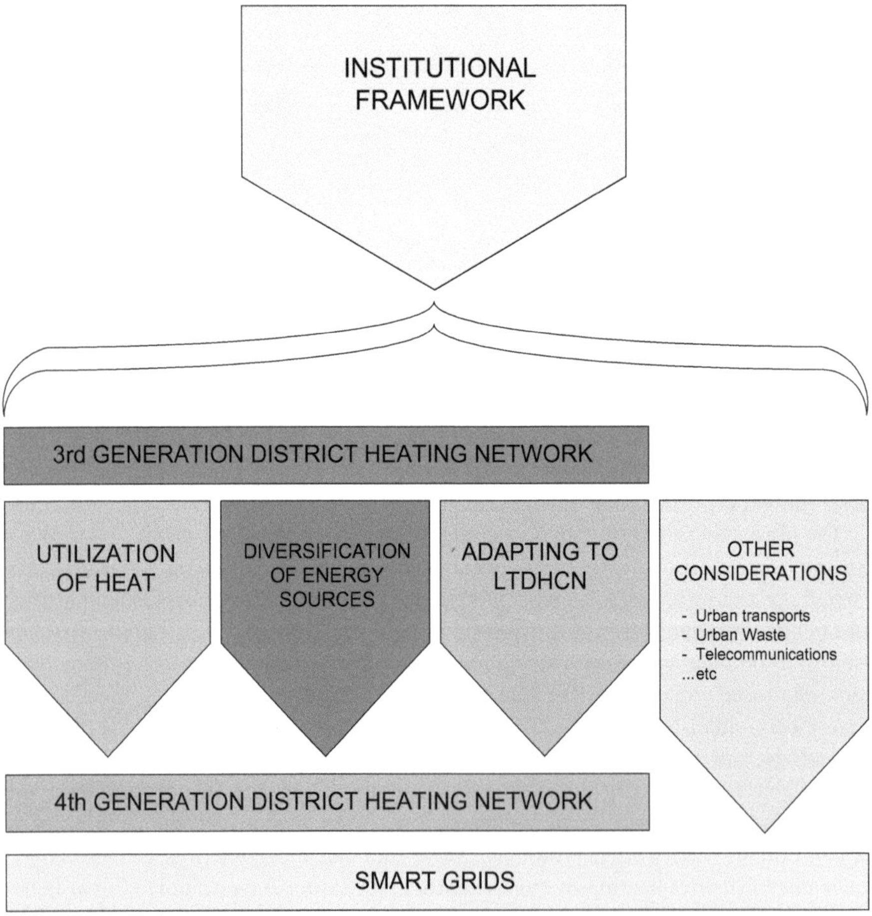

Fig. 5.14 Concept of grid infrastructure development.

A global strategy is required, integrating all actions and regulations concerning urban networks. Not only thermal network are included inside this procedure, covering the following aspects:

1. Suitable planning
2. Promoting green buildings already prepared and designed for smart grids and low-temperature heating networks
3. Encouraging strategic and political actions to motivate and adapt actual facilities into sustainable energy systems
4. Defining new operation and maintenance procedures, increasing efficiency and reducing costs

- Utilization of available heat:

1. Energy classification: Energetic evaluation and qualification of the buildings by estimating heat demand through the different seasons of the year.
2. Definition of archetypes, by creating groups with buildings with the same or just similar characteristics (Fig. 5.15).
3. Definition of temperature levels, defining at least three levels (high, medium, low, and return temperatures) and considering three-pipe technologies. Hence, a temperature hierarchy is defined, in order to provide high-temperature heat to high-demand consumers and medium to low temperature to green or better adapted buildings. This principle is known as an energy cascade and leads to a reorganization of the thermal networks through energy integration tools (Köfinger et al., 2017) (Fig. 5.16).
4. Extension of the new networks to integrate all possible heat sources, even recycling low-medium temperature effluents from industries and urban waste waters (Fig. 5.17).

   As the picture shows, dotted lines represent all possible heat that can be recovered from the different waste effluents generated from the buildings or factories. Of course, the last ones have different temperature levels, depending of what kind of productive process they have. Green buildings are able to use energy from low temperature levels (like water for heat pumps) and send effluents at low or ambient temperature; for example, return DH water from hot water production will be warmer than return DH water from heating.

- Diversification and use of new energy sources. Integration with fourth-generation DHN:

1. Conversion of thermal power plants to CHP plants and combined cycles, introducing new fuels like biomass, biogas, or urban waste (Fig. 5.18).
2. Integration of renewables with district heating networks (Fig. 5.19).

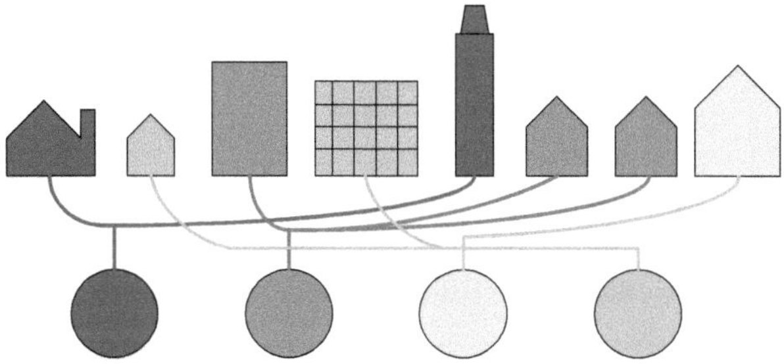

Fig. 5.15 Definition of archetypes.

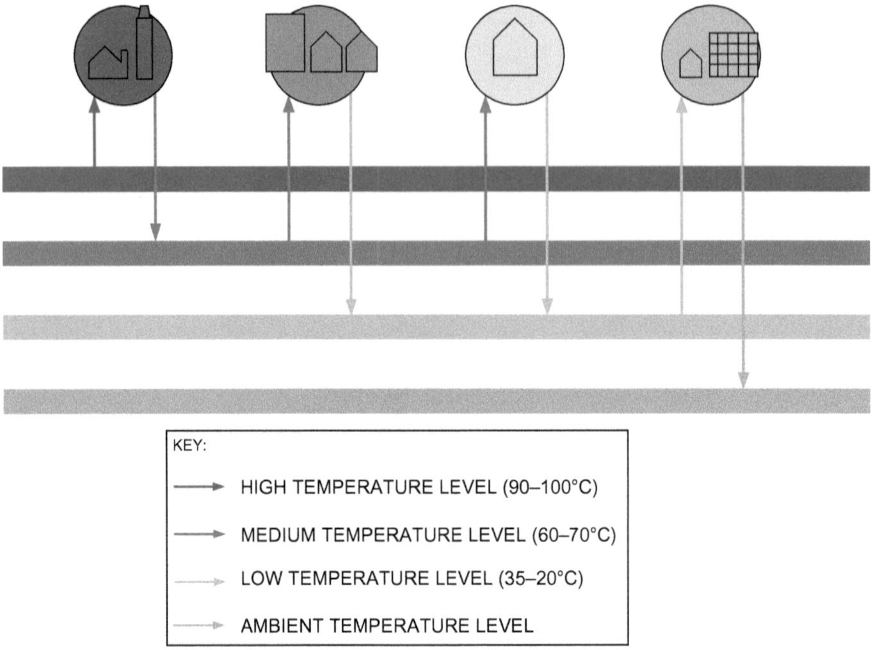

Fig. 5.16 Defining a cascade strategy for the new district heating networks.

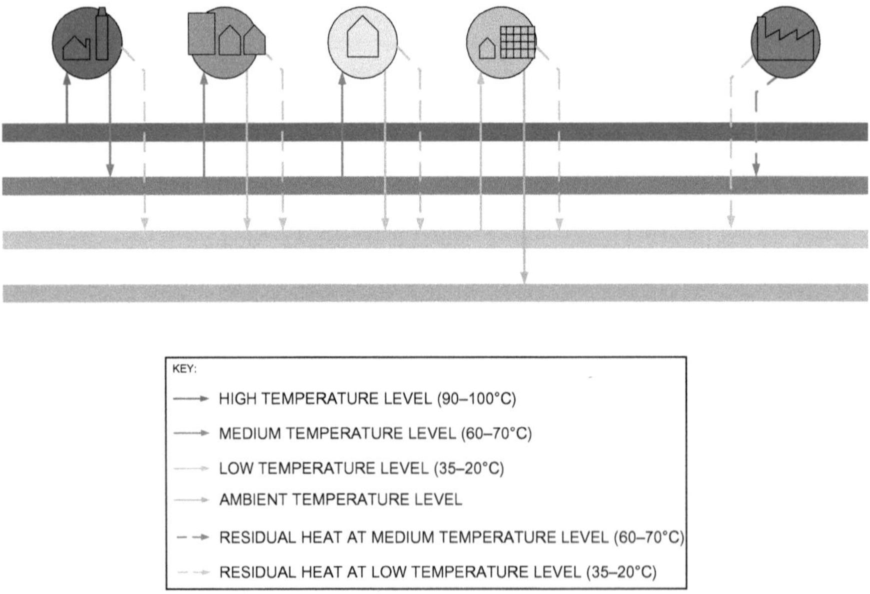

Fig. 5.17 Integration of industrial heat using four temperature levels.

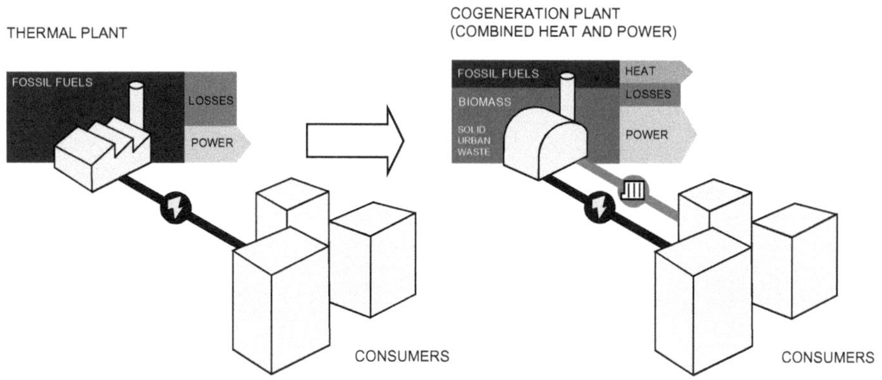

Fig. 5.18 Transition from thermal plants to CHP plants.

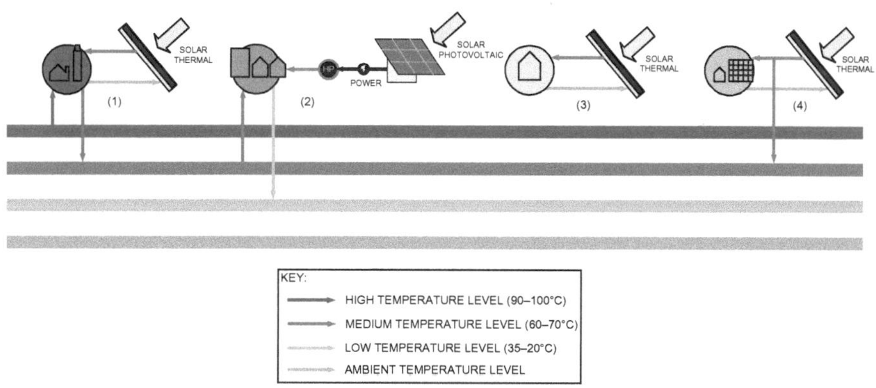

Fig. 5.19 Several ways to combine local renewables with DHN.

(1) Solar thermal in combination with DHN (consumer): For high-energy consumers, solar thermal alone is not expected to be sufficient to cover all thermal needs; so, the building is also receiving support from the thermal network.

(2) Solar photovoltaic (electric renewables in general) in combination with DHN: Via heat pumps, it is possible to use the electric work from renewables to increase locally the temperature of water coming from a low-temperature DHN.

(3) Individual solar thermal, self-sufficient buildings: Buildings with all thermal needs covered by solar thermal energy. No DHN required.

(4) Solar thermal in combination with DHN (prosumer): In this case, the prosumer is able to provide energy surplus to the network.

3. Integration of district heating networks with other grids. For more information, see Section 5.3 on multimodal networks (Fig. 5.20).

- Adaptation to low-temperature district heating networks (LTDHN):

1. Refurbished buildings can be connected to low-temperature networks with heat distribution by floor heating (Fig. 5.21).

2. Installation of localized heat pumps for green buildings (Fig. 5.22).

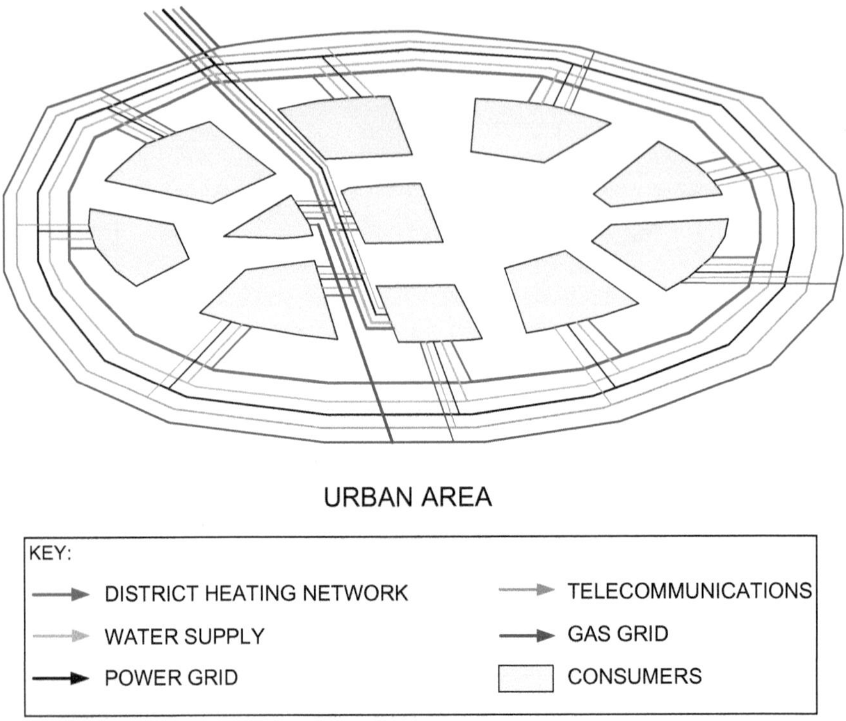

Fig. 5.20 Several grids integrated for the same urban area.

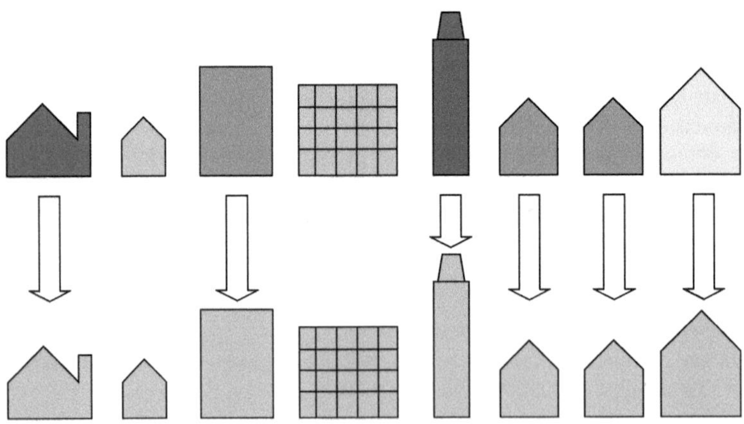

Fig. 5.21 Building refurbishment.

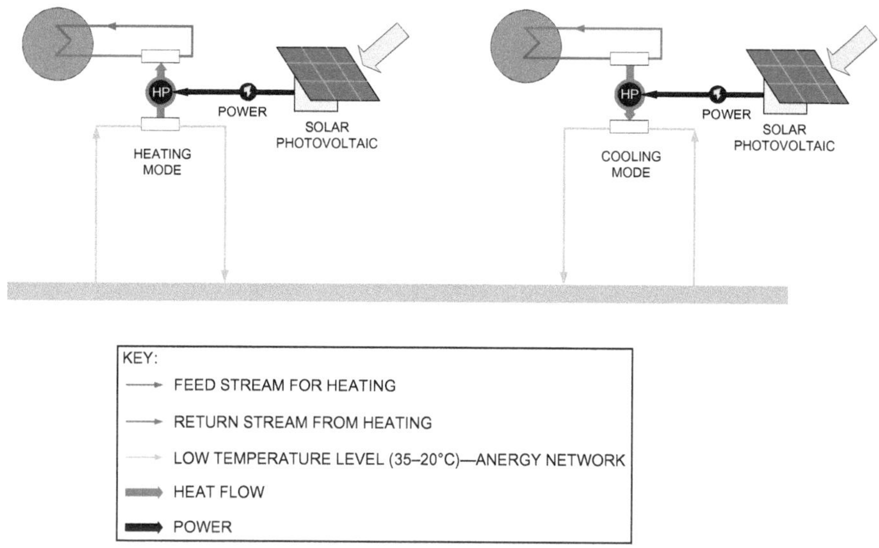

Fig. 5.22 Reversible heat pump configuration.

Heat pumps can increase network feed temperatures, requiring an anergy source (Fig. 5.10).

Fifth-generation district heating networks are ultralow-temperature anergy networks. This system would be a heat source for heat pumps and can be used to collect, store, and recycle heat from industries, waste water, etc. There is also a second mode, also for summer seasons, in which the anergy network works directly as a district cooling system (see Fig. 5.11). It means that the heat pump is switched off, and only a heat exchange is occurring, with the anergy network providing cooling (Fig. 5.23).

5.3 Thermo-hydraulic network models and tools

5.3.1 District heating network models

Two approaches can be found in network modeling: the model of a specific component, that is, substations, pipes, pumps, and heat generators, or system modeling of the entire district heating system. In this section, the second option is in the focus. The network system includes here pipes and all pipe connectors, pumps, and substations (both for the "consuming" and the "producing" sides).

Also, several media can be considered, mainly steam and water. When going toward the fourth generation of district heating (Fig. 5.13), steam is less and less used; in fact, some district heating systems are changing medium from steam to water. Thus, the only medium considered here is water.

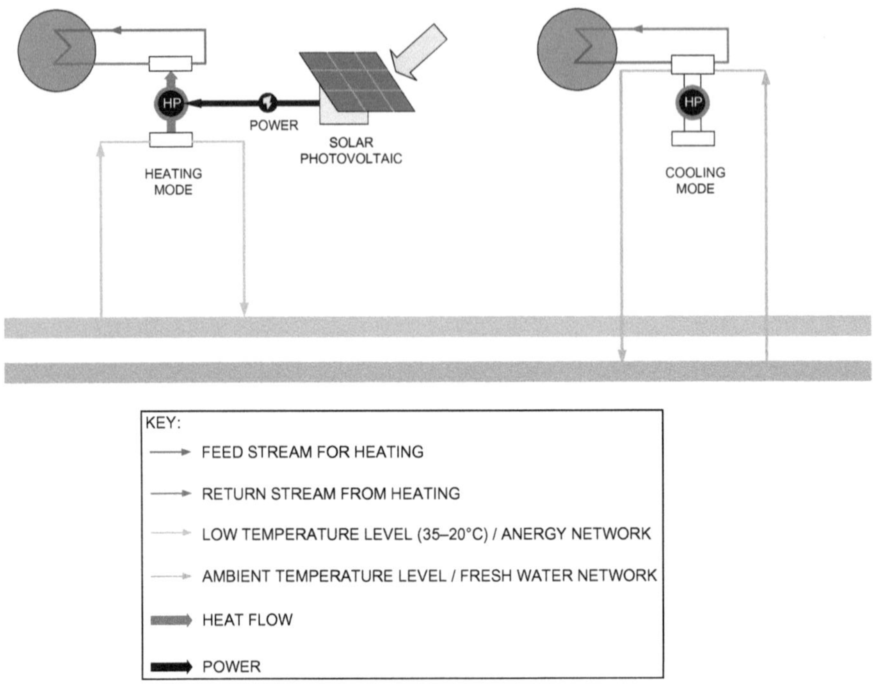

Fig. 5.23 Alternative configuration for reversible heat pumps.

A network model can be divided into two parts: the hydraulic model, where flows and pressures are calculated, and the thermal model, where the temperatures field is obtained.

Two main approaches can be defined for both parts following the bottom-up strategy: a steady-state and a dynamic one (Bøhm et al., 2002).

5.3.1.1 Steady-state approach

The steady-state approach is commonly used in DH mainly for network dimensioning (Gabrielaitiene et al., 2003) and optimization of the layout together with optimization techniques (Curti et al., 2000).

The model is divided into the two parts: first, the flow rates through every pipe and pressures at each node are calculated with constant water properties (density, viscosity heat capacity, and thermal coefficient). Then, the temperature field is obtained with a heat balance equation in every pipe, depending on the previously calculated flow ratios. The heat transfer is considered to be only radial, and the temperature does not change along the complete volume under study.

5.3.1.2 Dynamic approach

Temperature and velocity fluctuations appear when connecting unstable sources to the network, that is, waste heat or solar heat. This behavior is not possible to capture with a steady-state approach, and a dynamic one must be used (Gabrielaitiene, 2011).

Velocity fluctuations increase the probability of turbulences. Turbulent diffusion means higher heat exchange. Furthermore, the temperature field affects the water velocity because of the dependencies of the water properties on temperature. Hence, a *fully dynamic approach* must have both parts of the model, hydraulic and thermal coupled (Gabrielaitiene et al., 2002), compared a coupled with an uncoupled model with numerical experiments, and showed that not considering the coupling between thermal and hydraulic models leads to small errors in temperature predictions but are significant for pressure differences.

Nonetheless, the dynamics of the hydraulic part are much faster than those of the thermal part. Pressure waves travel at the speed of sound, while temperature waves do it with the water flow (the typical time steps for velocity and pressure calculations are between 0.5 and 2 s, while the time step for temperature can be minutes or even over an hour). That makes the changes in pressure to be less important from the operational point of view than the changes in temperature. A pseudodynamic approach, with a "high" time step of calculation (some minutes), leads to an uncoupled approach where the velocities are obtained from steady-state calculations and the temperatures dynamically (Bøhm et al., 2002). This is the approach most frequently used in engineering for network simulation (Pálsson, 2000).

Two methods are used to model the dynamics of a network based on the pseudodynamic approach, the element and the node methods (Pálsson, 2000).

The simulation with the element method is based on a temperature profile along a pipe. The heat losses in radial direction are calculated considering a single uniform temperature in water and pipe and also uniform temperatures in insulation and surrounding ground. It is a robust method and avoids numerical oscillations and divergence if the stability criterion, the Courant-Friedrichs-Lewy (CFL) condition, is satisfied. The CFL number is defined as $c = u\Delta t/\Delta x$ and must be lower than 1 to satisfy the criterion. On top, the element method suffers of artificial diffusion, which could result in abnormal smoothing of sharp temperature profiles. The artificial diffusion depends strongly on the CFL too and gets worse when the CFL number approaches to zero (Pálsson, 2000).

The node method bases the computation on a time history of temperatures and flows in one single point. Heat losses are computed by assuming different water and steel temperatures. This method does not take into account the heat capacity of the insulation and surrounding ground; thus, no dynamic changes in insulation temperatures are considered. The node method was described in 1997 by Benonysson (Wollerstrand, 1997).

The amplitude of the temperature waves together with the time for reaching the peak value are typical magnitudes to analyze in the dynamics of a DHCS. Both models have limitations on these predictions. On the one hand, the node method has been proved to be more accurate in predicting the times in temperature changes, but on the other hand, the amplitudes are better predicted with the element method (Gabrielaitiene et al., 2008). Furthermore, the computational costs seem to be also lower for the node method (Benonysson et al., 1995).

In conclusion, none of the two methods seems to be much better than the other, and both are widely used in modeling tools, even though the node method seems to be preferred in the last few years.

5.3.2 State of the art of commercial network modeling tools

A brief summary concerning three of the most important computer tools related to the design, modeling, and simulation of thermal networks is done.

- STANET®

  Software specialized in calculating and simulating both stationary and dynamic utility networks, including gas, water, electricity, wastewater, and district heating.

  It is a very versatile program based on a graphical interface in which the user is able to view, edit, and add all network elements, either manually or importing the necessary files (maps, plans, etc.) from various platforms such as GIS or CAD systems. The program is capable to receive Excel tables, text in *Asc*I format, ODBC, or directly DXF AutoCAD files.

  From the simulation point of view, it is possible to execute the program both for steady-state or stationary simulation (i.e., for a single instant of time) and for dynamic calculation (for an entire period of time), obtaining results of each variable as a function of time.

  The program also offers some possibilities beyond simulation, such as automated topology improvement, creation of subnetworks, and diameter optimization (stafu.de).
- Bentley® sisHYD

  Bentley sisHYD is a network design and modeling application that provides all necessary thermal and hydraulic calculations used for pipeline networks. Like STANET, Bentley sisHYD offers several possibilities for the construction of a database from which the simulation will be built, being able to import files from multiple platforms such as Excel, GIS, or AutoCAD. Of course, it also allows to draw and define the networks manually using its own graphical interface.

  It is also possible to do the calculations both in static and dynamic mode, generating all kind of reports and editable graphs, in order to analyze the behavior of the system.

  Optimization is also available here to calculate different scenarios in which the balance between electricity and heat generated according to current or predictive demand is optimized. Another possibility is the reduction of operational costs by topological redesign of the network. Source: (bentley.com).
- MATLAB and Simulink®

  It is considered one of the most popular platforms inside the scientific community, which has the following characteristics (mathworks.com):
  - Flexibility and versatility: There are several ways to build models for the same objective.
  - Relatively easy and affordable writing of physical and mathematical models in the form of functions that can be visualized as blocks with their corresponding inputs, outputs, and necessary parameters.

5.4 Energy and exergy balances to evaluate performance

The behavior of a thermal network can be defined as a set of calculated variables resulting from applying hydraulic, energy, and exergy balances. For example, to meet certain heat demand objectives, the system will adapt itself modifying certain variables such as district heating water mass flow or return temperature, which in turn will influence the hydraulics of the system, distributing heat from supply points to consumers.

Efficiency is defined as the ratio between supplied and incoming energy/exergy flows; this incoming energy may come either from a production stage or from an intermediate point in the network (heat exchanger, substation, etc.). Therefore, energy and exergy losses are considered as a good way to measure the performance of thermodynamic systems.

It seems logical to affirm that energy balance and energy efficiency must be taken into account when evaluating the efficiency of a given thermodynamic system, but why is an exergy balance included in these studies as well? Why is destroyed exergy important to evaluate the efficiency of a thermal system? The definition of exergy will answer these questions:

Exergy (in older usage, available work, and/or availability) of a system is the maximum theoretical possible useful work during a process that brings the system into equilibrium with a heat reservoir. When the surroundings are the reservoir, exergy is the potential of a system to cause a change as it achieves equilibrium with its environment. Using other words, exergy is the work that is available to be used. After the system and surroundings reach the equilibrium, the exergy is zero (Martínez, 1992).

In essence, the exergy refers to the potential utility of a particular energy source. It is also defined as the maximum theoretical work that could be obtained from the interaction between a certain thermodynamic system and the reference environment. If the state of a quantity of matter is different from that of the environment, there will be the possibility of producing work. As the system evolves toward the equilibrium with the environment, this possibility will be reduced (Rant, 1956).

The reference environment refers to a portion of the environment whose intensive properties do not change significantly as a result of any thermodynamic process. Therefore, it will be considered a simple compressible system with enormous dimensions with pressure P_0 and temperature T_0, uniform and constant.

Several aspects about exergy are as follows:

- In order to make an exergetic study, the boundaries of the thermodynamic system must be clearly defined.
- The value of the exergy cannot be negative. Exergy transfers associated with energy flows in transit (in the form of heat or work) may have one or the other sign depending on the meaning and nature of these flows.

In this way, exergy quantitatively determines the thermodynamic value of any energetic resource. For example, if fuel is burned generating a mixture of air and combustion products, although the energy associated with the thermodynamic system is the same, the exergy of the initial system is much greater, since the potential of obtaining useful work is greater before the combustion. Therefore, in contrast to energy, exergy is not preserved, but it is lost by the evolution toward the equilibrium state. Because of that, it is right to say "exergy destroyed" instead of "exergy losses."

Another example is the cooling water of thermal power plants. Although the plant transfers a large amount of energy to the water, it only raises its temperature by a few degrees above the temperature of the surroundings, so its potential usefulness to obtain work is almost zero, which means a really low associated exergy. From this, we can easily infer that the greater the temperature difference between a hot stream exchanging heat and a cold one, the greater the destruction of exergy, which means a greater "waste" from the exergetic point of view, even though the heat exchanger was perfectly insulated, which means zero heat losses (Moran and Shapiro, 2000).

As we will see later in Section 10.1.4, all these concepts will be useful to interpret the obtained results.

Exergetic content of the different forms of exergy is as follows:

- Exergy of a closed system:

$$Ex = m_{cv}\left[(U - U_0) + p_0(V - V_0) - T_0(S - S_0) + \frac{c^2 - c_0{}^2}{2} + g(z - z_0)\right] \tag{5.5}$$

where m_{cv} is the mass of the control volume, kg; T_0 is the reference temperature, K; p_0 is the reference pressure, Pa; U is the specific internal energy of the system, J/kg; U_0 is the system's specific internal energy at the reference's pressure and temperature, J/kg; V is the specific volumen of the system, m^3/kg; V_0 is the system's specific volume at the reference's pressure and temperature, m^3/kg; S is the specific entropy of the system, J/(kg K); S_0 is the system's specific entropy at the reference's pressure and temperature, J/(kg K); $\frac{c^2 - c_0{}^2}{2}$ is the kinetic energy term, J/kg; $g(z - z_0)$ is the potential energy term, J/kg; and Ex is the exergy of the system, J.

Neglecting the variations of kinetic and potential energy, the following expression remains:

$$Ex = m_{cv}[(U - U_0) + p_0(V - V_0) - T_0(S - S_0)] \tag{5.6}$$

- Exergy associated with a mass flow:

$$\dot{E}x = \dot{m}\left[(H - H_0) - T_0(S - S_0) + \frac{c^2 - c_0{}^2}{2} + g\,(z - z_0)\right] \tag{5.7}$$

where $\dot{m}$ is the mass flow, kg/s; H is the specific enthalpy of the system, J/kg; and H_0 is the system's specific enthalpy at the reference's pressure and temperature, J/kg.

Neglecting again the variations of kinetic and potential energy,

$$\dot{E}x = \dot{m}[(H - H_0) - T_0(S - S_0)] \tag{5.8}$$

This equation is very useful in the simulation calculations and will frequently appear throughout Section 10.1.4.

- Exergy transfer associated with a heat flow:

$$\dot{E}x_Q = \int_1^2 \left(1 - \frac{T_0}{T_F}\right) d\dot{Q} \tag{5.9}$$

This is the maximum usable theoretical work that could be obtained from the mentioned heat flow when it is exchanged with the environment; therefore, it is intimately related to Carnot's performance.

T_F is the external temperature at the external boundary of the thermodynamic system, K.

- Exergy transfer associated with a work flow:

$$\dot{E}x_W = \dot{W}_{useful,\max} = \dot{W} + \dot{m}\,p_0 dV \tag{5.10}$$

It is the maximum usable theoretical work from the work flow in a thermodynamic process. If the system changes its volume in this process, part of the produced work (W) must be used for the volume change at a pressure p_0.

If the system does not generate a volume change, as it happens with incompressible fluids, the transfer of exergy associated with the work will be equal to the work of the system itself, obtaining the following equation:

$$\dot{E}x_W = \dot{W}_{useful,\max} = \dot{W} \tag{5.11}$$

- Exergy efficiency:

In general, it is defined as the quotient between the exergy that leaves the thermodynamic system and is truly used for the user and the total exergy provided to the system:

$$\eta_{EX} = \frac{\text{useful Exergy}}{\text{total supplied Exergy}} \tag{5.12}$$

The exergy balances applied to different thermodynamic systems are shown below. The principal ones for this project are the following:

- Exergy balance for a closed thermodynamic system (Fig. 5.24):

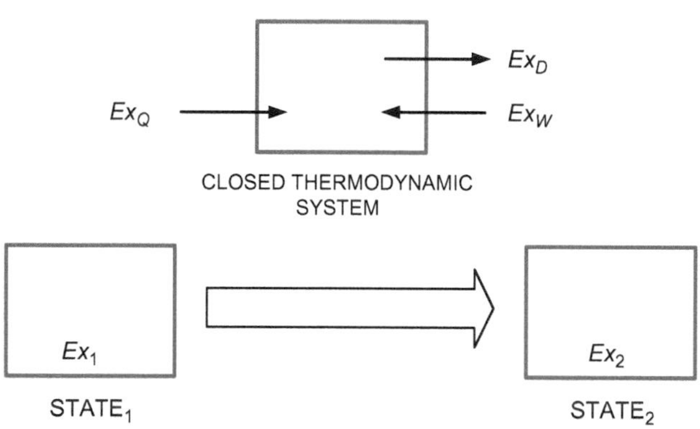

Fig. 5.24 Exergy balance for a close system.

State 1 is the initial state, and state 2 is the final state of a closed thermodynamic system that does not exchange mass, although it exchanges energy with the environment, fulfilling the following equations:

$$Ex_2 - Ex_1 = \int_1^2 \left(1 - \frac{T_0}{T_F}\right) dQ + [W + m_{cv}\, p_0 (V_2 - V_1)] - T_0\, m_{cv}\, \sigma \tag{5.13}$$

This is the exergy balance expression for a control volume. It shows how the exergy varies when the control volume evolves from one state to the other. Exergy increases in the following cases:

1. By receiving useful work, which is completely exergy
2. By receiving heat flow, which has only one part that contributes to the exergy

And it decreases by the internal entropy generation (Martínez, 1992).

σ is the internal entropy generation term, (J/(kg K)). It will give rise to the exergy destroyed term.

In simplified form,

$$Ex_2 - Ex_1 = Ex_Q + Ex_W - Ex_D \tag{5.14}$$

Solving Ex_D,

$$Ex_D = Ex_1 - Ex_2 + Ex_W + Ex_Q \tag{5.15}$$

Calculating exergy efficiency,

$$\eta_{EX} = \frac{Ex_2}{Ex_1 + \text{supplied Exergy}\,(Ex_Q, Ex_W)} \tag{5.16}$$

Ex_1 is the Exergy of the system in the initial state, kJ; Ex_2 is the exergy of the system in the final state, kJ; and $Ex_D = T_0\, m_{cv}\, \sigma$ is the term relative to destroyed exergy, kJ.

- Exergy balance associated with a nonsteady-state open thermodynamic system (Fig. 5.25):

The mass flow that enters the control volume might not be the same as the one that exits; therefore, the balance will be as follows (considering no variation with the volume):

$$\frac{dEx}{dt} = \sum_{j=1}^{n} \int_1^2 \left(1 - \frac{T_0}{T_{Fj}}\right) d\dot{Q}j + \dot{W} + \sum_{n\ inputs} (\dot{m}_{IN} ex_{IN}) - \sum_{n\ outputs} (\dot{m}_{OUT}\, ex_{OUT}) - T_0\, m_{cv} \frac{d\sigma}{dt} \tag{5.17}$$

with j as the subscript relative to each heat exchange point between the system and the environment and ex_{IN} and ex_{OUT} are the characteristic exergy values relative to the

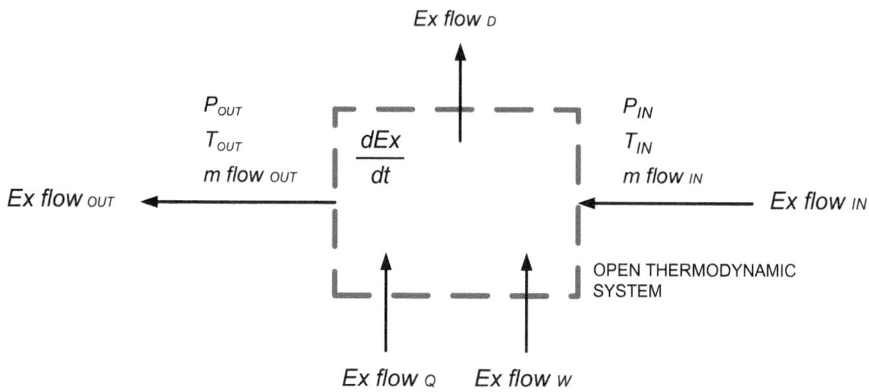

Fig. 5.25 Exergy balance for an open system (control volume). Nonsteady state.

pressure and temperature of each stream with respect to the reference values, defined by the following equations:

$$ex_{IN} = h_{IN} - h_0 - T_0(s_{IN} - s_0) \tag{5.18}$$

$$ex_{OUT} = h_{OUT} - h_0 - T_0(s_{OUT} - s_0) \tag{5.19}$$

To simplify calculations, the exergy flow associated with each stream is directly raised, just subtracting the input minus the output, resulting in the following expression:

$$\dot{E}x_{STREAM} = m_{STREAM\ IN}\, ex_{IN} - m_{STREAM\ OUT}\, ex_{OUT} \tag{5.20}$$

In the case of steady-state system, which will be explained later, the equality is written as follows:

$$\dot{m}_{STREAM\ IN} = \dot{m}_{STREAM\ OUT} \tag{5.21}$$

Hence, the following expression is defined for steady-state system:

$$\dot{E}x_{STREAM} = \dot{m}_{STREAM}\,(ex_{IN} - ex_{OUT}) \tag{5.22}$$

Expressing (5.17) in simplified form and clearing Ex_D,

$$\dot{E}x_D = \sum_{n\ inputs} \dot{E}x_{IN} - \sum_{n\ outputs} \dot{E}x_{OUT} + \dot{E}x_W + \sum_{j=1}^{n} \dot{E}x_Q - \frac{dEx}{dt} \tag{5.23}$$

Showing the equation in an easier way,

$$\dot{E}x_D = \dot{E}x_{IN} - \dot{E}x_{OUT} + \dot{E}x_W + \dot{E}x_Q - \frac{dEx}{dt} \tag{5.24}$$

Calculating exergy efficiency,

$$\eta_{EX} = \frac{\dot{E}x_{OUT}}{\dot{E}x_{IN} + supplied\,Exergy\left(\dot{E}x_Q, \dot{E}x_W\right)} \tag{5.25}$$

$\dot{E}x_{IN} = \sum_{n\,inputs} \dot{E}x_{IN} = \sum_{n\,inputs} (\dot{m}_{IN} ex_{IN}) =$ Exergy flow associated with mass flow inputs into the control volume (kW).
$\dot{E}x_{OUT} = \sum_{n\,inputs} \dot{E}x_{OUT} = \sum_{n\,inputs} (\dot{m}_{OUT} ex_{OUT}) =$ Exergy flow associated with mass flow outputs from the control volume (kW).
$\dot{m}_{STREAM\,IN} =$ Mass flow input to the control volume (kg/s).
$\dot{m}_{STREAM\,OUT} =$ Mass flow output from the control volume (kg/s).

Once all equations are theoretically described, this theory must be applied to all those main systems related to district heating networks, defining a set of expressions that constitute the exergetic method:

- Exergy balance for a heat exchanger:

The heat exchanger is considered as an open steady-state thermodynamic system, which exchanges matter and thermal energy. It is applied to all types of substations and, in general, to all kind of heat exchange points (Fig. 5.26).

T_0 the reference temperature = 8°C, corresponding to a typical average temperature of a trench where the DH pipes are installed, and T_F is the external temperature of the boundary that delimits the thermodynamic system = 13°C, corresponding to the DH substations where the heat exchangers are located, both in K.

The exergy balance applied to this system will be the corresponding to Eq. (5.28), following the sign criterion corresponding to Fig. 5.30.

$$\dot{E}x_D = \dot{E}x_{HS} - \dot{E}x_{CS} + \dot{E}x_W + \dot{E}x_Q \tag{5.26}$$

$\dot{E}x_{HS} =$ positive, inlet to the control volume.
$\dot{E}x_{CS} =$ negative, outlet from the control volume and it is useful to the user.
$\dot{E}x_W =$ zero, no work exchange between the control volume and the user.
$\dot{E}x_Q =$ positive, heat losses are leaving the control volume and are useless; therefore, they generate simple anergy, which is the unrecoverable part of the energy. Thus, $\dot{E}x_Q$ counts as a positive value because it is supposed to increase exergy destruction.

From Eq. (5.9) and considering that the heat exchanger is totally isolated, $\dot{E}x_Q$ is calculated:

$$\dot{E}x_Q = \int_1^2 \left(1 - \frac{T_0}{T_F}\right) d\dot{Q} = \left(1 - \frac{8 + 273.15}{13 + 273.15}\right) \times 0 = 0 \tag{5.27}$$

Finally, the exergy balance for the heat exchanger will be

$$\dot{E}x_D = \dot{E}x_{IN} - \dot{E}x_{OUT} \tag{5.28}$$

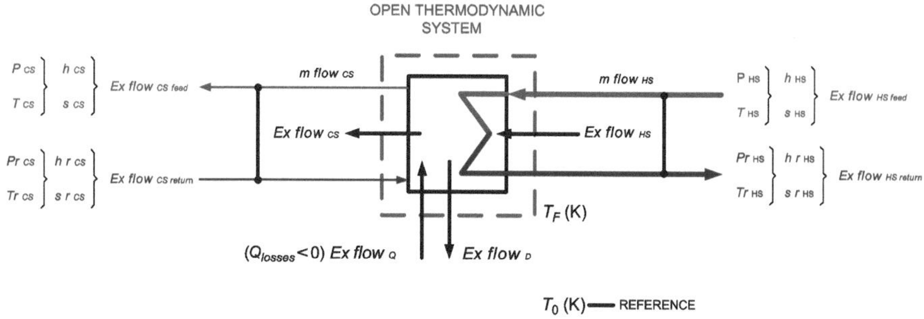

Fig. 5.26 Exergy method applied for a heat exchanger.

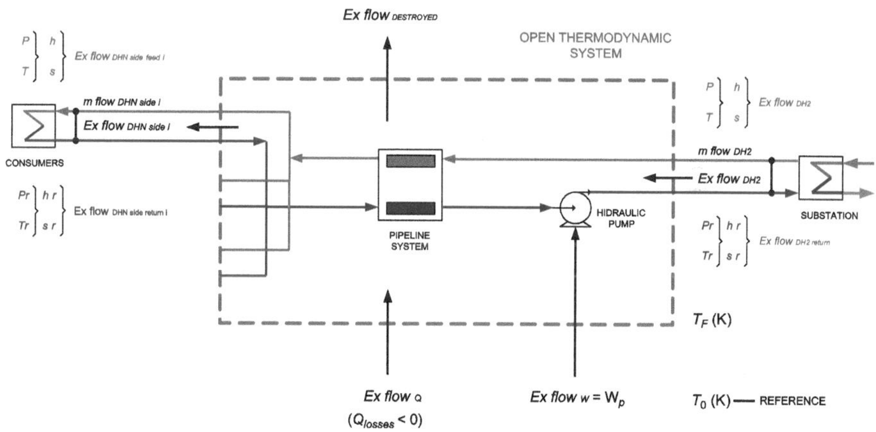

Fig. 5.27 Exergy method applied for a DHN's pipeline system.

Calculating exergy efficiency,

$$\eta_{EX} = \frac{\dot{E}x_{OUT}}{\dot{E}x_{IN}} \tag{5.29}$$

- Exergy balance for a district heating network:

The pipeline network is considered as an open thermodynamic system in a steady-state regime that exchanges mass and heat with the environment. It is applied to the entire pipe and branch segments located between substations (see Fig. 5.27).

T_0 the reference temperature = 8°C, corresponding to the average temperature of the trench where the DH pipes are installed, and T_F is the external temperature of the boundary that delimits the thermodynamic system = 8°C, corresponding to the DH substations where the heat exchangers are located, both in K.

The exergy balance applied to this system will be the one corresponding to Eq. (5.23), modified for a single mass input and several outputs, following the sign criterion corresponding to Fig. 5.27.

$$\dot{E}x_D = \dot{E}x_{DH2} - \sum_{i=1}^{n\ outputs} \dot{E}x_{DH\ SIDE\ i} + \dot{E}x_W + \dot{E}x_Q \tag{5.30}$$

$\dot{E}x_{DH2}$ = positive, it is an input to the control volume.
$\sum_{i=1}^{n\ outputs} \dot{E}x_{DH\ SIDE\ i}$ = it is an outlet from the control volume, and it is useful to the user.
$\dot{E}x_W$ = positive, it is an input to the control volume, as power for the hydraulic pump.

Again, based on Eq. (5.9) and considering the values of T_0 and T_F, $\dot{E}x_Q$ is estimated:

$$\dot{E}x_Q = \int_1^2 \left(1 - \frac{T_0}{T_F}\right) d\dot{Q} = \left(1 - \frac{8 + 273.15}{8 + 273.15}\right) \times \dot{Q}_{LOSSES\ DHN} = 0 \tag{5.31}$$

So, the expression for the entire pipeline system will be

$$\dot{E}x_D = \dot{E}x_{DH2} - \sum_{i=1}^{n\ outputs} \dot{E}x_{DH\ SIDE\ i} + \dot{E}x_W \tag{5.32}$$

Calculating exergy efficiency,

$$\eta_{EX} = \frac{\sum_{i=1}^{n\ outputs} \dot{E}x_{DH\ SIDE\ i}}{\dot{E}x_{DH2} + \dot{E}x_W} \tag{5.33}$$

See detailed analysis and the rest of the nomenclature and subindexes in Section 10.1.4.

5.5 System model thermal networks

Once the urban energy systems integrating a thermal network are known, a mathematical model can be set up. A large number of energy and exergy balances are required to be applied throughout the different facilities present in this type of systems. The number of equations to be considered and calculations to be solved becomes large as the complexity of the network increases. In fact, there are already several specialized softwares (see Section 5.3 in this chapter) designed for these tasks.

A detailed representation of the elements of the network (pipes, substations, and hydraulic pumps) and other elements inside the buildings like radiators or recirculation systems is needed.

For example, if the purpose is showing the effects of refurbishing all buildings changing from radiators to heated floors, first, it will be necessary to simulate a network with buildings with radiators installed and compare it with a second simulation in which these radiators are replaced by radiant floors. Therefore, the software used must be flexible and capable to construct each representative simulation.

Throughout this section, the different stages constituting the district network model are explained.

Fig. 5.28 explains the entire working methodology. The different events are happening in sequential order through MATLAB®. All consecutive stages of the process are described below:

- Reading and collecting the different input data to build and simulate the network (steps 1 and 2) and stored inside Matlab's internal database (workspace) (step 3). The source of these data comes from different formats, like Excel sheets, text, and XML files. The main and necessary data for the simulation are
  - hourly resolved yearly heat demand profiles.
  - separated yearly heat demand profiles for heating and hot water production (if possible).
  - external temperature profiles.
  - feed and return temperatures for each building and the general substation.
  - other parameters, like temperature regulation curves, pressure regulation, and equipment characteristics.
- Construction of the simulation using the previously mentioned structural data (step 4), including network topology, number of consumption points, number of pipes and connections' layout between pipes and substations, and other parameters like nominal diameters, width, and thermal properties of the insulating material.

  This step has in turn two substeps, hereafter described:
  - Library of blocks (Fig. 5.29):

    Each DHN component (pipes, buildings, internal installations of buildings, substations, control and regulation systems, hydraulic pumps, heat pumps, etc.) has its own model, represented by visualized blocks with inputs, outputs, and all necessary parameters used to solve the existing equations in the mentioned model.
  - Building the simulation (Fig. 5.30):

    The layout of the blocks extracted from the library is done according to what is specified in the structural data and assignment of the different parameters to their corresponding blocks.

- Indication by the user regarding whether there are separated heat demand data for heating and hot water profiles available (step 5). Depending if the answer is YES or NO, the model moves to one or other stage, as indicated in Fig. 5.28.
- If there are no separated heat demand profiles, the procedure continues to step 6, called calibration, which consists of adjusting the parameters of the simulation to obtain those separated profiles from the total profiles (heating + hot water).
- Then, the simulation is executed, obtaining all calculated temperatures, mass flows, and enthalpy profiles of the inlet and outlet streams for each building. Afterward, calculated and real profiles are compared (steps 7 and 8, validation). If the adjustment is acceptable, the model advances to the next step; if not, it comes back repeating step 6.
- In the next step (9), the simulation is correctly calibrated and run, generating pressure, temperature, enthalpy, and exergy profiles of all present streams at any point in the simulated network, from the substation to the last building. Mass, energy, and exergy balances are properly calculated, obtaining pressure and heat losses and destroyed exergy for all defined thermodynamic systems, explained in Section 10.1.4.
- All calculations mentioned in step 10 must be done not only for a single value but also for entire profiles in function of time (e.g., pressure-time or temperature-time), so that the

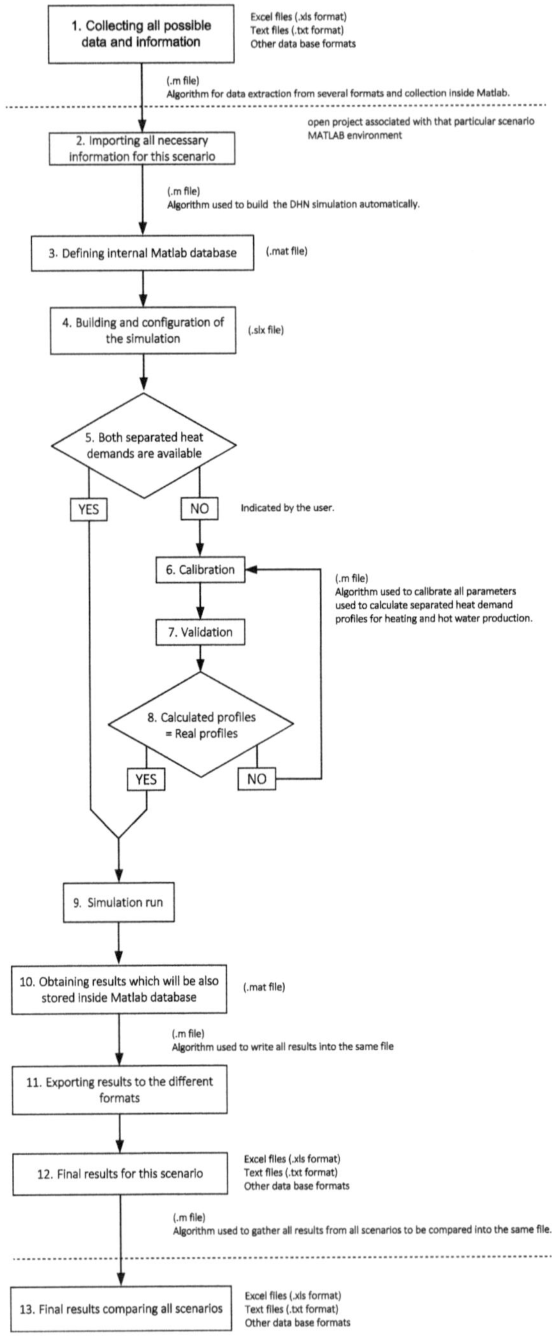

Fig. 5.28 Sequential process for the construction of a multistage DHN simulation for a given case study.

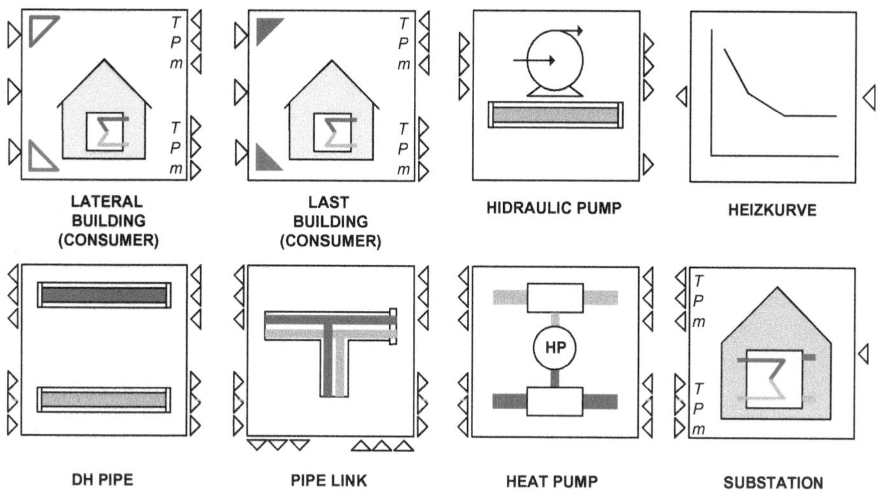

Fig. 5.29 Figurative representation of the block library.

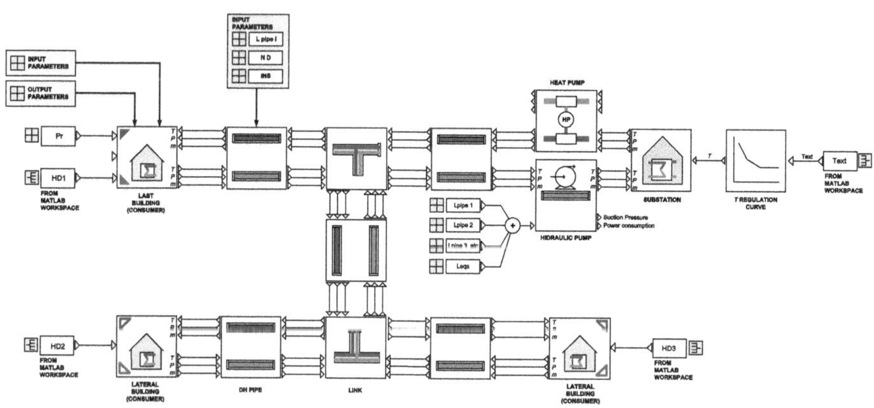

Fig. 5.30 Conceptual diagram representing the layout of the blocks for a small three-buildings network.

system outputs will be profiles and graphs that can be stored and exported to different formats, like Excel sheets, text files, or directly sent to a common database and distributed by the web service.

- Among steps 10, 11, and 12, results are obtained and written in several file formats. An Excel file is obtained for each scenario.
- In step 13, a final Excel file is obtained showing a comparison between results for all executed scenarios. Once all this material is obtained, a report can be prepared, as shown in Section 10.1.4 of this book.

In conclusion, the basic theoretical foundations needed for a district heating network analysis have been introduced. Different strategies are analyzed according to the urban areas to be studied, including the main available urban energy systems and how to use them. Likewise, theoretical notions related to the history, present and future situation of thermal networks, were shown, defining afterward a workflow methodology to be put into practice for the case study example studied in Chapter 10.

5.6 Multimodal networks

Nowadays, rapidly growing cities, increased energy consumption, and consequently increased impact on the ecological environment have promoted the usage of alternative energy sources, technologies with lower pollution rate, and optimization of energy production. As a consequence, urban utility networks become more complex and dependent on each other, and their operation, maintenance, and management become complex as well. However, the growing diversity and interconnection of urban utility networks lead to new challenges and to a new optimization potential. For example, by understanding and leveraging the dependencies between heat and power networks, heat production costs can be reduced, while power network peaks can be avoided. Likewise, network condition can be comprehensively observed in order to prevent failures and to analyze network restoration possibilities.

The main purposes of this work are

(1) to discuss the advantages of considering multimodal utility networks;
(2) to define and analyze interdependencies between networks such as electricity, heating, gas, and others; and
(3) to make first steps toward a novel modeling approach for multimodal utility networks of cities based on a combination of graph-theoretical models and optimization methods from data analytics.

Each city is covered by numerous urban infrastructures, such as electricity grid, gas network, and district heating and cooling, which distribute utilities across the city and supply them to residential, office, commercial, and other buildings. Although infrastructures differ from each other and use different facilities and technologies for producing and distributing utilities, their structures are similar: stations or plants for producing resources, distribution lines, and buildings as consumers. For example, the electricity grid uses power plants and transmission facilities to produce and transmit the electricity, whereas in water network, water pumps and pipelines are used to deliver water in buildings. We represent the city utility infrastructure on this macro level and represent the city as a set of buildings that are networked by different infrastructures such as the electricity grid and district heating and cooling.

Each building (or network node) belongs to one of three types:

i. A consumer, such as an office building
ii. A producer, such as a power or thermal plant
iii. A prosumer, such as a household with photovoltaic panels

The type of a node can vary depending on the utility network it belongs to; for example, an electric boiler transforms electricity into heating, so it is a consumer of electricity and a producer of heating. Such energy conversion systems represent tight interconnections between different urban utility networks that, taken together, constitute a multimodal utility network. A graph-theoretical representation of the city utility networks provides means to analyze these networks focusing on its structural components and corresponding properties. For instance, path, flow, and centrality analyses of the graph contribute to estimating cost efficiency of the network, recognizing bottlenecks, and evaluate vulnerability of the network. Supplying the graph model with additional static data such as connection lengths and flow capacities and dynamic data such as measurements or data from prediction models allows analyzing the network focusing on its dynamic processes. For instance, combining information provided by prediction and network models and analyzing this information contributes to optimizing network operation and redistributing resources efficiently in the network.

5.7 Overview on utility networks models

There are several approaches to represent urban utility networks and their interconnections; the most popular ways to depict interdependent utility networks are complex adaptive systems and agent-based models. Complex systems are often modeled as random graphs, such as the traditionally used Erdös-Rénui random graph or scale-free models (Albert and Barabási, 2002). As such, random graph models are used to represent utility networks in Holmgren (2006) or Brown (2007).

In Rinaldi et al. (2001), utility networks are represented as a population of interacting agents functioning under certain environmental conditions. The authors define types of dependencies between them, types of failures, and response behavior in case of failures. According to Kandiah and Rao (2008), water utility networks depend on six other utility networks and consist of several components such as water collection facilities or wastewater treatment systems. They propose to model the water network as an agent-based system and study possible events and failures that can affect its functioning.

However, recently, it has been recognized that topology, characteristics, and changes in real-world networks comply with more robust and strict organizing principles in comparison with random graphs (Albert and Barabási, 2002). Moreover, random graphs do not always reflect properties of real-world networks correctly. Watts and Strogatz (1998) compared the electric grid graph and a random graph with the same size and vertex degree and showed that although their path lengths are similar, the clustering coefficient for a real-world network was significantly higher. This leads to a multitude of recent work depicting utility networks explicitly. In Dueñas-Osorio et al. (2007), interdependent networks are represented as undirected graphs. The authors take water and electricity infrastructures as a case study and write interdependencies as a conditional probability of failure in the water network, given that a failure occurred in electricity network. Abdalla et al. (2007) model network interdependencies and interactions using geographic information system (GIS) data. In Winkler et al. (2001), a graph representation of infrastructure networks is used, however, under the assumption that there are no connections other than "producer-consumer" in the utility network. The authors analyze the resulting bipartite graph and study a set of metrics to assess network connectivity, efficiency, and stability against natural disasters. The authors of Chai et al. (2011) compare interactions between different utility networks and human-human interactions, represent interdependent networks as a social network graph, and apply analytic methods from graph theory to analyze density, degree centrality, and network efficiency.

Most of the research cited above studies the vulnerability of infrastructure networks in case of failures or natural disasters. However, there is much less work on analyzing the networks to improve and optimize its functioning. The authors of Collazos et al. (2009) present a predictive control system based on a mixed-integer linear approach for a small domestic energy system that supplies electricity, hot water, and heat to residential buildings. To ease computational efforts, Fazlollahi et al. (2014) introduces a k-means clustering approach to reduce the number of decision variables in the resulting optimization models, which normally depends on the amount of resources and utility networks.

In the following sections, we propose a flexible graph-based approach to model utility networks. In contrast to other approaches, we aim to capture interdependencies between the different utility networks and also to take into account storage facilities and prosumers that return excess energy back to the network.

5.8 Modeling of city infrastructure networks

We consider a city as a set of buildings connected by urban utility services, such as electric power cables, water pipelines, and other facilities. In this section, we present a graph-based model that formalizes this view. Our goal is to capture not only city utility network structures but also their interdependencies and interactions.

First, we consider a network for one resource, such as water or gas pipelines, district heating network, electricity grid, or any other utility. A distribution network is a system that is used to deliver resources from a producer to a consumer, for example, power grid, water pipelines, or district heating network. Graphs are a powerful and flexible way to represent all kinds of linked data; they are frequently used to represent urban infrastructures.

The network is represented as a directed graph $G\ (V, A)$, where producers and consumers form a set of vertices $V = P \bigcup C$ (i.e., the combination of P and C) where P denotes a set of producers, C denotes a set of consumers, and $S = P \bigcap C$ (i.e., the intersection set of P and C) denotes a set of prosumers. Distribution lines between vertices are represented as a set of arcs $A \subseteq V \times V$ (i.e., a subset A of the vertex sets) of the graph G. Note that for some resources, production means the actual generation, for instance, of electricity, whereas for resources such as water or gas, production means the generation of pressure.

A simple example of a utility network is shown in Fig. 5.31: an electricity grid with two producer nodes $P1$ and $P2$; three consumer nodes $C1$, $C2$, and $C3$; and one prosumer node $S1$.

These networks are not isolated from each other but interact with and depend one on one another. In general, there exist different types of interdependencies between

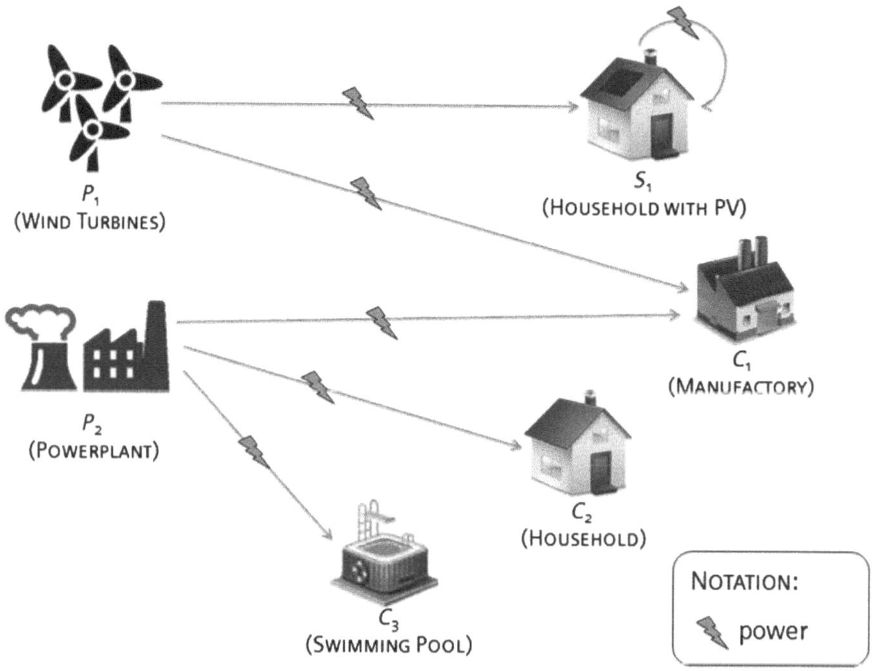

Fig. 5.31 An example of a utility network—electricity grid.

utility networks (Rinaldi et al., 2001; Kandiah and Rao, 2008), among which we distinguish the following:

(i) *Geographic interdependencies*, that is, elements of different utility networks either sharing the same facilities or being close to each other in terms of spatial proximity, such that changes or disturbance in functioning of one utility affect the other utility.

(ii) *Physical interdependencies*, that is, functioning of a utility network and its elements depending on the other utility network. The water system is physically dependent from the electric network, since water pumps and other facilities are electrically powered.

CHP systems that produce electricity and heat from one fuel are an example of both geographic and physical interdependencies: heat and electricity are produced from one fuel at the same place. A quantitative interdependency feature is a coupling order (Rinaldi et al., 2001)—a characteristic that describes whether utility networks are connected directly or through several intermediary networks. For instance, if utility network i is connected to the utility network j, which in turn is connected to network k, then although networks i and k are not connected directly, changes and events affecting network i might affect utility network k as well. In this case, networks i and k have a second order coupling. An example of such networks is natural gas and district cooling: heat produced from natural gas is used for heat driven chillers to generate cooling. Feedback loops are also possible, in case network k is connected further to other networks that are further connected to the initial network i.

Thus, we aim to identify and analyze such interdependencies in networks. We consider a *multimodal utility network*—a union of utility networks that includes all network graphs and, additionally, covers utility exchange between networks. The multimodal utility network is represented by a *multidigraph*—a directed graph that allows several arcs with the same source and target. Multiple arcs are introduced in order to avoid misconceptions, since two buildings in the network may be connected with distribution lines of different utilities. For example, having n utility networks $G_1(V_1, A_1)$, …, $G_n(V_n, A_n)$, the network of networks would be represented as a multidigraph $G(V, A)$, where $V = \cup_{i=1}^{n} V_i, A = \cup_{i=1}^{n} A_i \cup A_I$, A_I is an additional set of arcs, which describes interdependencies between utility networks $G_1(V_1, A_1)$, …, $G_n(V_n, A_n)$. In more detail, for each arc $a = (v_i, v_k) \in A_I$, the target vertex v_k belongs to the network G_k dependent from the network G_i containing the source v_i. Note that sets V_1, …, V_n are not disjoint: they share not only consumers, such as residential buildings, but also hybrid nodes that belong to several utility networks. We also define the *infrastructure dependency multigraph* $G_I(V_I, A_I)$, where the vertex set V_I consists of the utility networks $G_1(V_1, A_1)$, …, $G_n(V_n, A_n)$ themselves. In other words, graph G_I depicts interdependencies between utility networks.

However, the graph model defined above is a very coarse approximation of large complex utility networks: it does not cover most of its important characteristics, such as network load and flow characteristics. Therefore, G may be enriched by values or multiple values assigned to its vertices and arcs. Let Σ_A, Σ_V be finite alphabets of labels for arcs and vertices and l_A: $A \rightarrow (\Sigma_A)^p$, l_V: $V \rightarrow (\Sigma_V)^q$ be mappings assigning tuples of labels to arcs and vertices, where p and q are the numbers of attributes for arcs and vertices, respectively. Thus, a labeled multidigraph $G(V, A, \Sigma_A, \Sigma_V, l_A, l_V)$ represents the topology and properties of the urban utility network.

Some utility networks are highly interconnected with other network, whereas others are not. The main idea of our proposed modeling method is to take hybrid nodes and increasing exchange between city infrastructures into consideration for analysis and simulation. In the next section, we introduce possible applications of the multimodal utility network model in different use cases and analysis procedures (Fig. 5.32).

Apart from the topology represented by the network graph and its important characteristics assigned as labels to each vertex and arc, there typically exists also a large amount of time series information, such as sensor and smart meter measurements, energy production and consumption values, energy losses, flow temperatures, weather

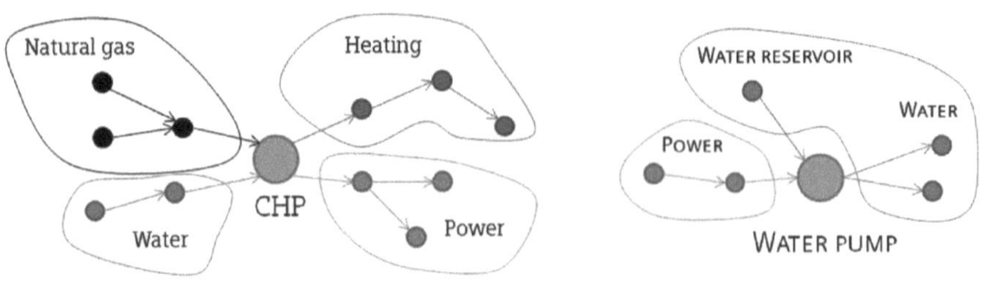

Fig. 5.32 Examples of hybrid nodes belonging to several horizontal levels: on the left a cogeneration plant and on the right a water pump.

conditions, and other dynamic data related to the daily functioning of the network. These data can be of different origin: real-world data or data generated by prediction models.

We suggest to bring more domain knowledge or "semantics" to the analysis of the numerical data and to extract more information by using the structure and topology properties of utility networks. Tasks that could be performed using such a combination of models include the following:

- Optimization-based predictive control of energy consumption
- Prediction of consumption for different utilities for consumers and their aggregations
- Exploration and assessment of the network vulnerability

In the following, we detail several use cases and show how the graph-theoretical model contributes to these tasks.

5.8.1 Optimization control

Network control and developing strategies of its functioning is one of the crucial directions of research in the area of smart cities. To reduce costs and fuel consumption and to increase network efficiency, improvements are essential in the conditions of modern society with constantly growing energy demands.

We suggest using the network model $G(V, A, \Sigma_A, \Sigma_V, l_A, l_V)$ to determine vertices and arcs that might require improvements, for instance, an overloaded transmission line or producer with high production rate without connection to storage facilities. While existing optimization and simulation models provide a strategy for optimizing utility flows for a more efficient usage of resources, information extracted from topology models can point at bottlenecks or directions of utility distribution. For instance, if a photovoltaic system installed on the roof of a prosumer building produces excess electricity during an extremely sunny day, the task of optimization is to determine a more profitable solution:

(i) Store for later usage depending on the appropriate facilities in a building.
(ii) Sell it to the network or neighbors, which can be determined by $G(V, A, \Sigma_A, \Sigma_V, l_A, l_V)$.
(iii) Convert into another utility, for example, use the electricity to heat water.

On the other hand, neighbors can choose what is better for them:

(i) Continue to buy electricity from the network.
(ii) Buy their own photovoltaic system.

In all these cases, path and flow analysis in the multidigraph $G(V, A, \Sigma_A, \Sigma_V, l_A, l_V)$ helps optimization to distribute the energies in the network cost and energy efficiently by the path, flow, and efficiency analysis of the multigraph.

5.8.2 Vulnerability

The interest in network vulnerability analysis increased significantly after recent large energy blackouts, when a single failure triggered shutdowns of major infrastructure networks that led to blackouts of cities and even countries. The most well-known examples are blackouts in North America (Andersson et al., 2005), Italy (Berizzi, 2004), and India (Lai et al., 2012).

There are two categories of failures that affect utility infrastructures: (i) *cascading* and (ii) *common-cause* failures (Kandiah and Rao, 2008). A failure is common cause if two or more utility networks are disrupted at the same time due to the same reason, such as natural or man-made disaster. This kind of failures is typically local and affects utility networks that are geographically interdependent. A failure is called cascading if it disrupts one utility network and initiates a chain reaction of further failures in dependent networks. This kind of failure may affect not only those networks that are directly physically dependent from initially disrupted network but also networks that have nth coupling order with the initial network and cause *nth order effects*.

Our graph-theoretical model of utility networks $G\ (V, A, \Sigma_A, \Sigma_V, l_A, l_V)$ can support study of cascading and common-cause failures: firstly, the dependency multidigraph $G_I\ (V_I, A_I)$ lends itself to the understanding of nth order coupling effects between networks and may be used to predict consequences of a blackout in a specified utility network. Further, studying connectivity and vertex degree in $G_I\ (V_I, A_I)$ contributes to a better understanding of failure propagation. For example, what happens if a vertex, arc, or set of vertices and arcs are down, that is, deleted from the network? How can network functioning be restored as soon as possible? We suggest using the multidigraph $G_I(V_I, A_I)$ to find bypasses between nodes and corresponding optimization models to redistribute the flows efficiently.

5.8.3 Planning future network expansions

The task of planning future network expansions is tightly connected to vulnerability and optimization tasks: suggestions concerning network expansions may involve strengthening vulnerable or critical elements of the network to ensure that in case of failure, it is possible to quickly restore the network functioning. Moreover, having information on bottlenecks in the networks and analyzing the possibilities to improve the efficiency of the network also produce corresponding suggestions for network improvement. For instance, will efficiency of network, costs, and distances of transferring utilities or other characteristics improve by building new transmission lines (adding a set of arcs) in the network $G(V, A, \Sigma_A, \Sigma_V, l_A, l_V)$?

5.8.4 Maintenance

Sensors and smart meters in networks produce huge amounts of measurement data. Processing these data to extract information regarding network functioning and compiling statistics is one of the essential tasks. It largely contributes to all other tasks, in particular to condition monitoring, optimization control, and vulnerability analysis.

Typically, the main goals of sensor data processing are (i) *descriptive*, that is, to describe what happens in the network (e.g., failures and unusual events) and summarize network functioning characteristics, and (ii) *diagnostic*, that is, to explain why certain events occurred (e.g., peaks and trends). In the latter case, it is crucial to know the location of sensors in the network and whether their measurements correlate or influence each other. That is the task of correlation and causality analysis in the utility network $G(V, A, \Sigma_A, \Sigma_V, l_A, l_V)$. Assigning to each arc or vertex information on sensors and meters installed there as well as links to the database with measurement tables allows easier combination of graph and exploratory analysis and achieving better results.

As shown above, the labeled multidigraph $G(V, A, \Sigma_A, \Sigma_V, l_A, l_V)$ representing the city infrastructure network can contribute to various tasks, apart from the area of vulnerability analysis, where graph representation of city networks are currently mostly used. The aim of our work is to develop analysis algorithms for multimodal utility network that combine the advantages of graph-theoretic algorithms, optimization models, and data analyses.

5.9 Analyzing the network

This section describes and details how multimodal networks can be analyzed.

5.9.1 Graph-theoretical algorithms

Graph theory provides us means to analyze these graphs focusing on its structural components and corresponding properties. Several common techniques for network analysis are described in the following:

Centrality analysis is used to identify the most central entities in your network. For instance, one of the metrics used is betweenness centrality—a measure of nodes centrality—which is equal to the number of shortest paths containing node v in the network:

$$C(v) = \sum_{i \neq v \neq j, i, j \in V_I} \frac{\sigma_{ij}(v)}{\sigma_{ij}} \tag{5.34}$$

where $\sigma_{ij}(v)$ is the number of shortest paths between nodes i and j and σ_{ij} is the total number of shorter paths between i and j. The measure shows importance of a node in the whole network. It can be normalized:

$$\text{norm}(C(v)) = \frac{C(v) - \min(C)}{\max(C) - \min(C)} \tag{5.35}$$

In this case, *norm*$(C(v))$ is between 0 and 1 and ranges from smaller to larger centrality. Apart from this metrics, in and out centrality is considered, since it is also important for dependencies between nodes. Indegree indicates the reliance of a node to other nodes, whereas outdegree shows how strongly the node is relied upon. However, in and out centrality can be considered in relation not only to nodes but also to node agglomerations or even the whole network.

Path analysis is used to identify all the connections between a pair of entities, determining the shortest paths between nodes, and estimating the graph resilience to removal of a node or of an edge.

Community detection is used to identify structural components on the basis of connectivity, such as strongly connected components. One of the metrics used is the clustering coefficient, which shows local importance of a node among neighbors of a node and interchangeability of a node. It is equal to the number of edges shared between neighbors divided by total possible number of connection between the neighbors:

$$Cl(v) = \frac{2e_v}{k_v(k_v - 1)} \tag{5.36}$$

where k_v is the number of neighbors of the node and v e_v is the number of edges they share.

Subgraph isomorphism is used to search for a pattern of relationships, useful for validating hypotheses, searching for improving the networks, and identifying the generic type of a graph.

Density of the network is used to indicate how knitted the network is Chai et al. (2011). It represents the measurement of the proportion of the actual number of links to the possible number of links between the nodes.

Efficiency of the network is used to represent the performance of transporting between nodes (Chai et al., 2011). It is determined using the formula:

$$E = \frac{1}{n(n-1)} \sum_{i \neq j \in V_I} \frac{1}{d_{ij}} \tag{5.37}$$

where d_{ij} is the shortest path between v_i and v_j.

5.9.2 Statistical algorithms

For an analysis of time series data, such as sensor or smart meter measurements and production values, various statistical methods can be used. There are several purposes for which statistics may be utilized that we consider below: (i) descriptive statistics, to summarize a dataset, describing its average values, variance, maximum and minimum values, and other basic parameters; (ii) data processing, used to prepare a dataset for further tasks, for example, for assessing quality of a dataset or detecting and imputing missing data; and (iii) analysis, to extract further knowledge from a dataset, for example, correlations with other values and results of statistical tests and hypotheses.

Descriptive statistics: Statistical methods that summarize and describe characteristics of a dataset are called *descriptive statistics* (Hoaglin et al., 1983). Consider a dataset $\{x_n\} = \{x_1, x_2, \ldots, x_n\}$, where $x_i \in R$. The main parameters describing $\{x_n\}$ are as follows:

- The *mean* or *expected value* of $\{x_n\}$ is an average of all sampled values and is a basic measure of the central tendency of $\{x_n\}$:

$$\overline{x} = \frac{1}{n}\sum_{1}^{n} x_i \tag{5.38}$$

- The *standard deviation* shows how much on average values in $\{x_n\}$ deviate from their mean:

$$\sigma = \sqrt{\frac{1}{n}\sum_{1}^{n}(x_i - \overline{x})2} \tag{5.39}$$

Additionally, the dataset $\{x_n\}$ is characterized by its minimum and maximum points and quartiles—three points in a data—which divide the dataset into four equal groups: (i) 25th percentile, which divides 25% of data from other 75%; (ii) 50th percentile that is a median $\tilde{x}$ of a dataset: Exactly half of data have lower value then $\tilde{x}$, and other 50% have greater value; and (iii) 75th percentile that is a middle between the median and the maximum value of a dataset.

There exist various estimators of these statistical parameters, which are robust to outliers and very helpful for analyzing data in the presence of outliers and noise. Typical robust methods for estimating the central tendency include the following:

- Median of a dataset, that is, 50th percentile
- *Trimmed* (or *truncated*) mean—calculation of a mean value discarding k% (usually from 5% to 25%) of data from each end. One of the examples of the truncated mean is interquartile range (IQM) that trims 25% of data from each end.
- *Trimean*—weighted average of three quartiles of data. For Q_1 denoted as 25th percentile, Q_2 denoted as median, and Q_3 as 75th percentile,

$$T = \frac{Q_1 + 2Q_2 + Q_3}{4} \tag{5.40}$$

Methods for robust estimation of variance include the following:

- Median absolute deviation (MAD)—calculated as median of absolute deviations from data median in dataset. For $\{x_n\}$,

$$\text{MAD} = median\left\{\left(\left|x_i - \tilde{x}\right|\right)\right\}, \tag{5.41}$$

- Average (or mean) absolute deviation is similar to MAD—a mean of absolute deviations of data from its mean. Generally, any measure of central tendency could be chosen for calculation of absolute deviations:

$$\text{AAD} = \frac{1}{n}\sum_{1}^{n}|x_i - M(\{x_n\})| \tag{5.42}$$

where M $(\{x_n\})$ represents mean, trimmed mean, median, or any other measure.

- IQR—difference between 75th and 25th percentiles of a dataset $\{x_n\}$, that is,

$$\text{IQR} = Q_3 - Q_1 \tag{5.43}$$

These are all well-known and widely used estimators of statistical parameters. Practically, there exist many more different estimators classified in broad classes (Hampel et al., 2011), but their usage is out of scope of this work.

Additionally, sometimes in order to detect long-term trends or changes in data, statisticians analyze and calculate basic statistical parameters of different subsets of data set, using a moving (or rolling and running) window. Given a dataset $x_n = \times 1, \times 2$, xn and natural number $k < n$, we analyze each subset generated by moving a k-sized window along the full dataset, that is, $\{x_1, x_2, \ldots, x_k\}$, $\{x_2, x_3, \ldots, x_{k+1}\}$, and so on. In total, there are $n - k+1$ such subsets of $\{x_n\}$.

For two datasets $x_n = x_1, x_2, \ldots, x_n$ and $y_n = y_1, y_2, \ldots, y_n$, we use the measures of covariance and correlation in order to analyze how they depend on each other. For $x = \{x_n\}$ and $y = \{y_n\}$, covariance is calculated:

$$\mathrm{cov}(x, y) = \frac{1}{n}\sum_{1}^{n} (x_i - \tilde{x})(y_1 - \bar{y}) \tag{5.44}$$

Covariance is zero if there is no dependence between x and y and nonzero if they have any dependence between them; the bigger absolute value of covariance shows greater interdependence. Normalized covariance is called correlation, and it changes between -1 and 1, where the value 1 indicates at perfect direct linear dependence between x and y (i.e., y increases if x increases and decreases when x decreases) and -1 indicates at perfect inverse relationship (anticorrelation). The closer the correlation value is to 1 or -1, the stronger the dependence between datasets, and similarly, correlation value closer to 0 indicates at weaker relationship between the datasets. There exist multiple formulas to measure correlation of values examined in Rodgers and Nicewander (1988), but the most familiar and widely used way to measure correlation is Pearson's correlation coefficient.

$$\mathrm{cor}(x, y) = \frac{\mathrm{cov}(x, y)}{\sigma_x \sigma_y} \tag{5.45}$$

where σ_x and σ_y denote standard deviations of $\{x_n\}$ and $\{y_n\}$ correspondingly. This formula is used further in this thesis.

In the multivariate case, that is, for a vector $x = (x_1, x_2, \ldots, x_k)T$, where each component xi is a dataset $\{x_i\} = \{x_{i,1}, x_{i,2}, \ldots, x_{i,n}\}$, the covariance matrix is a $k \times k$ matrix, whose (i and j) element is a covariance value between ith and jth elements of x. Similarly, correlation matrix is a $k \times k$ matrix, whose (i and j) element is a correlation value between x_i and x_j.

Data analytics methodologies are used to find, characterize, and analyze interdependencies between utility network. Data analytics (DA) is a wide interdisciplinary field having its theoretical and practical roots in computer science, artificial intelligence, and mathematics. The goal of data analytics is to inspect, process, and model raw data and draw conclusions from them. There are four major steps of data analytics, also shown in Fig. 5.1. The complexity of analysis and value of results increases with each step, as it is depicted in Fig. 5.1. Descriptive analytics summarizes data and provides a comprehensive description of operation condition, for example, average values of load in the network, detecting and visualizing changes in operation parameters over time. By means of diagnostic analytics, it is possible to derive the cause of a particular event, for example, determine a cause of gas turbine overheat in CHP plants. Predictive analytics forecasts events and operation condition in the nearest future, for example, predicting a peak load in the network. Prescriptive analytics provides different outcomes depending on our actions and recommends the optimal action, for example, predicting demand values and prescribing optimal configurations of production. Altogether, these four steps of data analytics serve to (i) inform, communicate information on the operation state; (ii) analyze, derive new knowledge and predictions; and (iii) act, prescribe actions based on predictions and performance goals.

Data analytic technologies are an important asset in industries, used to process large input of industrial data and extract or derive the new knowledge required according to the task. Descriptive and diagnostic analytic techniques are widely adopted in industry; predictive analytic techniques are still very rare (Fig. 5.33).

Data analysis is used to extract further knowledge from a dataset, for example, correlations with other values and results of statistical tests and hypotheses. For time series methods of a time series, analysis is mostly used. *Time series analysis* is a branch of statistics that provides methods for analyzing time series data in order to derive meaningful characteristics of data. Specifically, there are several common approaches for modeling time series data for analysis or forecasting. Models represent observations as a sum of several independent components, such as trend or season (cyclic reiterations). One of the most common time series model is the ARIMA model (Brockwell and Davis, 2002). Generally, ARIMA models combine autoregression, namely, fitting of data points to linear combination of previous data, and moving averages to estimate the next value.

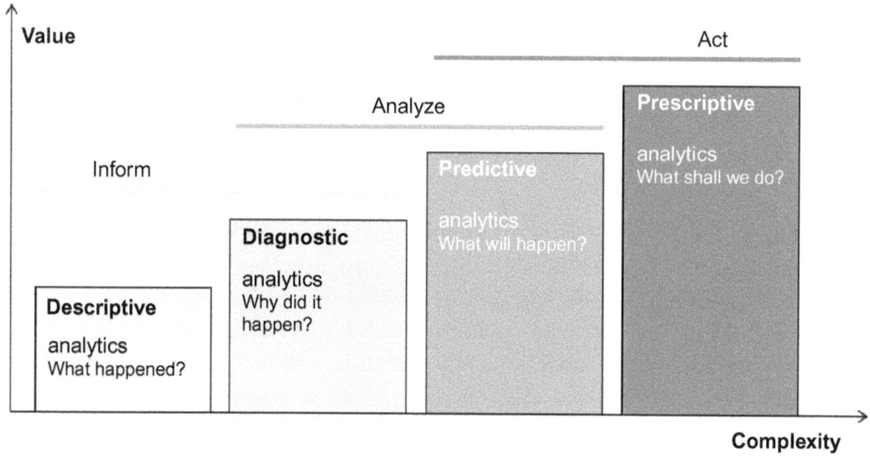

Fig. 5.33 Four steps of data analytics.

All statistical parameters defined in the previous section apply and can be calculated for time series data as well. Additionally, we define autocorrelation for a single time series dataset $\{x_n\} = \{x_1, x_2, \ldots, x_n\}$ as a correlation of x with time-shifted version of itself:

$$\mathrm{acf}(\tau) = \frac{1}{\sigma^2} \sum_{1}^{n} (x_i - \overline{x})(x_{i+\tau} - \overline{x}) \tag{5.46}$$

where τ is a time lag for which we shift initial time series to calculate the correlation between them. Thus, autocorrelation is expressed as a function of time lag, and it is equal to 1 for zero time lag. Additionally, autocorrelation is the so-called even function, that is, $\mathrm{acf}(\tau) = \mathrm{acf}(-\tau)$.

5.10 Application and use cases

5.10.1 Network data and models

Different kinds of interdependencies require different kinds of data to be able to detect them. Such, for geographic interdependencies, information on proximity and location of objects regarding each other is required, whereas for physical interdependencies, knowledge on utility exchange between network objects is necessary. Cyber interdependencies and telecommunication networks are not going to be considered.

To represent each relevant aspect, several models focusing on relevant aspects of case study cities are needed. For instance, typical city case study data include ESRI shapefiles containing (among other information) geographic coordinates on objects and network connections, which can theoretically be used for determining both geographic and physical interdependencies. However, this data format is not structure-oriented and is not reliable to use for determining utility exchanges between networks objects. Of course, using spatial queries, it is still possible to determine whether there is an electricity line between a substation and a building. But it is cumbersome and not efficient for analyzing connections within a district. On the contrary, graph models focus on a structure and connectivity in networks and allow more substantial and complex processing. Below the models and tools used in this work are discussed.

5.10.1.1 Spatial model

Spatial data are an important asset contributing to detection of geographic interdependencies in networks. The ESRI shapefile file format is one of the most common geospatial data format for GIS software. It consists of a collection of files that store geometry of objects and attribute columns for each geometrical object. Here, ESRI shapefiles provide a vector model of a city: building contours, network connections, and objects, which are depicted using the following main types of geometry objects:

- *Point*—a point in space represented by a pair of coordinates (x, y). In the context of utility networks, points might represent measuring devices, valves, pressure regulators, hydrants, and other equipment.

- *PolyLine* (or multiline)—an ordered set of points that form line segments. These segments may be connected or intersected and not connected or not intersected. In the context of utility networks, polylines represent conduits of utility networks, such as water pipes or electricity lines.
- *Polygon*—a geometric shape that is represented by sets of four points that form a closed nonself-intersecting loop. In the context of utility networks, polygons are buildings, power plants, substations, gutters, and other appliances and structures.

In this work, QGIS is used for data visualization and PostGIS database and PostGIS JDBC driver for Java for data processing and querying. The following information can, for example, be extracted:

- Geospatial characteristics of an object: coordinates, dimensions (lengths and perimeter), and area (Fig. 5.34)
- For several objects: proximity and arrangement (whether they are neighbors, located one inside the other, etc.)
- Filtering objects according to their attribute columns

5.10.1.2 Graph model

The graph model reflects the topology of networks, that is, connections between network objects, buildings, and other structures, which contributes to physical interdependencies detection. The network topology is extracted using spatial data as an input. Due to the lack of processing and analysis possibilities of existing solutions, such as pgRouting10, an algorithm was developed for converting spatial data to a graph. First, geometries of type line or polyline are read. For each such line, the algorithm creates (if not already existing) two vertices and creates an edge between them. These vertices are of the type "simple," which means that it is simply a point on the map marking intersections of lines. Further, the algorithm reads points and polygons from the shapefile. If there already exists a simple vertex with the same coordinates as the point or lying within the polygon, it is updated to the type "object" with the corresponding parameters, such as attributes from the shapefile table or geographic parameters, for example, perimeter and area. Otherwise, a new vertex of type

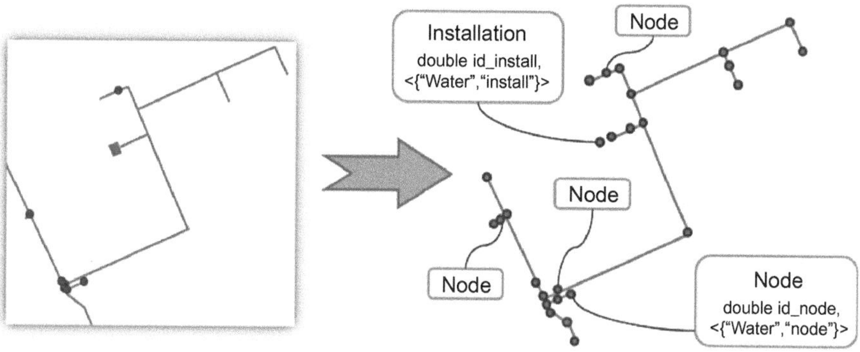

Fig. 5.34 Building graph model from spatial data.

Table 5.1 Summary of Geneva canton utility networks

	Electricity	Gas	Heating	Cooling	Water	Buildings	Σ
Conduit	173,868	196,803	6631	2096	265,252	–	664,650
Objects	97,953	99,894	1947	154	76,428	87,922	364,238

"object" is created. Fig. 5.2 provides an illustration to the procedure. The algorithm is implemented in Java using the JGraphT library. However, the resulting graph model does not always perfectly reflect real-world network connections. The accuracy and correctness depend a lot on the spatial file used and its correctness.

A graph model was built using Geneva canton GIS data on five utility networks: electricity, gas, water, district cooling, and district heating. Table 5.1 shows sizes of these networks and total number of buildings in the canton. The resulting graph built based on these networks has 664,650 edges and 846,057 vertices. Such large amount of vertices is explained by the branching of the networks and hence large amounts of "simple" vertices. Further, vertices that have edges belonging to different networks are called hybrid vertices. An example of a hybrid vertex is CHP appliances. Detection of such vertices contributes to identification of physical interdependencies between networks.

In order to explore the graph and to be able to find hybrid vertices and perform other operations, graph analysis algorithms are needed, for example,

- descriptive characteristics (number of vertices and edges, vertex degree, vertex centrality, and graph connectivity),
- component detection using breadth-first search, and
- path analysis (existence of the path, Dijkstra shortest path algorithm, and minimum spanning tree).

Also to be considered are depth-first search, flow and density analysis, and more efficient algorithms for efficient component detection (Lancichinetti and Fortunato, 2009) and shortest paths (Goldberg and Harrelson, 2005).

5.10.1.3 Numerical data

Numerical data for the network typically comprise characteristics of network objects that can be expressed by numbers, such as diameter of pipes, heating area for buildings, and year of installation for appliances. They can be included in attribute tables in the ESRI shapefile or be provided separately. These are static numerical data. Moreover, each network continuously produces numerical data on its operation, such as supply, demand values, and various parameters. These are typically time series or dynamic data, that is, they consist of sequences of values provided with a timestamp. These dynamic data are stored as tables in PostgreSQL database. Analyzing numerical data contributes to the detection of physical interdependencies.

Various analysis methods are applied to numerical data in order to extract meaningful statistics and other characteristics of the data. Two main stages of data analysis are as follows:

1. Exploratory data analysis (Hoaglin et al., 1983). It is an important stage of data processing used for understanding main characteristics of data and consisting of visual methods (i.e., charts, plots and other visualizations) and descriptive statistics (i.e., mean, variance, and robust measures). It provides a basis for further steps of data analysis, that is, supports the selection of appropriate statistical techniques, suggests hypotheses to be tested, and provides deeper understanding of data.
2. Modeling and data analysis algorithms (Han et al., 2011; Wei, 1994). Their purpose is to extract new knowledge from a dataset. For example, statistical inference is for testing hypotheses about data and includes various statistical tests, for example, tests on distributions and analysis of variance (ANOVA). Regression analysis is used for identifying relations among different variables. Time series analysis provides tools to analyze and forecast time series.

Exploratory data analysis methods (descriptive statistics, visualization, robust statistics, and running window methods) and model algorithms (hypothesis testing, regression models, and ARIMA models) are used in this work. These methods are implemented in the statistical computing language R.

5.10.1.4 Model integration

A Java program connects all above-described models and types of data and allows data exchange between them. Each object in the spatial model, each vertex or connection in the graph model, and each numerical data table are labeled with a unique identifier in order to combine analysis results from different perspectives. That enables the development of novel algorithms operating on all these models, orchestrating data exchange between them and providing more significant results that would not be possible to infer using one model only. Fig. 5.35 shows which types of data are considered and combined.

There are two main possibilities for model integration:

- Sequential operation—apply already existing analysis methods to each type of data and use results of one algorithms as an input or parameters for another algorithms.
- Simultaneous operation—apply more sophisticated algorithms taking fractions from different models as an input. This principle is illustrated in Fig. 5.36B.

5.10.1.5 Data quality assessment and improvement

Performing data analysis effectively and obtaining benefit from its results depend a lot on the quality of data used. Different data quality (DQ) criteria are defined to estimate how credible are results of data processing and analysis, such as completeness, free-of-error, credibility, and consistency. These criteria are very much dependent on the data purpose and the use case. However, real-world data do not always satisfy all the quality criteria. Below in this section, the results of DQ assessment for Geneva GIS data are shown, which are used in this work.

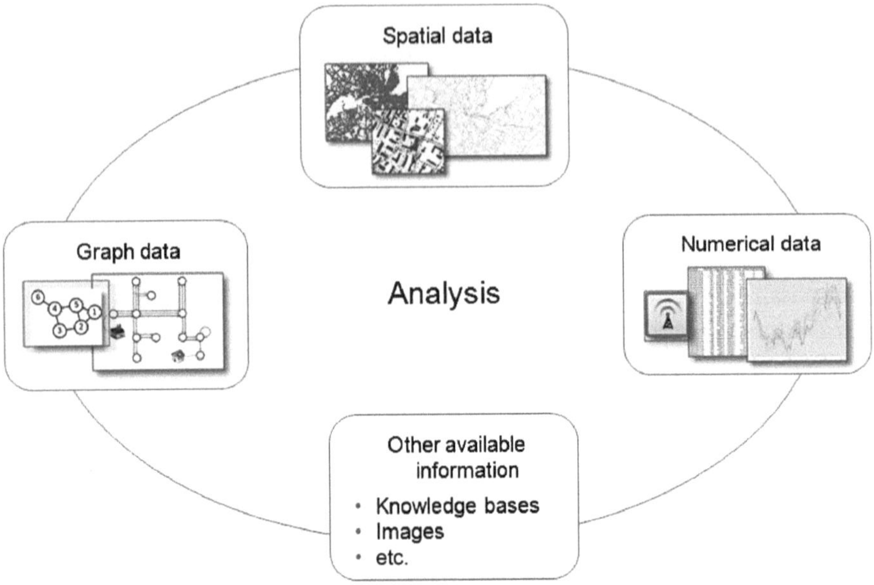

Fig. 5.35 Different types of data used in the approach.

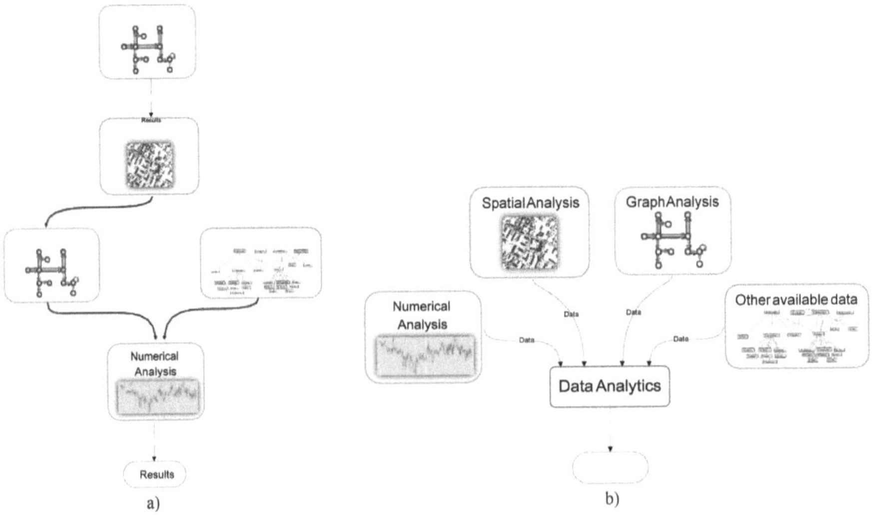

Fig. 5.36 (A) Sequential and (B) simultaneous approaches.

Data of a sufficient quality are crucial. GIS data are primarily used to build a graph model that reflects utility exchanges. During the data exploration phase, discussions with city partner showed that errors are mainly caused by data losses and errors during conversion between formats or domain-specific peculiarities. A domain-specific peculiarity is, for example, cables rolled around causing self-intersections of network connections, which are detected as errors during geometry validity check. Several common DQ issues are shown in Fig. 5.37. These DQ issues cause the resulting graph model to have incorrect structure. Missing data, for example, missing connections shown in Fig. 5.37A and B, cut the objects (appliances, buildings, etc.) off from the networks. Taking such data as an input for the Algorithm 1, building a graph results in a large graph component number, the loss of utility exchanges, and hence possibly imprecise identification of physical interdependencies. Similarly, missing objects might also cause inaccuracies in analysis. Self-intersections and missing intersections shown in Fig. 5.37C and D also reduce accuracy of the model construction. Missing intersections result in additional components cutoff from the network. It includes not only the particular connection but also all further connections and objects connected to it.

In order to build a graph model that reflects network structure as precise as possible, a procedure was elaborated that detects and corrects the following DQ discrepancies:

- Restoration of missing links between objects and network.
- Restoration of missing intersections (as show in Fig. 5.37D and described in Algorithm 3).
- Treatment of isolated objects (as shown in Fig. 5.37A).

There are four possibilities for treating isolated object: (i) not include them in the model, that is, ignore; (ii) include them as they are; (iii) create a connection to the nearest object; or (iv) create a connection to the nearest point. Each of these methods has its pros and cons. After tests, it could be shown that the latter two methods create an abundance of synthetic connections.

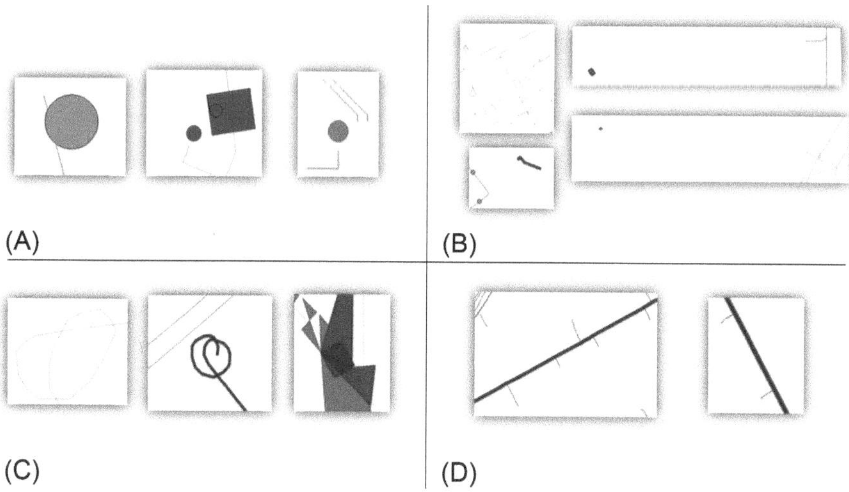

Fig. 5.37 Data quality issues in GIS data: (A) objects and connections are not linked correctly, (B) objects isolated from the networks, (C) self-intersections of connections and objects, and (D) missing intersection points for linked connections.

Therefore, it was decided to treat an isolating object conditionally on the location and proximity from other objects. Further, after completion of the graph model, it makes sense to omit nonobject vertices (or vertices of type "simple"). These vertices do not play any further role in the analysis, unless they mark a place where parameters of the network connection (e.g., diameter) have changed. A pseudocode for deletion of such vertices is shown in Algorithm 4 in Appendix.

5.10.1.6 Further specific use cases

One of the objectives of CI-NERGY Project is to determine city and industry needs for an analysis, resulting in further specific use cases.

5.10.1.7 Correlation and causalities

This topic is interesting for both industry and city partners and corresponds to the performance analysis study. Continuous operation of networks and their appliances is to a great extent affected by other networks depending on types of interdependencies between them. Changes of parameters in one network, such as pressure, temperature, or load, may affect parameters of another network. By determining physical and geographic interdependencies between utility networks, it might be possible to track changes and causes of changes in operational parameters of networks. However, it is necessary to include known physical laws into the analysis in order to avoid inference of trivial knowledge and use them to infer new information on the networks.

5.10.1.8 Competitivity and complementarity of networks

This topic is of a particular interest for cities and their energy companies and contributes to the network planning use case. The stakeholders are interested in which technology is better to use to provide energy, for example, how cogeneration competes with renewables. Further, for new or isolated districts and buildings, there often exists a question of how to deliver utilities to the customer, whether it is optimal to install new appliances or to connect to existing utility networks and use existing appliances. Moreover, different sources of energy can be used. For example, heating can be provided to the customer using waste heat, industrial heat, boilers, or other available energy sources. The Osmose Platform developed in IPESE laboratory at EPFL provides possible optimal solutions for a choice of energy source. The question is how these choices affect existing utility networks. For example, whether installation of new heat pumps affect electricity network and how, create bottlenecks or require further investments in electricity or heating network.

5.10.1.9 Condition monitoring and forecasting services

This topic is of a particular interest for industrial companies using business data analytics. Diagnostic and predictive analytics are of a particular interest. For example, if the temperature or supply of water used to cool a power plant changes, it may influence operation of the power plant. Predictive analytics is used to forecast those occurrences.

In summary, a novel graph-theoretic approach was presented for the representation of city infrastructure networks. The multigraph model captures not only utility network structure but also interdependencies and utility exchange between different networks. Graph models are mostly used for vulnerability analysis of city networks and preventing cascading failures, yet other applications of such models have not been considered until now. However, topological information of the network can be useful for various applications and use cases. It can also be integrated with other models or databases, for instance, binding arcs or vertexes with a database with measurement data regarding this vertex (building) or arc (distribution line). Further, we pursue our main project goal that is the extensive usage of topological information on the utility network acquired from the multidigraph for the network functioning analysis and optimization. In the future, our results will be evaluated on data provided by two large European cities Vienna and Geneva that serve as use case cities.

References

Abdalla, R., Tao, C.V., Cheng, Q., Li, J., 2007. A network-centric modeling approach for infrastructure interdependency. Photogramm. Eng. Remote. Sens. 73, 681–690.

Agemar, T., Weber, J., Schulz, R., 2014. Deep geothermal energy production in Germany. Energies 7, 4397–4416.

Albert, R., Barabási, A.L., 2002. Statistical mechanics of complex networks. Rev. Mod. Phys. 74, 47.

Andersson, G., Donalek, P., Farmer, R., Hatziargyriou, N., Kamwa, I., Kundur, P., et al., 2005. Causes of the 2003 major grid blackouts in North America and Europe, and recommended means to improve system dynamic performance. IEEE Trans. Power Syst. 20, 1922–1928.

Belson, K. and DePalma, A. (2007) Asbestos and Aging Pipes Remain Buried Hazards, The New York Times, 19th of July.

Benonysson, A., Bøhm, B., Ravn, H.F., 1995. Operational optimization in a district heating system. Energy Convers. Manag. 36 (5), 297–314.

Berizzi, A., 2004. In: The Italian 2003 blackout. Power Engineering Society General Meeting. IEEE. IEEE, pp. 1673–1679.

Bøhm, B., Koljonen, T., Larsen, H.V., Lucht, M., Sipilä, K., Park, Y., Wigbels, M., Wistbacka, M., 2002. Simple Models for Operational Optimization. Technical Report, Netherlands Agency for Energy and the Environment.

Brockwell, P.J., Davis, R.A., 2002. Introduction to Time Series and Forecasting. Taylor & Francis, US.

Brown, T., 2007. Multiple modeling approaches and insights for critical infrastructure protection. NATOSecurity Sci. Ser. D: Inform. Commun. Secur. 13, 23.

Chai, C.L., Liu, X., Zhang, W., Baber, Z., 2011. Application of social network theory to prioritizing oil & gas industries protection in a networked critical infrastructure system. J. Loss Prev. Process Ind. 24, 688–694.

Chen, J., 1994. The maximum power output and maximum efficiency of an irreversible Carnot heat engine. J. Phys. D. Appl. Phys. 27, 1144.

Cheng, C.Y., Cheng, C.K., 1995. Performance optimization of an irreversible heat pump. J. Phys. D Appl. Phys. 28, 2451.

Collazos, A., Maréchal, F., Gähler, C., 2009. Predictive optimal management method for the control of polygeneration systems. Comput. Chem. Eng. 33, 1584–1592.

Curti, V., von Spakovsky, M.R., Favrat, D., 2000. An environomic approach for the modeling and optimization of a district heating network based on centralized and decentralized heat pumps, cogeneration and/or gas furnace. Part I. Methodology. Int. J. Therm. Sci. 39 (7), 721–730.

Dueñas-Osorio, L., Craig, J.I., Goodno, B.J., Bostrom, A., 2007. Interdependent response of networked systems. J. Inf. Syst. 13, 185–194.

Energy foundation (FENERCOM), 2016. Region of Madrid, Guide of Solar Thermal Energy (Guía Sobre Energia Solar Térmica). .

Fazlollahi, S., Girardin, L., Maréchal, F., 2014. In: Clustering urban areas for optimizing the design and the operation of district energy systems. Proceedings of the 24th European Symposium on Computer Aided Process Engineering, vol. 33. Elsevier, pp. 1291–1296.

Gabrielaitiene, I. (2011) Numerical simulation of a district heating system with emphases on transient temperature behavior.

Gabrielaitiene, I., Bøhm, B., Sunden, B., 2008. Evaluation of approaches for modeling temperature wave propagation in district heating pipelines. Heat Transfer Eng. 29 (1), 45–56.

Gabrielaitiene, I., Kačianauskas, R., Sunden, B., 2003. Thermo-hydraulic finite element modelling of district heating network by the uncoupled approach. J. Civ. Eng. Manag. 9 (3), 153–162.

Gabrielaitiene, I., Sundén, B., Kačianauskas, R., 2002. Analysis of fluid flow and heat transfer in district heating pipelines using the finite element method. WIT Trans. Eng. Sci. 35.

Goldberg, A.V. and Harrelson, C. (2005) Computing the shortest path: a search meets graph theory (In Proceedings).

Hampel, F.R., Ronchetti, E.M., Rousseeuw, P.J., Stahel, W.A., 2011. Robust statistics: the approach based on inuence functions. Ser. Probab. Math. Stat., 114.

Han, J., Kamber, M., Pei, J., 2011. Data Mining: Concepts and Techniques: Concepts and Techniques. Elsevier.

Hoaglin, D.C., Mosteller, F., Tukey, J.W., 1983. Understanding Robust and Exploratory Data Analysis. vol. 3 Wiley, New York.

Höckner, E., 2014. Wien Energie—District Heating, Presentation for the 2nd Training Module, CI-NERGY European Project. Vienna.

Holmgren, Å.J., 2006. Using graph models to analyze the vulnerability of electric power networks. Risk Anal. 26, 955–969.

Howard, B., Saba, A., Gerrard, M., Modi, V., 2014. Combined heat and power's potential to meet New York City's sustainability goals. Energy Policy 65, 444–454.

Kandiah, V., Rao, P., 2008. Identifying and understanding the infrastructure interdependencies in water systems. West Indian J. Eng. 30.

Khan, M.A., 2010. In: The geysers geothermal field, an injection success story, division of oil, gas, and geothermal resources, 50 D street # 300, Santa Rosa, CA 95404 (USA). Proceedings World Geothermal Congress 2010, Bali, Indonesia, 25–29 April, p. 2010.

Kittel, C., Kroemer, H., 1980. Thermal Physics, second ed. W.H. Freeman, San Francisco.

Köfinger, M., Basciotti, D., Schmidt, R.R., 2017. Reduction of return temperatures in urban district heating systems by the implementation of energy-cascades. Energy Procedia 116, 438–451.

Lai, L.L., Zhang, H.T., Mishra, S., Ramasubramanian, D., Lai, C.S., Xu, F.Y., 2012. In: Lessons learned from July 2012 Indian blackout. 9th IET International Conference on Advances in Power System Control, Operation and Management (APSCOM 2012). IET, pp. 1–6.

Lancichinetti, A., Fortunato, S., 2009. Community detection algorithms: a comparative analysis. Phys. Rev. E. 80(5):056117.

Lund, H., Werner, S., Wiltshire, R., Svendsen, S., Thorsen, J.E., 2014. 4th Generation District heating (4GDH) integrating smart thermal grids into future sustainable energy systems. Energy 68, 1–11.
Martínez, 1992. Termodinámica Básica y Aplicada. Editorial Dossat.
Moran, M.J., Shapiro, H.N., 2000. Fundamentals of Engineering Thermodynamics. Wiley.
Pálsson, H., 2000. Methods for Planning and Operating Decentralized Combined Heat and Power Plants. Risø & DTU-Department of Energy Engineering (ET).
Rant, Z., 1956. Exergie, Ein neues Wort für technische Arbeitsfähigkeit. Forschung Gebiete Ingenieurwesens 22, 36–37.
Rinaldi, S.M., Peerenboom, J.P., Kelly, T.K., 2001. Identifying, understanding, and analyzing critical infrastructure interdependencies. IEEE Control Syst. 21, 11–25.
Rodgers, J.L., Nicewander, W.A., 1988. Thirteen ways to look at the correlation coefficient. Am. Stat. 42 (1), 59–66.
Watts, D.J., Strogatz, S.H., 1998. Collective dynamics of "small-world" networks. Nature 393, 440–442.
Wei, W.W.S., 1994. Time Series Analysis. Addison-Wesley Publ.
Winkler, J., Dueñas-Osorio, L., Stein, R., Subramanian, D., 2001. Interface network models for complex urban infrastructure systems. J. Inf. Syst. 2011 (17), 138–150.
Wollerstrand, J., 1997. District Heating Substations. Performance, Operation and Design. Thesis, Lund University, ISSN 0282-1990.

Further reading

Arreglé, J., 2002. Procesos y tecnología de máquinas y motores térmicos. Universidad Politécnica de Valencia.
Çomakli, K., Yuksel, B. and Çomakli, O. Evaluation of energy and exergy losses in district heating network, Appl. Therm. Eng., 1009–1017.
Directive 91/271/EEC, 1991. Directive 91/271/EEC of the European Council, 21/05/1991 About Urban Waste Water Treatment.
EnerAgen, 2007. Municipality of Pamplona, Thermal and Photovoltaic Energy Guide Inside Technical Edification Regulation Framework. .
Kotas, T.J., 1995. The Exergy Method of Thermal Plant Analysis. Krieger Publishing Company, Malabar, pp. 113–137. 197–204.
Linnhoff, B., 1989. Pinch technology for the synthesis of optimal heat and power systems. J. Energy Resour. Technol. 111(3).
McKendry, P., 2002. Energy production from biomass (part 3): gasification technologies. Bioresour. Technol. 83 (2002), 55–63.
Tipler, P.A., 1978. Modern Physics. Worth Publishers, Inc., New York, USA.

Open data-modeling standards for city-wide energy simulations 6

Giorgio Agugiaro
Smart and Resilient Cities Unit, Center for Energy, AIT—Austrian Institute of Technology GmbH, Vienna, Austria

6.1 Introduction

Many smart-city initiatives deal with improvement of integrated urban planning by means of comprehensive information and process modeling. The main idea is that standardization, harmonization, and integration of information contribute to creating the basis for well-informed city planning and management. Tackling modern urban challenges by means of integrated planning requires a deep understanding of the complex system interdependencies at urban level, in that all physical entities of a city, the citizens, and their mutual relations and processes need to be considered. For energy planning purposes, for example, a comprehensive knowledge of the demand and supply of energy resources, including their spatial distribution within urban areas, is therefore of utmost importance.

Precise and comprehensive knowledge about urban space, infrastructures, and underground features is thus required for simulation and analysis. In order to facilitate these applications, (geo-)information about functional, semantic, and topographical aspects of urban features, their mutual dependencies, and their interrelations is needed. Obtaining (and managing) such detailed information for each urban feature is however a complex and resource-demanding task, as it requires to gather, harmonize, integrate, and eventually manage heterogeneous data for a large quantity of objects for the whole city, therefore guaranteeing a city-wide coverage.

In the last decade, integrated semantic 3-D virtual city models have emerged from very specific application domains (often strongly related to 3-D visualization). They evolved to be a useful all-purpose tool allowing for the development of advanced geospatial applications, thanks to the inherent advantage of the integrated and harmonized information hub they represent.

As a matter of fact, the recent adoption of semantic 3-D city models goes hand in hand with the slow but steady evolution and development of new IT tools in the GIS community, ranging from the definition of standard data models, to "3-D-aware" database and editing tools, etc.

A semantic 3-D city model can represent coherently all city entities, spatially and (possibly) in 3-D, together with all relevant characteristics, relations, and dependencies. What is more, it should be based on open standards, first to reduce the effort in terms of data preparation and provision and secondly to offer clear data structures,

Urban Energy Systems for Low-Carbon Cities. https://doi.org/10.1016/B978-0-12-811553-4.00006-8

ontologies, and semantics to facilitate data exchange between different domains. In short, such a standard-based semantic 3-D city model would represent an "Urban Information Model" (UIM), analogous to the otherwise well-known Building Information Model (BIM). When dealing with modeling of the built environment, only a small number of open model standards exist, the common goal being to ease interoperability between heterogeneous software platforms. However, they are strongly dependent on the scale and application domain they are conceived for.

For single buildings within the Building Information Model (BIM) domain, the two most known and commonly adopted open standards are either Industry Foundation Classes (IFC) or green building XML (gbXML). The former is intended to describe building and construction-industry data. The latter is as an open schema for sharing building information between disparate building design and simulation software tools. Both data models cover both 3-D geometry and semantics.

When working at city scale, the number of choices with regard to open and standard data models comprises essentially only CityGML and (at least for Europe) the upcoming INSPIRE Data Specification on Buildings (which, in part, takes inspiration from CityGML itself). Although they share common concepts, the modeling granularity they offer is dependent on the scale and purpose, with CityGML being essentially conceived for the city and regional scale and INSPIRE for the regional/national/international level. At the same time, CityGML guarantees a functional overlap with the BIM domain, given its semantic and spatial granularity, which for the buildings reaches down till the room level. An exemplary overview is given in Fig. 6.1.

When it comes to open, energy-related data models, at building scale, the already mentioned gbXML is the most widely used data exchange format between different

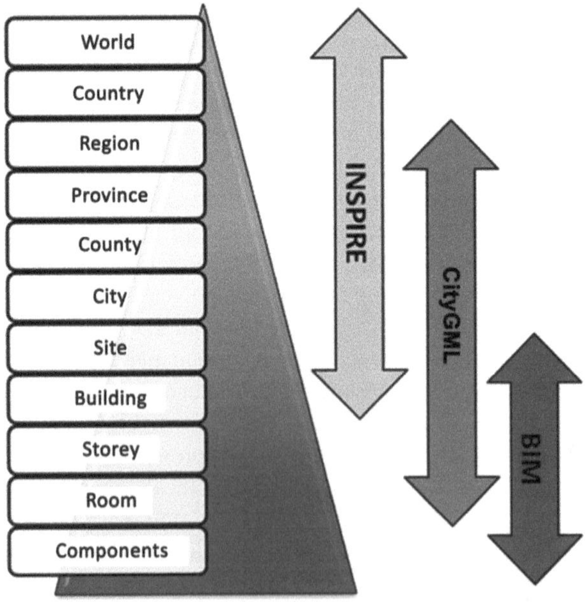

Fig. 6.1 Qualitative comparison of suitability of INSPIRE, CityGML, and BIM standards with regard to territorial scales.

modeling software products for energy analysis. However, at city scale, there is still no existing, complete, and published open standard data model as of 2018. INSPIRE allows indeed for a limited number of energy-related attributes for buildings, for example, connections to networks, energy performance, heating source, and system (INSPIRE-BUILDING). However, they are far too few to cover the needs of complex multidomain applications.

Even CityGML partially falls short: CityGML can in fact already store attributes with regard to the year of construction, the building class, and usage; however, not all energy-related attributes and features can be stored natively in a systematic and standard way.

Therefore, the existing tools for assessment of energy topics at urban scale often rely on proprietary, application-specific, and sometimes closed data formats, de facto nullifying data interoperability.

Nevertheless, in recent years, a growing number of new tools have been adopting CityGML as a common data model, given its open nature, its application-independent scope, and the possibility to extend it by means of the so-called application domain extensions (ADEs): depending on the specific needs, new features or properties can be simply added, hence greatly augmenting modeling capabilities offered by CityGML.

In the energy domain, two ADEs are of particular interest: the Energy ADE and the Utility Networks ADE. Both are currently in active development and are conceived to extend CityGML and allow a standardized way of representing, storing, and exchanging energy-related features and attributes, which are of relevance for urban energy models.

The Energy ADE focuses on the building, its physical properties, and the systems installed in it. It is not meant to cover urban centralized energy infrastructures, like district heating system or gas network, which are instead the focus of the Utility Networks ADE. Nevertheless, the Energy ADE allows for interfacing with the Utility Networks ADE through substation node objects. Both application domain extensions will be described more in detail in the following.

As a matter of fact, several applications adopting CityGML (and some of its ADEs) already exist and range from urban planning, noise mapping, augmented reality, and utility network management to energy simulation tools—just to name a few.

When it comes to Urban Information Modeling and energy-related topics, several research groups are focusing on a growing number of topics: Carrión et al. (2010) and Strzalka et al. (2011) have proposed algorithms to estimate space heating energy demand of buildings. A city-wide energy atlas, focusing on energy demand for residential buildings, was created within the project "Energy Atlas Berlin" (Krüger and Kolbe, 2012; Kaden and Kolbe, 2013). Other conceptually similar examples exist for the cities of Stuttgart, Hamburg, Karlsruhe, and Ludwigsburg in Germany; Lyon in France; Rotterdam in the Netherlands; and Trento in Italy (Nouvel et al., 2013; Bahu et al., 2013; Nouvel et al., 2015; Agugiaro, 2016b).

A state-of-the-art and up-to-date review of different applications of 3-D city models, including energy, can be found in Biljecki et al. (2015), although research is still going on. More recently, Monien et al. (2017) compare building modeling assumptions and methods for urban-scale heat-demand forecasting, and Romero Rodríguez et al. (2017) focus on the assessment of the photovoltaic potential at urban level based on 3-D city models, while Biljecki et al. (2017) deal with the generation of 3-D city models without elevation data.

The remaining of this chapter focuses on CityGML and provides a general description of the standard and of the two energy-related application domain extensions mentioned before.

6.2 CityGML

6.2.1 CityGML standard

3-D city models represent a good starting point to organize heterogeneous types of data and sensors within smart-city projects as they build a stable platform for information management, linking, and enrichment.

The City Geographic Markup Language, or CityGML, has been an Open Geospatial Consortium (OGC) standard since 2008 (CityGML-OGC), and it is conceived specifically as information and data model for storage, manipulation, presentation, and data exchange of semantic city models at urban and territorial scale (Gröger and Plümer, 2012). The term "city" is here broadly defined to comprise not only built structures but also elevation, vegetation, water bodies, "city furniture," and more.

CityGML defines a generic model describing geometry, topology, semantics, and appearance of 3-D objects in urban environments, with five possible levels of (geometric and semantic) detail. Generalization hierarchies between thematic classes are also included, as well as aggregations, relations between objects, and spatial properties.

CityGML is designed as an XML-based format. It is implemented as an application schema of the Geography Markup Language 3 (GML3), the extendible international standard for spatial data exchange and encoding issued by the OGC and the ISO TC211. CityGML is based on several standards from the ISO 191xx family, the Open Geospatial Consortium, the W3C Consortium, the Web 3-D Consortium, and OASIS.

The CityGML standard is presented in diagrams using the Unified Modeling Language (UML) static structure diagram. Fig. 6.2 shows an example of the UML notation and coloring scheme used in the CityGML specification document: here, for better reading, the yellow classes are associated with the CityGML building module, the blue classes are from the CityGML core module, and the green colored class depicts a geometry element defined by the GML3 model. From the abstract class _CityObject, that is, the base class of all thematic classes within CityGML's data model, the subclass _Site is derived and inherits all its properties (i.e., attributes and relations, not shown in Fig. 6.2 for better readability). The abstract class _AbstractBuilding thus inherits all properties from _Site and _CityObject and is further specialized into the Building and BuildingPart classes.

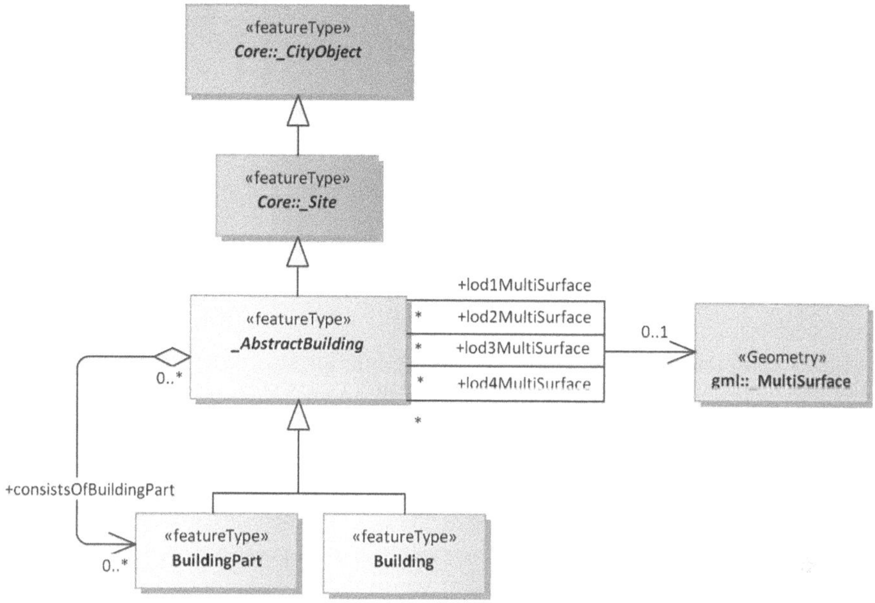

Fig. 6.2 Example UML diagram demonstrating the UML notation and coloring scheme used in the CityGML specification document.

6.2.2 *Main characteristics*

The CityGML data model is thematically decomposed into a core module and thematic modules. The core module comprises the basic concepts and components of the CityGML data model. Based on the core module, each thematic module covers a specific thematic field of virtual 3-D city models. CityGML introduces the following thematic modules: Appearance, Bridge, Building, CityFurniture, CityObjectGroup, Generics, LandUse, Relief, Transportation, Tunnel, Vegetation, and WaterBody. CityGML-compliant implementations may support any combination of extension modules in conjunction with the core module. Such combinations of modules are called CityGML profiles.

CityGML supports different levels of detail (LoDs). LoDs are required to reflect independent data collection processes with differing application requirements and facilitate efficient visualization and data analysis. In a CityGML dataset, the same object may be represented in different LoDs simultaneously.

The coarsest level LoD0 is essentially a 2.5-D digital terrain model (DTM) over which an aerial image or a map may be draped. Buildings may be represented in LoD0 by footprint (Fig. 6.3) or roof-edge polygons.

LoD1 comprises prismatic buildings with flat roof structures. In contrast, a building in LoD2 has differentiated roof structures and thematically differentiated boundary surfaces. LoD3 is conceived for architectural models with detailed wall and roof structures potentially including doors and windows.

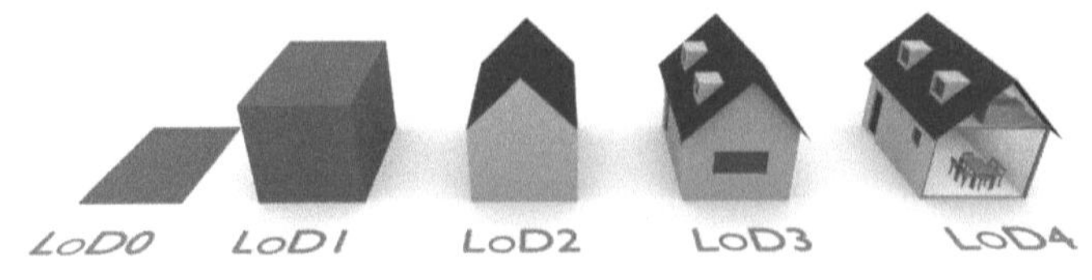

Fig. 6.3 The five LoDs of CityGML 2.0. The geometric detail and the semantic complexity increase, ending with the LoD4 containing indoor features.
From Biljecki, F., Ledoux, H., Stoter, J., 2016. An improved LOD specification for 3D building models. Comput. Environ. Urban Syst. 59, 25–37.

LoD4 completes a LoD3 model by adding interior structures for buildings. For example, buildings in LoD4 are composed of rooms, interior doors, stairs, and furniture.

More specifically, when it comes to buildings, data from topographical or cadastral maps are generally stored in LoD0, and depending on the available height information source, they are either extruded to generate the LoD1 geometries or used as one of the required input datasets, for example, together with 3-D point clouds (coming either from LIDAR or from photogrammetric dense matching) to generate LoD2 buildings fully or semiautomatically.

LoD3 models are generally not widely available at city scale, are mostly imported from computer-aided design (CAD) data, or require in any case a considerable effort to be generated. Alternatively, as this applies also to the LoD4 interior geometries, BIM models can be converted into CityGML keeping in mind the "scale range" overlap existing between CityGML and BIM (see Fig. 6.1).

As already mentioned, all LoD representations are nonexclusive: depending on the available data, a building might have different geometric representations associated to the same unique ID.

One of the most important design principles for CityGML is the coherent modeling of semantics and geometric/topological properties. At the semantic level, real-world entities are represented by features, such as buildings, walls, windows, or rooms. The description also includes attributes, relations, and aggregation hierarchies (part-whole relations) between features. Thus, the part-of-relationship between features can be derived at the semantic level only, without considering geometry. However, at the spatial level, geometry objects are assigned to features representing their spatial location and extent. So, the model consists of two hierarchies: the semantic and the geometric in which the corresponding objects are linked by relationships (Stadler and Kolbe, 2007). If both hierarchies exist for a specific object, they must be coherent. For example, if a wall of a building has two windows and a door on the semantic level, then the geometry representing the wall must contain also the geometry parts of both windows and the door.

In CityGML, all 3-D objects can have relations to external objects in other databases or datasets. The reference of a 3-D object to its corresponding object in an external data set is essential, for example, if additional data are required, for example, the name and address of a building's owner in a cadastral information system. In order to supply such information, each city object may refer to external datasets using the concept of *external reference*. Such a reference denotes the external information system and the unique identifier of the object in this system. The generic concept of external references allows for any city object an arbitrary number of links to corresponding objects in external information systems.

CityGML is designed as a universal topographical information model that defines object types and attributes that are useful for a broad range of applications. In practical applications, the objects within specific 3-D city models will most likely contain attributes that are not explicitly modeled in CityGML. Moreover, there might be 3-D objects that are not covered by the thematic classes of CityGML. CityGML provides two different concepts to support the exchange of such data: (a) generic objects and attributes and (b) application domain extensions (ADEs).

The concept of generic objects and attributes allows for the extension of CityGML applications, that is, any _CityObject may be augmented by additional attributes, whose names, data types, and values can be provided by an application without any change of the CityGML XML schema. Similarly, features not represented by the predefined thematic classes of the CityGML data model may be modeled and exchanged using generic objects.

Application domain extensions (ADEs) specify additions to the CityGML data model. Such additions comprise the introduction of new properties to existing CityGML classes (e.g., the number of habitants of a building) or the definition of new object types. The difference between ADEs and generic objects and attributes is that an ADE must be defined in an extra XML schema definition (XSD) file. The advantage of this approach is that the extension is formally specified. ADEs can be defined (and even standardized) by information communities that are interested in specific application fields. More than one ADE can be actively used in the same dataset. Further details on how to model a CityGML ADE can be found in van den Brink et al. (2014).

6.3 Energy ADE

6.3.1 Introduction

The possibility to extend CityGML by means of application domain extensions (ADEs) represents a useful characteristic of CityGML: depending on the specific needs, new features or properties can be simply added, hence augmenting de facto the modeling capabilities offered by CityGML. For this reason, an international consortium of urban energy simulation developers and users was started in May 2014 to develop an Energy ADE for CityGML.

The Energy ADE is conceived to extend CityGML 2.0 and allows for a standardized way of representing, storing, and exchanging energy-related features and attributes that are of relevance for urban energy models. It is intended to deal with heterogeneous data qualities, levels of details and urban energy model complexities (e.g., from monthly energy balance of ISO 13790 to subhourly dynamic simulations). It takes into consideration the INSPIRE Directive and the recent US Building Energy Data Exchange Specification (BEDES) (Nouvel et al., 2015).

The Energy ADE focuses on the building, its physical properties, and the systems installed in it. Nevertheless, it allows for interfacing with the Utility Networks ADE through substation node objects.

As of 2018, the development of the Energy ADE has reached its first stable version 1.0. However, from the first development versions onward, the Energy ADE has already been tested in a growing number of projects and coupled with different software solutions (e.g., CitySim (CITYSIM), Simstadt (SIMSTADT), and Teaser (TEASER)) directly or by means of interfaces, for example, to EnergyPlus (Agugiaro et al., 2015).

All resources (XSD schemas, documentation, and guidelines) are freely available on GitHub (ENERGY-ADE-GIT) or on the project wiki page (ENERGY-ADE-WIKI).

6.3.2 Main characteristics

The Energy ADE is structured in a modular way and consists of the following modules:

- The Core module
- The Construction and Materials module
- The Building Physics module
- The Occupant Behavior module
- The Energy System module
- The Supporting classes module

The Core module extends two classes of the CityGML base standard (_CityObject and _AbstractBuilding) with energy-related properties. It further provides connecting abstract classes to four of the abovementioned modules ("Occupant Behavior," "Material and Construction," "Energy Systems," and "Building Physics"), and finally, it defines the data types, enumerations, and codelists shared by the modules. With this structure, it is possible to avoid mutual dependencies between these modules.

The Construction and Materials module contains the physical characterization of building construction elements. It is applicable not only to simply store U- and g-values for walls and windows but also to model more complex, multilayered construction elements and their physical properties. In the Energy ADE, its main use is to characterize the geometric surfaces that build up the thermal envelope of a building.

The Building Physics module extends existing CityGML classes (e.g., _AbstractBuilding) and relates them to new "thermal" classes that are used to model the thermal zones of a building and their thermal boundaries and properties.

The Occupant Behavior module focuses primarily on the characterization of the building in terms of usage, occupancy, and facilities contained in it. It allows to define and model usage zones, that is, each zone of a building with homogeneous usage conditions and indoor climate control settings. Each usage zone is further characterized by means of attributes to describe its usage, schedules for heating, ventilation, and air conditioning (HVAC), internal gains, and floor area values. Additionally, this module allows to model building units, that is, subdivisions of a building each having its own lockable access from the outside or from a common area. Classes for facilities (domestic hot water production, electric appliances, and lighting) are also included in this module.

The Energy System module defines the energy systems (e.g., for conversion, distribution, and storage systems) to perform energy demand and supply analyses in a building. Several energy conversion systems can be represented, each one having both common parameters regarding technical properties and specific ones depending on the system type (e.g., solar thermal or PV systems, heat pumps, heat exchangers, and boilers).

Finally, the Supporting classes module contains several classes needed to represent series of time-dependent values, schedules, and weather or climate data. Such classes are used by most of the Energy ADE elements. The inclusion of temporal variables and the definition of their relevant metadata allow to model and store both simulation results and metering data. Moreover, different time steps, corresponding to different building simulation methods and metering systems, are possible, too.

6.4 Utility Networks ADE

6.4.1 Introduction

The Utility Networks ADE defines a topological network model facilitating sophisticated analyses and simulations on utility networks and supplying infrastructures. Included are network hierarchies of arbitrary depth, nesting of network components, and modeling of multimodal networks. Analogously to the Energy ADE and given the general interest for the level of development and the results achieved so far, an international consortium is carrying on the further development of the Utility Networks ADE (UTILITYNETWORK-ADE).

Several other data models and formats for representing utility networks exist. The most relevant ones in the geospatial domain are the INSPIRE Utility Networks model (INSPIRE, 2013a), which is based on the INSPIRE Generic Network Model (INSPIRE, 2013b); the ISO standard Industry Foundation Classes (IFC), which is predominantly used in Building Information Modeling; and the ESRI Geometric Network model. Another ISO standard that allows for representing utility networks is Synthetic Environment Data Representation and Interchange Specification (SEDRIS). SEDRIS focuses on the representation and exchange of synthetic environments and allows for modeling networks for electricity, water, and wastewater and for oil, gas, and chemicals. SEDRIS was developed for training simulation and is to date only applied in the military domain. Under the umbrella of the OGC, PipelineML, a GML-based data interchange standard for the exchange of pipeline data focusing on the oil and gas industry, is currently under development (PIPELINEML-OGC). In its current stage of development, the standard focuses on distribution components and 2-D geometries only; terminal elements such as pump stations are not considered; neither is a topological representation of networks. For a detailed discussion of these and other data models, please refer to Becker et al. (2011), Becker et al. (2012), and Halfawy (2010).

The Utility Networks ADE aims, on the one hand, at providing a common basis for the integration of the diverse models to ease joint analyses and visualization tasks but, on the other hand, also intends to overcome shortcomings of the abovementioned existing network models with respect to the following characteristics:

- They allow for the representation of heterogeneous networks, that is, not only for specific types of networks.
- They allow for a dual representation of network topography and topology and for a representation of topographical/graphic aspects (including 3-D) and of functional aspects.
- They allow for a hierarchical modeling of networks and subnetworks and of components and subcomponents and for modeling interdependencies between network features and city objects and between network features of different types of networks.

6.4.2 Main characteristics

Like the Energy ADE, also, the Utility Networks ADE is structured in a modular way and consists of the following modules:

- Core module
- Network components module
- Functional characteristics module
- Material module
- Network properties module
- Hollow space module

The Core module provides classes and concepts to model multiple different infrastructure networks, to embed the 3-D multiutility networks into the 3-D virtual urban environment and to relate them to each other. It also provides classes to achieve a dual (geographic and topological) representation of the same network. A network can be composed of subnetworks, and by means of modeling the external connectivity of network features, it is possible to make dependencies between networks of different commodity types explicit. Networks sharing the same urban space can therefore be modeled as intermodal networks by having explicitly modeled relations.

The Network components module contains the concrete representation of the entities within utility network. Specialized classes are defined for distribution elements, devices, protective elements, etc., to represent each distinct component of different utility network types. For example, specific classes exist for different distribution elements, depending on the type of the transported commodity (gas, electricity, water, etc.). Therefore, the modeling of utility network entities must take into consideration the functional level of network features and the transported commodity type of those features.

The Functional characteristics module allows for representing in detail the structure of a network together with all the discrete network components the network is composed of and, thus, for following the route of a commodity from source to sink and for determining the whole area the commodity is supplied to by the network, that is, the supply area. Due to the detailed modeling, it is easy to analyze which parts of the network itself and which city objects within the supply area are affected.

Sometimes, however, no information might be available with regard to the detailed modeling of a network supplying a certain area; in that case, it should still be possible to analyze the corresponding supply area and the concrete city objects located in that area. Therefore, the concept of the supply area is introduced: a supply area is defined as that geographic region a specific commodity is supplied to by a network. In other words, the supply area substitutes the network supplying that area, which, in turn, implies that the spatial extent of the supply area conforms to the spatial extent of the network it substitutes. More details can be found in Kutzner and Kolbe (2016).

In the Network properties module, the type and the characteristics of the transported commodity are modeled. The commodity type therefore is related to the network, as a (utility) network can transport, for example, gas, electricity, light, and water. Finally, the Material module provides classes to describe the material each network feature is made of, while the Hollow space module models, as the name suggests, hollow spaces, for example, in pipes transporting waste water pipes or storm water canals.

6.5 Existing tools

6.5.1 Overview

Several commercial and academic tools already support CityGML by providing import interfaces, export interfaces, or both. The 3-D City Database is a free and open-source 3-D geo database to store, represent, and manage virtual 3-D city models. It fully supports CityGML and is shipped with a tool for the import and export of CityGML models. It is based on the open-source Java class library and API for the processing of CityGML models citygml4j (CITYGML4J).

The commercial extract, transform, and load (ETL) tool Feature Manipulation Engine (FME) from Safe Software Inc., either as a stand-alone application or as part of the interoperability extension of ESRI's ArcGIS, has read and write interfaces for CityGML. The same applies to CAD tools as Bentley Map from Bentley Systems and to GIS tools like SupportGIS from CPA Geo-Information.

Many 3-D viewers provide read interfaces for CityGML: the Aristoteles Viewer from the University of Bonn, LandXplorer CityGML Viewer from Autodesk Inc., FZKViewer for IFC and CityGML from KIT Karlsruhe, and BS Contact from Bitmanagement Software GmbH that offers a CityGML plugin for the geospatial extension BS Contact Geo.

The number of companies providing commercial tools with CityGML capabilities include virtualCITYSYSTEMS (virtualcitySUITE), M.O.S.S. (NovaFactory), Fraunhofer IGD (CityServer3D), UVM (CityGrid3D), and 3DIS (CityEditor)—just to name some. This list of software tools is not exhaustive and steadily growing. The reader can refer to the official website of CityGML (CITYGML) and the CityGML Wiki (CITYGML-WIKI). All links can be found in the references section at the end of this document.

Given its relevance in terms of software maturity, the number of implemented features, and its free and open-source nature, the 3D City Database software package is described a bit more in detail in the following.

6.5.2 The 3D city database

The 3D City Database (3DCITYDB) is a free and open-source package consisting of a database schema and a set of software tools to import, manage, analyze, visualize, and export virtual 3-D city models according to the CityGML standard. The database schema results from a mapping of the object-oriented data model of CityGML 2.0 to the relational structure of a spatially enhanced relational database management system (SRDBMS). The 3DCityDB supports both the commercial SRDBMS Oracle (with "Spatial" or "Locator" license options) and the open-source SRDBMS Postgres with its spatial extension PostGIS.

3DCityDB makes use of the specific representation and processing capabilities of the SRDBMS regarding the spatial data elements. It can handle also very large models in multiple levels of details consisting of millions of 3-D objects with hundreds of millions of geometries and texture images. 3DCityDB is already in use in real-life production systems and in many research projects. For example, the cities of Berlin, Potsdam, Munich, Frankfurt, and Zurich all keep and manage their virtual 3-D city models within an instance of 3DCityDB.

Since 3DCityDB is based on CityGML, interoperable data access from user applications to the database can be achieved in at least two ways:

(1) By using the CityGML Import/Export tool or the basic Web Feature Service 2.0 to exchange the data in CityGML format.
(2) By directly accessing the database tables. It is therefore easy to enrich a 3-D city model by adding information to the database tables in some user application.

The 3DCityDB can be used to import a CityGML dataset and then access and work with the city model by directly accessing the database tables from application programs or ETL software.

The Import/Export tool also provides functionalities for the direct export of 3-D visualization models in KML, COLLADA, and glTF formats. A tiling strategy is supported that allows to visualize even very large 3-D city and landscape models in GIS or digital virtual globes like Google Earth or Cesium Virtual Globe.

The Import/Export tool comes with an API to create further importers, exporters, and database administration tools.

Starting from release 3.3.0, the 3DCityDB software package comes with a Cesium-based 3-D viewer called "3DCityDB-Web-Map-Client," which links the 3-D visualization models with online spreadsheets and facilitates interactive visualization and exploration of 3-D city models over the Internet within a web browser.

The provided software tools like the Importer/Exporter application are implemented in the Java language to be platform-independent. The tools run under Microsoft Windows, Linux, and Apple Mac OS X. High performance is achieved by exploiting multithreading on multiprocessor or multicore CPU systems.

6.5.3 *Extending the 3D city database*

With the Energy ADE and the Utility Networks ADE reaching a good level of maturity and stability, there has been a growing interest and need in having a suitable database implementation of the ADEs, ideally extending the already available 3DCityDB.

The current version of the 3DCityDB (and the set of accompanying software tools like the Import/Export tool) does not provide currently a generic solution for handling CityGML ADE elements in CityGML instance documents. Research work is indeed being carried out, for example, at the Technical University of Munich and by the 3DCityDB development team, to extend to 3DCityDB and the accompanying software tools to deal with any ADE and to automatically perform the mapping between the object-oriented model and the relational database model, in order to derive database schemas "on the fly." Further details can be read in Yao and Kolbe (2017).

In the meanwhile, manually derived database schemas for both ADEs were implemented and are already freely available as open-source software. The 3DCityDB extensions for the Energy and the Utility Networks ADEs are based on the 3DCityDB and are currently implemented for PostgreSQL only. As the overall goal of this extension is to facilitate direct connection and usage of the 3DCityDB relational database by users and applications programmers, several design and implementation decisions were taken according to the inspiring criteria briefly listed in the following:

(a) Define a nonconcurrent way of extending the 3DCityDB with multiple ADEs at the same time, for example, adopting naming prefixes to avoid potential homonymy conflicts.
(b) Build upon the existing objects of the 3DCityDB but keep the original ones untouched to guarantee the normal usage of the Importer/Exporter tool with standard "pure" CityGML instance documents.
(c) Try to stay as close as possible to the original design "style" of the 3DCityDB when it comes to tables, constraints, naming conventions, data types, etc.
(d) Follow the Energy ADE and the Utility Networks ADE as close as possible. This means also that tables and attribute names in the database are kept as close as possible to the original names indicated in the UML diagrams.
(e) Keep the number of tables in check, that is, grouping and merging multiple classes into one table—whenever reasonable.

All resources of the database implementation of the Energy and Utility Networks ADEs (DLL procedures, documentation, test data, etc.) can be downloaded from (3DCITYDB-ADE).

6.6 Conclusions

This chapter has provided a brief overview of the existing standards with regard to city-wide energy-related applications. Being the unique *open* standard able to cover most of the requirements in terms of urban energy modeling and simulation, CityGML and its main characteristics were described, before focusing on two of its application domain extensions (ADEs) that play a major role in the energy domain, namely, the Energy ADE and the Utility Networks ADE. The last part of the chapter introduces some of the most common tools to generate, store, and visualize CityGML-based datasets.

References

Agugiaro, G., 2016b. Energy planning tools and CityGML-based 3D virtual city models. Experiences from Trento (Italy). Appl. Geomatics 8 (1), 41–56.

Agugiaro, G., Hauer, S., Nadler, F., 2015. In: Coupling of CityGML-based semantic city models with energy simulation tools: some experiences.REAL CORP 2015 Proceedings/ Tagungsband, pp. 191–200.

Bahu, J.M., Koch, A., Kremers, E., Murshed, S.M., 2013. Towards a 3D spatial urban energy modelling approach. ISPRS Ann. Photogramm. Remote Sens. Spatial Inf. Sci. II-2/ W1, 33–41.

Becker, T., Nagel, C., Kolbe, T.H., 2011. Integrated 3D modeling of multi-utility networks and their interdependencies for critical infrastructure analysis. In: Kolbe, T.H., König, G., Nagel, C. (Eds.), Advances in 3D Geo-Information Sciences, Lecture Notes in Geoinformation and Cartography. Springer, Berlin, Heidelberg, pp. 1–20.

Becker, T., Nagel, C., Kolbe, T.H., 2012. Semantic 3D modeling of multi-utility networks in cities for analysis and 3D visualization. In: Pouliot, J., Daniel, S., Hubert, F., Zamyadi, A. (Eds.), Progress and New Trends in 3D Geoinformation Sciences, Lecture Notes in Geoinformation and Cartography. Springer, Berlin, Heidelberg, pp. 41–62.

Biljecki, F., Stoter, J., Ledoux, H., Zlatanova, S., Çöltekin, A., 2015. Applications of 3D city models: state of the art review. ISPRS Int. J. Geo-Inform. 4 (4), 2842–2889.

Biljecki, F., Ledoux, H., Stoter, J., 2017. Generating 3D city models without elevation data. Comput. Environ. Urban. Syst. 64, 1–18.

Carrión, D., Lorenz, A., Kolbe, T.H., 2010. Estimation of the energetic rehabilitation state of buildings for the city of Berlin using a 3D city model represented in CityGML. Int. Arch. Photogramm. Remote Sens. Spatial Inf. Sci. 38 (4/W15), 31–35.

Gröger, G., Plümer, L., 2012. CityGML – interoperable semantic 3D city models. ISPRS J. Photogramm. Remote Sens. 71, 12–33.

Halfawy, M.R., 2010. Municipal information models and federated software architecture for implementing integrated infrastructure management environments. Automat. Constr. 19 (4), 433–446.

INSPIRE, 2013a. D2.8.III.6 INSPIRE Data Specification on Utility and Governmental Services – Technical Guidelines. European Commission Joint Research Centre.

INSPIRE, 2013b. D2.10.1 INSPIRE Data Specifications – Base Models – Generic Network Model, Version 1.0rc3. European Commission Joint Research Centre.

Kaden, R., Kolbe, T.H., 2013. City-wide total energy demand estimation of buildings using semantic 3D city models and statistical data. ISPRS Ann. Photogramm. Remote Sens. Spatial Inf. Sci. II-2/W1, 163–171.

Krüger, A., Kolbe, T.H., 2012. Building analysis for urban energy planning using key indicators on virtual 3D city models – the Energy Atlas of Berlin. Int. Arch. Photogramm. Remote Sens. Spatial Inf. Sci. 39 (B2), 145–150.

Kutzner, T., Kolbe, T.H., 2016. Extending semantic 3D city models by supply and disposal networks for analysing the urban supply situation. In: Lösungen für eine Welt im Wandel, Dreiländertagung der SGPF, DGPF und OVG, 36. Wissenschaftlich-Technische Jahrestagung der DGPF. Deutsche Gesellschaft für Photogrammetrie, Fernerkundung und Geoinformation e.V., pp. 382–394.

Monien, D., Strzalka, A., Koukofikis, A., Coors, V., Eicker, U., 2017. Comparison of building modelling assumptions and methods for urban scale heat demand forecasting. Future Cities and Environment 3 (1), 2.

Nouvel, R., Schulte, C., Eicker, U., Pietruschka, D., Coors, V., 2013. In: CityGML-based 3D city model for energy diagnostics and urban energy policy support. Proceedings of IBPSA World. 2013, pp. 1–7.

Nouvel, R., Bahu, J.M., Kaden, R., Kaempf, J., Cipriano, P., Lauster, M., Haefele, K.-H., Munoz, E., Tournaire, O., Casper, E., 2015. In: Development of the CityGML application domain extension energy for urban energy simulation.Proceedings of the 4th International Conference of the International Building Performance Simulation Association (IBPSA).

Romero Rodríguez, L., Duminil, E., Sánchez Ramos, J., Eicker, U., 2017. Assessment of the photovoltaic potential at urban level based on 3D city models: a case study and new methodological approach. Sol. Energy 146, 264–275.

Stadler, A., Kolbe, T.H., 2007. Spatio-semantic coherence in the integration of 3D city models. Int. Arch. Photogramm. Remote Sens. Spatial Inf. Sci. XXXVI-2/C43.

Strzalka, A., Bogdahn, J., Coors, V., Eicker, U., 2011. 3D city modeling for urban scale heating energy demand forecasting. HVAC&R Res. 17 (4), 526–539.

van den Brink, L., Stoter, J., Zlatanova, S., 2014. Modeling an application domain extension of CityGML in UML. OGC Best Practice.

Yao, Z., Kolbe, T.H., 2017. In: Dynamically extending spatial databases to support CityGML application domain extensions using graph transformations. 37. Wissenschaftlich-Technische Jahrestagung der DGPF in Würzburg – Publikationen der DGPF. Band 26.

Web resources

3DCITYDB, http://www.3dcitydb.org.

3DCITYDB-ADE, https://github.com/gioagu/3dcitydb_ade.

CITYGML, https://www.citygml.org.

CITYGML4J, https://github.com/citygml4j.

CITYGML-OGC, http://www.opengeospatial.org/standards/citygml.

CITYGML-WIKI, http://www.citygmlwiki.org.

CITYSIM, http://citysim.epfl.ch.

ENERGY-ADE-WIKI, http://en.wiki.energy.sig3d.org.

ENERGY-ADE-GIT, https://github.com/cstb/citygml-energy.

INSPIRE-BUILDING, http://inspire.ec.europa.eu/documents/Data_Specifications/INSPIRE_DataSpecification_BU_v3.0.pdf.

PIPELINEML-OGC, http://www.pipelineml.org.

SEDRIS, http://standards.sedris.org.

SIMSTADT, http://www.simstadt.eu.

TEASER, https://github.com/RWTH-EBC/TEASER.

UTILITYNETWORK-ADE, http://www.citygmlwiki.org/index.php?title=CityGML_UtilityNetworkADE.

Further reading

Agugiaro, G., 2016a. Enabling "energy-awareness" in the semantic 3D city model of Vienna. ISPRS Ann. Photogramm. Remote Sens. Spatial Inf. Sci. III-4/W1, 81–88.

Biljecki, F., Ledoux, H., Stoter, J., 2016. An improved LOD specification for 3D building models. Comput. Environ. Urban. Syst. 59, 25–37.

Uncertainty assessment of building performance simulation

7

An insight into suitability of methods and their applications

Parag Wate[*,†], *Volker Coors*[*], *Marco Iglesias*[‡], *Darren Robinson*[§]
[*]Center for Geodesy and Applied Informatics, Stuttgart University of Applied Sciences, Stuttgart, Germany, [†]Department of Architecture and Built Environment, The University of Nottingham, Nottingham, United Kingdom, [‡]School of Mathematical Sciences, The University of Nottingham, Nottingham, United Kingdom, [§]Sheffield School of Architecture, University of Sheffield, Sheffield, United Kingdom

7.1 Introduction

Modeling and simulation of thermal energy flows in a building aims to understand and predict its thermal behavior and performance. The obtained performance results are employed for critical decision-making and energy-planning tasks, for instance, an evaluation of different energy efficient design options or estimation of energy savings potential of existing building stock. These models represent the underlying physical phenomena in varying levels of abstraction. Depending upon the scale of simulation and complexity of the model involved, the range of data sources that are required for feeding these models would vary largely. In case of preparing individual dynamic building performance simulation (BPS) models for multiple buildings, the model user is often faced with the scarcity of data either largely due to sparse or missing data related to building information, which has been a main hindrance in performing dynamic building simulation at urban scale. Moreover, the data acquisition, preparation, and computing of dynamic building performance for multiple buildings are time-consuming and error-prone tasks. Therefore, often the required building information is obtained based on basic data sources available from city municipalities (e.g., building register or cadastre) and the standards (e.g., IWU norm in Germany) and the building energy performance results are obtained using normative standards (e.g., DIN18599 in Germany) established by the regulation authorities. The methodology thus established is termed as a simplified energy balance approach for calculations of energy demands for heating and cooling in buildings.

The required input datasets are related to: building location and geometry, construction and household composition, weather and simulation parameter datasets. With the simplified approach, the model user arrives at the stage of virtually complete building information from the stage of scarce information but with caveats.

Urban Energy Systems for Low-Carbon Cities. https://doi.org/10.1016/B978-0-12-811553-4.00007-X

For instance, he/she might possess the required data to fill in the placeholders in preparing the BPS model but not necessarily the actual operational data, which are typically unknown and/or sometimes hard to obtain because of myriad reasons. At first place, the required data for simulation can be acquired at different levels of spatial abstraction and temporal resolution. Given that the limited resources at hand, only particular level of data abstraction or resolution can be accessed and it is often only the coarser level of datasets can be acquired and afforded. Sometimes inputs such as infiltration rate are not practically measurable for all the candidate buildings and never been thought of to collect them. Instead, the values for such datasets are assumed based on the heuristics as a function of benchmark or default data values from the available data sources.

In some cases such as the modeling of building geometry and short- and long-wave solar radiation exchange, the values of inputs are also the result of preprocessing and submodeling tasks. The possibility of uncertainties inherent in these submodeling tasks and their respective inputs cannot be ignored completely. On other hand, the availability of household composition data, such as number of occupants, demography, etc., varies between aggregated and detailed levels. Weather phenomena in itself are stochastic. Many simulation parameters are thus based on the assumptions derived from building physics, weather, and usage libraries or standards. Apart from that, the heterogeneity of data availability in the desired spatial and temporal resolutions, differing from one data source to another, results in feeding of interpolated data to the simulation models.

Even when all these typical data uncertainty problems have been understood, decision makers mostly still use one single value from performance predictions without any error bars or statistical distribution of results. Therefore, we argue that the uncertainty inherent in datasets is often not given due consideration. Hence we state that the consideration of the uncertainty problem in BPS would need a paradigm shift in simulation practices from the single-value assignment predictions to probabilistic predictions followed by the assessment of qualitative and quantitative impact of these uncertain input parameters on the simulation results.

This chapter is organized as follows: Section 7.2 on the existing literature summarizes the potential applications of uncertainty assessment approaches in BPS. The section of Section 7.2 in this context puts forth the problem and objectives that are going to be addressed in this research work. Section 7.3 on the methods presents the details of two uncertainty analysis methods: a qualitative measure called Morris method and a quantitative one on classical Monte Carlo (MC) techniques. The sections of Section 7.3 present the implementation details of these methods. Section 7.4 on the case study design formulates our case study in terms of specific method requirements to provide a proof of principle of our concept. Section 7.5 demonstrates how the uncertainty analysis framework has been applied to address the identified problem of uncertain geometry. The sections of Section 7.5 present the uncertainty characterization and corresponding structure assignment procedure. Section 7.6 presents results and discusses the comparative application of both the above-mentioned measures. Section 7.7 concludes the proposed research work and provides open directions for further research.

7.2 Literature review

7.2.1 Uncertainty assessment approaches in building performance simulation

The reviewed approaches can be potentially categorized by the following aspects:

1. quantified uncertain parameter, either an input or phenomenon;
2. uncertainty characterization;
3. uncertainty propagation through MC techniques;
4. applied uncertainty measures either qualitative, quantitative, or both; and
5. the type of simulation models used.

The initial studies were focused on the analysis of performance for different building design configurations. Lomas and Eppel (1992) considered standard weather and operating schedules for a typical building design example with a Gaussian distribution for the thermophysical properties. The outcome of a local sensitivity analysis (LSA)[1] measure called differential analysis revealed the linear nature of the thermal model, and the set-point temperature and window glazing *U*-value as the most influential inputs. De Wit (2001) and his related works (De Wit, 1997a,b) applied the parameter screening method called Morris (1991) (illustrated later in the separate section) and found out that the wind pressure coefficient and indoor air temperature were among the most influential factors. The uncertainties in all other parameters were assumed to be Gaussian distributed during their propagation and the resulted uncertainty in thermal comfort was evaluated in terms of coefficient of variation (= standard deviation divided by sample mean).

The uncertainty analysis study by MacDonald (2002) used a dynamic building thermal model using ESP-r software. The study was largely focused on the identification of suitable methods for the uncertainty representation while assessing individual and overall impacts of the parameters comprising of thermophysical properties, casual heat gains, and infiltration rates. A differential analysis was used to know the critical parameters affecting the resultant indoor temperature. The corresponding impacts of uncertainty in these parameters on the resultant indoor temperature were analyzed in context of a typical IT classroom building during early design stage in order to know the essential performance change requirements to avoid overheating during peak summer temperatures. The infiltration, occupant, and computer gains were identified as the most influential parameters while construction materials having little or no effect on the risk of overheating.

Mara and Tarantola (2008) conducted analysis of variance-based global sensitivity analysis (GSA)[2] on the results from a test-cell building at University of Reunion Island. The objective was to better characterize and understand the thermal behavior

[1] LSA is a qualitative screening measure in which an input parameter is varied in isolation, that is, in a one-at-a-time fashion keeping other parameters at their base level or datum.

[2] GSA is a quantitative measure where all the concerned input parameters are varied simultaneously in order to capture the impact of parameter interactions on the output quantity of interest.

of an actual building. The methodology employed two-level full factorial sampling design: one level called base level assumes the actual thermal characteristics and second level assumes perfectly adiabatic conditions with null heat transfer coefficients. The study enabled to investigate the impact of construction assemblies (wall, roof, ground, and window) in order to know the influence of variation in their thermophysical properties on the predicted indoor air temperature. A hourly computation of sensitivity indices revealed that the window and door during morning, the roof and west sidewall in the afternoon, and ground during nighttime were the most responsible for the variation in hourly indoor temperature. The results corroborated with the facts: During daytime, the diffuse solar radiation transmitted through the window contributes to the warming up of indoor air and hence, having larger impact during the day and while its effect decreases during night time. In case of the ground assembly to be influential during night time, the thermal capacity of the concrete stores the absorbed irradiation during daytime and the same is released during the night.

Mechri et al. (2010) analyzed the variance in energy performance of a typical office building against the six design variables—thermal envelope area to volume ratio, transparent area to envelope area ratio, exterior surface absorptance, orientation, external solar shading factor, and internal effective heat capacity of a building. The uncertainty analysis methodology was based on the MC technique, which produces a set of outcomes by conducting repetitive simulations at the random points in an input sample space, using an ISO standard 13790:2008 (ISO, 2008)-based quasisteady-simplified monthly method for building energy performance assessment. The study concluded with the observations that—transparent to envelope ratio being the most significant and the surface absorptance being the least significant factor in determining the heating and cooling energy requirements of the office building.

An alternative and a more rigorous approach to the quantification of changes in building energy performance due to the changes in design parameters during early design stage was proposed by Hygh et al. (2012). A multivariate linear regression model was developed based on the dataset constructed using the randomly chosen input parameter values and the resultant yearly energy consumption estimates obtained from each EnergyPlus, an existing building performance simulation (BPS) program, model instance within the MC simulation framework. The impact of total 27 building design parameters, of a medium-sized rectangular office building, related to building form, orientation, fenestration, shading, and thermophysical properties was investigated under four different climatic conditions of US continent. The study claimed to fit an excellent linear regression model to the EnergyPlus simulation results and suggested it as an effective instrument for decision support during early design stages. According to the analyzed sensitivity of energy loads to design parameters using standardized regression coefficients, the building area was observed to be the most influential design parameter with its orientation being the least and that too largely a similar trend across all the four climatic zones considered.

A further more systematic approach by Spitz et al. (2012) suggested step-by-step methodology to identify the most influential parameter and as well as to determine a simulation uncertainty band. A simulation model of an experimental house was defined in EnergyPlus and the comparison was made between measured and simulated data. The first step involved knowing of 10 most important parameters out of selected 139 parameters using LSA measure. In the next steps, the repetitive model runs were performed at the parameter sampled values of these 10 parameters to compute global sensitivity indices (explained later in the separate section). The applied GSA measure revealed electric heating system capacity, internal gains, heat exchanger efficiency, infiltration, insulation thickness, and conductivity having highest impact on the room air temperature.

7.2.2 Problem statement and objectives

The application of available methods under uncertainty analysis framework has been demonstrated in the past to address the parameter uncertainty in early building design stages. The current literature is limiting in assessing the impact of uncertain geometry arising from the building modeling and reconstruction procedures. In addition, the full potential of the existing uncertainty analysis methods also lies in understanding the simulation model behavior, which has not been presented before. We therefore attempt to accomplish the following objectives:

- To identify, characterize, and assign the structure for the representation of uncertain building geometry in an uncertainty analysis framework.
- To present a comparative study on the application of both qualitative and quantitative measures for assessment of the uncertainty in building energy performance arising from the building modeling procedures.

7.3 Methods

Complex prognostic models involve an integration of deterministic and stochastic processes whose unknown combined behavior results in an addition of inherent model uncertainty to the predictions of the quantity of interest. Apart from that, depending upon the nature and availability, the model input data sources prone to measurement errors and scaling issues are also responsible for the contribution of either systematic or random uncertainty to the model predictions. As a consequence, the veracity and confidence in the prediction capability of complex models is adversely affected.

Uncertainty and sensitivity analysis techniques offer a pragmatic treatment to uncertainty in model predictions by providing understanding of the model behavior. The model input-output relationship can be investigated to reveal the additivity, linearity, and monotonic behavior of the underlying model. On one hand, uncertainty analysis quantifies the amount of uncertainty in model output due to uncertainties in its inputs. While on the other hand, sensitivity analysis provides measures to decompose the total uncertainty in model output into respective sources of uncertainty in model inputs (Saltelli et al., 2004). The treatment aids modelers in different ways:

1. to reassure the robustness of model against assumptions made;
2. to provide guidance on prioritization of model improvement efforts and subsequent model simplifications by knowing influential inputs;
3. to identify critical regions in input space responsible for desired target behavior; and
4. to assess reliability and know justification for the extreme model behavior.

We classify these techniques into qualitative and quantitative measures.

7.3.1 Qualitative measures

Qualitative measures consist of an experiment design and a screening method. The experiment design is a sampling procedure to select a small number of points from the probability distributions of input parameters in such a way that they uniformly represent the regions of input sampling space. The simulation model is executed at these evenly distributed sample points. Then, the screening method is employed to identify the most influential inputs or the groups containing influential inputs. In screening methods, the indicators about the influence of input x_i on model output y are determined at the local point in an input space. These indicators convey the model behavior in terms of total and interaction effects. While total effect accounts for an overall influence of input on output, an interaction effect determines the higher-order influence involving interactions of x_i with other inputs due to model nonlinearity and nonadditivity. The improved screening methods provide indicators to detect the model monotonicity.

7.3.1.1 Morris design and method

Morris design is a type of One-At-Time (OAT) design, where the changes in input values are represented by the finite levels within an interval. OAT is a local method where model output is computed by changing the value of only one input between the consecutive pair of simulation runs. The local measure of incremental ratios such as an elementary effect is computed at each of these levels accounting for the whole input space. The measures are averaged out to give the single-effect independent of the selection of sample points. Morris (1991) proposed two sensitivity measures based on an elementary effect to determine whether the effect of x_i on y is negligible, linear and additive or nonlinear and nonadditive. An elementary effect and measures are defined as:

k independent inputs x_i, $i = 1, 2, \ldots, k$, define k-dimensional input space discretized into a p-level grid Ω. For a given input x_i, the elementary effect is defined as:

$$EE_i = \frac{[y(x_1, x_2, \ldots, x_{i-1}, x_{i+\Delta}, \ldots, x_k) - y(x_1, x_2, \ldots, x_k)]}{\Delta} \tag{7.1}$$

where Δ = multiple of $\frac{1}{p-1}$ for $p \geq 2$

$X = (x_1, x_2, \ldots, x_k)$ is transformed to a point $(X + e_i\Delta)$ for each input $i = 1, \ldots, k$ and e_i is a vector of zeros but with the unity value at its ith component. The random sampling of a vector X from the grid Ω on input space is used to generate the distribution of elementary effects F_i for every ith input. If p is even, Δ is assumed to be $\frac{p}{2(p-1)}$ with number of effects EE_i in F_i is equal to $p^{k-1}[p - \Delta(p - 1)]$.

The mean μ_i and standard deviation σ_i estimates of distribution F_i represent sensitivity measures. While mean μ_i measures an overall influence of input on the output, σ_i measures the collective effects of an input due to nonlinearity and/or interactions with other inputs. In the simplest design, a computation of a single elementary effect requires two sample points, that is, the economy of design $= \frac{1}{2}$ so for r EEs per k inputs result in $2rk$ model executions. The economy of design can be improved by building r random trajectories, that is, computing r EEs per k inputs from $(k + 1)$ sample points where each trajectory provides k EEs, one per input. So with efficient Morris design, the total simulation runs required are $r(k + 1)$ with the economy of design $= k/(k + 1)$. Fig. 7.1 illustrates a trajectory for $k = 2$, $p = 4$, and $\Delta = \frac{2}{3}$. Here, the trajectory $r = 1$, $(k + 1) = 3$ sample points and model runs $= 3$ at $X^{(1)}[1/3, 1]$, $X^{(2)}[1, 1]$ and $X^{(3)}[1, 1/3]$. For the case of nonuniform distributions, input is sampled in the quantiles space [0, 1] of the distributions and then actual input values for simulation models are obtained using the inverse function method.

The Morris method in the current form is prone to type II errors[3] —failing to identify an influential input. When the underlying model exhibits a considerable nonmonotonous response, a mean μ_i would measure a negligible overall influence due to the cancelling effects of the opposite polarity elements. Alternatively, Morris suggested to study both the μ_i and σ_i measures simultaneously using a graphical representation in (μ, σ) plane. Although if the μ_i would measure negligible, but σ_i would certainly capture the variability in elementary effects. However, this approach has been found to be impractical in case of large models having multiple outputs.

Campolongo et al. (2007) defined an improved Morris measure μ_i^* as the mean estimate of distribution G_i of absolute values of an elementary effect $|EE_i|$. μ_i^* measure is

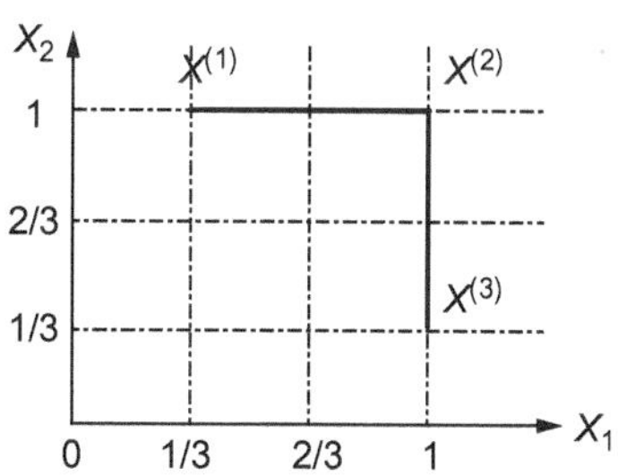

Fig. 7.1 Illustration of trajectory in input space when number of inputs $k = 2$.

[3]Let, Null hypothesis H_0: Input is noninfluential. Alternative hypothesis H_a: Input is influential. Type II error occurs when one accepts a null hypothesis when it is actually not true.

suitable for the models with several uncertain inputs and multiple outputs. This improved model independent measure is robust against type II errors providing realistic ranking of the inputs. An efficient Morris design-based generation of trajectories is susceptible to a nonuniform representation of an input space due to the small fixed number of random trajectories. This leads to biased estimates of the sensitive measures when inputs involved are highly uncertain. Campolongo proposed selection of r trajectories from the set of random number of trajectories in a way that they are optimally dispersed in an input space. A high number of different trajectories M are generated using efficient Morris design discussed earlier. The optimal r trajectories are selected from M trajectories by maximizing geometric distances between them. According to the experiments conducted in the literature, a selection of $p=4, \Delta=\frac{2}{3}$ and $r = 10$ has been considered as an optimum for highly dispersed representation of the input space. The measures are given as:

$$\mu_i = \frac{1}{r}\sum_{j=1}^{r} EE_i^j \tag{7.2}$$

$$\mu_i^* = \frac{1}{r}\sum_{j=1}^{r} |EE_i^j| \tag{7.3}$$

$$\sigma_i^2 = \frac{1}{r-1}\sum_{j=1}^{r} (EE_i^j - \mu_i)^2 \tag{7.4}$$

where $i = 1, 2, \ldots, k$.

The measures μ_i and μ_i^* represent average and absolute average of elementary effects, while σ_i represents their deviation from the mean effect. Using the settings as mentioned for Morris sampling design, the input sample values can be obtained from the characterized distributions.

7.3.2 Quantitative measures

Quantitative measures aid to know by how much a given uncertain input is important than the other among their rankings. Quantitative measures belong to a class of GSA techniques. Unlike qualitative measures, the effect of input x_i on an output y is quantified by exploring the entire space of uncertainty in the inputs. The samples are drawn from the probability distributions considering that the inputs are independent of each other. A multidimensional averaging is performed to decompose the quantified output uncertainty as individual and total uncertainty contributed by the respective inputs. An estimation of the quantitative measure is a two-step procedure: (1) uncertainty propagation and analysis; and (2) computation of sensitivity estimator for quantitative ranking. Roy and Oberkampf (2011) had described a predictive uncertainty analysis framework employed for uncertainty estimation due to different uncertain sources in scientific computing. The employed framework is based on the classical MC technique for uncertainty propagation and quantification.

7.3.2.1 Classical MC technique

The numerical methods for uncertainty propagation aid in mapping of k-dimensional unit hypercube input space represented by N sample points to a scalar or functional model output. The ultimate aim is to quantify the uncertainties in model outcomes arising from uncertain inputs by estimating an unconditional variance of the model outcome by evaluating the model at the sampled points in an input space. Lee and Chen (2009) had classified and compared the performance of representative uncertainty propagation methods for estimation of first (mean)- and second (variance)-order moments with different combinations of function examples and inputs. The MC simulation-based method (Metropolis and Ulam, 1949; Hammersley and Handscomb, 1964) was found to be advantageous.

The MC method involves the evaluation of the multidimensional integral of a square-integrable function $f(x_i)$ over a k-dimensional unit hypercube. In principle, the process draws samples from an output distribution by model evaluations at these sampled points in order to obtain an approximate MC estimate for the numerical integral. Considering the law of large numbers, an estimate converges to the true value of the integral with an estimate error $= \frac{\sigma}{\sqrt{N}}$. Uncertainty analysis is followed by computation of sensitivity indices for the quantitative ranking of uncertain sources. Uncertainty decomposition is the process by which model output uncertainty is apportioned to respective uncertainty of influential sources that cause it. Sensitivity index defines the degree of responsiveness of model output to its uncertain sources. The index multiplied by an uncertainty measure such as variance of input probability distribution gives the amount of quantitative variance contributed.

Sobol (1993) proposed an MC-based evaluation of multidimensional integral of a square-integrable function $f(x_i)$ and the decomposition of functional terms over a k-dimensional unit hypercube. For a two-input variable case, the output variance contributed by interaction of inputs X_1 and X_2 is given as:

$$V = V_1 + V_2 + V_{12} \tag{7.5}$$

where the unconditional variance V is readily computed from the values obtained from evaluations of function f at the sampled points and

$$V_1 = V(\mathbb{E}[Y|X_1]) \tag{7.6}$$

$$V_2 = V(\mathbb{E}[Y|X_2]) \tag{7.7}$$

$$V_{12} = V(\mathbb{E}[Y|X_1, X_2]) - V(\mathbb{E}[Y|X_1]) - V(\mathbb{E}[Y|X_2]) \tag{7.8}$$

The consistency of previous variance decomposition equation can be validated by equating the summation of individual and interaction variance terms to the computed unconditional variance. In principle, an unconditional variance represents the variance contribution of first-, second-, and higher-order effects from the uncertain sources. Dividing both sides by unconditional variance V results in estimates of first- and higher-order global sensitivity indices.

In two-input variable case,

$$S_1 + S_2 + S_{12} = 1 \tag{7.9}$$

In general,

$$\sum_i S_i + \sum_i \sum_{j>i} S_{ij} + \sum_i \sum_{j>i} \sum_{l>j} S_{ijl} + \cdots + S_{123\ldots k} = 1 \tag{7.10}$$

To reduce the problem of dimensionality and hence number of evaluations, Homma and Saltelli (1996) proposed a total effect index S_{T_i} defined as the summation of first-order effect and all the higher interactions involving only input X_i. The total effect is also used for detecting nonadditivity in the model where a collective higher-order interactions would reflect the larger value than that of first order in the total effect index.

$$S_{T_i} = \frac{\mathbb{E}[V(Y|X_{\sim i})]}{V(Y)} = 1 - \frac{V[\mathbb{E}(Y|X_{\sim i})]}{V(Y)} \tag{7.11}$$

For instance, in case of three input models, the total effect sensitivity index S_{T_1} due to X_1 alone is given as:

$$S_{T_1} = S_1 + S_{12} + S_{13} + S_{123} \tag{7.12}$$

7.4 Case study designs

For the purpose of testing of the methodology to uncertainty quantification, we consider two separate case study designs for qualitative and quantitative methods, respectively. For both case study designs, the weather data for Geneva, Switzerland (46 degrees 15′ N, 6 degrees 08′ E, elevation 412 m) has been used for conducting simulations.

7.4.1 Qualitative method: Case study design 1

The two typical roof shape geometries, flat and gabled are considered (Fig. 7.2). In case of flat roof geometry, since eaves and ridge heights are the same, there are five input factors—building position/origin, length, orientation, width, and height to be considered for the analysis. While there are six input factors with split of eaves and ridge height in case of gable roof shape geometry. We consider the optimum assumption of highly dispersed representation of the input space that can be obtained with the number of trajectories $r = 10$, number of levels $p = 4$, and $\Delta = \frac{2}{3}$ for the Morris design. With this design

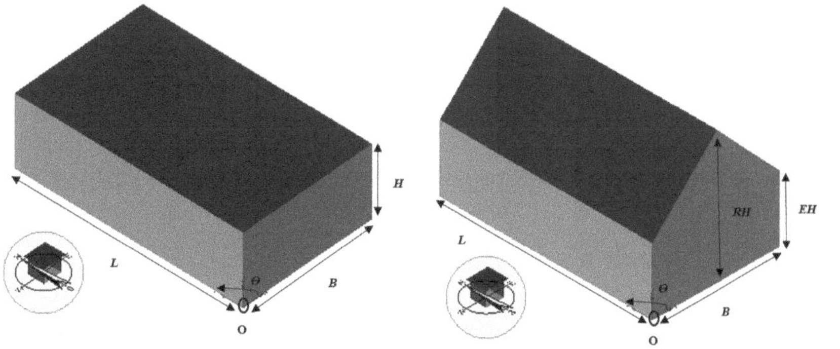

Fig. 7.2 Flat and gabled roof-shaped building geometry.

settings, the total of 60 and 70 representative sample points are obtained from the five- and six-dimensional input spaces of flat and gable roof geometry, respectively. The sampled values obtained from the characterized probability distributions of these uncertain factors have been used to generate the uncertain geometry for each simulation. The characterization of uncertainty in geometry input factors has been explained later in Section 7.5.1. The uncertain geometry generation algorithm has been described in Wate et al. (2016). The simplified thermal model based on DIN 18599 standard for the building energy demand calculations in Germany is executed for each obtained uncertain geometry and the annual thermal energy demand (kWh/m^2) values are obtained. Then, the Morris measures (μ_i,μ_i^*,σ_i) are computed and (μ_i^*,σ_i) plots are obtained for the screening analysis.

7.4.2 Quantitative method: Case study design 2

A typical office configuration with shoe box building geometry has been considered (Fig. 7.3). The same geometry parameters as in the first design of flat-roof geometry have been varied and an uncertainty in energy predictions is quantified. Here the

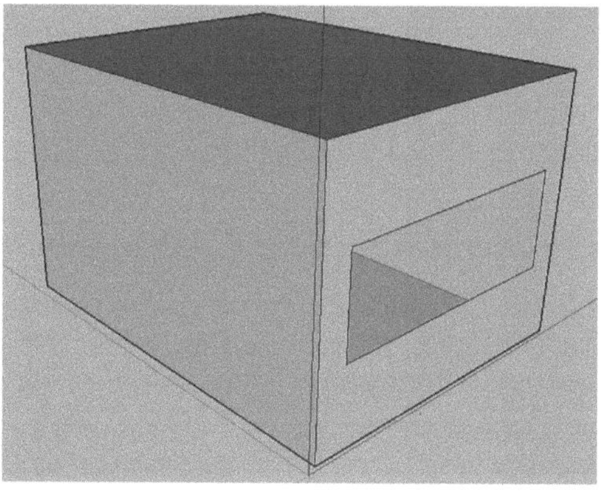

Fig. 7.3 Building geometry of hypothetical office.

uniformly distributed Sobol low-discrepancy sequence (Caflisch, 1998) of geometry parameters has been generated. The samples are generated by exploring the entire uncertainty space, typical as in case of GSA methods. Then, the simulations are conducted using building energy demand prediction model in EnergyPlus (Crawley et al., 2001) building simulation software as a "systemic" black box for each uncertain geometry generated using sampled values for a typical cold week simulation period.

7.4.3 Implementation details

7.4.3.1 Qualitative measure

The concept is implemented in Stuttgart (2015) platform as an independent workflow. SimStadt allows the creation of workflows consisting of the workflow steps where each workflow step performs a defined task and passes its results to the next consecutive workflow steps. The steps perform the tasks of importing building geometry dataset, preprocessing, and model execution. Morris design and screening measures (μ_i,μ_i^*,σ_i) are computed using the "sensitivity" package in R software for statistical computing. We have introduced our algorithm as "Uncertain Geometry Creation" workflow step between import and preprocessing steps and thereby configured a new "Uncertain Geometry Impact Analysis" workflow in SimStadt.

7.4.3.2 Quantitative measure

The first- and total-order indices are computed using the "soboljansen" estimator available in "sensitivity" package in R software. For higher convergence of sensitivity indices' values, the sample size = 5000 is set which resulted into 35,000 simulation runs of simulation model for the selected "soboljansen" estimator. The parametric runs of EnergyPlus model are conducted in a batch process using Python scripts. Then the graphs showing first- and total-order effects of geometry parameters are plotted.

7.5 Application of the uncertainty analysis framework

The presented uncertainty analysis framework has been employed to demonstrate its potential application for the consideration of building geometry uncertainties. Toward this end, we present the step-by-step procedure from geometry uncertainty identification and characterization to its qualitative and quantitative decomposition.

7.5.1 Uncertainty characterization: Building geometry

Uncertainty characterization is a procedure of assigning a mathematical structure to the sources of uncertainty in order to describe the state of knowledge about the uncertainty being considered (Roy and Oberkampf, 2011). The virtual geometry representation of the building is reconstructed from the different data measurements such as surveyed ground control points, terrestrial (El Meouche et al., 2013), and airborne LiDAR cloud points (Verma et al., 2006). The reconstruction procedure may produce a rudimentary form of building representation deviating from the reality, typically in spatial dimensions, internal and external envelope details. This section presents a general overview of the approaches in building modeling and then focuses on the evaluation of two common types of uncertainties—planimetric and vertical in building parameters, arising from the building model reconstruction process.

The continuous improvements in pulse and scan rates of the airborne laser scanners (ALS) (i.e., from Optech, Leica, and Reigel companies) have resulted in an increase in density of the ground point measurements. The acquisitions of high-density datasets allow national mapping agencies to perform large-scale automated mapping and detection of ground objects (e.g., building, vegetation, power cables) at reasonable processing costs. For over more than a decade, the progress in mapping has led to the development of advanced methods and techniques particularly in building object detection and the virtual 3D geometry reconstruction (Vosselman, 1999; Vosselman et al., 2005; Dorninger and Pfeifer, 2008) for different applications (Biljecki et al., 2015). For the reconstruction at urban scale, the building model is derived using model and data-driven approaches ranging from basic ones proposed by Maas and Vosselman (1999) to advanced ones described in Haala and Kada (2010) and Huang et al. (2013). The very first step in the model reconstruction is the building roof plane extraction. Although complex roof forms exist, they are simply the different combinations of one or more typical roof shapes belonging to either flat, gambrel, or hipped roof family shown in Fig. 7.4. Kada and McKinley (2009) presented a similar approach to decompose the building footprints of complex roof forms into nonintersecting cells and then examining the typical roof shapes extracted from the off-terrain LiDAR data points belonging to each cell. The approach was demonstrated by reconstructing the 3D city models for East Berlin and Cologne in Germany.

So, given the set off-terrain LiDAR points, the realistic typical roof shapes can be detected based on the general characteristics and rules. The rules govern the azimuth and slope characteristics of the horizontal and oblique ridge edges (highlighted in red

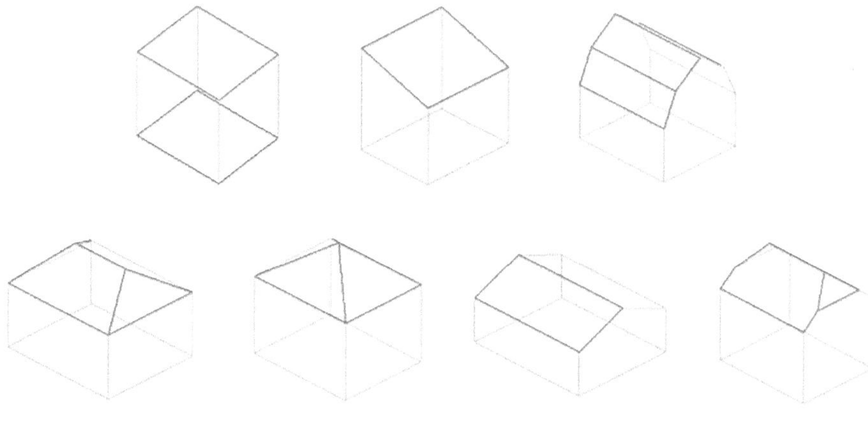

Fig. 7.4 Building roof models. *Top row*: flat and gambrel family, *bottom row*: hipped roof family.

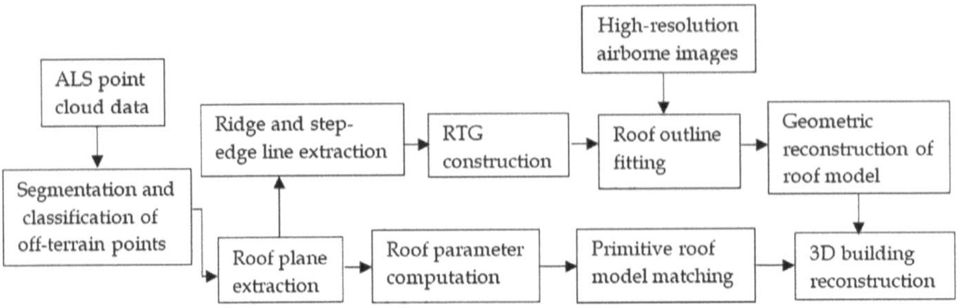

Fig. 7.5 Routine data and model-driven workflow in 3D building reconstruction.

[in bold black in print version] in Fig. 7.4). Perera (2015) discussed the orientation and height thresholds on the adjacent pair of planes for the flat and oblique roof plane extraction. Fig. 7.5 gives an overview of the routine workflow in a building model reconstruction process. The process differs in terms of how the final geometry of a roof model is reconstructed. Maas and Vosselman (1999) introduced the model and data-driven approaches for the typical and complex roof shapes, respectively. However, later, the model-driven approach has been adapted by introduction of cell decomposition technique (Kada and McKinley, 2009) and by defining the rules for merging of typical roof shapes (Huang et al., 2013).

The first common part of the workflow starts with the segmentation of planar faces from the LiDAR point measurements followed by filtering of off-terrain points for the prospective roof planes based on the height thresholds (Sampath and Shan, 2008). Then, the rule-based roof plane extraction can be achieved by fitting the local regression plane equation. In case of top-down model-driven approach (lower part in Fig. 7.5), the prospective roof parameters typically—position, contour, and shape, are estimated from the extracted roof plane. Finally, the best possible roof shape match is obtained by searching the primitive roof model library (Huang et al., 2013). Unlike model-driven approach, the roof model in data-driven approach (upper part in Fig. 7.5) is reconstructed by representing the relationship between plane and ridgeline features in a roof topology graph (Verma et al., 2006; Elberink, 2009). The ridgeline, that is, the intersection line of adjacent roof planes, is extracted by deciding on the distance threshold between the common boundary points. The roof outlines are detected by tracing the boundary points using segment contouring (Vosselman, 1999) and hierarchical least square fitting (Sampath and Shan, 2007) or line straightening methods (Kwak, 2013). Further, the roof feature regularization rules are applied using additional boundary information from the image data in order to present the refined 3D building model for the applications.

7.5.1.1 Uncertainty issues

The various uncertainty issues in building modeling process have been highlighted and a mathematical structure has been proposed to characterize it. It is to be noted that the uncertainty has been considered as a broader term to represent the ambiguities resulting from the errors occurring at different stages in the reconstruction workflow.

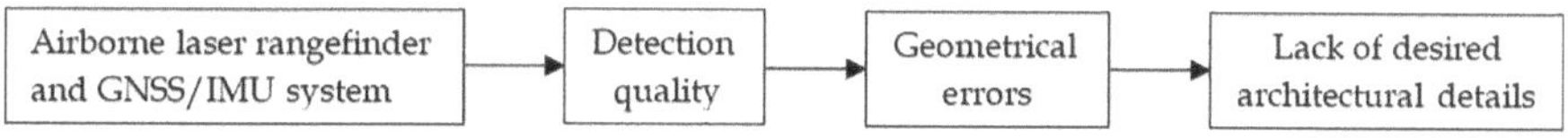

Fig. 7.6 Uncertainty chain in building model reconstruction workflow.

Fig. 7.6 points at the possible sources of uncertainties in workflow starting with acquisition systems to reconstruction techniques. Airborne laser systems consist of measurement and geo-referencing components. The ground data-measuring instrument—laser rangefinder estimates the ground range and elevation covered by the laser pulse reflected from the target. The Global Navigation Satellite System and Inertial Measurement Unit provide the position and attitude data during the acquisition. The suppliers of laser scanning systems specify the overall error budget for measured range and elevation values depending upon the flying height. This could be further augmented by data strip offsets and sensor noise errors due to unstable platform configurations during the acquisition. The typical error values can be found in Petrie (2011). A usual practice is to reflect the overall error budget into the quantifiable terms of planimetric and height uncertainties. In case of laser scanning surveys, Crombaghs et al. (2002) and Vosselman (2008) assess the impact of earlier stated system error factors on height and planimetric accuracies, respectively. Whereas, Vosselman and Maas (2001) and Rentsch and Krzystek (2012) provide a summary of the systematic errors due to the sensor calibration and strip adjustment factors. The absence of data or data gaps due to tree clutter or poor reflection from the window or water features in the scene may result into low or varying laser point density. Based on these datasets, a more bottom-up data-driven approach may reconstruct such roof structures, which could hardly depict the scene reality.

The reconstruction techniques add another layer of uncertainty to the geometric representation of the building. Broadly, it relates to the roof plane detection and the geometrical errors while roof modeling. Segmentation of planar roofs can be achieved using different techniques: 3D Hough transform (Vosselman et al., 2001), surface growing (Dorninger and Nothegger, 2007), and random sample consensus (RANSAC) (Tarsha-Kurdi et al., 2007). The principle is based on the proximity and distance to plane constraints for the neighboring points of the seed surface. A random seed is selected at locally smooth patches and then grown iteratively till the neighboring points fall within the distance thresholds. Then, the filtering criteria is applied to classify the building objects from other nonbuilding objects in the scene. Fig. 7.7 illustrates the roof modeling from the segmented roof planes as well as the geometrical errors occurring at the ridgeline, building outline, and boundary vertices.

Over the past decade, a number of building reconstruction methods have been developed. The approaches reported varying quality measures to demonstrate the reliability of their methods. It would be less plausible to evaluate the common uncertainties from the results of these techniques without a common measure. Moreover, each method used different datasets representing varying complexity of the urban form. Rottensteiner et al. (2012) presented an exhaustive study on the comparison of 3D building reconstruction results based on the five ISPRS benchmark study areas

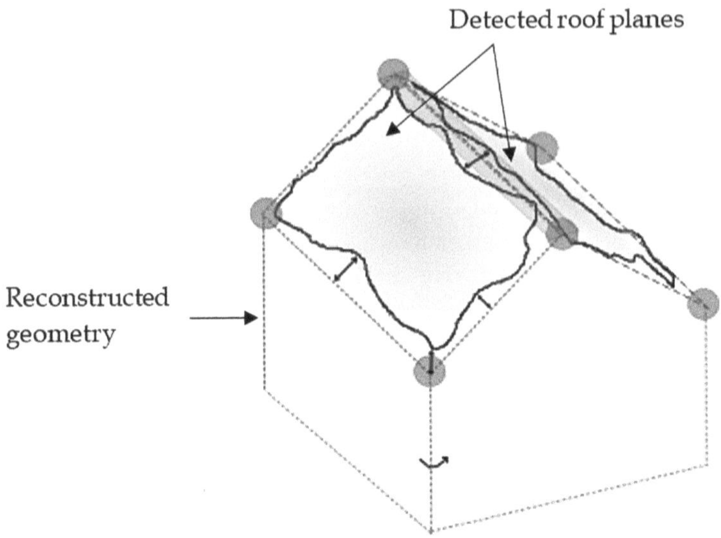

Fig. 7.7 Uncertainty illustration: Reconstruction of building envelope and roof geometry.

in Vaihingen, Germany and Toronto, Canada (Table 7.1). For the building reconstruction comparison, the reference 3D building models were generated with the detailed roof models without overhangs and façade details in the level of detail LoD2 CityGML standard. The researchers were requested to submit the results of their reconstruction techniques for the purpose of comparison with the reference models based on these datasets. Overall, the results of 14 different 3D building reconstruction methods were obtained (Rottensteiner et al., 2014).

The objective of the comparison study was to evaluate the segmentation quality and the geometrical accuracy of the reconstructed 3D roof polygons. The planimetric

Table 7.1 ISPRS benchmark test datasets for 3D building reconstruction

Study area (mean point density p = 4 points/m^2)	Description
Vaihingen, Germany	
Area 1	Historic buildings with complex shapes, roads, and trees
Area 2	High-rise residential buildings surrounded by trees
Area 3	Residential area with detached houses
Toronto, Canada	
Area 4	Mixture of low- and high-rise buildings with complex roof structure
Area 5	Cluster of typical American-style high-rise buildings

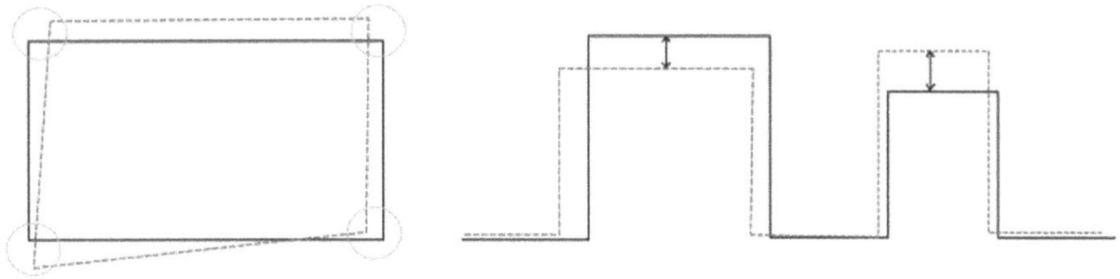

Fig. 7.8 Position (*left*) and vertical (*right*) uncertainty. Reference (*black*) and reconstructed (*red* [*dashed line* in print version]) roof plane and DSM.

accuracy of the roof boundary polygons was measured by estimating the 2D Euclidean distances between the reconstructed and the corresponding reference polygon vertices. Then, the root mean square error (RMSE) of distance was calculated per polygon. On similar lines, the vertical accuracy was estimated as the RMSE of height differences between the reference and the corresponding reconstructed roof planes. The height differences were measured by comparing two synthetic digital surface models (DSM) generated from the 3D building models (Fig. 7.8). The RMSEs can be expressed as

$$RMS_{XY} = \sqrt{\frac{\sum_{i=0}^{n} d_i^2}{N}} \tag{7.13}$$

$$RMS_Z = \sqrt{\frac{\sum_{i=0}^{k} z_i^2}{K}} \tag{7.14}$$

where d_i is the 2D Euclidean distance, N is the number of vertices of a polygon, z_i is the height difference, and K is the number of roof polygons in given scene.

Rottensteiner et al. (2014) presented the evaluation results for the three study areas in Vaihingen dataset. A total of 592 roof planes were processed by all the 14 techniques. The average values of planimetric and vertical uncertainties are reported in Table 7.2. The evaluation study aided in establishing the general relation between the involved geometrical uncertainties and the ALS point spacing. It is reasonable to consider such relation to be valid given the fact that nearly all the results from the reconstruction methods agree on approximately the same value of RMSEs.

Table 7.2 Mean value for uncertainties from the evaluation of building reconstruction results

Uncertainty type	**Mean value (m)**	**Relation to point spacing** $\Delta = \frac{1}{\sqrt{p}}$
Planimetric	≈0.8	$\approx 1.5 \times \Delta$
Vertical	≈0.35	$\approx \Delta$

7.5.1.2 Uncertainty indicators and structure assignment

The indicators of uncertainty in building geometry can now be defined to reflect the final impact of planimetric and vertical uncertainties arising from the reconstruction process. It is to be noted that although the focus is on the typical roof models from the flat or hipped roof family but as a starting point in uncertainty evaluation it is reasonable to select the simple roof primitives. Moreover, the simple roof primitives are the basic unit of complex roof shapes and the roof shapes derived as the combinations of basic roof primitives or they actually represent the majority of the buildings in urban scene. The indicators belong to three categories namely position, contour, and shape, where position and contour reflect the planimetric uncertainty and shape represents the vertical uncertainty. Following notions given by Huang et al. (2013), position = origin, orientation; contour = length, width and shape = ridge, eaves height, hip depth have been adopted. Fig. 7.2 describes the indicators of uncertainty denoted as origin O, orientation θ, length L, width B, ridge height RH, and eaves height EH.

The next task is to assign a probability structure in order to characterize the planimetric and vertical uncertainties in these indicators. Then, the assignment would allow to sample the uncertain values for the indicators and as well as to generate the samples of uncertain geometry for Monte Carlo uncertainty propagation. The uncertainty evaluation data presented by Perera (2015) for the three study areas of ISPRS benchmark test dataset of Vaihingen, Germany have been used. The Weibull distribution is fitted to both planimetric and vertical uncertainties' evaluation data (Fig. 7.9A and B). The shape and scale parameters are calculated using maximum likelihood estimator (Table 7.3). The closeness between the probability distributions of evaluation and fitted data can be observed in Figs. 7.10A and B and 7.11A and B. The uncertainty in position and contour indicators can be characterized by planimetric uncertainty distribution, and the shape indicators by vertical uncertainty. The other study by Vosselman (2012) on planimetric and height accuracy measures also reported approximately the similar probability distribution structure for the observed uncertainties.

Although the limited evaluation data have been analyzed for the characterization of uncertainty, the procedure could be generalized by establishing the relation between the parameters of the distribution and the ALS point spacing given that the sufficient evaluation data are available. It is observed that small values of uncertainty occur more often than the large ones, which is justifiable given the fact that the occurrence of large values could only be the worst case scenario. Thus, it restricts the shape parameter of Weibull distribution to approximately lie between 0 and 3 m in order to preserve the positive skewness of the distribution. At the same time, the scale parameter could be made proportional to the ALS point spacing which would control the spread or width of the bell-shaped distribution, where the high ALS point density (i.e., less ALS point spacing) would result into smaller width of the bell shape (i.e., low uncertainty), which is also justifiable given the fact that the higher point density results into the better reconstruction quality.

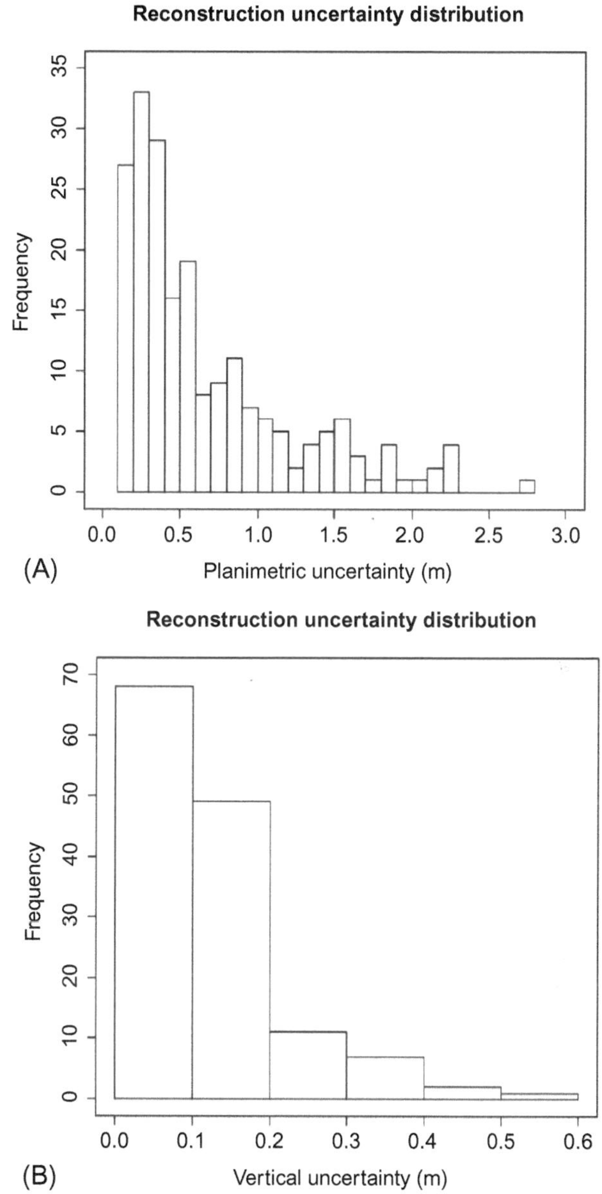

Fig. 7.9 Frequency distribution of planimetric and vertical uncertainties arising from the building reconstruction process (Perera, 2015).

Table 7.3 Shape and scale parameters of Weibull distribution

Distribution parameter	Shape (k)	Scale (λ)
Planimetric	1.31	0.74
Vertical	1.16	0.13

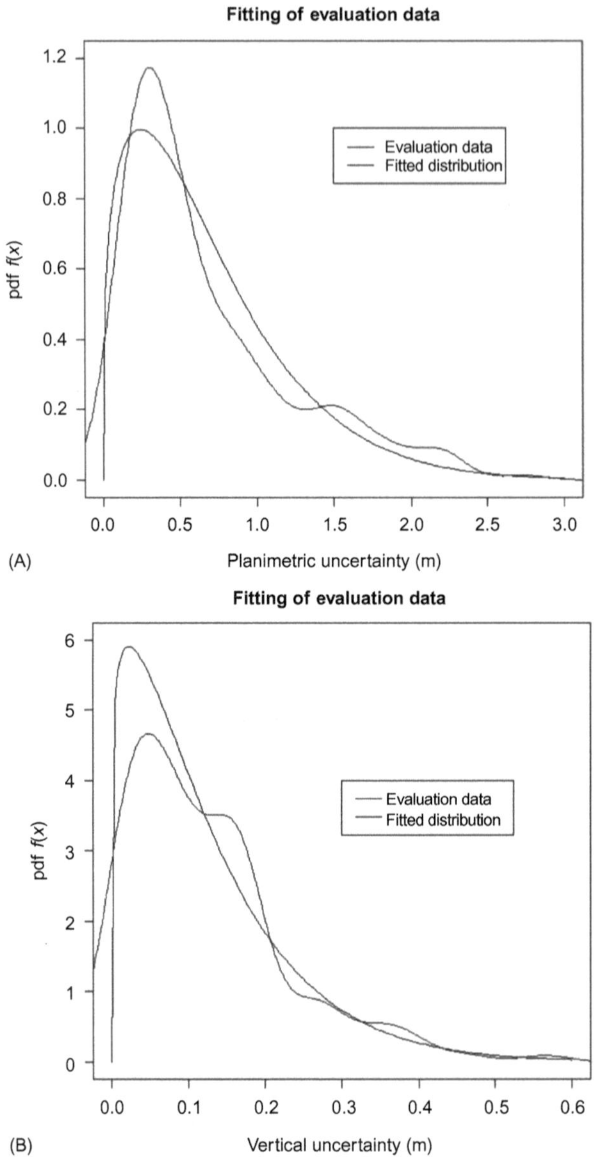

Fig. 7.10 Uncertainty evaluation data fitting using Weibull distribution for planimetric and vertical uncertainties.

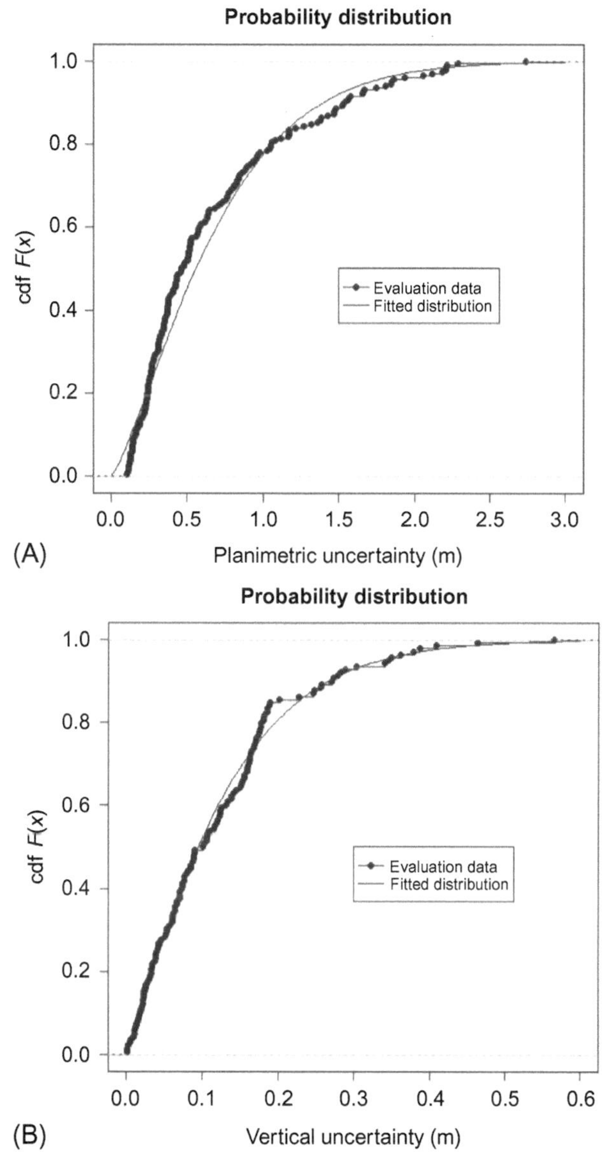

Fig. 7.11 Cumulative distribution functions for planimetric and vertical uncertainties.

7.6 Results and discussion

7.6.1 Qualitative measures and case study design 1

Morris measures (μ_i,μ_i^*,σ_i) are used to understand the model behavior as well as to compute the qualitative ranking of influential geometrical parameters. Sanchez et al. (2014) introduced an indicator ($\frac{\sigma_i}{\mu_i}$) to detect the underlying model behavior. For a given input factor i:

1. Linear ($0 \le \frac{\sigma_i}{\mu_i} \le 0.1$): Deviations of elementary effect are so small that all the effects are very close to their mean or constant depicting linear response with respect to factor i.
2. Monotonic ($0.1 < \frac{\sigma_i}{\mu_i^*} < 0.5$ and $\mu_i = \mu_i^*$): Most of the effects have same sign (+ve/−ve) leading to monotonically increasing or decreasing effect of factor on the model output.
3. Almost monotonic ($0.5 < \frac{\sigma_i}{\mu_i^*} < 1$ and $\mu_i \cong \mu_i^*$): Although there are high deviations in effects, but most of them are either highly +ve or −ve.
4. Nonlinear, nonmonotonic, and nonadditive ($\frac{\sigma_i}{\mu_i^*} > 1$ and $\mu_i \neq \mu_i^*$): The elementary effects deviate highly from the mean effect having nonlinear and/or interactive effect on the model output.

For qualitative ranking of the parameters, μ_i^* measures the influence and σ_i accounts for the effect of interactions between the input parameters on the model output. (μ_i^*,σ_i) and ranking graphs are plotted (see Tables 7.4 and 7.5; Figs. 7.12A and B and 7.13A and B).

Table 7.4 Estimated qualitative measures (flat roof-type case)

Factors	μ	μ*	σ	$\frac{\sigma}{\mu^*}$	Is μ* = abs (μ)?
Origin X	−12.14	25.16	29.04	1.15	No
Origin Y	−4.69	40.2	56.5	1.41	No
Length	13.57	32.93	52.67	1.6	No
Orientation	−28.13	54.38	60.19	1.11	No
Width	30.53	40.5	57.13	1.41	No
Height	−26.06	47.53	78.52	1.65	No

Table 7.5 Estimated qualitative measures (gabled roof-type case)

Factors	μ	μ*	σ	$\frac{\sigma}{\mu^*}$	Is μ* = abs (μ)?
Origin X	13.79	29.29	37.89	1.29	No
Origin Y	−8	39.68	58.62	1.48	No
Length	−12.57	28.48	33.82	1.19	No
Orientation	33.5	63.16	70.46	1.12	No
Width	9.98	20.24	25.47	1.26	No
Eaves height (EH)	20.57	31	45.05	1.45	No
Ridge height (RH)	0.54	36.27	54.3	1.5	No

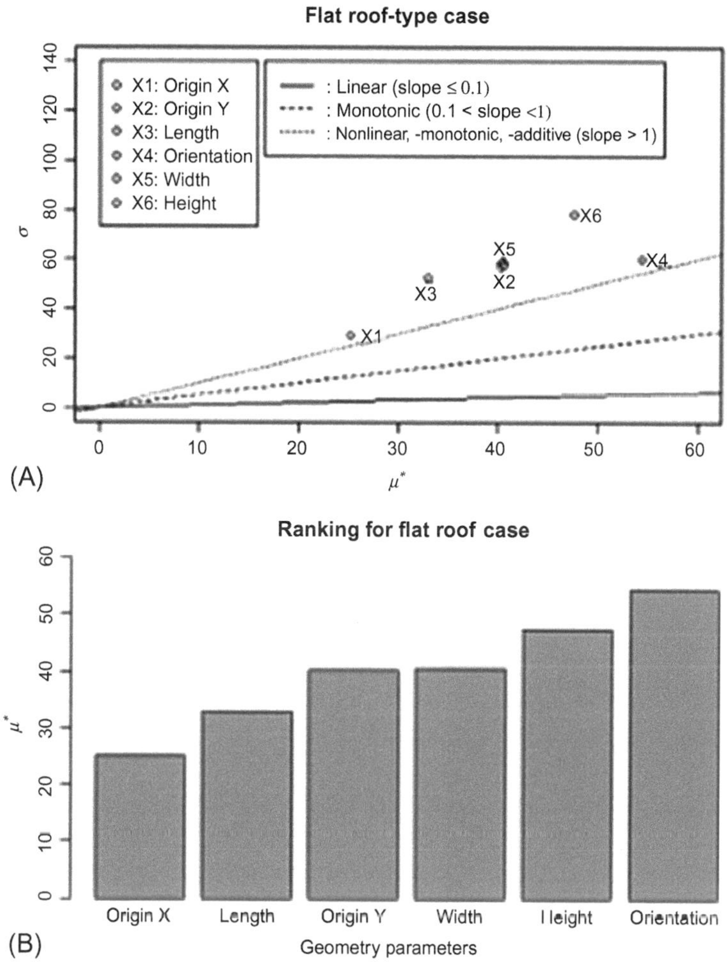

Fig. 7.12 Flat roof type case: (μ_i^*, σ_i) plot and ranking of influential parameters.

For flat roof-type case, the model exhibits highly nonlinear, nonmonotonous, and nonadditive behavior indicated by consistent high values of $\frac{\sigma_i}{\mu_i^*}$. The results state that the building orientation and height are the most influential parameters affecting the annual heating demand values. It is also observed that the building height and orientation have higher interaction effect on the model output.

Similar results are observed for gabled roof-type case. Here, also the input parameters have highly nonlinear, nonmonotonous, and nonadditive impact on the model response ($\frac{\sigma_i}{\mu_i^*} > 1$) with approximately same sequence for the individual and interaction effects.

From the building modeling literature, it can be understood that the orientation and height parameters are more accurate to extract than others are. Because usually they

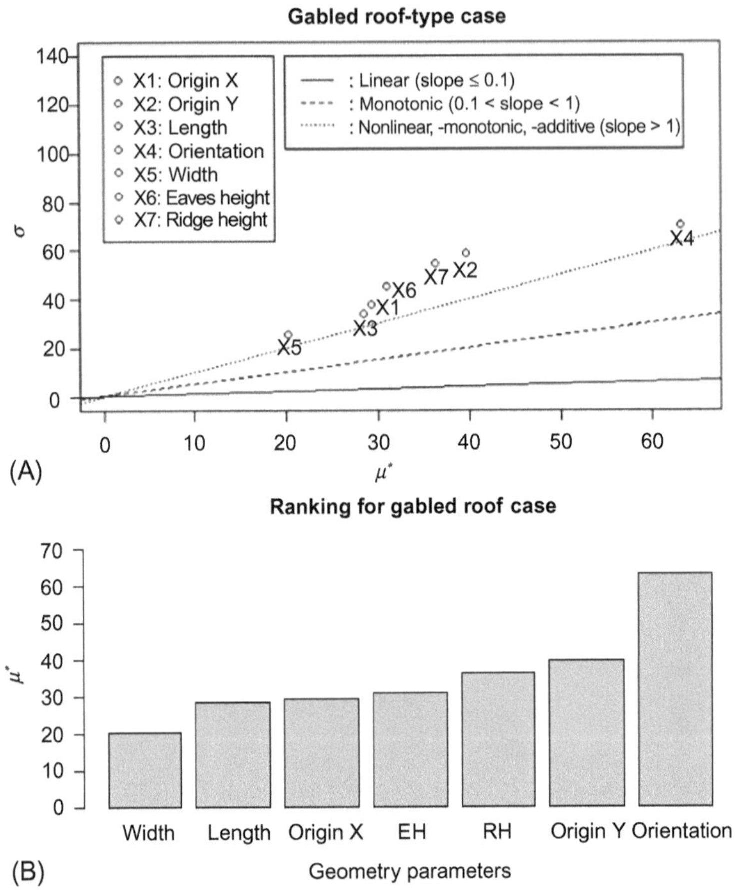

Fig. 7.13 Gabled roof type case: (μ_i^*, σ_i) plot and ranking of influential parameters.

are obtained from the intersection of roof planes derived from the large number of LiDAR point measurements. However, in case of annual energy demand predictions, they are found to be the most influential ones, essentially indicating that even a small amount uncertainty present in them can cause substantial variation in energy predictions. In other words, a large proportion of uncertainty energy demand predictions can be reduced by employing near accurate techniques for roof plane and ridgeline extraction.

7.6.2 Quantitative measures and case study design 2

In this section, we present the application of quantitative measure for addressing of the uncertain geometry problem. So far, the identified and characterized geometry uncertainty in last sections has been propagated using classical Monte Carlo simulation techniques. The propagated uncertainty in energy demand can now be quantified

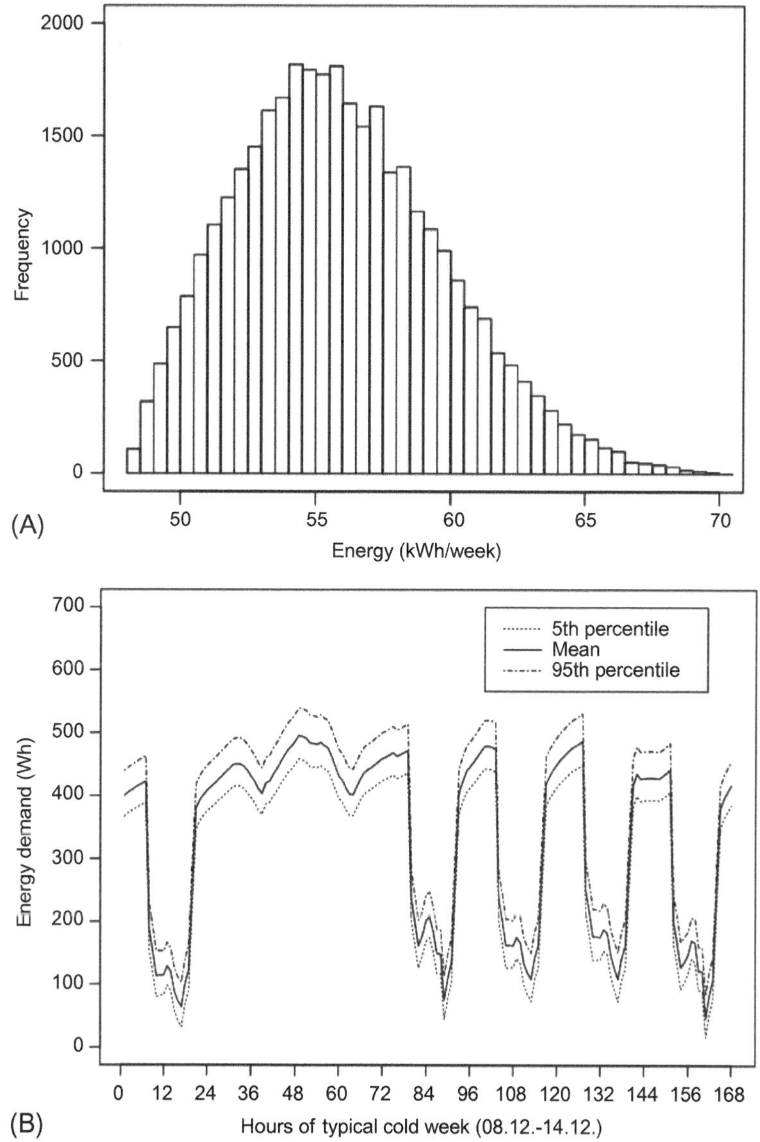

Fig. 7.14 (A) Heating energy demand distribution and (B) hourly variation of energy demand for a typical cold week.

and decomposed into the geometry parameters that caused it. This section presents results that are being observed while understanding the underlying relationship between the model inputs and output.

Fig. 7.14A describes the frequency distribution of heating energy demand over the period of typical cold week. With the amount of geometry uncertainty considered, the

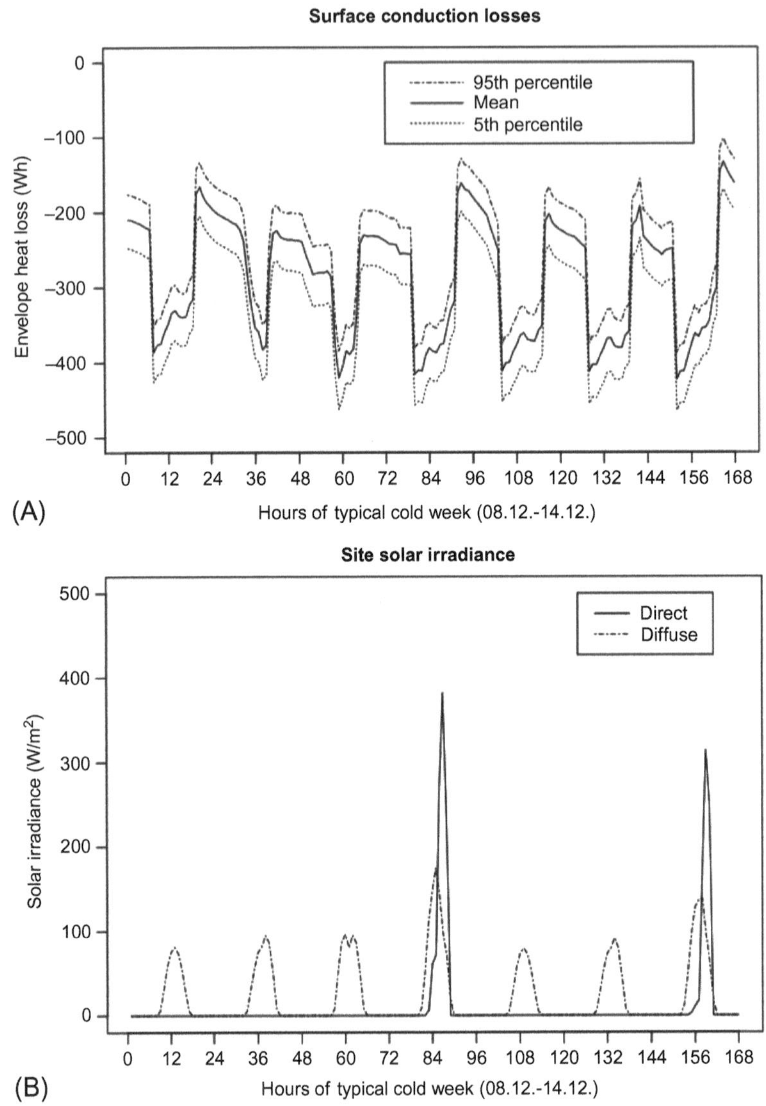

Fig. 7.15 (A) Conduction losses through building envelope and (B) solar irradiance at site.

heating energy demand lies in range of 48–64 kWh/week, which represents a more realistic estimate for establishing of the building energy efficiency strategy than the single deterministic value. It can be observed that the variation in energy demand remains almost constant at both the low and higher values of energy demand (Fig. 7.14B). So the variation is almost homoscedastic in nature.

The variation in conduction losses through the building surfaces also reveals approximately the similar homoscedastic nature (Fig. 7.15A). With the kind of problem formulation in the present study, only the indirect impact of radiation exchange on the energy demand is reported. Because the variation in geometry parameters only alters the amount of conduction losses as well as the solar heat gains due to the variation in building surface areas. Due to the presence high amount of solar irradiance during the noon time on 4th and 7th day of the week (Fig. 7.15B), a substantial decrease in heating energy demand has been observed because of the solar heat gains dominating the conduction losses.

Fig. 7.16B gives the estimation of hourly computed sensitivity indices. It is observed that the main and total effects for the parameters are almost the same, meaning that there is negligible impact of higher-order interactions between the uncertain

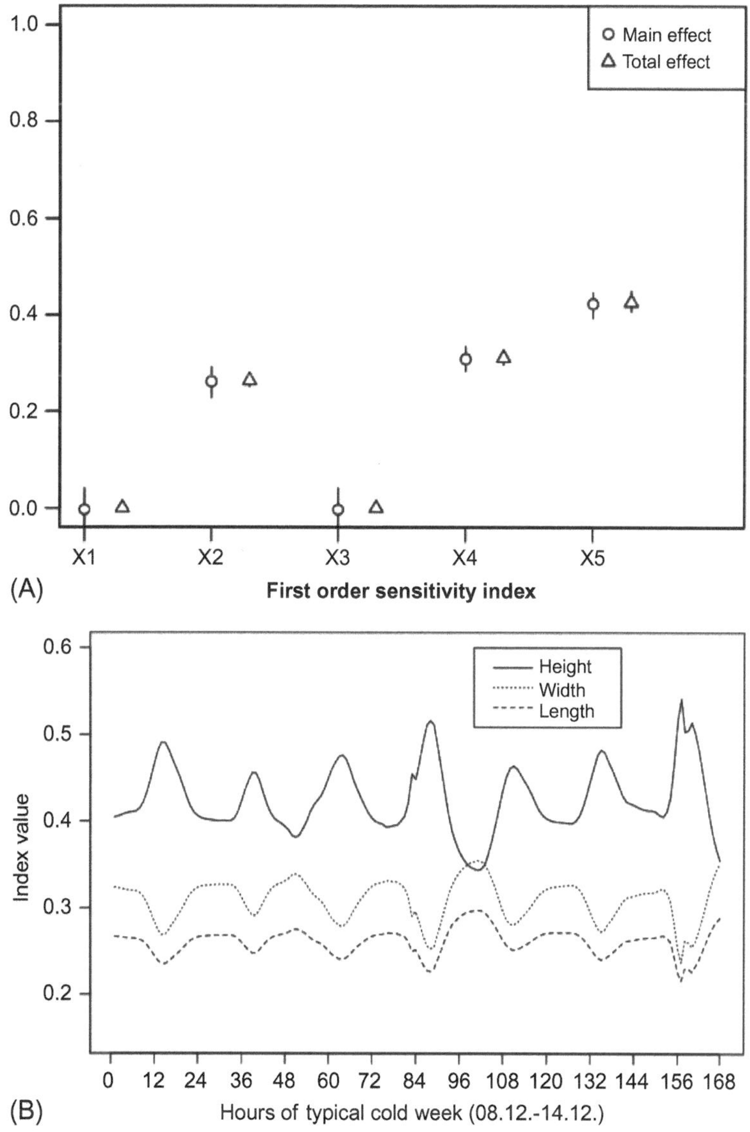

Fig. 7.16 (A) Main and total effect where X1 (origin/position), X2 (length), X3 (orientation), X4 (width), and X5 (height); (B) temporal variation of first-order sensitivity indices.

parameters on the energy demand. The building height, length, and width are observed as the influential parameters and height being the most influential among them. It points out that the vertical uncertainty being the important factor even though the discrepancy observed is relatively small (equal to ALS point spacing) than the planimetric uncertainty (1.5 times the ALS point spacing). It invites for further research into the improvement of algorithms for building height estimation during reconstruction process.

7.7 Conclusions and future work

The research work on qualitative measures has introduced the mechanism to account for the uncertainty arising from building geometry modeling and assessed its impact on the energy demand calculations. The methodology has proposed a probabilistic simulation method for the inclusion of building geometry data uncertainties. The study has also attempted to address the impact of uncertain geometry problem by establishing both qualitative screening and quantitative methods. On the one hand, at the level of urban scale simulation, it is essential to know the important geometry parameters so that the available resources can be judicially allocated in order to derive more precise values, while on other hand it aids suitable building representation in performance prediction software still giving plausible results. Considering the context of study (e.g., simulation model used, type of input varied), the building orientation and height were assessed as the most influential parameters affecting the energy demand calculations. The building height was reported to be relatively the most important one by both uncertainty quantification methods.

The methodology will be further extended to quantify the uncertainties arising from the deterministic and stochastic phenomena involved in the prediction of energy performance of buildings in their urban context. A comparative study of simulation results will be conducted while considering the different modeled levels of building geometry abstraction. The comparison could be used to answer the question of how detailed the geometry should be acquired in order to obtain sufficiently reliable simulation results.

Acknowledgments

The authors gratefully acknowledge the European Commission for providing financial support during the conduct of research under the FP7-PEOPLE-2013 Marie Curie Initial Training Network "CI-NERGY" project with Grant Agreement Number 606851.

References

Biljecki, F., Stoter, J., Ledoux, H., Zlatanova, S., Çöltekin, A., 2015. Applications of 3D city models: state of the art review. ISPRS Int. J. Geoinf. 4 (4), 2842–2889.

Caflisch, R.E., 1998. Monte Carlo and quasi-Monte Carlo methods. Acta Numer. 7, 1–49.

Campolongo, F., Cariboni, J., Saltelli, A., 2007. An effective screening design for sensitivity analysis of large models. Environ. Model Softw. 22 (10), 1509–1518.

Crawley, D.B., Lawrie, L.K., Winkelmann, F.C., Buhl, W.F., Huang, Y.J., Pedersen, C.O., Strand, R.K., Liesen, R.J., Fisher, D.E., Witte, M.J., et al., 2001. EnergyPlus: creating a new-generation building energy simulation program. Energy Build. 33 (4), 319–331.

Crombaghs, M., Elberink, S.O., Brugelmann, R., de Min, E., 2002. Assessing height precision of laser altimetry DEMs. Int. Arch. Photogramm. Remote Sens. Spat. Inf. Sci. 34 (3/A), 85–90.

De Wit, M.S., 1997a. Identification of the important parameters in thermal building simulation models. J. Stat. Comput. Simul. 57 (1–4), 305–320.

De Wit, M.S., 1997b. Influence of modeling uncertainties on the simulation of building thermal comfort performance. In: Building Simulation, vol. 5. International Building Performance Simulation Association, Prague, Czech Republic.

De Wit, M.S., 2001. Uncertainty in Predictions of Thermal Comfort in Buildings (Ph.D. thesis), TU Delft, Delft University of Technology.

Dorninger, P., Nothegger, C., 2007. 3D segmentation of unstructured point clouds for building modelling. Int. Arch. Photogramm. Remote Sens. Spat. Inf. Sci. 35 (3/W49A), 191–196.

Dorninger, P., Pfeifer, N., 2008. A comprehensive automated 3D approach for building extraction, reconstruction, and regularization from airborne laser scanning point clouds. Sensors 8 (11), 7323–7343.

El Meouche, R., Rezoug, M., Hijazi, I., Maes, D., 2013. Automatic reconstruction of 3D building models from terrestrial laser scanner data. ISPRS Ann. Photogramm. Remote Sens. Spat. Inf. Sci. (1), 7–12.

Elberink, S.O., 2009. Target graph matching for building reconstruction. Proc. Laserscanning 9, 49–54.

Haala, N., Kada, M., 2010. An update on automatic 3D building reconstruction. ISPRS J. Photogramm. Remote Sens. 65 (6), 570–580.

Hammersley, J.M., Handscomb, D.C., 1964. General principles of the Monte Carlo method. In: Monte Carlo Methods. Springer, pp. 50–75.

Homma, T., Saltelli, A., 1996. Importance measures in global sensitivity analysis of nonlinear models. Reliab. Eng. Syst. Saf. 52 (1), 1–17.

Huang, H., Brenner, C., Sester, M., 2013. A generative statistical approach to automatic 3D building roof reconstruction from laser scanning data. ISPRS J. Photogramm. Remote Sens. 79, 29–43.

Hygh, J.S., DeCarolis, J.F., Hill, D.B., Ranjithan, S.R., 2012. Multivariate regression as an energy assessment tool in early building design. Build. Environ. 57, 165–175.

ISO, 2008. ISO 13790: Energy Performance of Buildings—Calculation of Energy Use for Space Heating and Cooling. International Standard Organisation.

Kada, M., McKinley, L., 2009. 3D building reconstruction from LiDAR based on a cell decomposition approach. Int. Arch. Photogramm. Remote Sens. Spat. Inf. Sci. 38 (Part 3), W4.

Kwak, E., 2013. Automatic 3D Building Model Generation by Integrating LiDAR and Aerial Images Using a Hybrid Approach (Ph.D. thesis), University of Calgary.

Lee, S.H., Chen, W., 2009. A comparative study of uncertainty propagation methods for black-box-type problems. Struct. Multidiscip. Optim. 37 (3), 239.

Lomas, K.J., Eppel, H., 1992. Sensitivity analysis techniques for building thermal simulation programs. Energy Build. 19 (1), 21–44.

Maas, H.G., Vosselman, G., 1999. Two algorithms for extracting building models from raw laser altimetry data. ISPRS J. Photogramm. Remote Sens. 54 (2–3), 153–163.

MacDonald, I.A., 2002. Quantifying the Effects of Uncertainty in Building Simulation (Ph.D. thesis), University of Strathclyde Glasgow, Scotland.

Mara, T.A., Tarantola, S., 2008. Application of global sensitivity analysis of model output to building thermal simulations. Build. Simul. Int. J. 1 (4), 290–302.

Mechri, H.E., Capozzoli, A., Corrado, V., 2010. USE of the ANOVA approach for sensitive building energy design. Appl. Energy. 87 (10), 3073–3083. ISSN 0306-2619.

Metropolis, N., Ulam, S., 1949. The Monte Carlo method. J. Am. Stat. Assoc. 44 (247), 335–341.

Morris, M.D., 1991. Factorial sampling plans for preliminary computational experiments. Technometrics 33 (2), 161–174.

Perera, G.S.N., 2015. Automated Generation of 3D Building Models From Dense Point Clouds and Aerial Photos (Ph.D. thesis), Technical University Dresden.

Petrie, G., 2011. Airborne topographic laser scanners. Geoinformatics 14, 34–44.

Rentsch, M., Krzystek, P., 2012. Automatic reconstructed roof shapes for LiDAR strip adjustment and quality control. ISPRS Ann. Photogramm. Remote Sens. Spat. Inf. Sci. 3, 117–122

Rottensteiner, F., Sohn, G., Jung, J., Gerke, M., Baillard, C., Benitez, S., Breitkopf, U., 2012. The ISPRS benchmark on urban object classification and 3D building reconstruction. ISPRS Ann. Photogramm. Remote Sens. Spat. Inf. Sci. 1 (3), 293–298.

Rottensteiner, F., Sohn, G., Gerke, M., Wegner, J.D., Breitkopf, U., Jung, J., 2014. Results of the ISPRS benchmark on urban object detection and 3D building reconstruction. ISPRS J. Photogramm. Remote Sens. 93, 256–271.

Roy, C.J., Oberkampf, W.L., 2011. A comprehensive framework for verification, validation, and uncertainty quantification in scientific computing. Comput. Methods Appl. Mech. Eng. 200 (25–28), 2131–2144.

Saltelli, A., Tarantola, S., Campolongo, F., Ratto, M., 2004. Sensitivity Analysis in Practice: A Guide to Assessing Scientific Models. John Wiley & Sons, West Sussex, England.

Sampath, A., Shan, J., 2007. Building boundary tracing and regularization from airborne LiDAR point clouds. Photogramm. Eng. Remote Sens. 73 (7), 805–812.

Sampath, A., Shan, J., 2008. Building roof segmentation and reconstruction from LiDAR point clouds using clustering techniques. Int. Arch. Photogramm. Remote Sens. Spat. Inf. Sci. 37, 279–284.

Sanchez, D.G., Lacarrière, B., Musy, M., Bourges, B., 2014. Application of sensitivity analysis in building energy simulations: combining first-and second-order elementary effects methods. Energy Build. 68, 741–750.

Sobol, I.M., 1993. Sensitivity estimates for nonlinear mathematical models. Math. Model. Comput. Exp. 1 (4), 407–414.

Spitz, C., Mora, L., Wurtz, E., Jay, A., 2012. Practical application of uncertainty analysis and sensitivity analysis on an experimental house. Energy Build. 55, 459–470.

Stuttgart, H., 2015. Urban energy simulation platformAvailable from: http://www.simstadt.eu/en/index.html Accessed on 13.09.2018.

Tarsha-Kurdi, F., Landes, T., Grussenmeyer, P., 2007. Hough-transform and extended RANSAC algorithms for automatic detection of 3D building roof planes from LiDAR data. In: ISPRS Workshop on Laser Scanning 2007 and SilviLaser 2007, vol. 36, pp. 407–412.

Verma, V., Kumar, R., Hsu, S., 2006. 3D building detection and modeling from aerial LIDAR data. In: 2006 IEEE Computer Society Conference on Computer Vision and Pattern Recognition, vol. 2, pp. 2213–2220.

Vosselman, G., 1999. Building reconstruction using planar faces in very high density height data. Int. Arch. Photogramm. Remote Sens. 32 (3; SECT 2W5), 87–94.

Vosselman, G., 2008. Analysis of planimetric accuracy of airborne laser scanning surveys. Int. Arch. Photogramm. Remote Sens. Spat. Inf. Sci. 37 (3a), 99–104.

Vosselman, G., 2012. Automated planimetric quality control in high accuracy airborne laser scanning surveys. ISPRS J. Photogramm. Remote Sens. 74, 90–100.

Vosselman, G., Maas, H.-G., 2001. Adjustment and filtering of raw laser altimetry data. In: Proceedings of OEEPE Workshop on Airborne Laserscanning and Interferometric SAR for Detailed Digital Terrain Models, Stockholm, Sweden.
Vosselman, G., Dijkman, S., et al., 2001. 3D building model reconstruction from point clouds and ground plans. Int. Arch. Photogramm. Remote Sens. Spat. Inf. Sci. 34 (3/W4), 37–44.
Vosselman, G., Kessels, P., Gorte, B., 2005. The utilisation of airborne laser scanning for mapping. Int. J. Appl. Earth Obs. Geoinf. 6 (3–4), 177–186.
Wate, P., Coors, V., Robinson, D., Iglesias, M., 2016. Qualitative screening method for impact assessment of uncertain building geometry on thermal energy demand predictions. Int. Arch. Photogramm. Remote Sens. Spat. Inf. Sci. 42, 127–134.

Model integration: A SOA approach

8

*Preston Rodrigues**
*Hochschule für Technik Stuttgart, Stuttgart University of Applied Sciences Schellingstr, Stuttgart, Germany

8.1 Introduction

Over the last several years Information and Communication Technology (ICT) has progressed at an exponential rate. Thanks to this improvement, it is now possible to store, process, and exchange large amounts of data seamlessly. Furthermore, with decreasing hardware and software prices it is relatively easy to purchase storage and time critical computational resources (AWS, 2016; GCP, 2016). This has encouraged the scientific community to look for solutions beyond conventional simulations by addressing and incorporating a multidisciplinary approach. Toward this initiative, the urban energy domain is currently proposing simulation strategies that integrate information from different sources and domains. These information sources can be building registers, census, cadastre records, and geospatial datasets. However, to integrate information from different sources the data has to be organized in a structured way and semantic support has to be provided.

CityGML,[1] an application independent city information model, has shown a lot of promise within the geospatial domain. A considerable asset of CityGML is its flexibility. This ensures an object in a city can be modeled with different Level of Detail, thereby enabling the virtual city objects to adapt to the scale of local urban surroundings. Furthermore, geospatial data of most cities are now readily available as open data in CityGML compliant virtual 3D city models (Berlin 3D—Download Portal, 2015; 3D City Model of New York City, 2015). However, a critical drawback of CityGML is that it does not include specific elements that can used by other domains in its core specification. Thanks to the concept of Application Domain Extensions (ADE) in CityGML it is now possible to incorporate domain-specific entities. For example, Utility Network ADE (2017), which allows modeling of gas, wastewater, district heating/cooling networks, and electricity grids with 3D topographic city objects, or Energy ADE (2017), which extends CityGML by a number of energy-related features and attributes in order to perform energy simulations at city scale but with the granularity that reach down to a single building. This has empowered building energy experts with an unique opportunity to use their knowledge to target a specific energy field and model its behavior (e.g., the impact of varying loads and demand response on the low-voltage electricity

[1] See https://www.citygml.org/.

Urban Energy Systems for Low-Carbon Cities. https://doi.org/10.1016/B978-0-12-811553-4.00008-1

grid or energy conversion systems in distribution networks). This evolution is envisioned toward a broader goal to achieve sustainable urban energy management.

Urban energy management is an integrated approach. Its key objective is to provide a balanced sustainable energy solution to predict current and future urban energy needs. This approach has paved the way for many physical and numerical models (static and dynamic) with a focus on building energy demand, demand response, energy storage, district heating and cooling networks, and supply strategies (Boyer et al., 1996; Clarke, 2001). Energy domain experts have transformed their knowledge about physical models into efficient energy simulation systems. The developed energy simulation systems are computational software programs, tools, and platforms providing different use-case scenarios under certain boundary conditions. Thus we have efficient simulation tools and software operating independently to provide desired results. Nonetheless, in order to have accurate and realistic results it is highly advantageous to integrate them as a distributed simulation environment. However, these tools in general are stand-alone software programs supporting heterogeneous data formats making them difficult to interoperate.

With the soaring popularity of the Internet, service-oriented architecture (SOA) has been gaining a lot of momentum. SOA is an enabler that exposes software functionalities as *Web Services*. As shown in Fig. 8.1, SOA is made up of three core entities, namely; Service Provider, Service Consumer, and Service Registry.

Service Provider is an entity that is willing to provide a software functionality. Service Consumer is an entity that needs to use a software functionality. Service Registry is an entity that enables Service Provider to publish its software functionality and allows Service Consumer to search for a software functionality that it requires to fulfill a task. Each of these entities relies on key standardized technologies (i.e., SOAP, WSDL, UDDI). Simple Object Access Protocol (SOAP) (Gudgin et al., 2007) defines a messaging standard based on Extensible Markup Language (XML) and, with the help of Hypertext Transfer Protocol (HTTP), provides a communication protocol for accessing *Web Services*. Web Service Description Language (WSDL) (Christensen et al., 2001) is used to describe the web service access interface. Universal Discovery, Description and Integration (UDDI) (Clement et al., 2004) is a registry that allows advertisement and discovery of web services thereby providing the opportunity to dynamically bind a *Web Service* at runtime.

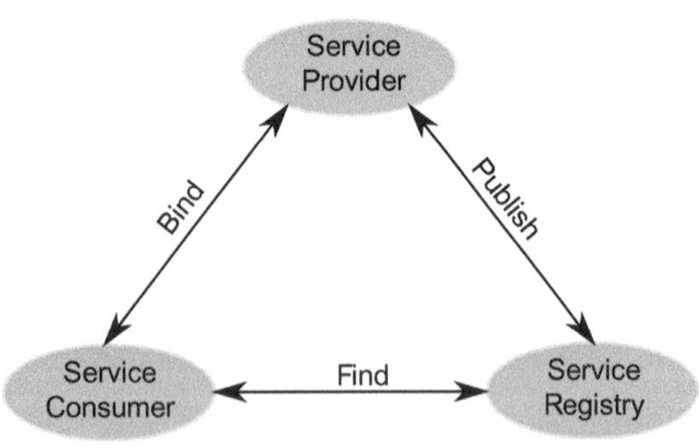

Fig. 8.1 Service-oriented architecture.

8.2 Distributed urban energy infrastructure

Currently, urban simulation tools focus on modeling and simulating specific entities of the urban domain (e.g., buildings). However, apart from buildings, other urban entities (e.g., bridges, vegetation, roads) should also be considered to achieve sustainable urban energy management. The heterogeneity within the urban energy domain makes it quite difficult to simulate all entities of an urban environment with a single simulation tool. This necessitates an integrated framework. A framework designed from scratch could be a way forward. However, it is a time and resource-consuming activity. Furthermore, the entire process may end up with redundant and erroneous implementations of similar functionalities. Additionally, as modeling experts (using efficient tools) are located at different places, it would be very helpful if they can use their preferred tools. A feasible alternative could be to extend the useful (valid, tested, and well documented) models implemented by experts and encapsulate them into software component through open interfaces.

As shown in Fig. 8.2, an open interface (control and data) acts as a communication point between simulation tools. The data interfaces enable exchange of information and functionalities during run-time simulation (i.e., common data models, processing time requirements), while on the other hand, the control interface manages the run-time simulation operations (e.g., start, pause, resume, stop) among connected simulation tools. Open interfaces can potentially be used to connect different simulation tools to address complex urban simulation scenarios. The resulting chain of connected tools can be characterized as a distributed urban simulation infrastructure. A distributed urban simulation infrastructure is not only the need of the hour, but also highly advantageous. Such a simulation infrastructure can be beneficial at three different levels, namely; *computational*, *decision*, and *business*.

Computational level. The sharing of data resources and functionality in a distributed computing environment could exploit the high-performance computing capabilities of processors and hardware to predict future energy scenarios at high granular scale.
Decision level. An integrated system consisting of individual expert's knowledge from their respective domain could facilitate urban energy planners and managers to assess the overall and specific primary energy requirements at varying spatial scales.
Business level. An integrated system could bring new business opportunities for building equipment manufactures and energy supply companies by bringing simulation results as services depicting energy-saving potential and comfort living conditions to their customers.

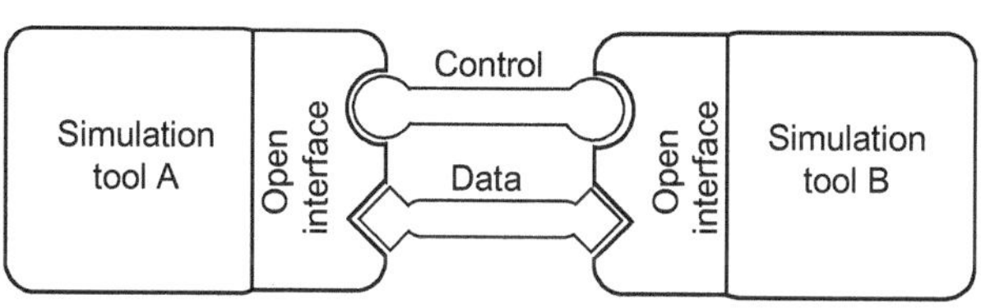

Fig. 8.2 Open interface.

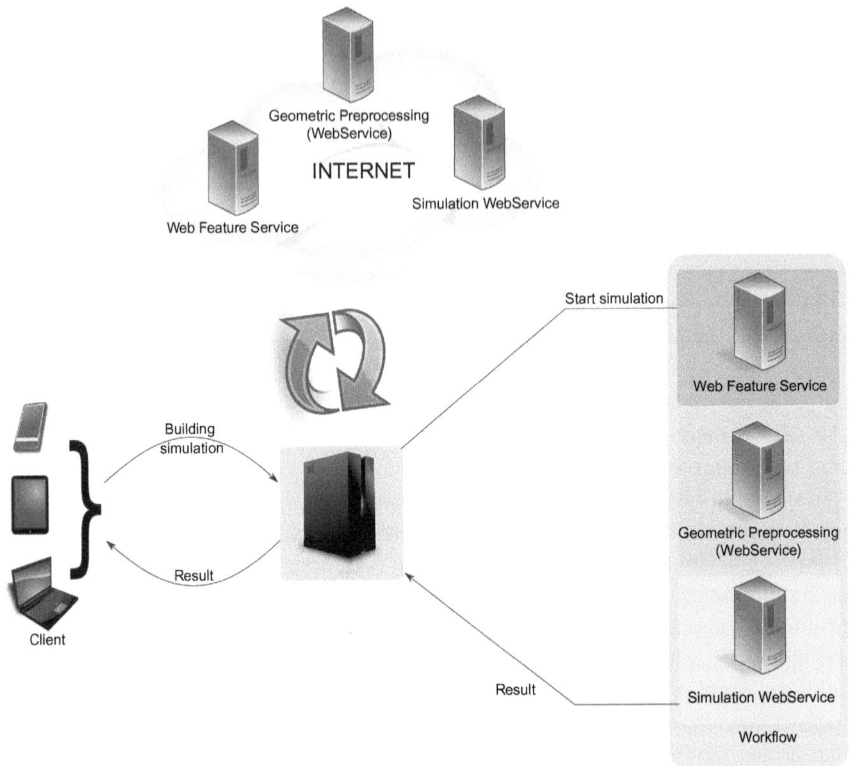

Fig. 8.3 Distributed GIS infrastructure.

Traditionally, Geographic Information System (**GIS**) were stand-alone systems used to store, query, visualize, and analyze geographically referenced information. However, with recent advancement in ICT it is now possible to store, process, and exchange large amounts of data seamlessly in a distributed environment. This has enabled the GIS community to rely on well-known industry concepts of open interfaces. Furthermore, the use of *Web Services* provides the necessary flexibility to access data and computing resources by different GIS entities. Moreover, it enables interoperability among distributed GIS resources. Fig. 8.3 illustrates a typical *Web Services*-based distributed GIS infrastructure. Realizing the numerous advantages of *Web Services*, the Open Geospatial Consortium (OGC) is leading and coordinating the efforts of international organizations and enterprises to develop interoperable GIS services to access distributed geospatial resources. Toward this effort, a series of standard-based interface specifications are already available, namely; Web Feature Service (WFS), Web Map Service, Web Coverage Service, Catalogue Service for the Web, and Web Processing Service (WPS). Furthermore, to enable seamless integration of OGC services, WPS is designed to act as a glue. It is used to orchestrate complex geo-computation processes and their execution mechanism along with their associated inputs and outputs.

8.3 Concept and use case

In order to highlight the importance of distributed urban energy simulation infrastructure and its advantages, we take an example to estimate solar heat gains for a building. Fig. 8.4 illustrates the coupling of modeling and processing. As shown, to estimate solar heat gains for a building, a shortwave irradiation simulation algorithm is used with the 3D building geometry as input. However, at urban scale, the task becomes highly challenging. In order to estimate solar heat gains at urban scale we need to address three key complexities; (1) *data processing and exchange*, (2) *modeling*, and (3) *computational*.

Data processing at an urban scale requires automated extraction and processing of building geometry parameters. Furthermore, the extracted data need to be modeled into a compatible format required by a simulation program. This is necessary as it will ensure that the generated data model can be easily fed to any simulation algorithm. *Modeling* complexity involves practical considerations of adjacency and proximity between buildings to account for detailed modeling of shadowing effects, while *computational* complexity relates to graphics hardware available for processing of large urban scenes.

In order to realize our use case, we exploit the features of SOA. Using *Web Services*, we will show how to couple processing and modeling functionalities on-the-fly and run a simulation targeting the urban scale. The following sections provide a brief overview about our concept. Energy simulations are accomplished using SOA-based *workflows*. A *workflow* is made up of many *workflow steps*. We have defined three default *workflow steps* namely; `input`, `preprocessing`, and `simulation`. Each *workflow step* is responsible to trigger the next *workflow step* in the chain.

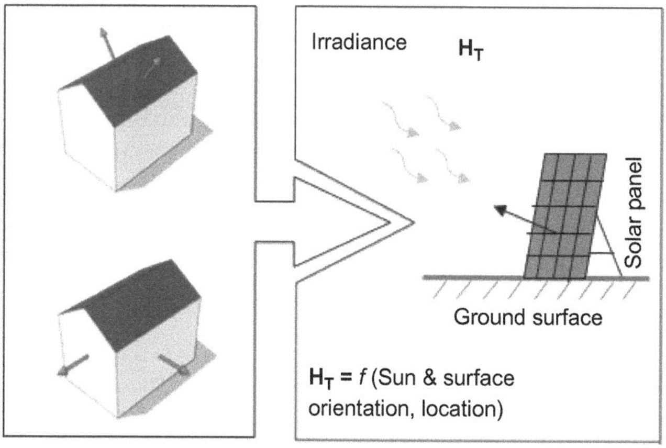

Fig. 8.4 Coupling between modeling and processing.

8.3.1 `Input` step

The `input` *workflow step* reads a CityGML model and extracts building information from *CityObject*. This information is then mapped into an internal data structure. The internal data structure assigns unique IDs to each wall and maintains the accuracy of buildings' geometric information with the help of vertices, edges, and polygons. This step is essential to remove redundant information and provide spatio-semantic coherent data structure for energy simulation. Furthermore, the coherency and quality assurance of geometry data aids in reducing the uncertainty of simulation results. On successful completion of the `input` *workflow step*, the `preprocessing` *workflow step* is triggered.

8.3.2 `Preprocessing` step

The `preprocessing` *workflow step* has a flexible design. This flexibility allows user-defined modules to be integrated as plugins. By default we provide two plugins, namely; *geometric* and *physics*. As the name suggests the *geometry plugin* computes geometric parameters such as surface area, orientations, gross volume, building heights, and window-to-wall ratios (Alam et al., 2014). On the other hand building physics-related parameters like construction type are derived from the *physics plugin*. Depending on a particular simulation requirement either or both plugins can be enabled.

8.3.3 `Simulation` step

On successful completion of `preprocessing` *workflow step*, `simulation` *workflow step* is triggered. Unlike `preprocessing`, which can integrate user-defined plugins, the `simulation` *workflow step* can integrate existing simulation tools. To showcase the flexibility of our platform and its advantage as a distributed energy simulation platform, we have coupled the software environment INSEL into the `simulation` *workflow step*. INSEL (2016) provides a graphical programming language for the simulation of renewable energy systems and buildings. Furthermore, it uses a template-based system to execute simulations. Using INSEL programming language, users can implement customized simulation templates on-the-fly and execute them. Key application fields of INSEL include, but are not limited to, solar photovoltaics and solar thermal. The following section explains in details the implementation of our proposed solution.

8.4 Implementation

As described in Section 8.3, to evaluate the solar potential of surfaces, we need to calculate irradiance value on each surface. This is needed, for example, for space heating or cooling demand, as solar irradiance is transmitted through glazings with varying orientations. Irradiance of surfaces is calculated with the help of a radiation processor. Using INSEL we have created a radiation processor template, it calculates monthly average daily irradiance (W/m^2) incident for a given orientation (surface azimuth and tilt). Optionally, to address weather conditions in a particular urban quarter we also provide as input the geo-coordinates of the building. Fig. 8.5 shows a snapshot of INSEL monthly irradiation template. As shown, the template provides unique place holders for all the required inputs.

Fig. 8.6 illustrates the flow of data at runtime. As shown, the `Data Input` *workflow step* loads 3D data from the CityGML file. The extracted 3D data are passed as an input to the `Preprocessing` *workflow step* where the geometric plugin calculates the geometric parameters. The output of the `Preprocessing` *workflow step* is passed as an input to the simulation workflow step. Each of the *workflow steps* is connected via *Web Services.*

One of the key design objectives of this proof-of-concept prototype is to describe and automate geospatial workflows using web service technology and to provide an easy to handle user interface. Toward this objective we have designed the SOA-based platform. The platform enables an end user to define a custom geospatial workflow. In order to design a custom geospatial workflow, an `End User` has to complete the following tasks:

- Choose a 3D city data source
- Choose a preprocessor to process the 3D data source
- Choose a simulation template

However, in order to complete these tasks, an `End User` needs to see the 3D object and select it. Visualization is an effective method not only to display data but also to support complex analysis, information extraction, and knowledge discovery. Dynamic visualization features like view controls (zooming, panning, and layering) and 3D

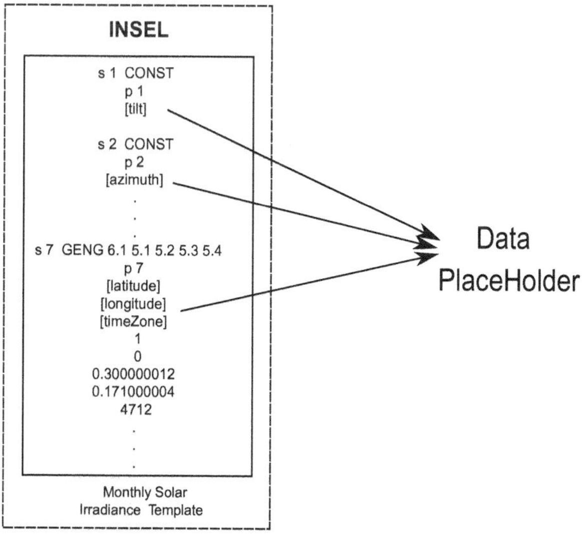

Fig. 8.5 INSEL template.

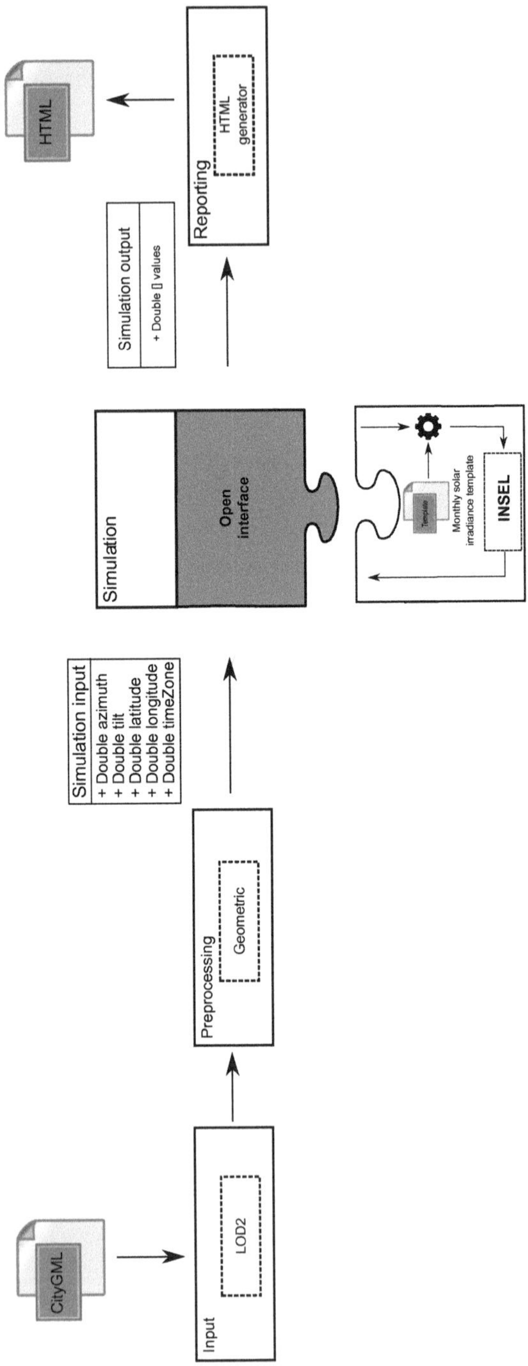

Fig. 8.6 Runtime data flow.

fly-through help users clearly recognize and interact with 3D objects. For instance, viewing a building from different view points shows the complexity of building architecture. Over the years, several open source and commercial software have been developed to visualize 3D objects. Some of the well-known software include SKYLINE (TerraExplorer) (2017), NASA WORLD WIND (2017), ARCGIS (2017), GOOGLE EARTH (Hello, Earth, 2017), and CESIUM. For the purpose of this implementation we use CESIUM (2017) for visualizing and selecting 3D objects. However, any of the aforementioned software can be integrated with the platform.

Each of the aforementioned tasks is designed as an individual *workflow step*. Fig. 8.7A shows the first task `Data Input` *workflow step*, which enables an `End User` to select a 3D building. In this task, the `End User` has to complete several subtasks. First, the `End User` has to select the *city*. Second, the `End User` has to decide whether he/she prefers to do a simulation on a *Single Building* or a *Region* and choose the appropriate option accordingly. And finally, choose a 3D building by simply clicking on the building he/she wishes to consider for simulation. In this example, the `End User` chooses to do a simulation on a *Single building*. At this point, the *gmlID* of the selected building is displayed to confirm the `End User` choice. Once this task is complete the `End User` should proceed toward the next task `Preprocessing`.

The `Preprocessing` *workflow step* is used to calculate geometric parameters for the selected building. It takes a *CityGML* files as an input and generates several output formats namely; *CSV*, *Building-to-Zone*, and *Building_storey-to-Zone*. Using the *gmlID* selected from the previous task, a *CityGML* data file is retrieve from a data store (3DCityDB[2]) using WFS. As shown in Fig. 8.7B, the `End User` chooses *Building-to-Zone*. The output of the `preprocessing` *workflow step* is passed as an input to the `Simulation` *workflow step*. In this example (see Fig. 8.7C), the `End User` chooses *Irradiation on tilted surface*. This simulation template does not require any further user inputs as all the data required for this template to run are acquired from the previous *workflow step*. Finally, the result of the simulation is shown as a plot (Fig. 8.7D). Alternatively, the resulting values can be displayed on the selected 3D building as well.

8.5 Conclusion

Climate change, air pollution, and poverty are some of the challenges faced by cities all over the world. With energy being at the heart of many of these issues, urban planning and energy planning need to go hand in hand. Urban energy modeling along with simulation and optimization methods appears to be a key enabler in finding a promising solutions toward a sustainable urban energy management.

In this chapter, we tried to show how we can integrate different urban domains with the help of ICT. We introduced a SOA for urban energy simulation. Our approach uses the concept of open interfaces. As proof-of-concept a prototype was developed to illustrate solar irradiation calculation as web service. The prototype demonstrated

[2] See http://www.3dcitydb.org.

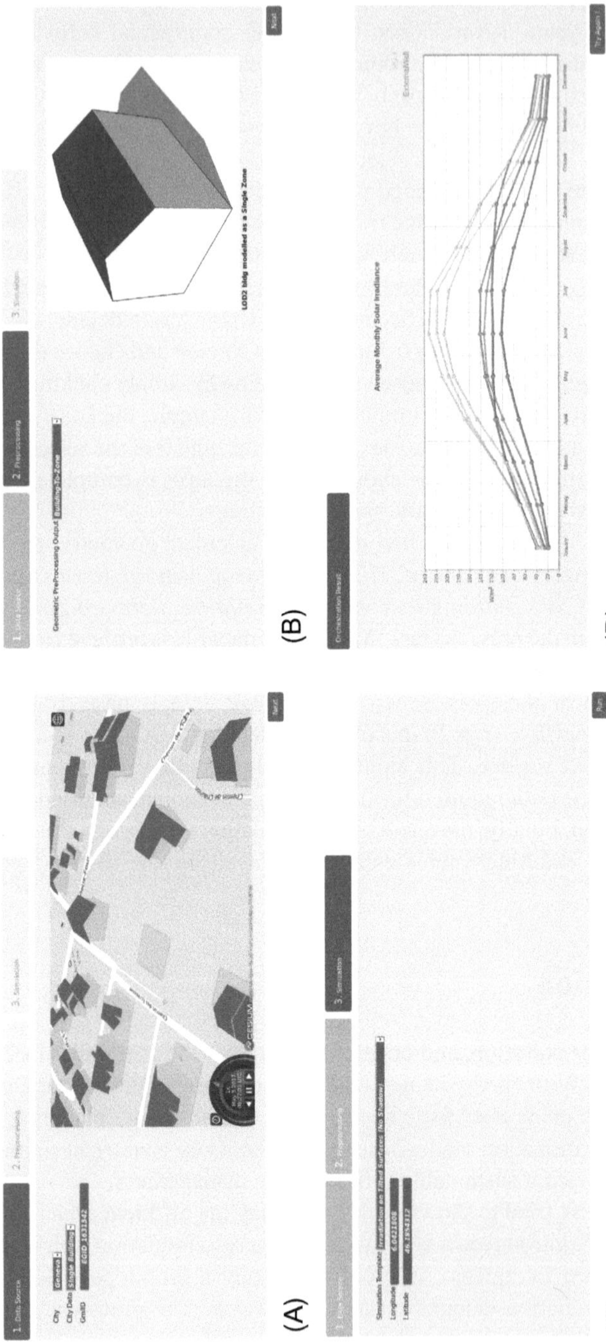

Fig. 8.7 SOA-based platform. (A) Data Input; (B) Preprocessing; (C) Simulation; (D) Result.

the feasibility of our approach by invoking the INSEL simulation engine with an INSEL template generated based on `End User` inputs. The flexibility of generating a dynamic INSEL template is highly suitable to carry out solar potential analysis in urban environment. Since a given urban façade possesses arbitrary tilt and azimuth, which can be directly fed to the web service in order to have a quick primary assessment of solar potential of urban surfaces. While our solution provides a way forward, it only supports CityGML v2.0 standard-based models. In future we plan to support CityGML v3 and Inspire standard.

References

3D City Model of New York City, 2015. https://www.gis.bgu.tum.de/projekte/new-york-city-3d/.

Alam, N., Wagner, D., Wewetzer, M., von Falkenhausen, J., Coors, V., Pries, M., 2014. Towards automatic validation and healing of CityGML models for geometric and semantic consistency. In: Innovations in 3D Geo-Information SciencesSpringer, New York, NY, pp. 77–91.

ArcGIS Mapping Without Limits, 2017. https://www.arcgis.com/features/index.html.

AWS, 2016. Amazon Web Services. https://aws.amazon.com/.

Berlin 3D—Download Portal, 2015. http://www.businesslocationcenter.de/en/downloadportal.

Boyer, H., Chabriat, J.-P., Grondin-Perez, B., Tourrand, C., Brau, J., 1996. Thermal building simulation and computer generation of nodal models. Build. Environ. 31 (3), 207–214.

CESIUM, 2017. https://cesiumjs.org/.

Christensen, E., Curbera, F., Meredith, G., Weerawarana, S., 2001. Web Services Description Language (WSDL) 1.1. W3C.http://www.w3.org/TR/2001/NOTE-wsdl-20010315.

Clarke, J.A., 2001. Energy Simulation in Building Design. Routledge, Abingdon.

Clement, L., Hately, A., von Riegen, C., Rogers, T., 2004. Universal Description Discovery & Integration (UDDI) version 3.0.2. OASIS. http://uddi.org/pubs/uddi-v3.0.2-20041019.htm.

Energy, A.D.E., 2017. http://en.wiki.energy.sig3d.org.

GCP, 2016. Google Cloud Platform. https://cloud.google.com/.

Gudgin, M., Hadley, M., Mendelsohn, N., Moreau, J.J., Karmarkar, A., Lafon, Y., 2007. Simple Object Access Protocol (SOAP) version 1.2 part 0: Primer, second ed. W3C.http://www.w3.org/tr/2007/rec-soap12-part0-20070427/.

Hello, Earth, 2017. https://www.google.com/intl/en_uk/earth/.

INSEL, 2016. http://www.insel.eu/index.php?id=301&L=1.

NASA World Wind, 2017. https://worldwind.arc.nasa.gov/.

Skyline—TerraExplorer for Web, 2017. http://www.skylineglobe.com/SkylineGlobe/corporate/Default.aspx?

Utility Network ADE, 2017. http://en.wiki.utilitynetworks.sig3d.org.

Part Two

Case Studies and Applications

The case study cities of Geneva and Vienna

9

Jean-Loup Robineau, Giorgio Agugiaro†*
*Industrial Processes and Energy Systems Engineering Group, Ecole Polytechnique Fédérale de Lausanne, Rue de l'Industrie, Sion, Switzerland, †Smart and Resilient Cities Unit, Center for Energy, AIT—Austrian Institute of Technology GmbH, Vienna, Austria

9.1 Introduction

Two European cities—Geneva in Switzerland and Vienna in Austria—have been chosen as test case studies within the project CI-NERGY due to their ambitious sustainability goal. This chapter is divided into two parts, each containing the description of either city. First, a general description is provided, followed by an analysis of the available data about the overall energy demand, the available resources (renewable and nonrenewable), and the existing and planned energy infrastructure. Successively, the overall goals of each city in terms of urban and energy planning are briefly presented, together with the main actors and the principal planning tools adopted. Finally, some information is given about the available data used in the project CI-NERGY. Of each city, a short description of the specific case studies is given at the end of either section.

9.2 Geneva

9.2.1 General description

Geneva[1] is a city in Western Switzerland located at the southwestern tip of Lake Geneva (see Fig. 9.1). It is the capital and most populous municipality in the canton of Geneva. With a population of 201,164 inhabitants in 2015, it is also the second city in Switzerland, after Zürich. The canton has a population of 490,578, with an increase of 10% over the past 10 years, and covers an area of 282 km^2 (the population density is around 1700 inhabitants/km^2). Geneva is at the center of a transnational agglomeration named "Greater Geneva" that extends to parts of the neighboring canton of Vaud (district of Nyon) and France (parts of Haute-Savoie and Ain). It represents a population of 946,000 and an area of 2000 km^2.

Geneva is surrounded by two mountain chains, the Jura to the north and the Alps to the southeast, and has two main rivers flowing through it, the Rhône and the Arve, which meet in Geneva City. Fig. 9.2 shows a satellite view of Geneva.

[1] Statistics can be found on the website of Cantonal Statistics of Geneva, http://www.ge.ch/statistique/.

Urban Energy Systems for Low-Carbon Cities. https://doi.org/10.1016/B978-0-12-811553-4.00009-3

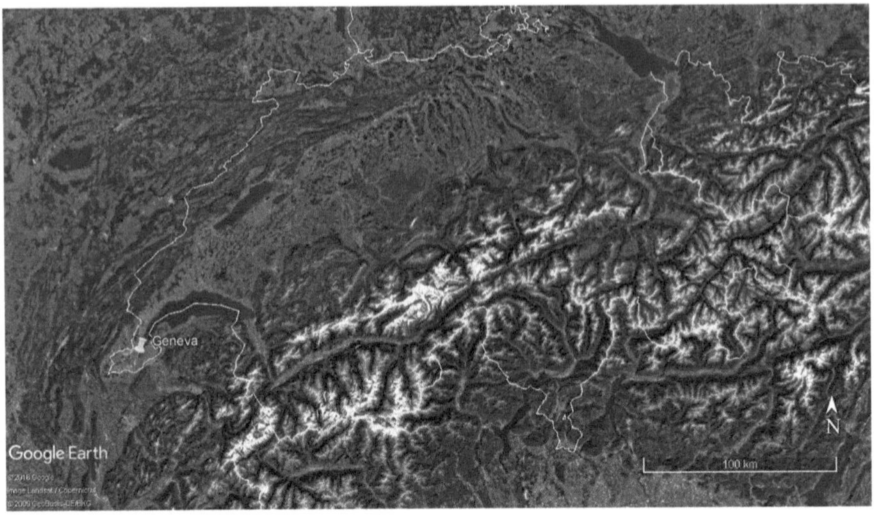

Fig. 9.1 Location of Geneva in Switzerland.
Source: Google Earth.

Fig. 9.2 Satellite image of Geneva.
Source: Google Earth.

The climate of Geneva is temperate oceanic, with cool winters (average temperature of 1.5°C in January) and pleasantly warm summers (average temperature of 20.2°C in July). Precipitation is adequate (1005 mm yearly) and relatively well distributed throughout the year. Besides, it has an average of 1828 h of sunshine a year

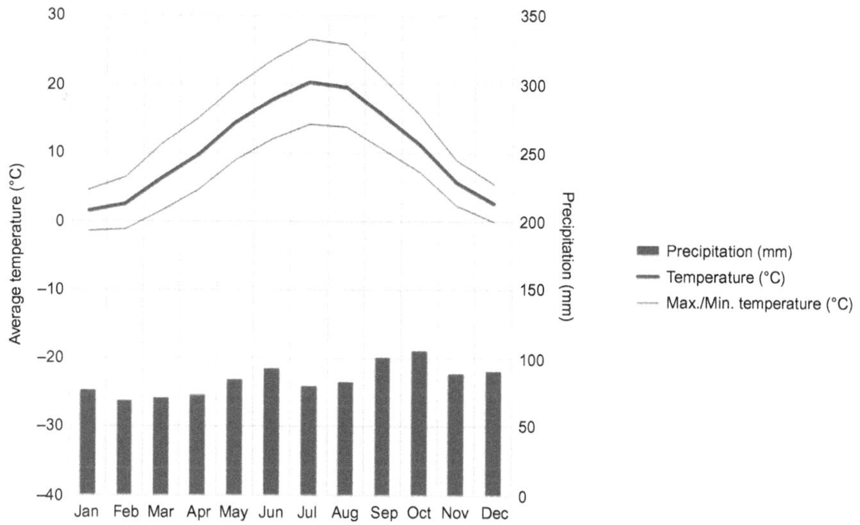

Fig. 9.3 Climatic data for Geneva (Genève-Cointrin station; 1981–2010) (MeteoSuisse, 2016).

(MeteoSuisse, 2016). The monthly precipitation and temperatures are represented in Fig. 9.3. The number of heating-degree days in Geneva was 2658 for the heating period (September–May) during the year 2010 (Khoury, 2014).

Geneva is a center of trade and finance and home to several international organizations' headquarters, such as the United Nations. The service sector is by far the most significant in Geneva in terms of employment (87% of the workforce in 2013). Precision machinery and instruments are other branches where the economy of the Canton of Geneva is strong (e.g., watchmaking). There are also some major chemical industries, including transnational cosmetic groups Givaudan and Firmenich. Besides, agriculture is also commonplace in the hinterlands of Geneva, particularly wheat and wine. Thirty-eight percent of the cantonal area is reserved for agricultural use, whereas woodland represents only 12%.

9.2.2 Energy demand, resources and infrastructure

According to the Cantonal Statistics, the total energy delivered in Geneva in 2015 was 35,314 TJ. This includes the delivery of electricity, natural gas, fuel oil, and other liquid fuels (mostly for mobility) but excludes solid fuels and aviation fuels. The evolution of the share of each energy carrier between 1980 and 2015 is represented in Fig. 9.4. While the total energy delivered has remained more or less stable over the past 35 years, there has been a progressive substitution of fuel oil with natural gas, which has led to a decrease in the CO_2 intensity of the energy demand from 217 to 166 g/kWh in that same period.

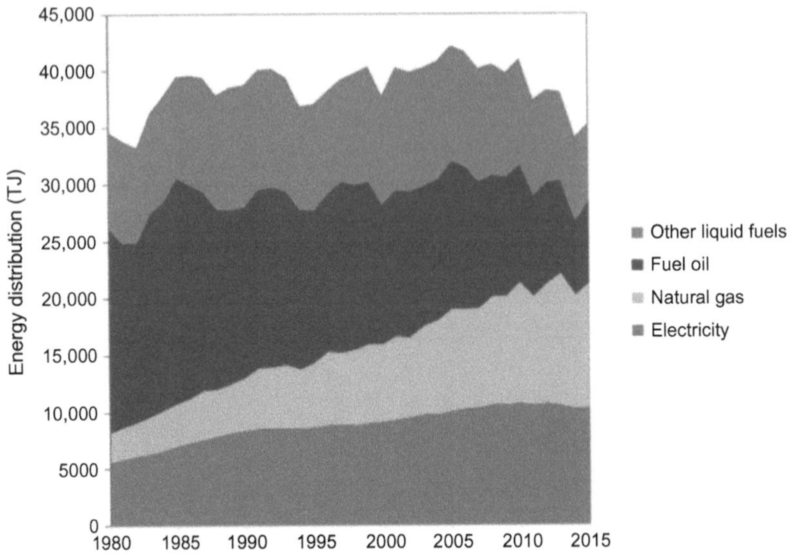

Fig. 9.4 Evolution of share of energy carriers in Geneva between 1980 and 2015. *Source*: Cantonal Statistics of Geneva.

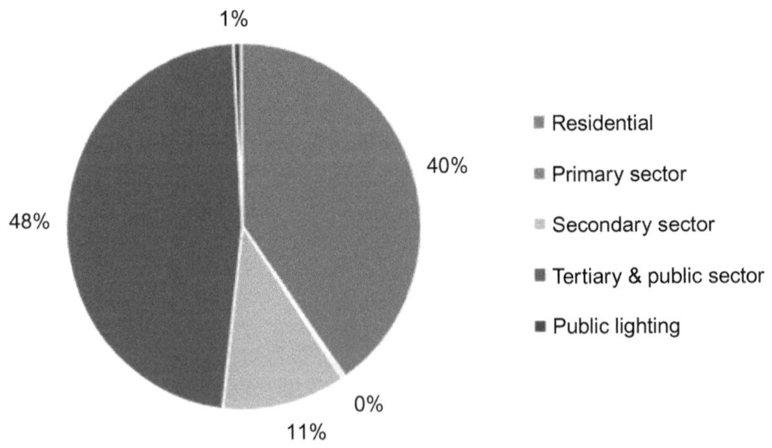

Fig. 9.5 Share of electricity consumption by end-user type. *Source*: Cantonal Statistics of Geneva.

The share of electricity consumption by end-user type is shown in Fig. 9.5. The majority of the demand (88%) comes from the residential, tertiary, and public sectors. Most of the gas consumption is used for space heating and domestic hot water, mostly through direct use in buildings (88% of total gas consumption) or through a district

Table 9.1 Reference energy surface (SRE) in million m^2 per building category (Khoury, 2014)

<table>
<tr><th></th><th>Individual houses</th><th>Houses with two dwellings</th><th>Buildings with >3 dwellings</th><th>SRE per category</th><th>Total SRE in Geneva</th></tr>
<tr><td colspan="6">With residential use</td></tr>
<tr><td>Exclusive residential</td><td>4.20</td><td>0.43</td><td rowspan="2">19.30</td><td rowspan="3">26.15</td><td rowspan="3">40.85</td></tr>
<tr><td>Residential with annexe usage</td><td>0.32</td><td>0.17</td></tr>
<tr><td>Residential with annexe usage</td><td colspan="3">1.72</td></tr>
<tr><td colspan="6">Without residential use</td></tr>
<tr><td>Nonresidential buildings</td><td colspan="3">14.7</td><td>14.7</td><td></td></tr>
</table>

heating network (10%). All commercial and industrial installations of 120 kW or higher represent only 2% of the total gas consumption, while gas use for domestic cooking and mobility fuel are negligible.

Statistics on the building stock in Geneva canton can be found in Jad Khoury's PhD thesis (Khoury, 2014). In 2010, the building stock in Geneva was made up of just over 47,000 buildings totaling around 41 million m^2 of reference energy surface. Almost two-thirds of this surface (26 million m^2) comes from buildings with residential use, either exclusively or partially. The exact breakdown of surfaces is given in Table 9.1. The reference energy surface (SRE) is defined in the Swiss norm SIA 416/1 as being the sum of all the floor surfaces, above- or belowground, which are included in the calculation of the building envelope and have a heating or cooling need.

The residential building stock can be divided into three major construction periods: the period before 1946, the period between 1946 and 1980, and the period since 1981. The year 1946 marked the beginning of the postwar economic boom and reconstruction period, during which quick-to-build and inexpensive apartment complexes were the norm, and there was little concern for energy savings. In 1980, the first building norm imposing a minimum U-value for new constructions was enacted in Switzerland, as a consequence of the oil crises in 1973. Regarding the construction period of multifamily residential buildings (i.e., with three dwellings or more), 30% of the 12,602 buildings were built prior to 1946 (27% of the SRE), 42% between 1946 and 1980 (49% of the SRE), and 28% between 1981 and 2010 (25% of the SRE).

Due to their predominant share, low isolation level, and the fact that they have reached their renovation period, postwar residential constructions (built between 1946 and 1980) represent the main focus of building energy rehabilitation programs. Indeed, they represent 5240 buildings (including mixed-use buildings) totaling almost 100,000 dwellings (54% of the total), and their heated surface adds up to 9.4 million m^2, representing almost half of the residential building stock in Geneva.

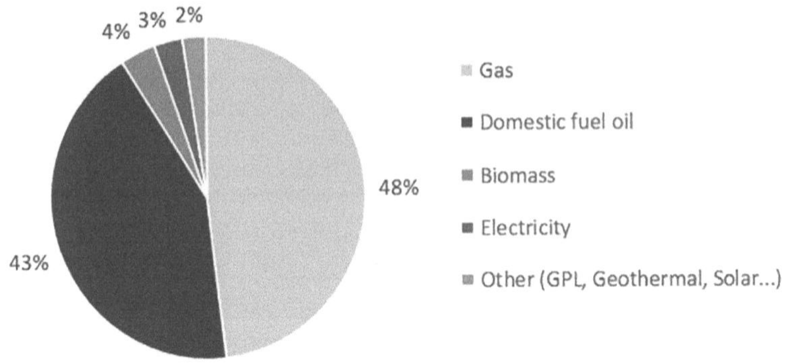

Fig. 9.6 Share of heating installation type (in number) for residential buildings.

Calculation of the heating consumption index (IDC[2]) is mandatory in Geneva since 2011 for nonresidential buildings and buildings with five dwellings or more (Canton de Genève, 2016). The IDC is the annual energy required for space heating and domestic hot water per heated surface (SRE). For these building categories (and a few smaller residential buildings), the heat demand in 2012 amounts to 12,876 TJ according to the data available[3] in SITG.[4] However, this demand corresponds to a SRE of 26,778,241 m^2 and does not include all buildings, either because it is not mandatory (small residential) or because the data are missing. According to the available data, the average heat demand (space heating and domestic hot water) for individual houses, buildings with two dwellings, and buildings with three or more dwellings is 154, 151, and 140 kWh/m^2, respectively.[5]

Heating demand is still dominated by gas and domestic fuel oil systems (see Fig. 9.6). Indeed, out of 37,769 residential installations that were surveyed,[6] 91% use gas or domestic fuel oil. There is a small share of biomass and electricity heating systems, while other system types (GPL, geothermal, solar, coal, and DHN) represent negligible shares. In this context, the potential of demand side management of the grid using the building's thermal load is very limited, without replacement of existing boilers.

A thorough evaluation of renewable resources in Geneva was carried out by Jérôme Faessler in his PhD thesis (Faessler, 2011). This included biomass (including biowaste), hydrologic (both hydrothermal and hydroelectric), geothermal, solar, wind, waste heat, and solid waste resources. Unless otherwise mentioned, the data presented in this section were taken from Faessler's thesis.

Biomass energy resources fall into three main categories: wood, agriculture, and organic waste. For agriculture, only by-products are considered. The total energy

[2] "Indice de dépense de chaleur" in French.

[3] Data obtained from layer SCANE_INDICE_MOYENNES_3_ANS in SITG platform (http://ge.ch/sitg/).

[4] SITG is an online sharing platform of geospatial datasets relative to the canton of Geneva.

[5] Data for buildings with less than three dwellings are not very reliable because it only concerns a small sample of buildings of these categories.

[6] Data obtained from layer SCANE_CHAUDIERE in SITG platform (http://ge.ch/sitg/).

potential from biomass for Greater Geneva has been evaluated at around 425 GWh.[7] This number includes the part that is already valorized and the part that is not yet used. As 308 GWh are already valorized, the remaining potential is around 117 GWh, mostly from natural wood (43 GWh) and agricultural by-products (49 GWh). This potential is very limited as it only represents 0.5% of the final energy consumption of the agglomeration. Nevertheless, the development of this sector is justified as it has low CO_2 emissions (Office de l'environnement, 2012).

The hydrologic resource can be exploited in two ways: by using its sensible heat (hydrothermal) or by using its gravitational potential to produce electricity (hydroelectricity). Three sources are considered: the Lake Geneva (for hydrothermal only) and the Rhône and Arve rivers (for hydrothermal and hydroelectric). The accessible hydrothermal potential (for heating or cooling) amounts to 9244 GWh (where only 13% is currently used, mainly for cooling down industrial processes). There is therefore a high unexploited potential. However, apart from the case of direct cooling, the heat is not directly usable, and a heat pump consuming electricity is required to harvest this potential. In his PhD thesis, Pierre-Alain Viquerat carried out a detailed study on the use of the lake as a hydrothermal source, including environmental impacts (Viquerat, 2012). On the other hand, the hydroelectric potential on the two main rivers represents a total of 2591 GWh, out of which 93% is already used, leaving little room for expansion.

Solar energy can be valorized in two ways in an urban context: through production of heat (solar thermal) or production of electricity (solar PV). As only one type of solar panel can be installed on a given surface, these two technologies are not complementary, but rather in competition with one another. To handle this aspect, different scenarios were considered in, for each of which the potential was evaluated. In a 100% PV scenario, the potential was calculated to be 616 GWh for the canton of Geneva. In a scenario where there is a mix of PV and solar thermal to cover some of the domestic hot water demand (1 m^2 per capita), the potential for heat and electricity is 227 and 560 GWh, respectively. Currently, only 8 GWh are used for each usage, leaving major room for expansion.

The geothermal potential of Geneva has also been extensively studied. The potential for different geothermal technologies was thoroughly evaluated in (Geotechnique appliquee Deriaz S.A., Centre d'hydrogéologie et de géothermie, and Jules Wilhelm, 2011). In this report, the potential of vertical geothermal probes was evaluated to be at least 1200 GWh for heat demand (corresponding to 20% of the total heating demand of the canton) and 1000 GWh for cooling demand. Moreover, this technique can be used for seasonal storage, although the exact potential is difficult to evaluate. The use of geothermal foundation piles in buildings could provide up to 30 GWh for both heating and cooling. There are several low depth aquifers that are available for geothermal exploitation in Geneva. The potential is 110 and 80 GWh, respectively, for heating (with a heat pump) and cooling (direct use). Deep aquifers (i.e., 2–3 km deep) can provide heat at a sufficient temperature to not require a heat pump for space

[7] Available energy per year, unless otherwise stated.

heating and domestic hot water. Moreover, it can be used to produce electricity through organic Rankine cycles (ORC). However, the potential is more difficult to evaluate due to a lack of knowledge on the geology at those depths. An initial estimate could be around 200 GWh direct heat (corresponding to 20 GWh electricity). Finally, enhanced geothermal systems (EGS) is a technology in which a fluid is injected in a deep fractured layer of rocks through a well and then collected back in another well after having retrieved heat at a high temperature (>150°C) from the rocks. There is a lot of uncertainty regarding the development of this technology, but a potential of heat production of 2400 GWh is envisaged within 30 years.

The wind power potential for the agglomeration of Greater Geneva is very low because of the dense settlements on the one hand and protected natural zones on the other hand. Taking this into account, the majority of the areas suitable for wind farms would be outside of the canton boundaries (i.e., district of Nyon in Vaud canton). Hence, wind power is not expected to play a role in the local energy mix, although development at the national scale would improve the CO_2 impact of electricity used.

The available waste heat from industries in Geneva has been estimated by Laurent Gilliard in his work (Gilliard, 2015), where he used a database of the manufacturing industries and simple energy models for each sector. It appears that the total heat recovery potential amounts to 67 GWh, which only represents 3% of the total final energy input to the industries (2092 GWh). He also showed that 90% of that heat is available between 25°C and 55°C, while only 9% is available at temperatures higher than 55°C. This is however lower than the potential evaluated by Faessler, 320 GWh, which also includes heat produced from waste incineration and heat recovered from wastewater (not considered by Gilliard).

Lastly, the potential for electricity production from waste incineration amounts to 150 GWh (corresponding to the 220,000 t of municipal solid waste that can be incinerated). However, this is the existing situation, and therefore, no more energy can be expected from here. Moreover, the wastewater treatment plant can generate biogas (42 GWh in lower heating value).

9.2.3 City goals, actors and tools

The energy strategy of the canton, delegated to the Cantonal Office of Energy (OCEN), is based on two instruments, as explained in the following (Service cantonal de l'énergie, 2016):

- The general energy conception (CGE) defines the long-term energy strategy considering the evolution of the international and national contexts. It is submitted to the Great Council at least once per mandate.
- The energy directive plan (PDE) uses the orientations set out by the CGE and translates them into quantified objectives and establishes a list of actions necessary for their accomplishment. It must be approved by the state counsel.

The postulate on which the energy strategy is based is twofold:

- First and foremost, the canton should align its vision on the confederation's ambition of a 2000 W society, which consists in reducing the primary energy consumption to a level corresponding to an average power of 2000 W per person over a year and limiting the CO_2 emissions to 1 t per person and per year, without reducing the quality of life, by 2050. This level of CO_2 emissions corresponds to a consumption of 500 W, meaning that the energy supply should be 75% renewable.
- Another aspect that is put forward is to not be supplied by nuclear energy.

To achieve this goal, a milestone was set for 2035, which consists in reducing the primary energy consumption per person by 26% compared with 1990 levels. It also stipulates that the CO_2 emissions should be reduced by 33% between 2005 and 2035 (Novatlantis, 2006).

The PDE defines the general objectives and sets the path to take to achieve them. The canton, as the central actor of this plan, delegates whenever possible responsibilities to appropriate partners and facilitates the implementation of actions by the latter. Some of the objectives are therefore transferred to those partners who will take actions in their respective domains to achieve those goals. The main actors of this plan are as follows:

- The Swiss confederation and the cantons, including all the agencies and organizations associated to the Swiss energy program
- Public authorities at the service of the state (City of Geneva and other municipalities)
- Economic actors (companies) and professional associations
- Real estate actors (building owners, building operators, and construction companies)
- Energy producers and distributers, in particular SIG[8]
- Organizations and associations, in particular "Genève à 2000 W"

Several legal and planning documents are used in Geneva to control urban and energy planning. First of all, a set of legal documents (laws, regulations, ordinances, and norms), presented in Table 9.2, define the general constraints of urban development. Urban and energy planning documents complement those laws by defining an actual road map to follow and providing recommendations (see Table 9.3). Those planning documents are available for different decision levels and development stages, each one being a local translation of the goals and actions set out at the closest higher scale. For urban planning, the cantonal directive plan defines the overall goals of the canton in terms of urban development, followed by communal directive plans for each municipality that define the local actions and goals. At the lower scale, neighborhood directive plans are used for the development of new districts. Moreover, energy planning is specifically addressed in the energy directive plan, which is further broken down at lower scales with the territorial energy concept.

9.2.4 Data

Geneva has obtained a large variety of data from the canton concerning all domains, providing public authorities, private companies, and citizens with valuable information that supports, among others, the decision-making process of urban and energy

[8] SIG, service industriel de Genève. Multiutility service company in Geneva (http://www.sig-ge.ch/).

Table 9.2 **Main legal documents regarding construction and urban planning in Geneva (Cajot et al., 2016)**

Name	Type	Description
LGZD	Cantonal law	Development zone law (Loi générale sur les zones de développement du 29 juin 1957)
LCI	Cantonal law	Construction and installation law (Loi sur les constructions et les installations diverses du 14 avril 1988)
RCI	Cantonal regulation	Regulation on the construction and installation law (Règlement d'application de la loi sur les constructions et les installations diverses du 27 février 1978)
LEN	Cantonal law	Energy law (Loi sur l'énergie du 18 septembre 1986)
REN	Cantonal law	Regulation on the energy law (Règlement d'application de la loi sur l'énergie du 31 août 1988)
OBP	National ordinance	Ordinance on noise protection (Ordonnance du 15 décembre 1986 sur la protection contre le bruit)
LaLAT	Cantonal law	Application of the federal spatial planning law (Loi d'application de la loi fédérale sur l'aménagement du territoire du 4 juin 1987)
LAT	National law	Spatial planning law (Loi fédérale sur l'aménagement du territoire du 22 juin 1979)
SIA	Norm	Set of standards and requirements for planning and construction in Switzerland (Société Suisse des ingénieurs et architectes)

planning. In particular, they have acquired numerous georeferenced datasets, many of which are publicly available on the SITG platform.[9]

SITG is an online platform for georeferenced data of Geneva. It includes over 500 "layers." Around two-thirds of the data are open data and can be downloaded in different file formats (CSV, SHP, KML, etc.), the rest being either private data that requires authorization from the data provider or data that can only be visualized online.

There are many datasets of interest on the SITG platform for studies regarding energy. First of all, there are two fundamental layers that allow to link many datasets together: the building footprint layer and the address layer.

The building footprint layer gives the footprint of each building in Geneva using a polygon geometry and many useful information regarding the building, such as building usage, construction period, and number of floors. Most importantly, each building is identified by a unique building identifier named EGID. Any dataset or information about buildings is usually associated to this EGID.

[9]Link to SITG platform, http://ge.ch/sitg/.

Table 9.3 Main urban and energy planning documents in Geneva (Cajot et al., 2016)

Name	Type	Description
PDCn	Cantonal directive plan	Document that coordinates confederation, cantons, and surrounding regions. It sets strategic goals and measures to reach them (PDCn—plan directeur cantonal)
PDCom	Communal directive plan	Document providing a global vision of a commune's development over 10–15 years, coordinating with canton and neighboring communes. Contains a concept and synthesis map and measures to reach the objectives (PDCom—plan directeur communal)
PDQ	Neighborhood directive plan	Document depicting a future neighborhood in the midterm. Sets the projects principles, but not the details of layout and construction. Contains layout alternatives, objectives, and synthetic map and measures to reach objectives (PDQ—plan directeur de quartier)
PDE	Energy directive plan	Document that sets the energy policy goals on the canton level and establishes desired shares of energy resources, including RES (PDE—plan directeur de l'énergie)
CET	Territorial energy concept	Approach and document on the territorial scale that organizes and coordinates actors to reduce energy needs by developing energy efficient infrastructure and promoting the use of local energy sources (CET—concept énergétique territorial)

The address layer gives all the addresses in the canton of Geneva. They are represented as points in the spatial database and each have a unique address identifier called IDPADR. Each address has a number of characteristics and is associated to a building through the EGID number. Many data are associated to the address, especially energy-related data such as energy consumptions billed to clients connected to the networks. As a consequence, data that are originally linked to the IDPADR can also be linked to the corresponding building EGID. It is to be noted that there can be several addresses (IDPADR) linked to one building (EGID), but the reverse is not true.

Energy-related data exist in different forms. A heating consumption index (IDC) is available for all nonresidential buildings and buildings with five dwellings or more, since 2011. It gives the yearly heating consumption per square meter year by year and averages over the last 3 years. These data are openly available as a layer in SITG and are directly linked to the building EGID. In addition to that, yearly consumption data of each client on the electricity, gas, water, and district heating grids can be provided by network operators upon request, provided that personal data protection measures are put in place (such as signing an NDA). These data can be linked to addresses through the IDPADR and hence with a specific building.

On the other hand, more detailed demand profiles are not easily available. Detailed electric load profiles at 15 min timesteps are available for almost 7000 clients that are on the nonregulated market (out of around 274,000), and the number is increasing every year. However, these are mostly "big" consumers such as industries, office and administrative buildings, or large residential complexes. Regarding the district heating networks, many substations are equipped with teletransmission devices monitoring and saving hourly measurements: input/output temperature, pressure, flowrate, and thermal load. However, the number of buildings connected to the district heating network is currently limited.

Statistical population data are also available at address level from the cantonal statistics office OCSTAT. In particular, it includes the population at each address, which is a useful parameter for building energy demand models (both data-driven and physical models). However, these data are protected by the LStat B 4 40 law and cannot be shared with partners outside of Switzerland, except for addresses that have three or more inhabitants (RStat B 4 40.01).

9.2.5 Case studies

Geneva has ambitious goals in terms of energy efficiency, renewable energy penetration, and reduction of CO_2 emissions, which are aligned to the 2000 W society. To go in this direction, numerous urban development projects are currently in progress or planned. These include both the construction of new districts (greenfield developments) and the rehabilitation of existing areas (brownfield developments). Within the context of the CI-NERGY project, two case study areas were chosen (Fig. 9.7). The first one, a greenfield development—"Les Cherpines" district—was chosen

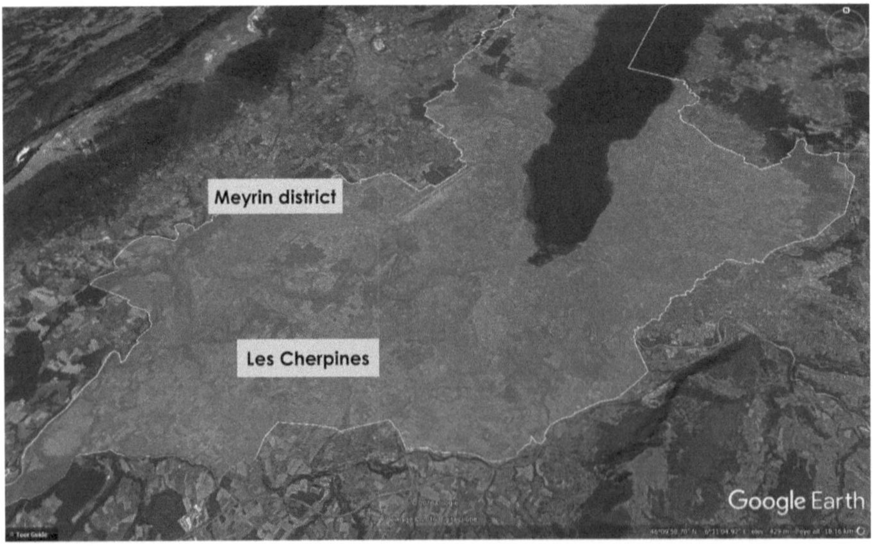

Fig. 9.7 Case study site location in Geneva.

Fig. 9.8 Bird's-eye view of Les Cherpines site (upper zone) and ZIPLO industrial zone (lower zone).

and will be used throughout this book to demonstrate the methodologies that were developed and that are presented in the following chapters.

Les Cherpines is an ongoing urban development project located in the south of Geneva, in the outskirts of the main urban center (see Fig. 9.8). Identified in 2001 as a suitable urban development site, a rural zone has been modified in 2011 to receive a mixed-use "ecodistrict" that will host 3000 dwellings and 2500 jobs on 0.6 km^2 of land. A bird's-eye view of the area is shown in Fig. 9.9.

Planning-wise, the "strategic" neighborhood master plan of Les Cherpines was released in 2013, and more detailed localized master plans will be available in early 2016. Implementation will begin in late 2017 with the tram and first dwellings. It will last until 2030, developing step by step the 12 "urban pieces" (or lots) that constitute Les Cherpines.

Energy-wise, the project aims to become a positive energy district, by using local resources and synergies between activities. Targets include covering 75% of energy supply with renewable sources and a threefold reduction of per capita consumption of future dwellers compared with current Swiss values. Main energy strategies involve geothermal and waste heat recovery with interseasonal storage from the neighboring industrial zone, ZIPLO.

In addition to this greenfield development, a case study of an existing area was also considered, especially for validation purposes. The district around Rue de la Prulay in Meyrin, a municipality in the northeast of Geneva (see Fig. 9.7), was chosen for this purpose because of the good data availability. The area mostly consists of residential apartment blocks constructed between the 1960s and 1980s. A bird's-eye view of the area is shown in Fig. 9.9.

Fig. 9.9 Bird's-eye view of the district in Meyrin.

9.3 Vienna

9.3.1 General description

Vienna is the capital and largest city of Austria, and one of the nine states of Austria. It is Austria's primary city, with a population of about 1.8 million (2.6 million within the metropolitan area, >20% of Austria's population), with an average population density of about 4000 inhabitants/km^2. Vienna is the cultural, economic, and political center of Austria. In 2001, the city center was designated as a UNESCO World Heritage Site.

Vienna is the seventh largest city by population within city limits in the European Union and is the second city to Berlin in terms of German speakers. Vienna is composed of 23 districts (in German, *Gemeindebezirke*).

Vienna is located in northeastern Austria (Fig. 9.10), at the easternmost extension of the Alps in the Vienna basin, close to the borders of the Czech Republic, Slovakia, and Hungary. The city spans both sides of the Danube river and covers an area of c.830 km^2 distributed as follows: 415 km^2 city, 395 km^2 land, and 19 km^2 water. Elevation ranges from 151 to 524 m above sea level. Although Vienna is quite compact and Austria's smallest federal province by far, >50% of its territory is covered by green spaces (Fig. 9.11).

In the last 20 years, Vienna has been a fast-growing city: in 2013, Vienna's population had increased by almost 200,000 people as compared with 2000; as of 2017, the city has reached approximately 1.86 million inhabitants. According to forecasts, this trend in population growth will continue in the coming years: around 2030, Vienna will reach and very probably surpass the 2-million-citizen mark. For the entire Viennese agglomeration area, which comprises the adjacent districts of Lower Austria and

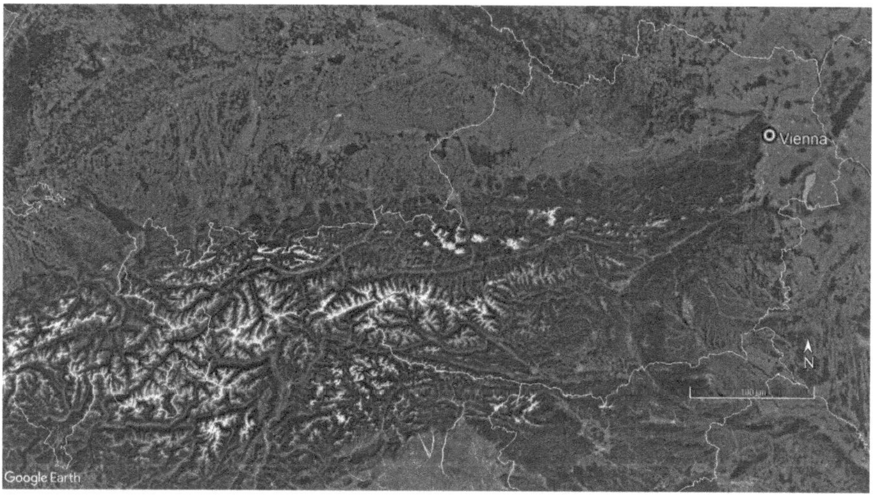

Fig. 9.10 Location of Vienna in Austria.

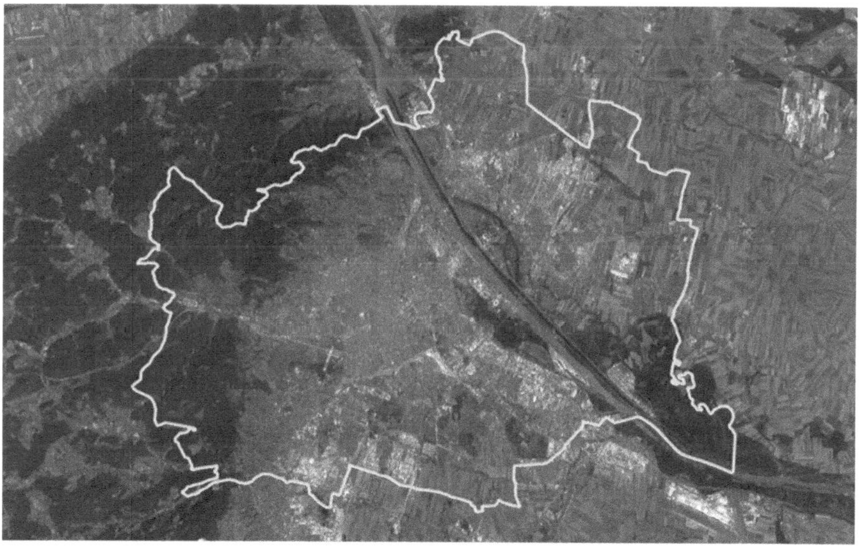

Fig. 9.11 Satellite image of Vienna with city boundaries.
Source: Google Earth.

parts of northern Burgenland, the population growth until 2030 is expected to equal >400,000 inhabitants.

Vienna lies within a transition of oceanic climate and humid continental climate and features an oceanic climate. The city has warm summers with average high temperatures of 24–33°C, with maxima exceeding 38°C and lows of around 17°C. Winters are relatively dry and cold with average temperatures at about freezing point.

Spring and autumn are mild. Precipitation is generally moderate throughout the year, averaging 550mm annually, with considerable local variations.

Industries are located mostly in the southern and eastern districts. The inner city is situated away from the main flow of the Danube but is bounded by the Danube canal. Vienna's second and twentieth districts are located between the Danube canal and the Danube river. Across the Danube and in the southern areas are the newest parts of the city (districts 21, 22, and 23).

9.3.2 Energy demand, resources and infrastructure

According to the yearly Energy Report[10] issued in 2016 by the Municipal Department 20 (Energy Planning), which contains the latest available data (from 2014) and upon which most of the section is based, Vienna imports about 88.2% of the energy, mostly in form of gas (36.4%), oil (34.3%), electricity (12.7%), renewables (3.2%), and district heating (1.4%). The remaining 12.8% is generated within the city boundaries from renewables (8.6%) (e.g., hydropower, wind, and PV), combustible waste (3.8%), and ambient heat (0.5%). c.1% is the exported energy (Fig. 9.12).

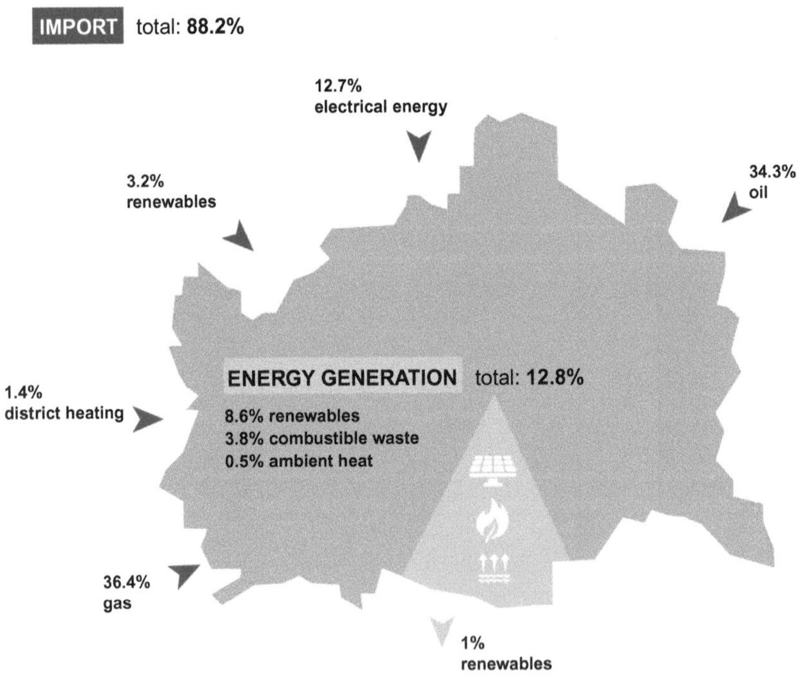

Fig. 9.12 Distribution of energy generation and imports in Vienna by source. Image source (Vienna Energy Report 2016).

[10]https://www.wien.gv.at/kontakte/ma20/publikationen/index.html#studien.

For 2014, the gross inland energy consumption was 40.65 TWh. There is a clear dominance of fossil energy sources (for 2014, c.36% natural gas and 32% fuels) in the city's energy mix. Approximately, half of the natural gas is converted into electricity and district heating. Fuels and mineral oils (petrol, diesel, and other petroleum products), on the other hand, are used nearly exclusively by the largest consumption sector, namely, transportation. Overall, losses amount to over 20 TWh or 48% of gross inland energy consumption. These losses occur during different phases of the energy flow, with approximately 3.85 TWh lost before arriving at the end user (conversion losses, transport losses, etc.) and approximately 15.65 TWh losses in end-user consumption. Therefore, the global energy consumption of Vienna in 2014 (i.e., after 3.85 TWh conversion and transport losses) was 36.79 TWh, subdivided approximately as follows: 7.02 TWh (19%) gas, 8.19 TWh (22%) electricity, 5.80 TWh (16%) district heating, 2.02 TWh (6%) renewables, 12.96 TWh (35%) fuels, and 0.79 TWh (2%) liquid energy sources. A more detailed view of the energy flowchart of the city of Vienna is given in Fig. 9.13.

The subdivision of the global energy consumption according to usage by sector yields the following values: private households 10.80 GWh (29%), productive sector and agriculture 3.02 GWh (8%), services 8.82 GWh (24%), and traffic and transport 14,157 GWh (38%).

In terms of end use, the global energy consumption is subdivided as follows: space heating 12.97 GWh (35%), process heat 5.13 GWh (14%), and power and lighting 18.96 GWh (52%). Taking into account the energy losses at end users (38.5%), the consumption of useful energy in 2014 was 21.15 GWh.

Table 9.4 presents the gross inland energy consumption in Vienna, subdivided according to the energy carries.

A comparison of the 2014 values (normalized by the number of inhabitants) with previous years is given in Table 9.5.

In terms of annual final and primary energy consumption per person, the trends are presented in Tables 9.6 and 9.7, respectively, while Table 9.8 gives some numbers in terms of CO_2-equivalent tons per capita. The emissions calculated using the KliP balance calculation method are the basis for all calculations for Vienna's climate protection programs (KLiP I and KLiP II).

Vienna has a large share of old buildings. Refurbishing them requires solutions that reduce energy consumption and carbon dioxide emissions considerably. The potential for reduction is particularly high in blocks of several houses. More specifically, when it comes to the building sector in a fast-growing city like Vienna, several measures have already been adopted in the past 20 years with regard to the building rehabilitation (especially in the municipal and social housing sector) or will further be improved in the near future. For existing buildings, a comprehensive rehabilitation is planned to reduce the energy consumption for space heating and cooling and domestic hot water production by 1% per person per year. For all newly built buildings (as well as additions and refurbishments), cost-optimized zero-energy building standards will be enforced from 2018 to further boost energy efficiency.

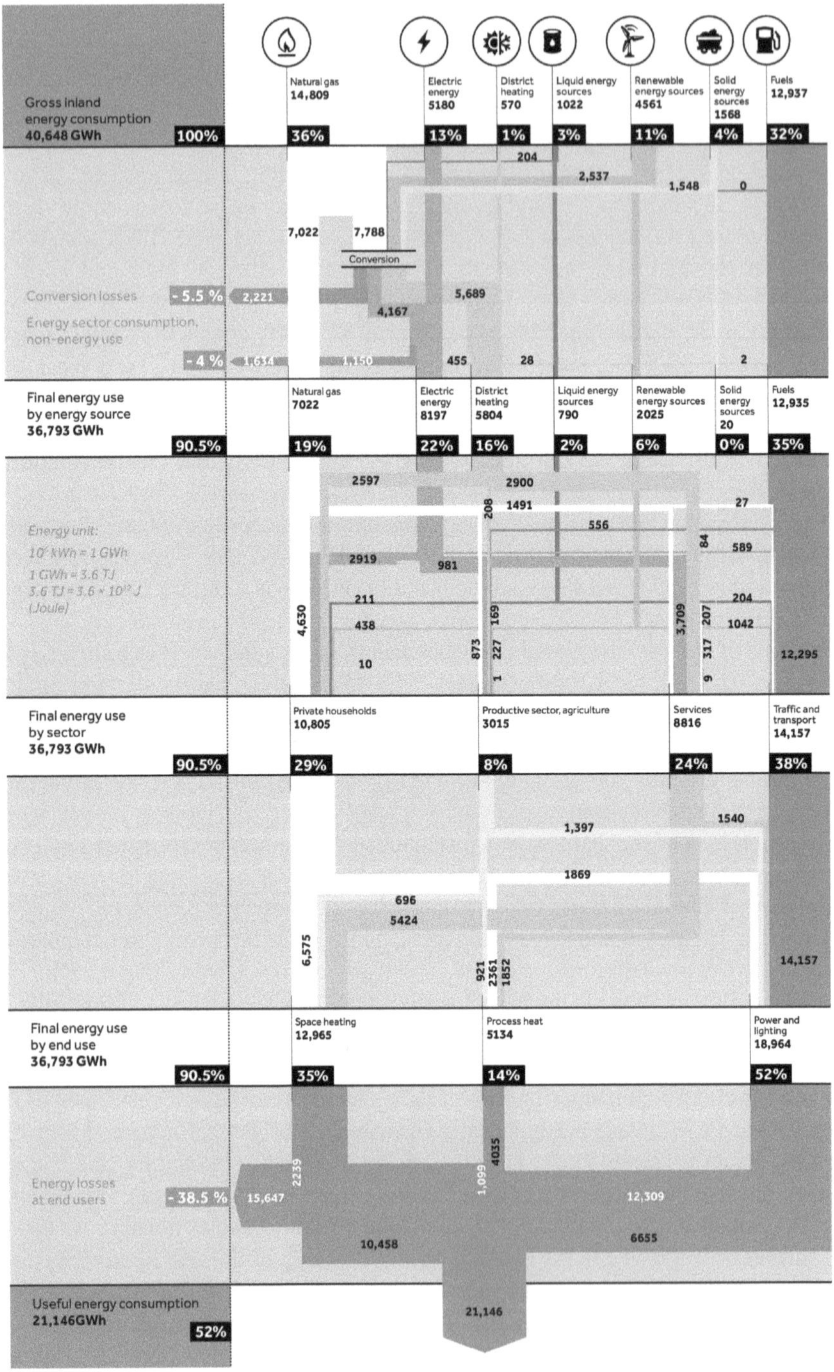

Fig. 9.13 Energy flowchart of the city of Vienna. Image source (Vienna Energy Report 2016).

Table 9.4 **Gross inland energy consumption in Vienna, subdivided according to the energy carries**

GWh/a	1995	2000	2005	2010	2013	2014	Variation (% since 1995)
Combustible renewables	917	1341	1121	3057	2949	2918	218.0%
Combustible waste	1044	937	1404	1461	1514	1550	48.5%
Wood	332	312	347	292	439	353	6.3%
Electricity	3364	3076	2045	1388	4808	5180	54.0%
District heating	426	644	413	463	632	570	33.9%
Gas	18,218	18,287	22,851	22,777	17,250	14,809	−18.7%
Coal	300	190	87	22	22	18	−94.1%
Oil	13,987	13,746	13,481	15,263	14,185	13,959	−0.2%
Ambient heat	50	72	89	140	173	183	267.2%
Hydroelectric	–	1271	1087	1117	1167	1063	
Wind and PV	–	1	7	14	41	44	
Sum	38,639	39,878	46,931	45,994	43,180	40,648	5.2%

Table 9.5 **Final energy consumption per person in Vienna**

kWh/person	1995	2000	2005	2010	2013	2014	Variation (% since 1995)
Private households	7032	6992	6915	7036	7114	6116	−13.03%
Productive sector and agriculture	2587	2070	2048	1983	1871	1706	−34.04%
Services	5057	5032	5758	6203	5321	4990	−1.33%
Traffic and transport	6255	7316	9130	8481	8206	8013	28.11%
Sum	20,931	21,680	24,211	23,704	22,513	20,825	−0.51%

In January 2016, Vienna introduced a database for the energy performance certificates,[11] in which all EPCs issued for Viennese buildings must be registered. The aim of the registration is to improve quality assurance and to ensure that the data in the certificates are correct and up-to-date.

[11]WUKSEA, Wiener unabhängiges Kontrollsystem für Energieausweise.

Table 9.6 Annual final energy consumption per person in Vienna

kWh/ person/a	1995	2000	2005	2010	2013	2014	Variation (% since 2005)
Final energy consumption	18,486	20,931	21,680	24,211	23,704	22,513	−13.98%

Table 9.7 Annual primary energy consumption per person in Vienna as continuous power rating

W/person/a	1995	2000	2005	2010	2013	2014	Variation (% since 2005)
Primary energy consumption	2833	2940	3286	3172	3062	2831	−13.8%

Table 9.8 Annual CO_2-equivalent emission per person in Vienna

$tCO_{2,eq}$/ person/a	1995	2000	2005	2010	2013	2014	Variation (% since 1995)
Emissions (according to KliP method)	3.8	3.6	3.1	3.2	2.8	2.8	−26.1%

New buildings are planned and built according to very high standards of energy efficiency (low-energy standard and its successors). Energy and heating systems are always included in these considerations as well. According to the EU Energy Performance of Buildings Directive (EPBD) of 2010, the zero-energy standard will be mandatory for all new building types, with cost optimization being taken into account in defining and updating all related requirements. In addition to new buildings, great attention is also paid to building rehabilitation. Above all, in municipal and cooperative housing construction, great advances were made in this field over the past 20 years. Table 9.9 contains the trend of the annual final energy consumption per person for space and water heating and air conditioning in Vienna.

The detailed share of energy sources in final energy consumption for space and water heating and air conditioning in Vienna is shown in Table 9.10.

Table 9.9 Final energy consumption for space and water heating and air conditioning per person in Vienna

kWh/person/a	2005	2010	2013	2014	Variation (% since 2010)
Final energy consumption	10,014	10,405	9628	8277	−20%

Table 9.10 Share of energy sources in final energy consumption for space and water heating and air conditioning in Vienna

Share in final energy consumption	2005	2010	2013	2014	Variation (% since 2005)
Combustible waste	0.1%	0.0%	0.0%	0.0%	−91.6%
Electric energy	9.5%	9.3%	9.5%	10.2%	6.6%
District heating	32.7%	37.8%	38.8%	39.3%	20.2%
Gas	44.5%	43.2%	42.1%	40.8%	−8.4%
Coal	0.5%	0.1%	0.1%	0.1%	−77.1%
Oil	9.2%	4.9%	3.2%	3.5%	62.0%
Renewables total	3.5%	4.6%	6.3%	6.1%	77.3%

In terms of adoption of renewable energies, an important goal is the increased use of waste heat and ground heat as sources of energy, with some showcase projects already carried out in the 17th district.

In addition, Vienna has favorable wind conditions, particularly in the northern and northeastern parts of the 21st and 22nd districts and on the southern edges of the 10th, 11th, and 23rd districts. A wind-potential map is available for the use of small wind turbines in the Vienna region.

By 2020, Vienna's wastewater treatment plant will be modified to double as a green power plant. Gas produced from sewage sludge fermentation will be burnt in gas-powered cogeneration plants. Currently, the wastewater treatment plant uses approximately 63 GWh of electricity per year, making it one of the city's largest energy consumers. The new power plant will produce 78 GWh annually, thus covering the entire electricity demand of the wastewater treatment plant and feeding the excess electricity in the public grid and providing power for additional 31,000 Viennese households. In addition, further 82 GWh/a of heat will be generated. Approximately half will be consumed on site and the other half fed into the district heating network, providing heating and hot water for some 5000 households.

Since March 2016, the city has been subsidizing solar heating systems, heat pumps in residential buildings that utilize ambient heat, systems for the thermal use of groundwater and ground heat, and seasonal heat storage systems (for waste heat and renewable energy) that help balance the load between the different times of production and use of heat.

Vienna is also planning to increase the e-mobility expected in the coming years. For this purpose, a network of electric-vehicle charging stations will be built in the next years. New charging technologies, improved vehicles, and sinking production costs make this the ideal time to start building a city-wide network of publicly accessible charging stations.

9.3.3 City goals, actors and tools

Thanks to the long-standing urban planning tradition of Vienna, several planning tools have been adopted over the course of the years to cope with the challenges tied to sustainable urban development.

In 2006, Vienna adopted the Urban Energy Efficiency Program[12] (SEP) with the objective to create a strategic framework for energy efficiency measures by 2015 and to reduce the expected increase in energy consumption from 12 to 7%. The final report on the full SEP implementation period (the period from 2006 to 2015) showed that the objective was more than achieved. In 2006–15, the documented project-related energy savings amounted to approximately 155 GWh per year. If the undocumented energy savings (e.g., in federal buildings and in businesses) are included, the total savings can be assumed to be above the SEP target of 180 GWh. SEP successfully established the importance of energy efficiency in Vienna and showed that energy savings are possible everywhere. Since the adoption of SEP in 2006, the conditions for energy efficiency policy have progressed considerably at both the European and national levels.

In 2011, the Smart City Wien Initiative was launched. Based on a broad stakeholder process, the development of the strategy begun in 2013 and has led, in 2014, to the adoption of the Smart City Wien framework strategy,[13] which builds upon and extends the previous Urban Energy Efficiency Program. The Smart City Wien framework puts further emphasis on sustainable energy supply by means of conservation and intelligent use of resources. It defines targets for different areas, such as energy efficiency, renewables, infrastructure, mobility, and buildings. One of the overall goals is to decrease the final energy consumption per capita in Vienna by 40% by 2050 (compared with 2005), and at the same time, the per capita primary energy input should drop from 3000 to 2000 W, that is, to 48 kWh per day, as envisioned by the so-called 2000 W society.

The Smart City Wien framework strategy spans its time horizon to 2050, since the necessary and often fundamental changes in the fields of energy, mobility, or construction are long-time processes. It lays down multiple ambitious objectives that will contribute to reshape Vienna as a research and business hub and to preserve all-important social achievements, protecting at the same time the national and international competitiveness of the Austrian capital.

[12] https://www.wien.gv.at/stadtentwicklung/energie/pdf/sep-programm.pdf.

[13] https://smartcity.wien.gv.at/site/en/the-initiative/framework-strategy/.

The strategy focuses on the intention of preserving and further evolving the city as a livable, socially inclusive and dynamic space for future generations, that is, saving resources to massively reduce CO_2 emissions and reducing dependency from non-renewable resources. At the same time, Smart City Wien wants to uphold and further increase Vienna's high quality of living and social participation, thus finally leading to a change based on innovation, active organization, and, where necessary, the development of new forms of public and private service delivery.

The goals of the Smart City Wien framework strategy regarding energy and resources can be summarized in the following points:

- Reduction of the per capita greenhouse gas emissions by at least 35% by 2030 and 80% by 2050 (from 1990 levels).
- Increase of energy efficiency and decrease of final energy consumption per capita by 40% by 2050 (from 2005 levels), reduction of the per capita primary energy input from 3000 to 2000 W.
- In 2030, over 20% and, in 2050, 50% of the gross energy consumption in Vienna will be covered from renewable sources.
- Reinforcement of the CO_2-free modes of transportation (walking and cycling), maintaining the high share of public transport and decreasing motorized individual traffic in the city to 20% by 2025, to 15% by 2030, and to markedly <15% by 2050.
- By 2030, the largest possible share of motorized individual traffic is to be shifted to public transport and nonmotorized types of traffic or should make use of new propulsion technologies (e.g., electric vehicles). By 2050, all motorized individual traffic within the municipal boundaries is to operate without conventional propulsion technologies.
- Cost-optimized zero-energy building standards for all new structures, additions and refurbishment from 2018/2020, and further development of future supply systems toward even better climate protection levels.
- Comprehensive refurbishment activities involve the reduction of energy consumption of existing buildings for space heating/cooling/water heating by 1% per capita and year.

Building upon Smart City Wien framework strategy, several other planning tools have been developed in the recent years, such as the Energy Strategy 2030 of the City of Vienna, the Climate Protection Program KliP, the Security of Supply Plan including the Renewable Energy Action Plan, the Urban Energy Efficiency Program, and the new Urban Development Plan STEP2025[14] and the Mobility Concept derived from it.

In particular, the Urban Development Plan STEP2025 sketches a picture of the city of tomorrow that combines social, economic, and environmental objectives. The document formulates strategic guidelines that will serve as an orientation for concrete implementation projects of coming years. STEP2025 therefore has a predominantly strategic character and provides an overview of measures to steer Vienna's development in key fields like building stock, housing, green and open spaces, economy, and infrastructure. The goal is to preserve what is valuable, renew what is obsolete, and transform the outdated.

If, on the one hand, inner-city urban development areas and existing urban expansion potentials already allow to meet the estimated demand for new building land until

[14] https://www.wien.gv.at/stadtentwicklung/studien/pdf/b008379a.pdf.

2025, on the other hand, it is of crucial importance to define how the urban developments will be carried out.

First, existing urban quarters need to be further development and enhanced in order to become full-fledged parts of the city with urban, mixed-use structures. For this purpose, structural measures are already being taken, for example, to upgrade the network of green spaces and to create strong urban subcenters. New inhabitants will be "adsorbed" in new urban quarters currently under development, brownfield and underused sites, and in already built-up areas.

According to the STEP 2025, out of the 120,000 new housing units expected to be built by 2025, c.55% will come from new structures erected on areas identified for potential development, 10% from modified use or conversion of existing buildings, 27% from further development of existing building stock, and the remaining 8% from areas so far not considered for development.

9.3.4 Data

Several heterogeneous datasets are available for the municipality of Vienna. Many of them are available as open government data (OGD) through the OGD portal.[15] It offers spatial and nonspatial data covering a huge number of topics. For spatial data, Web Feature Services and Web Map Services are also offered.

This section presents only a selection of datasets that were collected for the purpose of the CI-NERGY project. Although they consist mostly of OGD data, some non-publicly available datasets were kindly provided by the municipality of Vienna. In the following, the latter are identified by an asterisk (*). Most of the spatial data are georeferenced according to the MGI/Austria Gauss-Kruger East projection (EPSG code, 31256), which is the one generally used by the city of Vienna. The height reference is the so-called Wiener Null, set at 156.68 m above (Adriatic) sea level. In the following, the data sources will be classified into spatial and nonspatial ones, for better reading.

The spatial datasets include the following:

(1) The *Mehrzweckkarte* (MZK) is the official point and line-based map of Vienna, containing all most important objects (buildings, streets, etc.). From the MZK, the *Flächen-MZK* (F-MZK) is obtained, that is, a 2-D polygon-based map containing nearly 2.5 million geometries, where objects are classified into 50 classes (e.g., buildings, traffic areas, water bodies, and parks). For buildings, the *Bezug* code is also given. The Bezug identifies man-made features such as buildings, although it cannot be used as primary ID. If entities of the MZK are added or significantly updated for some reason, a new Bezug code is issued. Metadata information about the latest updates is available for each shapefile in which the F-MZK can generally be downloaded, however not at single-feature level.

(2) Another vector-based product derived from the MZW is the so-called Baukörpermodell, that is, a prismatic representation of all buildings in Vienna, in which the extrusion of the building polygon has been performed between the digital terrain model and the rain gutter height of the object. Nevertheless, some geometries with overhangs exist as well. The

[15] www.data.wien.gv.at.

Baukörpermodell contains approximately 665,000 objects and is available in two formats: as a set 3-D DXF files and as a set of 2-D shapefiles. In each DXF file, every layer corresponds to a Bezug and contains the corresponding triangulated geometric object(s); however, no other semantic information is available. In each shapefile, instead, each 2-D polygon is defined by means of the attributes Bezug, the FMK-derived class, the DTM height value, and the rain gutter height and, where required, the height of the lowest surface between DTM and rain gutter.

(3) For the whole city, a set of CityGML files containing buildings modeled according to the CityGML level of detail 2 (LoD2). The buildings are modeled starting from the Lidar and photogrammetric DTM data and the F-MZK (extracted in 2012) and contain information about the Bezug code. During the CI-NERGY project, the whole CityGML dataset for the city of Vienna was not publicly available yet, as internal data quality and consistency checks were being carried prior to the official publication. Nevertheless, the whole dataset for Vienna was kindly delivered by the municipality of Vienna and used to explore and test the data.

(4) A polygon-based vector map representing the land use for the whole city of Vienna. Polygons are classified using three hierarchical levels with growing degree of detail.

(5) A set of several point-based vector maps containing information about building names, building use/function (e.g., schools, theaters, commercial centers, kindergartens, swimming pools, public offices, and hospitals).

(6) A set of vector-based maps representing administrative boundaries for the city, the districts, and the building blocks, each with its unique ID.

(7) Point-based vector datasets with the installed photovoltaic and solar panel installations. The datasets contain information about the installed power and the surface of the solar collectors and the year of installation.

(8) *A set of raster maps used for the solar cadastre of Vienna, at 0.5×0.5 m geometric resolution, and consisting in monthly and yearly values of global solar radiation on the roof surfaces, and the corresponding values in terms of potential yield for photovoltaic and solar panel installations (the latter delivered as vector-based maps).

(9) A set of raster maps obtained from the solar cadastre of Vienna and representing three classes of roof suitability for installation of solar or photovoltaic panel systems.

(10) A vector-based map representing the geothermal potential of the underground at different depths.

(11) A raster-based map representing the groundwater thermal potential.

(12) A vector-based map of the wind-potential suitability zones.

(13) A CSV file containing all addresses of Vienna and some ancillary information about the building type/use and the coordinates corresponding to each address.

(14) Line-based shapefiles containing the layout of the gas and district heating network. Specific attributes like the pipe diameter or the construction materials are however not present. In addition, the geometries are not connected correctly from a topological point of view.

The nonspatial datasets used consist of the following:

(15) *An ESRI Geodatabase-based dataset containing in a number of table information for each building such as the following:

- The net floor area classified into different classes (residential, commercial, offices, industry, etc.)
- Number of floors above- and belowground

- Year or period of construction
- The (unclassified) net floor area for each floor and the type of floor (cellar, ground floor, attic, etc.)
- Address and address ID, to link the records to the spatially defined address dataset
- Further (relatively scarce) information regarding the presence of ventilation, heating, domestic hot water preparation systems, etc.

(16) *A Microsoft Excel file containing the Wiener Wärme Kataster (WWK) data, that is, the "Vienna Heat Cadastre," which contain many attributes among which are as follows:
- Address IDs, for example, to link the records to the spatially defined address dataset
- Estimated values for annual and monthly energy demand and energy consumption for space heating, domestic hot water, and electricity

(17) An XML-based file containing information about all *Gemeindebauten* (social housing buildings) in Vienna, such as the following:
- Address and address ID
- Name of the social housing building
- Year of construction and year of refurbishment
- Number of households
- Name of the architect(s)
- Some historical and architectonic information

(18) *For a limited number of residential buildings (64), a Microsoft Excel file containing information about the number of flats in the building, their floor net area, and the precise address with door number for each flat.

It must be stressed that a considerable number of further datasets are already available over the Open Government Data Wien platform. These datasets are surely worth to be integrated into a city model; however, they will not be treated in this paper as they are—for now—out of scope.

Unlike in Geneva, there is no unique ID system in Vienna to clearly identify a single building and to link it to its addresses or to other datasets, which as a matter of fact are often linked to an address information. Hence, the data integration process could not be carried out completely automatically.

Several other useful data are contained in the Energy Report, which is published by the Energy Department of the city of Vienna every year and contains data and analyses regarding the global energy balance and a number of parameters and indicators for the overall energy performance of the city.

9.3.5 Case studies

Given its heterogeneity in shape, structure, and characteristics, the district of Meidling was deemed adequate to represent a good test case (Fig. 9.14). Meidling spans an area of approximately 8.2 km^2; it lies just 5–7 km southwest of the city center and is located nearby the well-known Schönbrunn palace. It is a heavily populated urban area (c.90,000 inhabitants, i.e., c.11,000 inhabitants/km^2) not only with many residential buildings of greatly varying size and typology but also with large recreational areas and parks (Fig. 9.15). It can be approximately divided into two main parts: the northeastern one is characterized by a heavily developed urban residential texture, while the southwestern one is a more mixed (industrial and light residential) area, which then gradually continues southward to become the 23rd district.

Meidling represents also an important transportation hub, as the homonymous train station lies quite in the middle of the district and serves as an important connecting hub to many bus, tram, and rapid transit (S-Bahn and underground) lines between the city and the surrounding suburbs.

Given the relevance of the building sector in the overall set of energy-related goals defined by the Smart City Wien framework, project CI-NERGY focused in Meidling particularly on the identification of inefficient (residential) buildings, on the assessment of their energy performance and on the definition of some refurbishment scenarios. In addition, a bottom-up approach was adopted, in order to be able to scale up to block/district scale (whenever required) but keeping the granularity at the single-building level.

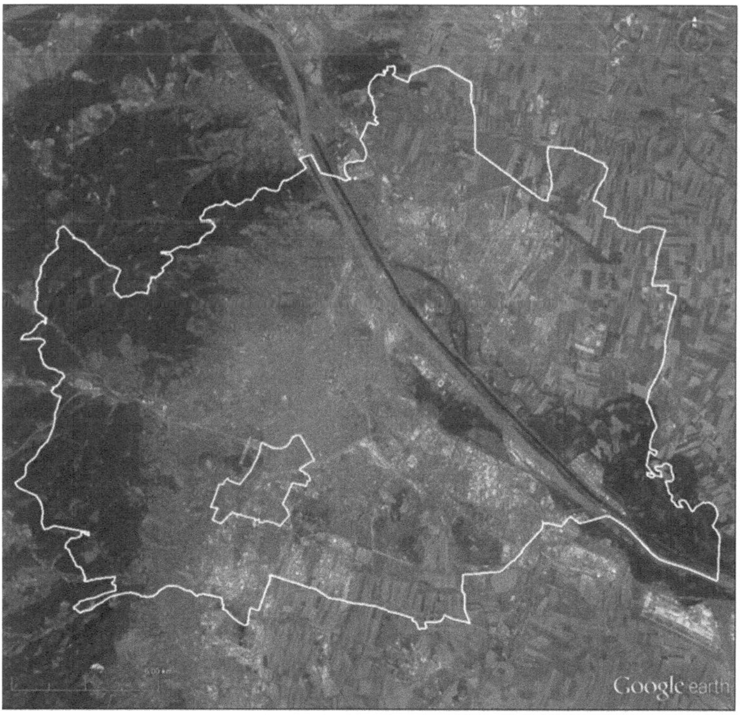

Fig. 9.14 Case study site location in Vienna.

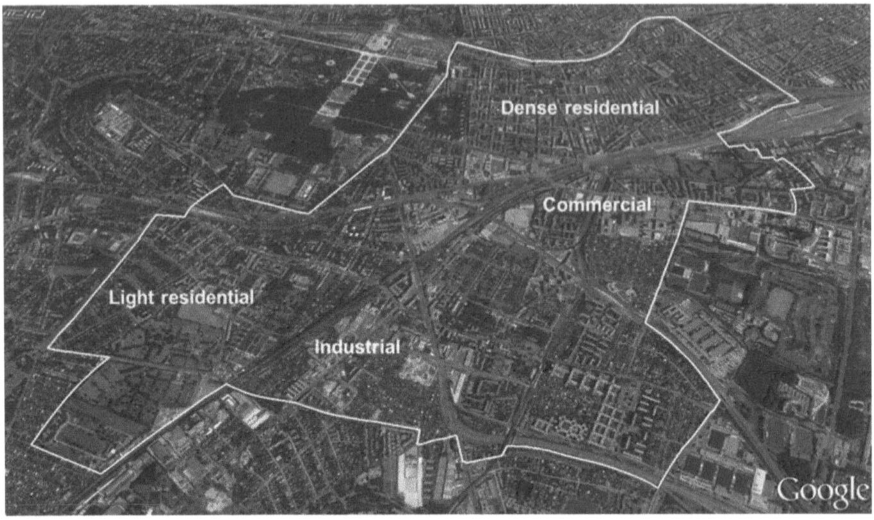

Fig. 9.15 The district of Meidling in Vienna.

9.4 Conclusions

This chapter gives a general introduction to two European case study cities—Geneva in Switzerland and Vienna in Austria—which were chosen as test case studies within the project CI-NERGY due to their ambitious sustainability goal. A general description of both cities was followed by an analysis of the available data about the overall energy demand, the available resources, and the existing and planned energy infrastructure. In addition, the overall goals of each city in terms of urban and energy planning were briefly presented. Finally, some information was given about the available data used in the project CI-NERGY. For each city, a short description of the specific case studies was given at the end of each section.

References

Cajot, S., Schüler, N., Peter, M., Page, J., Koch, A., Maréchal, F., 2016. Establishing Links for the Planning of Sustainable District (in Zürich).

Canton de Genève, 2016. B 4 40—Loi Sur La Statistique Publique cantonale(LStat). http://www.ge.ch/legislation/rsg/f/s/rsg_B4_40.html. (Accessed 26 October 2016).

Faessler, J., 2011. Valorisation Intensive Des Énergies Renouvelables Dans L'agglomération Franco-Valdo-Genevoise (VIRAGE) Dans Une Perspective de Société À 2'000W. Univ. Genève, Genève. http://archive-ouverte.unige.ch/unige:17272.

Geotechnique appliquee Deriaz S.A., Centre d'hydrogéologie et de géothermie, and Jules Wilhelm, 2011. Evaluation du potential géothermique du canton de Genève—PGG. Rapport Final. Genève.

Gilliard, L., 2015. Inventory of Energy Efficiency Resources of Industrial Processes of the Canton of Geneva. Semester projectEPFL.

Khoury, J., 2014. Rénovation Énergétique Des Bâtiments Résidentiels Collectifs: État Des Lieux, Retours D'expérience et Potentiels Du Parc Genevois. Univ. Genève, Genève. http://archive-ouverte.unige.ch/unige:48085.

MeteoSuisse, 2016. Diagrammes Climatiques et Normes Par Station: Genève-Cointrin. http://www.meteosuisse.admin.ch/home/climat/passe/normes-climatologiques/diagrammes-climatiques-et-normes-par-station.html?station=gve. (Accessed 5 October 2016).

Novatlantis, 2006. Conception Générale de l'Energie 2005–2009.

Office de l'environnement, 2012. Concept de valorisation énergétique de la biomasse. Rapport Final. Genève.

Service cantonal de l'énergie, 2016. Plan Directeur Cantonal de L'énergie 2005–2009. (Genève).

Viquerat, P.-A., 2012. Utilisation des réseaux d'eau lacustre profonde pour la climatisation et le chauffage des bâtiments; bilan énergétique et impacts environnementaux: Etude de cas: le projet GLN (Genève-Lac-Nations) à Genève. Univ. Genève, Genève.

Further reading

Cajot, S., 2015. Geneva "Cherpines" Project. A Common Case-Study in CINERGY. Internal Report. CI-NERGY Project.

Energie, 2016. Suivi Énergétique Des Bâtiments | Energie. http://ge.ch/energie/suivi-energetique-des-batiments. (Accessed 10 October 2016).

Tableau historique, 2014. B 4 40.01—Règlement D'exécution de La Loi Sur La Statistique Publique cantonale(RStat). https://www.geneve.ch/legislation/rsg/f/s/rsg_B4_40p01.html. (Accessed 26 October 2016).

Vassilakou, M., 2016. Energy! Ahead—Energy Report of the City of Vienna. Vienna City Administration/Municipal Department 20—Energy Planning.

Applying modeling and optimization tools to existing city quarters

10

*Mario Potente Prieto**, *Pilar Monsalvete Álvarez de Uribarri*[†], *Giovanni Tardioli*[‡,§]
*Technical University of Vienna (TU WIEN), Vienna, Austria, [†]Sustainable Energy Research Center (zafh.net), University of Applied Sciences, Stuttgart, Germany, [‡]Integrated Environmental Solutions Ltd., R&D, Dublin, Ireland, [§]School of Mechanical and Material Engineering, University College Dublin, Dublin, Ireland

10.1 Energy conversion and district heating systems in Vienna[1]

The Viennese energy system is based on centralized generation with several energy generators feeding a thermal network, defined as primary network. As shown in Fig. 10.1, in the federal district of Vienna and surroundings there are several energy sources, ranging from thermal and cogeneration plants, eolic and photovoltaic parks, and hydroelectric power plants. The latter are very popular in Austria and are located in the federal state of Niederösterreich, which surrounds the Viennese capital district and owns rich forest areas and areas of high wind speed. The diversification of energy allows an efficient utilization of the resources depending on location and season of the year. With an active energy efficiency policy in Vienna, there are currently 21,000 energy consultations per year, 3 million tonnes of CO_2 saved every year, and green electricity guaranteed for 660,000 people.

10.1.1 Evolution of the district heating network in Vienna

Initially, in 1969 no proper network existed. Only some decentralized heating plants were built providing small local networks in Hütteldorfer straße, Inzersdorf, Kagran, and Großfeldsiedlung. In addition, two thermal plants (Arsenal and Hofburg) and the first waste incineration plant (Flötzersteig) were built, none of them connected to the small thermal network. In 1981 the second Spittelau incineration plant started, creating the central branch of the eastern side of the Danube. The built network around the Kagran thermal plant extends along the western margin, connecting both sides of the river. Eight years later, many of the decentralized plants are dismantled, some of them reconfigured as combined cycles, constituting eight heat and power plants totally integrated with the district heating network. The third waste incineration plant is created.

[1] See Höckner (2014).

Urban Energy Systems for Low-Carbon Cities. https://doi.org/10.1016/B978-0-12-811553-4.00010-X

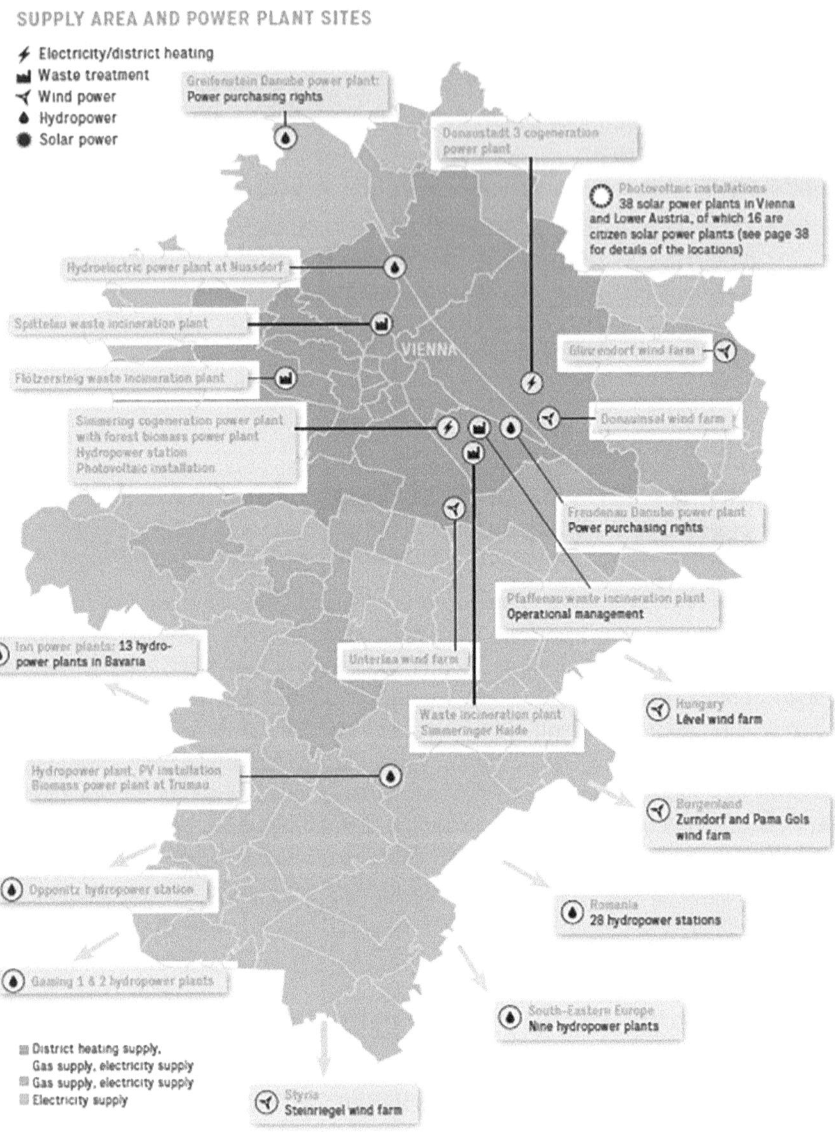

Fig. 10.1 General supply area and power plant sites around Vienna federal district (Höckner, 2014).

A heat recovery plant utilizes excess heat from the OMW refinery located in Schwechat, 15 km from Vienna. In 2003, two more plants were built: Inzersdorf and Donaustadt, which allow extending the network on both sides of the southwestern Danube, covering a length of 509 km for the primary network and 475 km for the local secondary networks, reaching all consumption points (dwellings, offices, hospitals, etc) (Figs. 10.2 and 10.3).

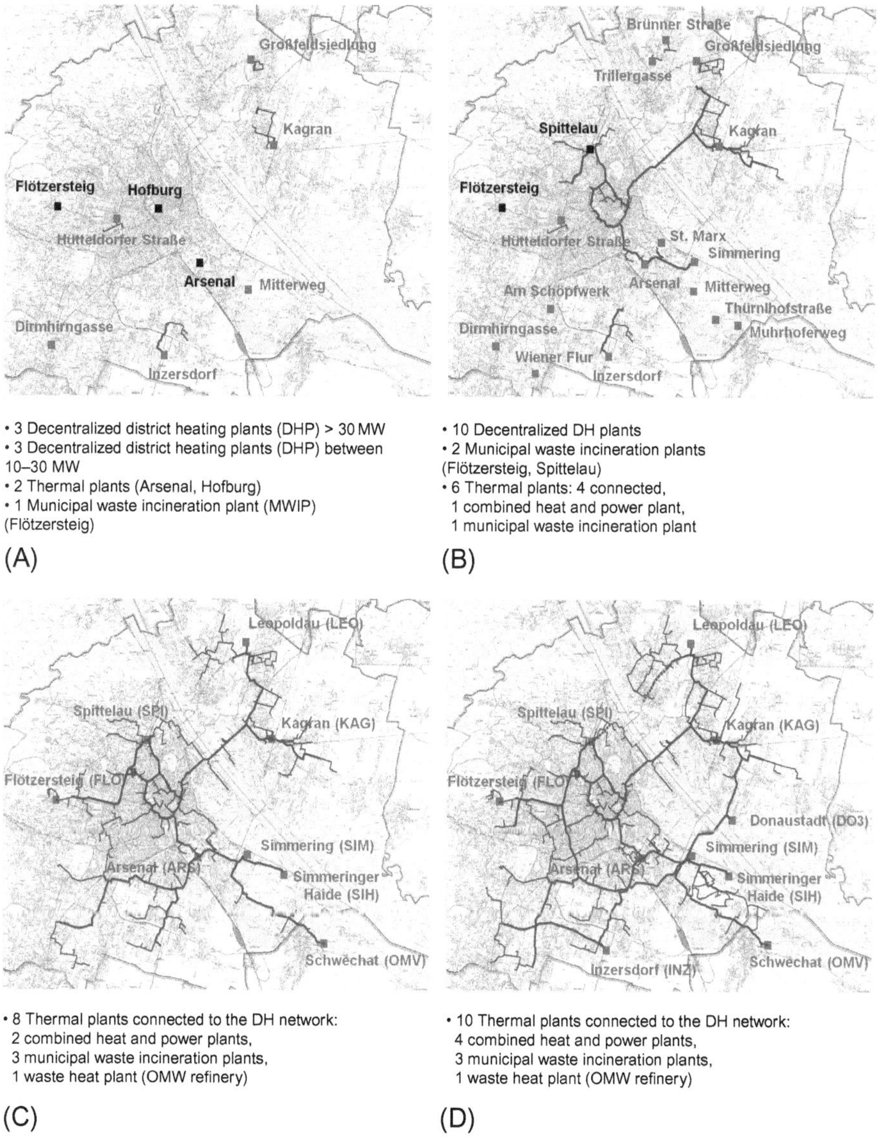

Fig. 10.2 (A) Year 1969. (B) Year 1981. (C) Year 1989. (D) Year 2003 (Höckner, 2014).

There are currently 11 operating thermal plants interconnected along a network covering 1192 km. It serves 336,902 users and 6602 major customers through 3758 substations and 5261 transfer stations. The total installed capacity is 3285 MW, reaching a 36% share in the heat market.

Operation. In the following lines, some of the most important Viennese thermal power plants are described. All of them are actually working linked to the district heating network.

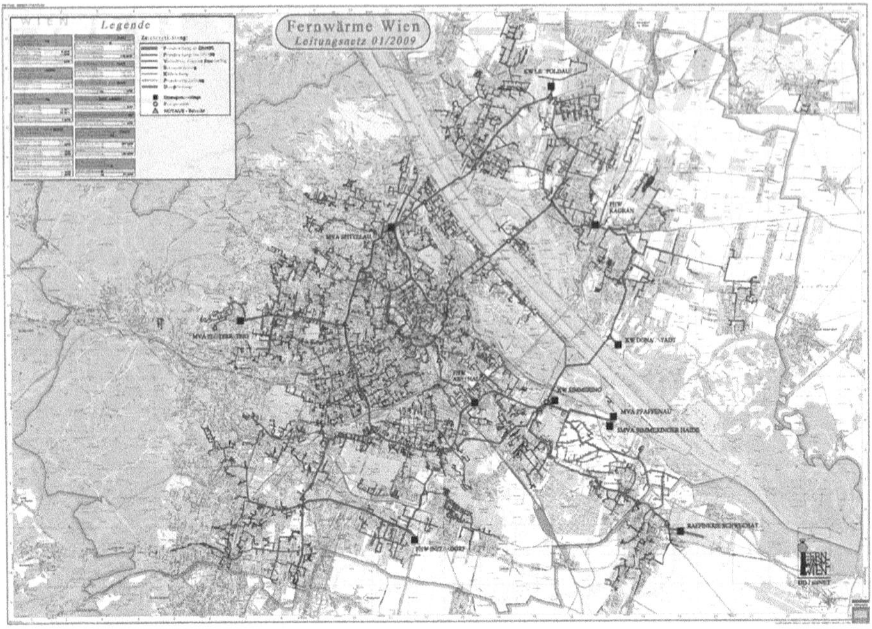

Fig. 10.3 Actual district heating network map (Höckner, 2014).

In summary:

- Energy generation, installed capacities in 2013:
  - Electricity: approx. 2100 MW
  - Heat: approx. 3300 MW
- Energy production in 2013:
  - Electricity: approx. 4000 GWh
  - Heat: approx. 6000 MW
- Share of energy generation from renewables:
  - Year 2012: 18.5%
  - Year 2013: 25%
  - Year 2030: 50%

10.1.2 Thermal power plant operation for district heating

In the following, some of the most important Viennese thermal power plants are described. All of them are actually working linked to the district heating network) (Fig. 10.4).

Combined heat and power plants (CHP). Those are the most technologically advanced, since their technology is based on Rankine cycles, producing electricity and heat simultaneously. The ratio between both can be modified according to the needs (Fig. 10.5).

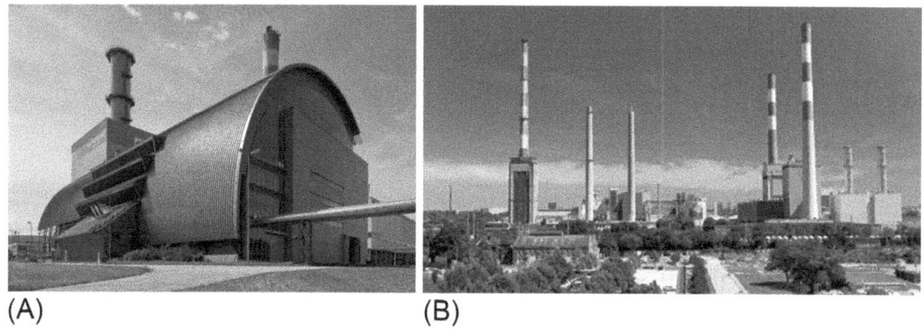

Fig. 10.4 (A) CHP plant Donaustadt 3. (B) CHP and biomass plant Simmering (Höckner, 2014).

Fig. 10.5 (A) MWIP and PL DH plant Spittelau. (B) MWIP Flötzersteig. (C) MWIP Pfaffenau (Höckner, 2014).

(A) (B)

Fig. 10.6 (A) PL DH plant Arsenal. (B) PL DH plant Arsenal (Höckner, 2014).

Municipal waste incineration plant (MWIP). Garbage is collected and stored to be burnt mostly for heat production. They use boiler technology and require a combustion gas treatment system, in order to avoid the expulsion of harmful particles into the atmosphere (Fig. 10.6).

Peak load district heating plants (PL DH) or boilers (B). These are exclusively dedicated to meet heat consumption peaks appearing at certain times of the year, increasing high temperature water flow. They operate using fossil fuels, especially natural gas or fuel. They are relatively small and their operation is not very complex. Table 10.1 shows total installed capacity as well as the total power generated (MW) for each of the operating plants and energy source used.

As shown in Table 10.1 and Fig. 10.7, most of the installed power corresponds to CHP plants and PL DH plants (44.7% and 44.6%, respectively).

In Fig. 10.7 the right column represents the installed capacity; it means that the maximum amount of produced heat obtained from the different thermal plants if they were at full load, while the column on the left reflects the percentages of actual utilization of the different types of plants for the year 2012/2013, which was especially cold in the city of Vienna (Fig. 10.8).

Here we can clearly distinguish the produced heat distribution during the winter/cold season between 2012 and 2013, depending on the used energy source.

The maximum consumption values occur in the months of January, February, March, and April, partially satisfied by the Peak Load District Heating Plants (PLDHP). At certain times it is better resorting to boilers since their start-up and operation is faster and more flexible than CHP plants.

Once that the water is heated in the thermal power plants, it is sent to the main piping network called primary network, whose temperature varies between 90 and 150°C at pressure levels higher than atmospheric pressure. The hydraulic pumps are located after the plants keeping water pressure at least at 5 bar, and, due to the geodetic situation of several buildings, it could reach 15 bar or even more. At the same time, control and security systems are inhibiting steam production or cavitation anywhere in the network.

The pipes are isolated and made of steel to be able to support the pressure. The primary network works as a source and at the same time energy sink, being connected to the different substations, which are partially linked to secondary networks. Consequently, the water inside has a high mass flow and temperature.

Table 10.1 Breakdown of heat generation: installed capacity (Höckner, 2014)

Energy source	Plant	Power (MW)	
Waste treatment	MWIP Spittelau	58	240
	MWIP Flötzersteig	51	
	MWIP Pfaffenau	75	
	Simmeringer Haide	54	
	Biogas	2	
Biomass	CHP Biomass Simmering	37	37
Industrial waste heat	CHP OMV refinery	168	173
	IWH Henkel	5	
Combined heat and power	CHP Donaustadt	250	1370
	CHP Leopoldau	170	
	CHP Simmering 1	450	
	CHP Simmering 2	150	
	CHP Simmering 3	350	
Peak load DH plants	B Leopoldau	230	1464
	B Spittelau	340	
	B Arsenal	340	
	B Kagran	200	
	B Inzersdorf	340	
	B Wilhelminenspital	14	
		Total installed capacity =	3284

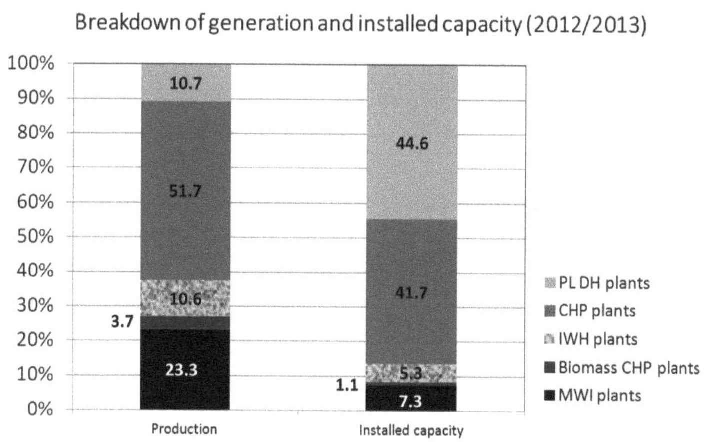

Fig. 10.7 Breakdown of heat generation and installed capacities (2012/2013) (Höckner, 2014).

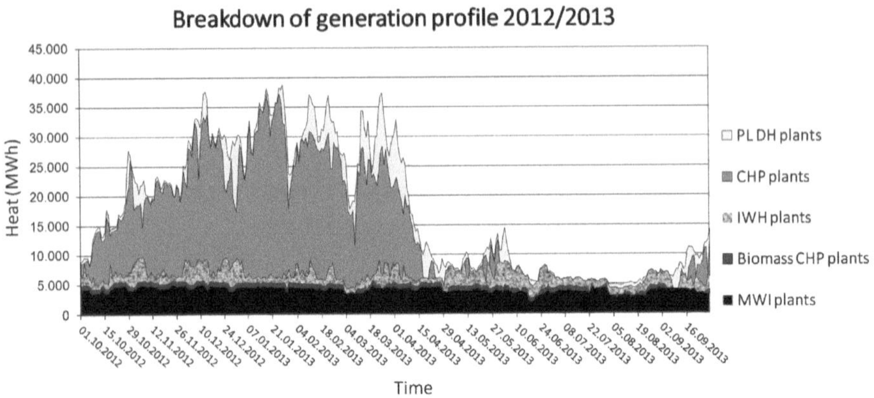

Fig. 10.8 Breakdown of heat generation through the year (2012/2013) (Höckner, 2014).

10.1.3 Advantages and disadvantages of the Viennese district heating network

From the ecological point of view, there are benefits obtained when choosing a thermal network instead of individual installations:

- Comparing network connected and nonconnected buildings, approximately 1.9 million tons of CO_2 emissions are avoided every year by district heating in Vienna.
- District heating in Vienna is ~75% more efficient than comparable heating systems.
- Without district heating the primary energy demand would be 42% higher. Greenhouse gas emissions would increase by 52%.

District heating systems have several weak points, especially considering the economic point of view:

- High gas prices and low electricity prices make gas fired CHPs uneconomic.
- Investing into DH network expansions is often not economically viable.
- DH network expansions focus on areas with high densities of heat demand. Extending 'the network to low population areas is not affordable.
- Existing trends toward decentralized heat production, for example, gas fired boiler houses combined with solar heat.

10.1.4 Possible improvements

Three general actions are considered, in order to improve the operation and increase the efficiency of the thermal network. Some of them are already developed or in progress:

Upgrading from boiler to CHP technologies, improving internal Rankine cycles, with the aim of increase versatility and energy efficiency. In short, the idea is introducing new ratios of heat production versus electricity and vice versa, depending on the energy prices and the needs of the moment.

- Reducing the return temperature generates two immediate effects:
  1. By maintaining a constant mass flow of district heating water, the temperature difference increases, raising in turn the supplied heat flow. This option is unrealistic from the operational point of view, because it is easier to satisfy peak heat demands increasing mass flows than manipulating temperatures.
  2. Considering a constant heat flow, as the return temperature is reduced, the flow of district heating water can be reduced, saving electricity for the hydraulic pumps.

  In order to achieve this temperature reduction, buildings must be adapted to be able to provide the same heating and hot water services at lower temperature. In this case, actions like refurbishing the envelope of the buildings and installing radiant floors replacing radiators must be taken into account. Radiant floors offer a greater area than radiators, which allow the reduction of the temperatures of the circulating hot water.

  Other technologies such as heat pumps and renewable energies in combination with district heating also allow this reduction. Besides, reducing the return temperature level implies reducing the temperature of the whole network, decreasing heat losses.
- Combining the actual district heating network with other technologies: for example installing decentralized heat pump systems, increasing temperatures in certain sections of the district heating network. Such systems utilize anergy networks and residual heat sources like wastewater, or medium/high temperature industrial effluents and require the consumption of electrical energy transformed into heat, which is known as "Power to Heat technology."
- In addition, there are other options such as the inclusion of energy storage tanks, which are useful to meet the heat demand peaks at certain times of the year, and even the generation of energy cascades, creating different temperature levels depending on the energetic efficiency of the buildings.
- There are actual projects related to geothermal energy, using the available heat contained in the underground water (for more information, see Section 10.5.1)

10.2 Actual situation in the urban district in a block of buildings within the urban district of Vienna Meidling: District heating network typologies and reference system

In the next section a case study urban district is described consisting of a set of buildings located in a residential area within the district of Vienna-Meidling.

This area was named as "Girasolstrasse," which is a fictitious name due to data privacy and confidentiality.

There are different uses of buildings, which define their characteristics. They are mostly residential homes, which include heating and domestic hot water (DHW) services, while others such as office blocks, churches, and others are only equipped with heating service. This distinction is important since not all buildings have the same purpose, so it is necessary to classify them for their correct modeling and simulation.

All buildings are connected to the local district heating network (secondary, in this case), although not all residents inside those buildings are clients of that network. Only connected apartments will be considered for our study. The case study consists of 21 buildings with 18 substations as the point of heat exchange (not mass) between the secondary local heating network and the thermohydraulic network inside the dwellings. Mainly plate heat exchangers are used, which are reduced size devices, compact, recommended for water as heat carrier with temperatures below 100°C. In addition, it is necessary to take into account the district substation (popularly called GUFO, by its name "Gebietsumformerstation" in German), which is thermally connected to the primary network and determines the temperature of the local network.

Note that number of buildings and substations are not coincident; this is because there are primary buildings, which have a heat exchanger linked to the network and secondary buildings that directly receive the heated water from the primary buildings (see Fig. 10.9). In addition, there are blocks defined as "slaves," which have no substation and are part of the same branch as their correspondent primary buildings.

In the other hand, all pipes present in the local grid must be classified as well. Basically there are two principal pipes, feed (H) and return (C), appearing from the general substation and progressively ramified, shaping a "Tree" configuration. If a branch is ramified into others a new letter is added, beginning from B and following alphabetic order (e.g., Fig. 10.10).

Using this methodology all buildings and pipes are named and classified. This is important to build the simulation. In addition, it has been decided to give a name and a number to each building (see Table 10.5).

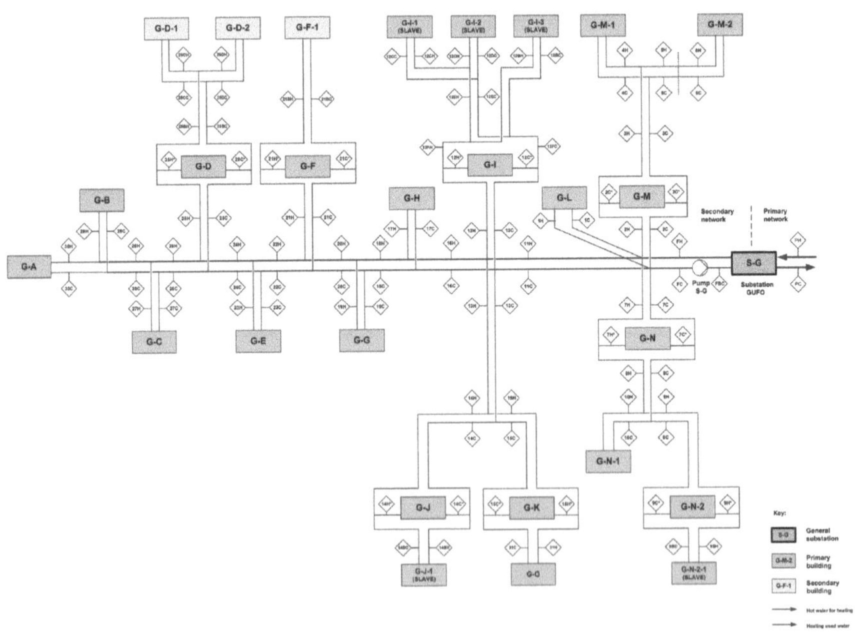

Fig. 10.9 General diagram of Girasolstrasse district heating network.

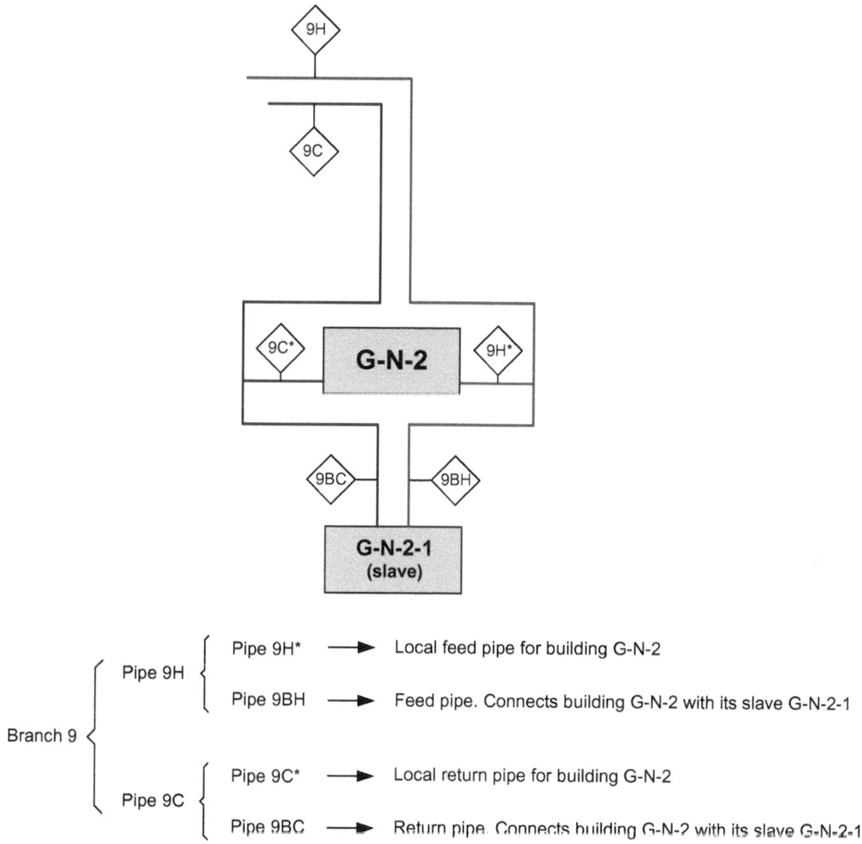

Fig. 10.10 Detail of buildings G-N-2 and G-N-2-1.

As we progress in the thermodynamic study of this type of networks, we realize the importance of the consumption points, since they are the final receivers of the heat transported in the pipes, generating possible energy and exergy losses, as we will analyze later.

However, not all flats sharing the same building are connected to the district heating network (Fig. 10.11).

There are three different possibilities or combinations to meet the needs of heating and hot water depending on the energy source used. Namely:

- Modality 1: Users of the district heating network for both services, having either radiators or radiant floors installed.
- Modality 2: Users of the district heating network but only for heating. DHW is produced by electric/gas boilers.
- Modality 3: Users not connected to the district heating network.

In practice, only those users connected to the district heating network will be taken into account, so that the apartments are not simulated as isolated elements, but the building as a whole object or unit.

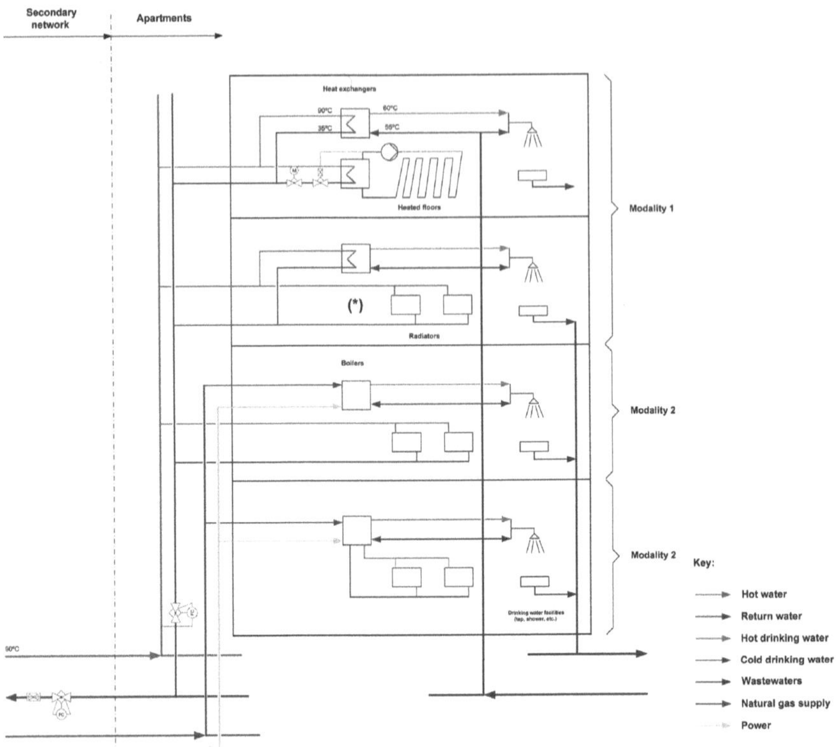

Fig. 10.11 General diagram for standard residential buildings. *Only in some cases, a heat exchanger is installed between the DH network and radiator's pipeline system.

Fig. 10.12 shows the hydraulic network installed in most of the buildings, in which two branches are distinguished; one for heating and another for domestic hot water (DHW). The heating branch has no heat exchanger installed except for some special cases, like floor heating systems or if the allowed pressure of radiators is too low. However, it will be taken into account for all cases when carrying out the simulation and calculations, as will be explained later.

The main function of the heat exchangers is preparing the water for the service temperature by exchanging heat with the water coming from the district heating network, which is particularly necessary for the hot water branch since it is potable water. This water cannot be the same or be mixed with the water coming from the network.

In addition, the heat exchanger is very useful to maintain the pressure of the internal circuit of the building independent of the network's pressure, which is much higher. This is important especially for old buildings whose facilities and materials cannot afford certain pressure values.

The goal of the control and instrumentation system is to control the temperature of the radiator's circuit. In our case we use the adapted Recknagel-Schramek-Sprenger

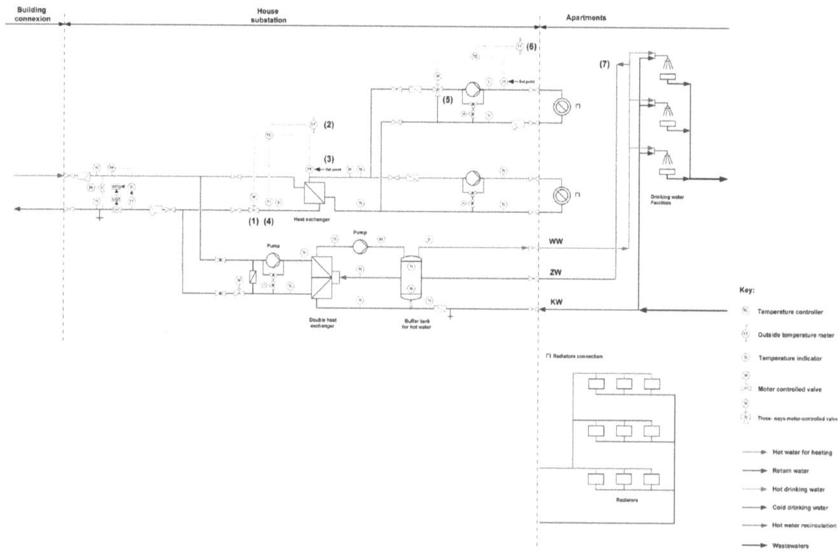

Fig. 10.12 Simplified diagram of a residential building (heat exchanger considered in the heating branch).
(FERNWÄRME WIEN GmbH, 2009. Technical Documentation TR-SZT.)

method (Recknagel et al., 2006), which calculates this temperature according to a fixed and constant value of comfort temperature in the rooms and the temperature measured outside the building.

Consequently, two control levels are used:

- Control system in the heat exchanger: The controller manipulates the motor-driven valve (1) controlling the returning flow of DH water to the network depending on the three temperatures measured in the following places:
  - outside the building (2)
  - radiator's feed (3)
  - return flow from the heat exchanger to the district heating network (4)
- Control system after the heat exchanger: Higher precision temperature control system by mixing part of the return water with the supply water in the radiator circuit through of a motor-driven mixing valve (5). Again, measuring and registering the outside temperature (6) is the key to carry out the control. It is important to emphasize how the recirculation system from the last floor works (7), assuring instant DHW to all residents.

The substation is the place where all necessary devices for energy exchange between primary and secondary networks are installed. The objective is to produce heating and hot water to cover the consumer's needs. A substation is usually a room located in the basement of one of the buildings or just next to them. Each substation normally contains two heat exchangers installed in parallel (Fig. 10.13).

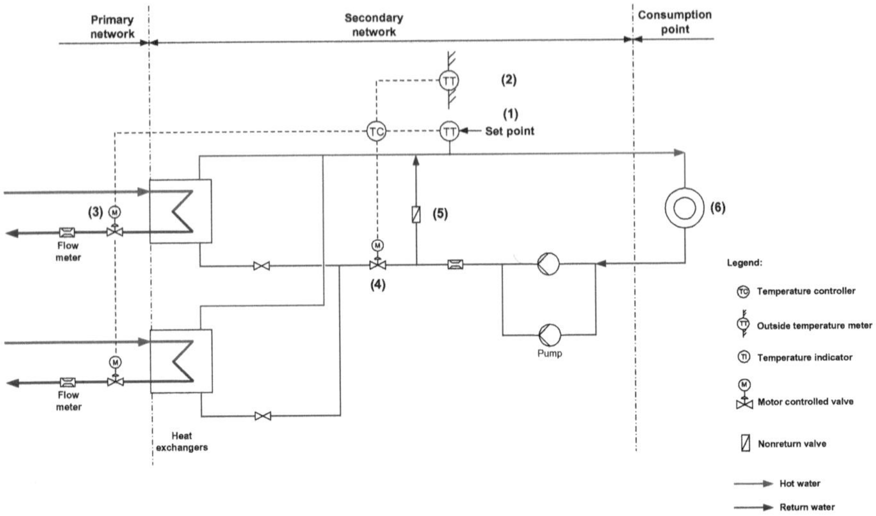

Fig. 10.13 Simplified diagram of the GUFO substation.
(FERNWÄRME WIEN GmbH, 2009. Technical Documentation TR-SZT.)

The control system is similar to the other ones previously explained, setting up two loops with the following elements:

(1) Temperature indicator: It is a part of the heat exchanger control system within the substation. It has a fixed set point depending on the outside temperature (2), according to the heating curve ("Heizkurve" in German), which will be defined afterwards. A heating curve is defined for both primary and secondary networks.
(3) Principal control level: The valve regulates high-temperature water mass flow from the primary network depending on the deviation from the set point value.
(4) This valve is connected to each substation controlling system, depending on the temperature shown at point (1) feeding more or less cold water into the hot stream. When it has 0% opening, then more cold water passes through (1) decreasing hot water supply temperature.
(5) Consumption points: buildings, dwellings, etc.

Regarding to temperature regulation, a linear mathematical function is used, through which it is possible to calculate the temperature of the feed streams for both DH networks depending on the external temperature. Each one has its own assigned heating curve.

For the primary side (Fig. 10.14):

- Highest value = 145°C, corresponding to the coldest external temperature condition: – 15°C.
- Lowest value = 80°C, corresponding to 12°C of ambient temperature. Above this ambient temperature, the value 80°C keeps constant.
- Applying linear regression shows R value = 0.999, thus we consider that data points are describing a linear function for both periods.

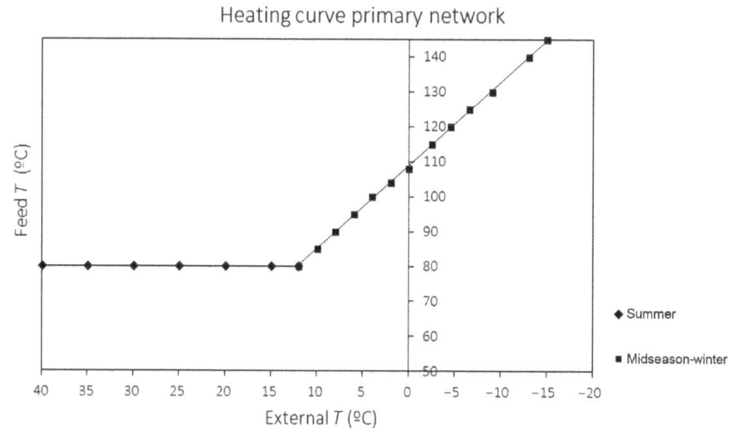

Fig. 10.14 Heating curve, primary network.
(FERNWÄRME WIEN GmbH, 2013. Technical Documentation TR-TAB.)

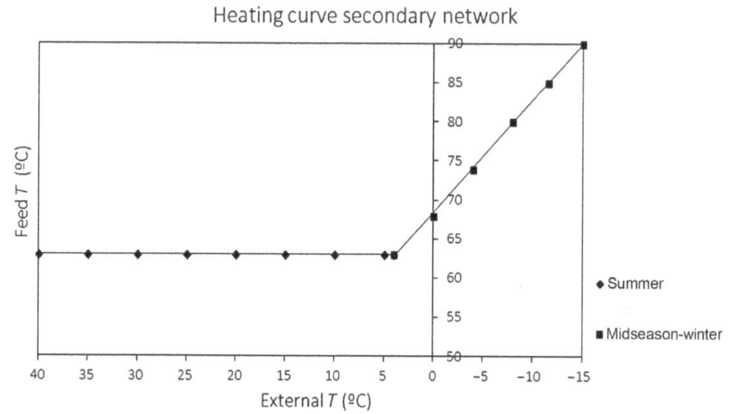

Fig. 10.15 Heating curve, secondary network.
(FERNWÄRME WIEN GmbH, 2013. Technical Documentation TR-TAB.)

For the secondary side (Fig. 10.15):

- The highest value is 90°C, corresponding to – 15°C ambient temperature.
- The lowest value is 63°C, corresponding to an external temperature of 4°C. From there onward the value of 63°C is constant.

Other important considerations at operation level, related with temperature management (FERNWÄRME WIEN GmbH, 2013):

– *Flow temperatures.* The minimum flow temperatures specified in the heating curves are guaranteed by the district heating and must be used for dimensioning. The maximum flow temperatures at the secondary side can temporarily happen and must always be managed by the customer. On the primary side, very small volume flows can happen, leading to temperature fluctuations in the secondary flow.

- *Design temperatures from radiators.* For newly constructed objects with a specific heating demand ≤50 kWh/m²a (energy certificate class B or better), feed temperature to radiators is limited to 60°C flow temperature.

  Respect to distribution losses for energy efficient heating systems, this results in a standard design of 60/40. Similar values are also a prerequisite for the economic operation of condensing boilers and heat pumps, corresponding to the state-of-the-art technologies.
- *Return temperatures from radiators.* The return temperatures to be adhered depend on the building substance. Generally, the lowest possible design temperature is desirable. Higher temperatures are only permitted for existing properties with poor thermal insulation (historical buildings, old residential public dwellings, etc.). The used characteristics are the transfer coefficient "*U*-value" as an average value of the building envelope and the annual heating demand according to the energy certificate (as a measure of the quality of the thermal insulation).
- *Return temperatures from DHW branch.* The specified maximum return temperature does not apply to the pure recirculation operation. For this time without hot water extraction and no storage charge, a return temperature of up to 3 K higher than the circulation temperature is allowed.
- *Storage tanks.* By using a storage tank, the return temperature of the DHW branch system should not be raised. Basically, a buffer should dampen the daily, house-side load peaks. In addition, the charge must not be made at the time of the daily peak performance.

Concerning Legionella issues, it is important to know that from a certain outdoor temperature (usually around 16°C) the temperature should never fall below 60°C. Legionella bacteria are able to grow in water systems in contact with air at medium to high temperatures, specially if the water is stored, remaining in a tank or any similar configuration, obeying Austrian regulation (ÖNORM B 5019, 2011; ÖNORM EN 806, 2011).

Regarding how pressure regulation works in this kind of local networks, our case study is clearly a tree structure, with the substation as starting point followed by with a main stream (FH, which divides off in such a way that in practice every consumption point has its own branch, one stream for feed and the other one for return. This hydraulic system and its components are easily explained in Fig. 10.16.

The minimum pressure of the system is fixed (Figs. 10.16 and 10.17, point 1, $P_{SET\ POINT} = 3$ bar), is fixed using an unpressurized tank with pump and overflow rate.

In the next step, the main hydraulic line is established, it links the impulsion point of the pump (Figs. 10.16 and 10.17, point 2, $P_{DISCHARGE}$) with the last building of the network (Figs. 10.16 and 10.17, point 6, $P_{FEED\ LAST\ BUILDING}$). In this way, the length of the pipe to be taken into account when calculating pressure losses ($\Delta P_{FEEDLINE}$) is defined.

In between, there is a short pipe called FBC connecting the hydraulic pump with the heat exchanger (Figs. 10.16 and 10.17, points 2 and 3), giving rise to certain pressure

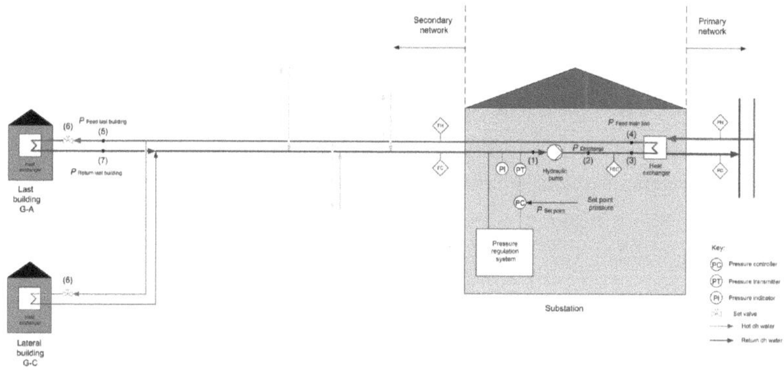

Fig. 10.16 (A) Hydraulic system of the secondary network. (B) Modeling intersections in the secondary network.

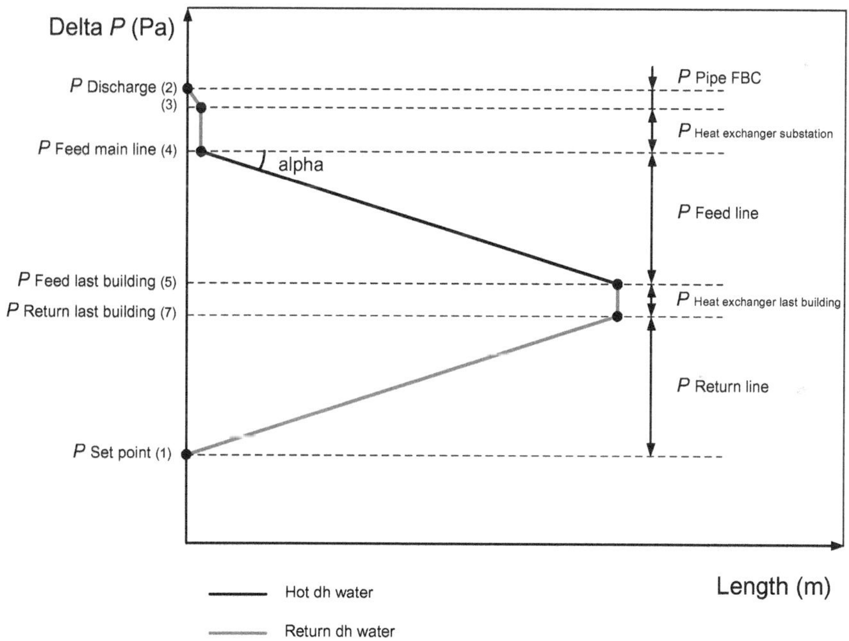

Fig. 10.17 Calculation of the hydraulic balance.

losses ($\Delta P_{PIPE\ FBC}$). The mentioned heat exchanger also generates pressure losses (Figs. 10.16 and 10.17, points 3 and 4; $\Delta P_{HEAT\ EXCHANGER\ SUBSTATION}$).

The water flows through the entire length of the line reaching the last point consumption point (Figs. 10.16 and 10.17, point 5, $P_{FEED\ LAST\ BUILDING}$), going into the substation of the building and losing pressure as it crosses through the heat exchanger and the control system (Figs. 10.16 and 10.17, point 7, $P_{RETURN\ LAST\ BUILDING}$), generating $\Delta P_{HEAT\ EXCHANGER\ LAST\ BUILDING}$.

Table 10.2 Limit operation values for Girasolstrasse local network

Maximum pressure (bar)	3.9
Minimum pressure (bar)	3
Maximum network temperature (°C)	90
Minimum network temperature (°C)	63
Maximum outside temperature (°C)	25.6
Minimum outside temperature (°C)	−12.31

On the way back the fluid crosses the pipes in the opposite direction, returning to point 1. Concerning the water pressure at the consumers side, a valve is set and regulated depending on each building's needs (Figs. 10.16 and 10.17, point 6).

To calculate intersections' pressure and determine the normal circulation of the fluid, the model considers compensation valves installed on the lateral pipes (Fig. 10.16B, points 1 and 2). The pressure in point 1 equals pressure in points 1a and 1b, since the pressure of each branch is equal to the pressure of the main stream. The same situation does not happen with the return branch. At point (2B) the compensation valve adjusts the pressure of the lateral branch until it is equal to the pressure of the main line (point 2A), in order to ensure the water to flow in the right direction. The following diagram shows how to calculate the discharge pressure $P_{DISCHARGE}$ generated by the hydraulic pump, considering all pressure losses previously mentioned (Fig. 10.17).

This diagram shows how to calculate the discharge pressure $P_{DISCHARGE}$ generated by the hydraulic pump, considering all pressure losses previously mentioned. The hydraulic balance is obtained:

$$P_{DISCHARGE} = P_{SET\ POINT} + \Delta P_{RETURN\_LINE} + \Delta P_{HE\_LAST\_BUILDING} + \Delta P_{FEED\_LINE} + \Delta P_{PIPE\_FBC} + \Delta P_{HE\_SUBSTATION} \quad (10.1)$$

$$\alpha = \frac{\Delta P}{L} \quad (10.2)$$

The angle α (alpha) represents the relation between pressure losses and length for each element.

To conclude this section, the considered limit operation values used for the calculations are shown, taking into account that water is all the time maintained in liquid phase (Table 10.2).

10.3 Scenarios for optimizing the energy system in a block of buildings

Given the scope and operation of the thermal network in the city of Vienna, and more specifically in the district where the case study is located (Meidling) and the case study itself, we proceed to explain the hypotheses studied. Each of them constitutes a new configuration of the network for the aforementioned case study, which implies changes such as the replacement of some technologies by others, the addition of new devices connected to the network or directly removing certain installations (Figs. 10.18–10.25).

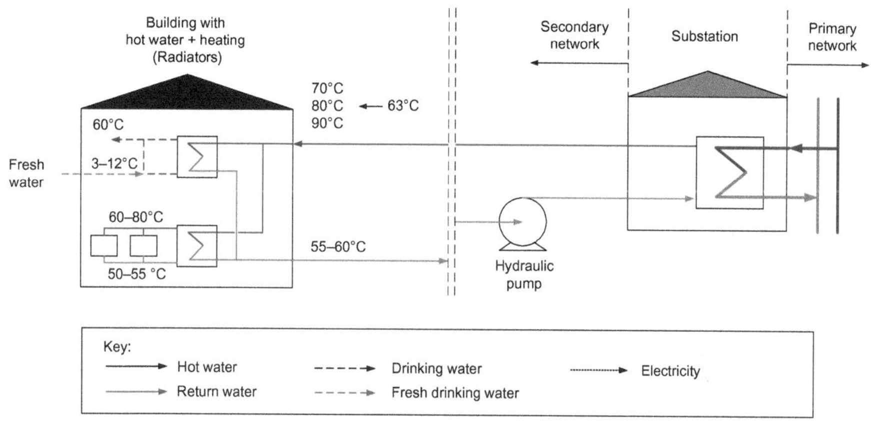

- It is necessary to know the current situation to have a reference with which to compare. Buildings and networks are simulated without any modification.
- Existing heating facilities: radiators.
- Temperature regulation of district heating network curve
- Hot water production service following Austrian technical regulation ÖNORM B 5019, which defines 60°C at consumer´s side, which means around 63°C from the DH network side.

Fig. 10.18 SCENARIO 1. Actual situation—reference case.

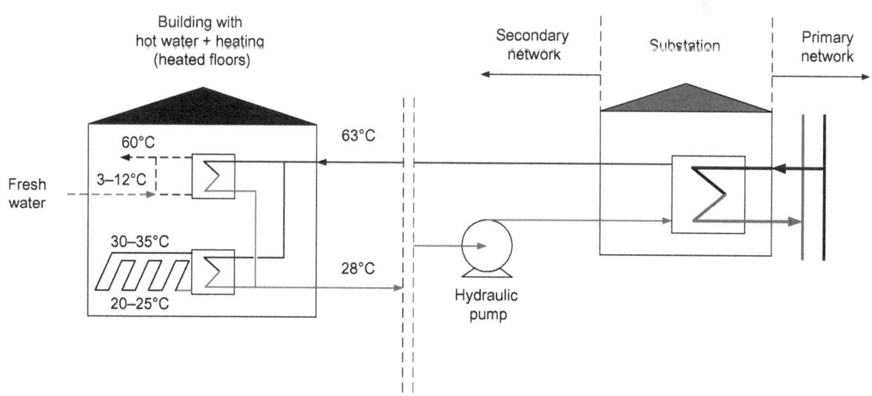

- Substitution of radiators by heated floors for all the buildings
- Temperature regulation of district heating network constant at 63°C for all the year.
- Hot water production service remains the same; Austrian technical regulation ÖNORM B 5019 defines 60°C a consumer´s side, which means around 63°C from the DH network side.

Fig. 10.19 SCENARIO 2. Basic refurbishment.

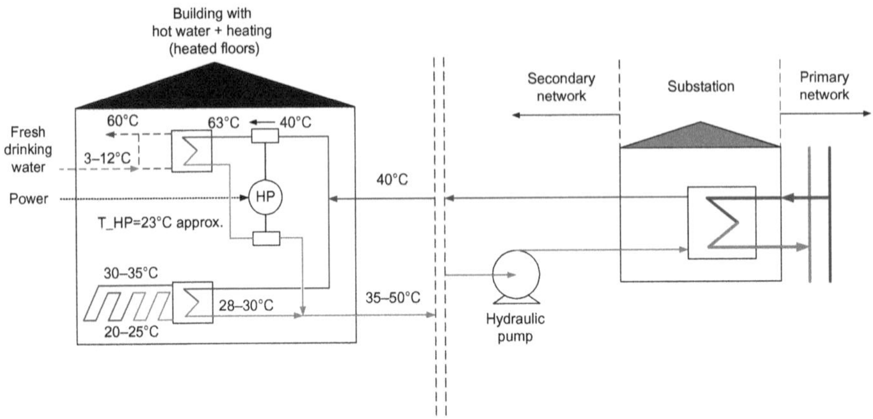

- Substitution of radiators by heated floors for all the buildings.
- District heating network water changes to 40°C for all the year (Low Temperature District Heating Network)
- Hot water production service remains the same; Austrian technical regulation ÖNORM B 5019 defines 60°C at consumer´s side, which means around 63°C from the DH network side.
- In order to satisfy hot drinking water service, the temperature of network water needs to be increased from 40 to 63°C, therefore, a heat pump must be simulated. This heat pump will be connected to the feed DHN water stream of the hot water heat exchanger.

Fig. 10.20 SCENARIO 3. Basic refurbishment LTDHN.

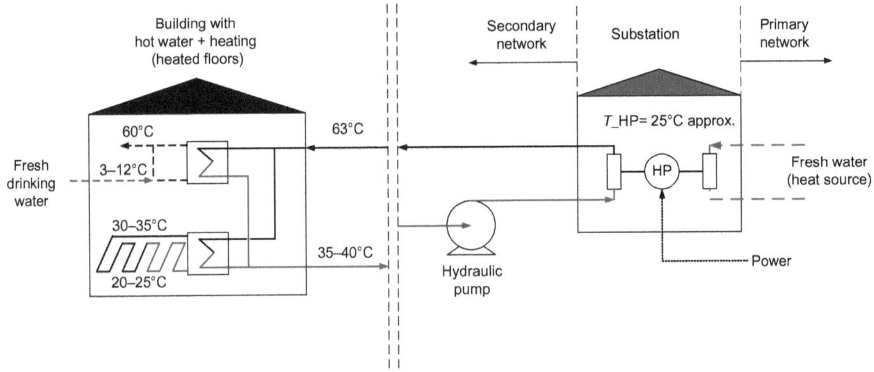

- Substitution of radiators by heated floors for all the buildings.
- District heating network changes to 63°C, constant for all the year.
- Hot water production service remains the same; Austrian technical regulation ÖNORM B 5019 defines 60°C at consumer´s side, which means around 63°C from the DH network side.
- Replacement of the heat exchanger of the general substation by a central heat pump intended to give service to the entire group of buildings, raising return water temperature to the above-mentioned value of 63°C. This scenario does not usethe primary network, but instead an anergy grid (river water, ground...etc.) to feed the heat pump.
- In the simulation, the heat pump block is adapted for all kind of heat-source fluids, to be specified by the user.

Fig. 10.21 SCENARIO 4. Central heat pump with refurbishment.

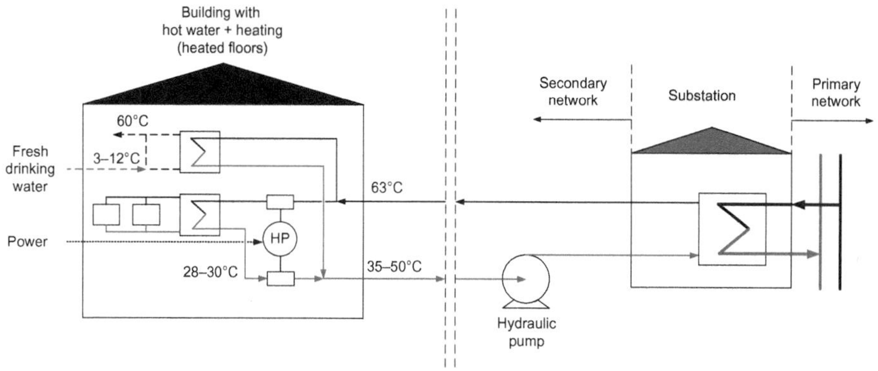

- District heating network changes to 63°C, constant for all the year.
- Hot water production service remains the same; Austrian technical regulation ÖNORM B 5019 defines 60 °C at consumer´s side, which means around 63 °C from the DH network side.
- To satisfy high radiator temperatures for non refurbished Viennese buildings an increasing DHN water temperature from 63 to 90 ° C is required. Therefore, a heat pump connected to the feed DHN water stream of the heating heat exchanger will be simulated.

Fig. 10.22 SCENARIO 5. Individual heat pumps for heating without refurbishment.

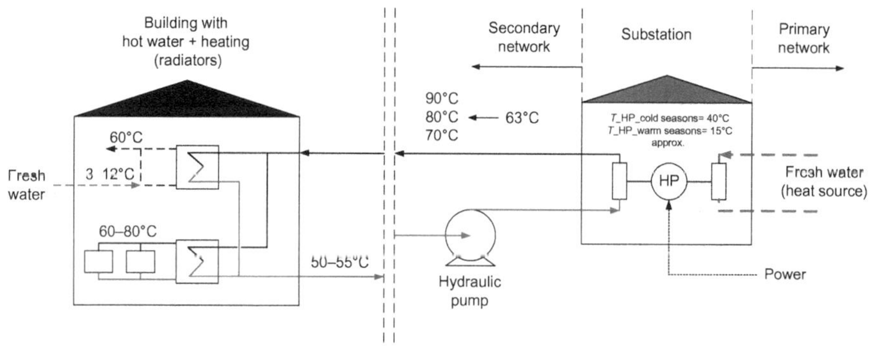

- Existing heating facilities: radiators.
- DHW curve remains the same as scenario 1.
- Hot water production service remains the same; Austrian technical regulation ÖNORM B 5019.
- Replacement of general substation´s heat exchanger by a central heat pump intended to give service to the entire group of buildings, raising return water temperature to the above-mentioned range of values between 63 and 90°C. This scenario does not use the primary network, but an anergy grid (river water, ground...etc.).
- In the simulation, the heat pump block is adapted for all kind of heat-source fluids, to be specified by the user.

Fig. 10.23 SCENARIO 6. Central heat pump without refurbishment.

Each scenario has been modeled with its corresponding structure of the buildings and the network, as well as its input parameters and other characteristics. There are eight scenarios; therefore, eight simulation runs whose results are shown and explained in Section 10.4.

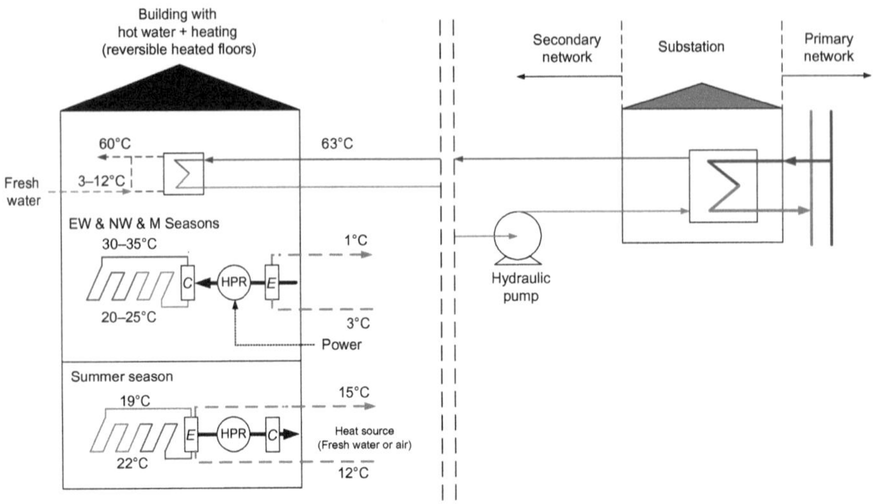

- Existing heating facilities: heated/cooled floors:
- District heating network changes to 63°C, constant for all the year.
- Hot water production service remains the same; Austrian technical regulation ÖNORM B 5019 is applied.
- To satisfy heated floors needs for both warm and cold seasons, a reversible heat pump is installed. This kind of device will be programmed to use different kinds of heat sources, like water or air, in order to heat the floors in winter and cool them in summer.

Fig. 10.24 SCENARIO 7. Refurbishment installing reversible heat pumps for heating/cooling.

10.4 Results and discussion

Terms and indexes are given in Tables 10.3 and 10.4.

- Defining the thermodynamic systems

Due to the high complexity of the thermal networks and their large number of variables and parameters, it is necessary to delimit the boundaries of the thermodynamic systems and their corresponding matter, energy and exergy balances. Thus, each variable is located and defined.

From the figures shown in Section 10.3, it is possible to synthesize all operation variables for all possible combinations, extending those diagrams into Figs. 10.26 and 10.27.

In order to better understand how the network operates, as well as finding deficiencies and analyzing possible improvements, it is necessary to separate the entire thermodynamic system into three independent subsystems, with their corresponding links and connections:

- Equations involved for both energy and exergy balances:
  - Global energy balance: It has been assumed that the heat exchangers in the buildings and in the substation are totally isolated; thus, heat losses are negligible. However, although the pipes are isolated and buried, there are certain heat losses, which must be taken into consideration.

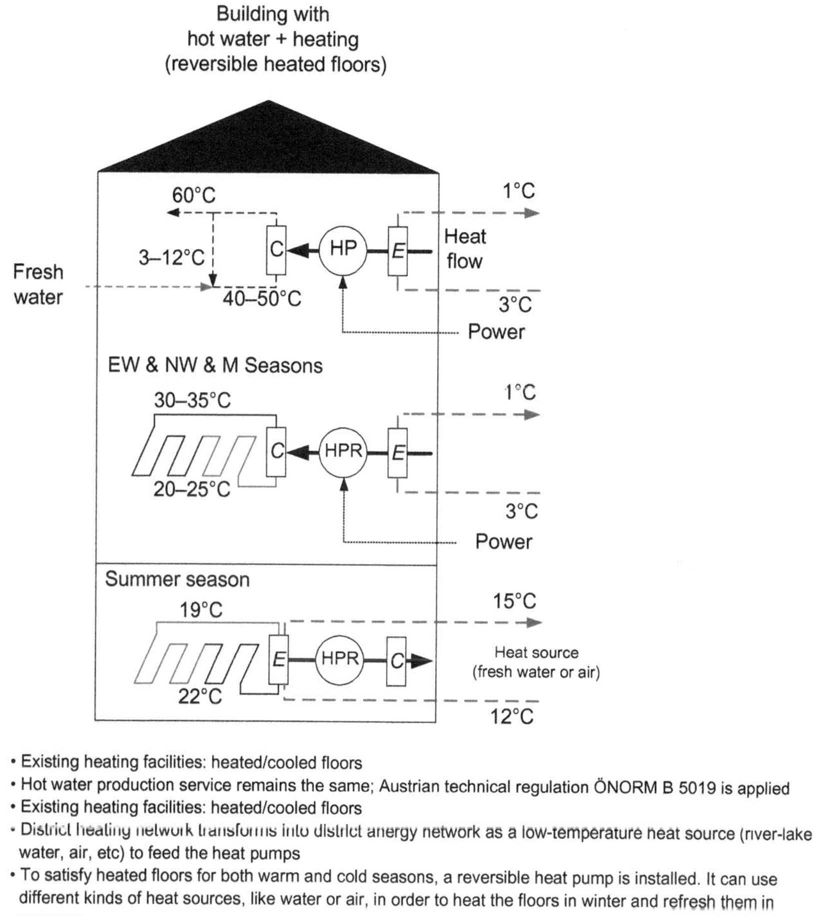

Fig. 10.25 SCENARIO 8. Refurbishment with no DH.

Table 10.3 Glossary of terms

C_p	kJ/kg K	Calorific capacity at constant pressure
COP	ad	Coefficient of heat pump's performance
$\dot{Ex}$	kW	Exergy flow
h	kJ/kg	Characteristic enthalpy at certain pressure and temperature
$\dot{h}$	kW	Enthalpy flow
$\dot{m}$	kg/s	Mass flow of DHN water
$\dot{Q}$	kW	Heat flow

Continued

Table 10.3 Continued

s	kJ/kg K	Characteristic entropy at certain pressure and temperature
T	°C	Temperature
ΔT	K	Temperature difference
$\dot{W}$	kW	Electricity flow
$\dot{W}p$	kW	Hydraulic pump's electricity flow
η	ad	Yield

Table 10.4 Glossary of subindexes

0	Relative to the reference values
AN	Anergy network
BUILDINGS	Relative to building's thermodynamic system
CARNOT	Related to theoretical heat pump's efficiency
client side	Relative to consumer side, inside the building
COND.	Heat pump's condenser
CS	Cold side of the heat exchanger
DH side	Relative to district heating network side, outside the building
DH1	Primary network
DH2	Secondary network
EN	Relative to the energy
env	Relative to the environment
EV	Heat pump's evaporator
EX	Relative to the exergy
feed	Relative to feed streams
H	Relative to heating branch
HE	Relative to the heat exchanger
HP	Relative to heat pumps
HS	Hot side of the heat exchanger
HW	Relative to hot water production branch
i	*n*-element identificator for each building
in	Relative to inlet streams
NETWORK	Relative to the network's thermodynamic system
out	Relative to outlet streams
p	Relative to district heating network side, outside the building
post HP	Relative to any stream after the heat pump
r	Return line

Table 10.4 Continued

rad	Relative to radiators cycle inside the building = heating branch
rad cs	Relative to radiators cycle inside the building, supply line
rad cr	Relative to radiators cycle inside the building, return line
REAL	Related to real heat pump's efficiency
return	Relative to return streams
s	Relative to consumer side, inside the building
SUBSTATION	Relative to the general substation's thermodynamic system

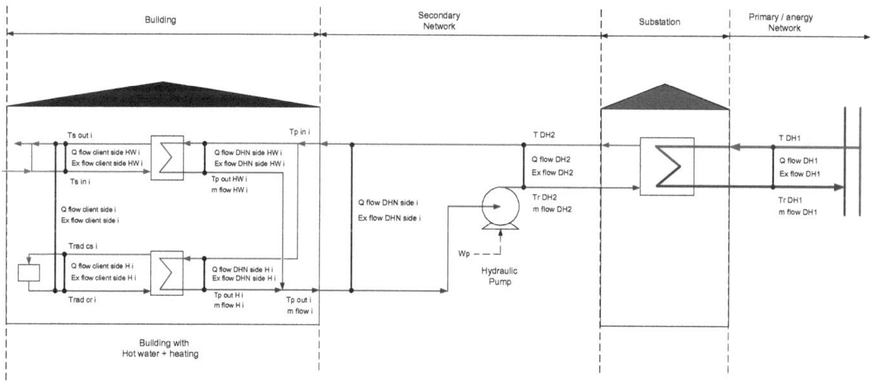

Fig. 10.26 Detailed schema showing the main operation and thermodynamic variables.

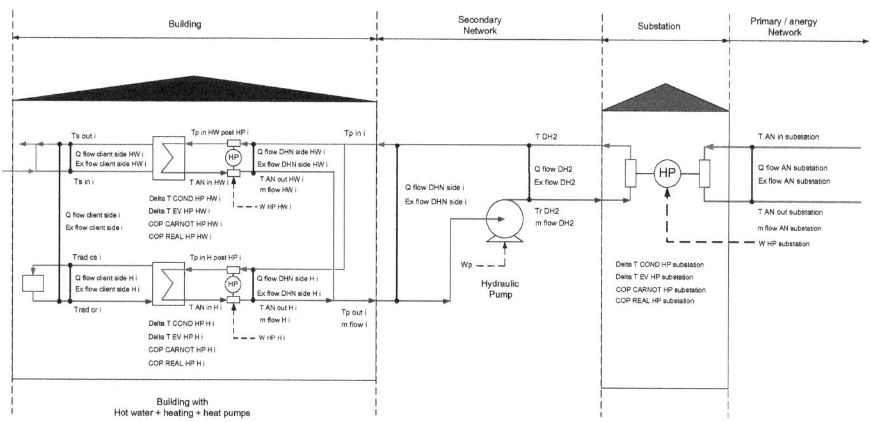

Fig. 10.27 Detailed schema showing the main operation and thermodynamic variables, including heat pumps connected to the network.

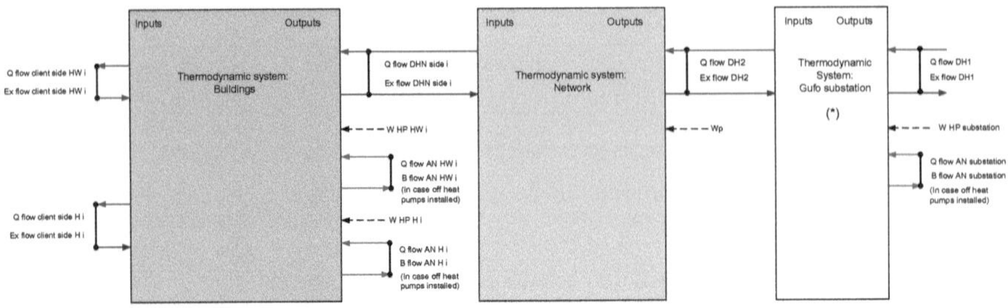

Fig. 10.28 Diagram showing the three thermodynamic systems in which the network is divided. *Although the subsystem of the substation is defined and appears in the figure, it has been taken into consideration in the calculations only in a partial way.

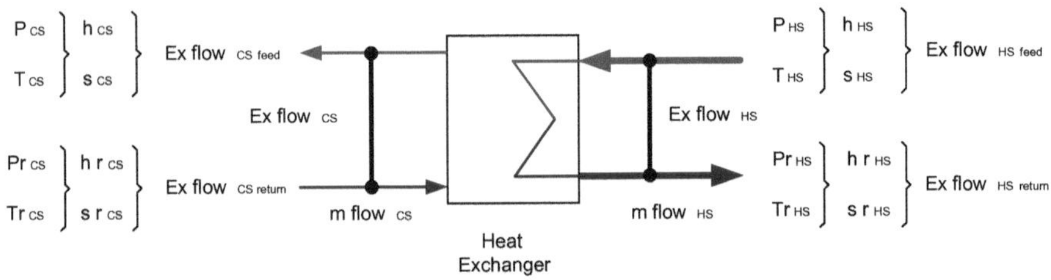

Fig. 10.29 Exergy balance applied to the heat exchanger.

Therefore, only Eqs. (10.3), (10.4) are obtained from Fig. 10.28.

Thermodynamic system: Network.

$$\dot{Q}_{losses\ NETWORK} = \dot{Q}_{DH2} + W_p - \sum_{i}^{n} \dot{Q}_{DHN\ side\ i} \tag{10.3}$$

$$\eta_{ENERGY\ NETWORK} = \frac{\sum_{i}^{n} \dot{Q}_{DH\ side\ i}}{\dot{Q}_{DH2} + W_p} \tag{10.4}$$

– Global exergy balance: As explained in Section 5.4 about the exergy method, exergy losses through any heat exchanger can be calculated by applying an individualized balance for each device, as shown in Fig. 10.29.

After applying the exergetic method, the following expressions are obtained:

$$\dot{E}x_{HS\,feed} = \dot{m}_{HS}[(h_{HS} - h_0) - T_0(s_{HS} - s_0)] \tag{10.5}$$

$$\dot{E}x_{HS\,return} = \dot{m}_{HS}[(h_{r\,HS} - h_0) - T_0(s_{r\,HS} - s_0)] \tag{10.6}$$

$$\dot{E}x_{HS}i = \dot{E}x_{HS\,feed} - \dot{E}x_{HS\,return} \tag{10.7}$$

$$\dot{E}x_{CS\,feed} = \dot{m}_{CS}[(h_{CS} - h_0) - T_0(s_{CS} - s_0)] \tag{10.8}$$

$$\dot{E}x_{CS\,return} = \dot{m}_{CS}[(h_{r\,CS} - h_0) - T_0(s_{r\,CS} - s_0)] \tag{10.9}$$

$$\dot{E}x_{CS}i = \dot{E}x_{CS\,feed} - \dot{E}x_{CS\,return} \tag{10.10}$$

$$\dot{E}x_{losses\,HE} = \dot{E}x_{HS} - \dot{E}x_{CS} + \left[\left(-\dot{Q}_{losses\,HE}\right)\left(1 - \frac{T_0}{T}\right)\right] \tag{10.11}$$

The variables enthalpy (h) and entropy (s) depend only on pressure and temperature, and more specifically on temperature, since the heat transfer fluid is kept every moment in a liquid phase. Hence, any pressure variation practically does not affect the values of those state functions. Through the thermodynamic tables of water (Schmidt, 1982), it is possible to determine h and s if pressure and temperature are known.

Subsequently, there are always temperature differences, which are able to generate exergy losses through enthalpy and entropy variables, no matter if heat flow is the same on both sides of the heat exchangers. Therefore, exergy losses are taken into account for the three systems studied, and $Q_{losses\ HE}$ are equal to zero, canceling one of the terms of Eq. (10.11).

After this clarification, Eqs. (10.12)–(10.21)), 10.22 are obtained, based on Fig. 10.28, and attached to their corresponding thermodynamic systems:

Thermodynamic system: Substation.

$$\dot{E}x_{losses\,SUBSTATION} = \dot{E}x_{DH1} - \dot{E}x_{DH2} \tag{10.12}$$

$$\eta_{EX\,SUBSTATION} = \frac{\dot{E}x_{DH2}}{\dot{E}x_{DH1}} \tag{10.13}$$

Thermodynamic system: Network.

$$\dot{E}x_{losses\,NETWORK} = \dot{E}x_{DH2} + W_p - \sum_i^n \dot{E}x_{DHN\,side\,i} \tag{10.14}$$

$$\eta_{EX\,NETWORK} = \frac{\sum_i^n \dot{E}x_{DHN\,side\,i}}{\dot{E}x_{DH2} + W_p} \tag{10.15}$$

Thermodynamic system: Buildings.

$$\begin{aligned}\dot{E}x_{losses\,BUILDINGS} &= \sum_i^n \dot{E}x_{DHN\,side\,i} - \sum_i^n \dot{E}x_{client\,side\,HW\,i} - \sum_i^n \dot{E}x_{client\,side\,H\,i} \\ &\quad + \sum_i^n \dot{W}_{HP\,HW\,i} + \sum_i^n \dot{B}_{AN\,HW\,i} + \sum_i^n \dot{W}_{HP\,H\,i} + \sum_i^n \dot{B}_{AN\,H\,i}\end{aligned} \tag{10.16}$$

$$\eta_{EX\,BUILDINGS} = \frac{\sum_i^n \dot{E}x_{client\,side\,HW\,i} - \sum_i^n \dot{E}x_{client\,side\,H\,i}}{\sum_i^n \dot{E}x_{DHN\,side\,i} + \sum_i^n \dot{W}_{HP\,HW\,i} + \sum_i^n \dot{E}x_{AN\,HW\,i} + \sum_i^n \dot{W}_{HP\,H\,i} + \sum_i^n \dot{E}x_{AN\,H\,i}} \tag{10.17}$$

$$\dot{E}x_{losses\,BUILDINGS\,HW} = \sum_i^n \dot{E}x_{DHN\,side\,HW\,i} - \sum_i^n \dot{E}x_{cl.\,side\,HW\,i} + \sum_i^n \dot{W}_{HP\,HW\,i} + \sum_i^n \dot{E}x_{AN\,HW\,i} \tag{10.18}$$

$$\eta_{EX\ BUILDINGS\ HW} = \frac{\sum_i^n \dot{E}x_{client\ side\ HW\ i}}{\sum_i^n \dot{E}x_{DHN\ side\ HW\ i} + \sum_i^n \dot{W}_{HP\ HW\ i} + \sum_i^n \dot{E}x_{AN\ HW\ i}} \quad (10.19)$$

$$\dot{E}x_{losses\ BUILDINGS\ H} = \sum_i^n \dot{E}x_{DHN\ side\ H\ i} - \sum_i^n \dot{E}x_{client\ side\ H\ i} + \sum_i^n \dot{W}_{HP\ H\ i} + \sum_i^n \dot{E}x_{AN\ H\ i} \quad (10.20)$$

$$\eta_{EX\ BUILDINGS\ H} = \frac{\sum_i^n \dot{E}x_{client\ side\ H\ i}}{\sum_i^n \dot{E}x_{DHN\ side\ H\ i} + \sum_i^n \dot{W}_{HP\ H\ i} + \sum_i^n \dot{E}x_{AN\ H\ i}} \quad (10.21)$$

- Validation of the simulation: The goal is to demonstrate how the used models are able to generate acceptable results, which are adjusted with real measured values. This action ensures a realistic, reliable, and applicable simulation, which can be adapted to other case studies. The compared variables are those related to buildings, measured in DHN real installations during the years 2012 and 2013.

  The results of the validation for 3 of the 21 buildings studied are shown in Figs. 10.30—10.32; each graph distinguishes four periods, described from left to right: extreme winter (year 2012), moderate winter (year 2013), midseason (spring-autumn), and summer. All of them are 48-h segments, with the first 24 h corresponding to a working day and the other 24 h to a weekend. In total 8 days are displayed in hourly resolution. In this way a representative profile for each period of the year is obtained.

  The readjustment of the heating curve improves significantly the accuracy of the feed temperature, immediately improving the return temperature.

  For a more accurate and close adjustment to the measured values, the parameters of the heat exchangers have been recalibrated, as well as those relating to the distribution of both demands, hot water production and heating, which are not measured separately.

  These adjustments also improve the accuracy of the calculation for mass flow and enthalpy flow, as shown in Figs. 10.33–10.38).

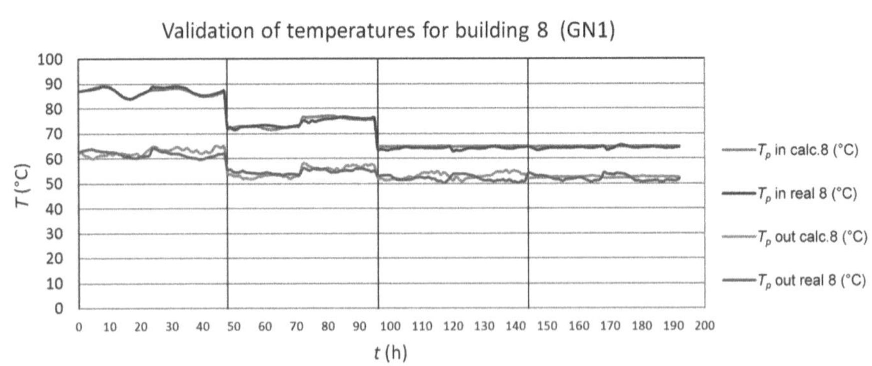

Fig. 10.30 Comparison between real and calculated temperatures for building 8.

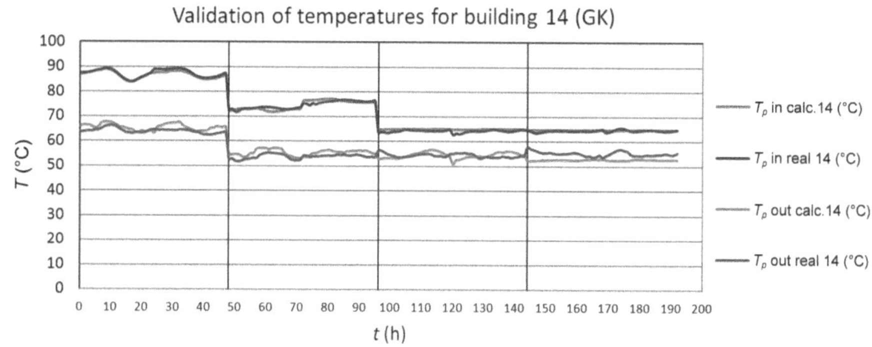

Fig. 10.31 Comparison between real and calculated temperatures for building 14.

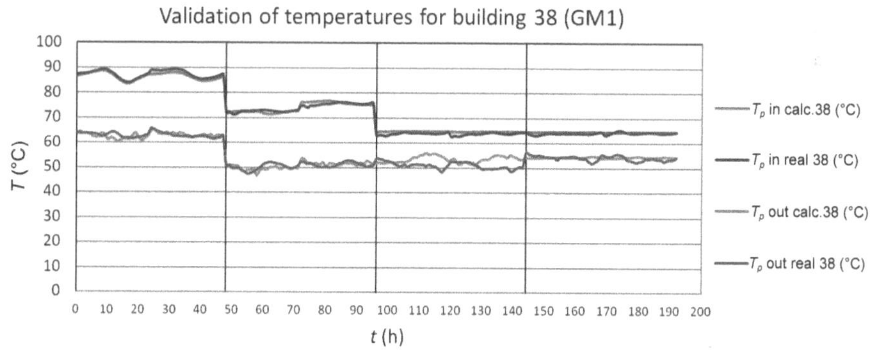

Fig. 10.32 Comparison between real and calculated temperatures for building 38.

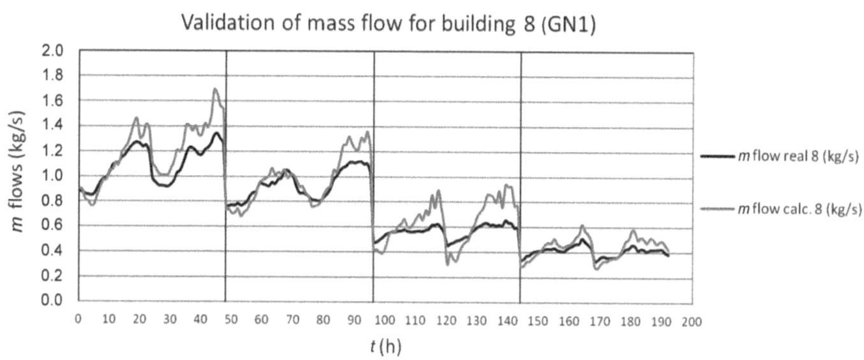

Fig. 10.33 Comparison between real and calculated mass flows for building 8.

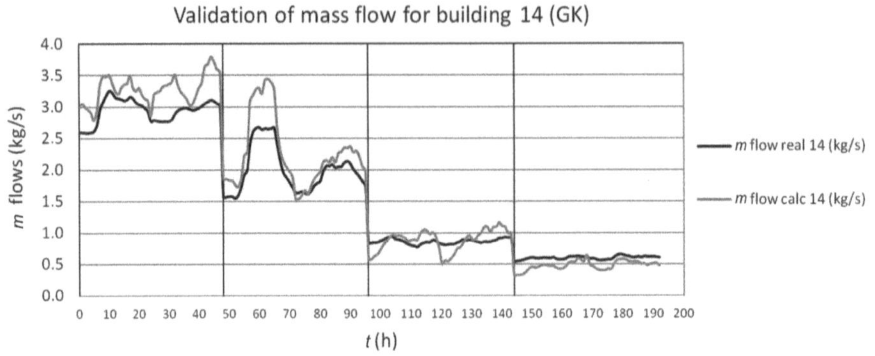

Fig. 10.34 Comparison between real and calculated mass flows for building 14.

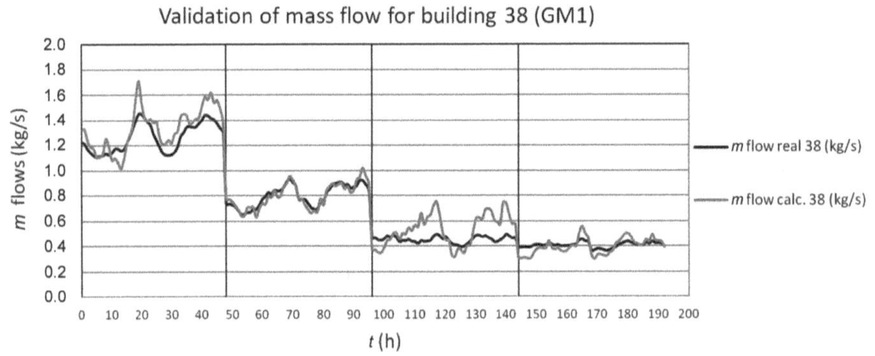

Fig. 10.35 Comparison between real and calculated mass flows for building 38.

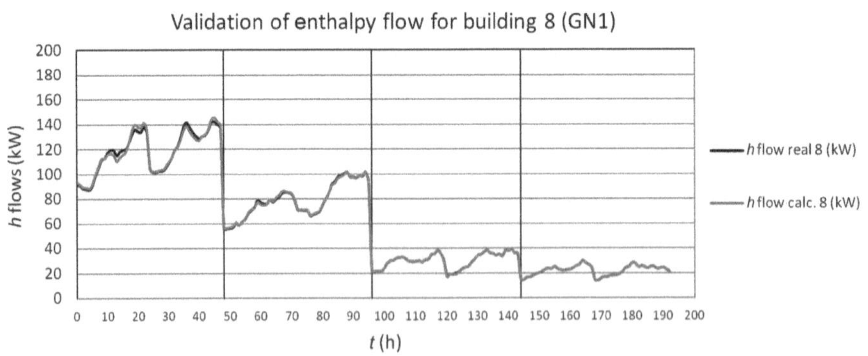

Fig. 10.36 Comparison between real and calculated enthalpy flows for building 8.

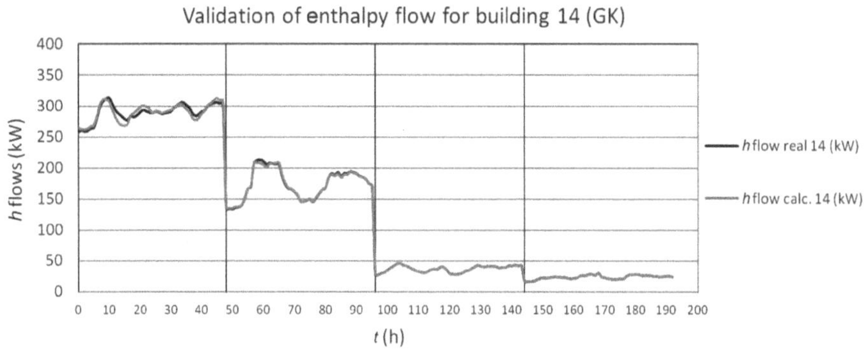

Fig. 10.37 Comparison between real and calculated enthalpy flows for building 14.

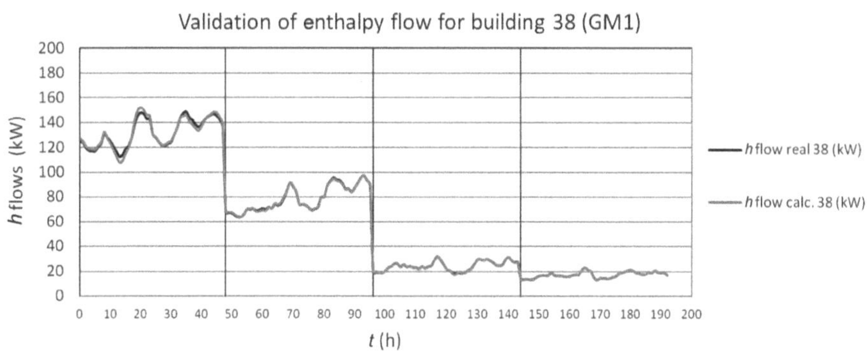

Fig. 10.38 Comparison between real and calculated enthalpy flows for building 38.

On the whole the simulation is realistic enough to now calculate scenarios. In the following, some results of the reference SCENARIO 1 are shown, defining how the performance of the actual network is. All variables and profiles are grouped according to the thermodynamic system where they come from.

- Operation variables, evolution of network's pressure (Fig. 10.39): The pressure immediately after the hydraulic pump is much higher than the calculated pressure in the last building and at the suction point of the pump. This large difference is due to the hydraulic balance (see Fig. 10.17 and Eq. 10.1). To calculate the pressure in the last building it is only necessary to add ΔP return line to the basic pressure (Pr prepump), while at impulsion point, the terms ΔP feed line + ΔP heat exchanger substation + ΔP building facilities must be added.

  For winter and extreme winter days, energy and heating consumption are higher, hence a greater flow of water circulating through the network must be driven, leading to higher pressures in comparison with other periods of the year.
- Operation variables, evolution of network's temperatures (Fig. 10.40): The graph shows how important heat losses along the pipes are. The pipes are installed throughout a buried ditch just below the pavement, conferring them some protection and insulation, keeping a constant ambient temperature throughout the year (between 5 and 8°C, in this study 8°C had been considered, constant throughout the year).

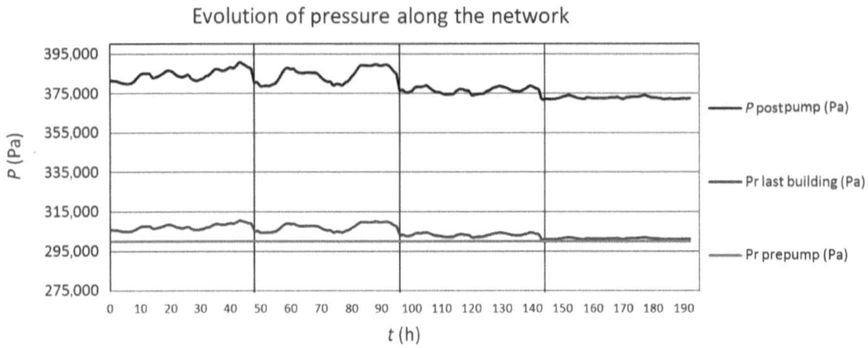

Fig. 10.39 Evolution of pressure throughout the network.

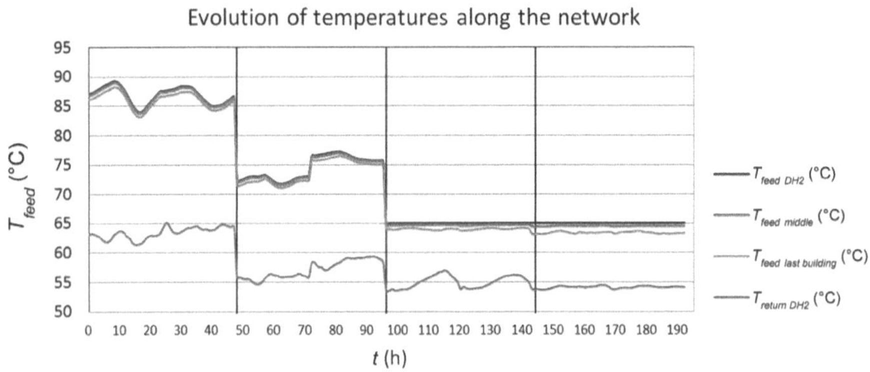

Fig. 10.40 Evolution of the different temperatures throughout the network.

- Operation variables, evolution of total mass flow (Fig. 10.41): Total DHN water mass flow and consumption peaks are higher for cold seasons and lower for the summer season. The frequency of peaks and troughs is similar throughout the seasons. The DHN's water flow for building 2 (GN1, mdot 2) (see Table 10.5) is shown, to have an idea of the flow for an individual building compared to the flow for the entire neighborhood. In addition the mass flow of network water is shown of which a hypothetical heat pump would extract energy. Since in the current situation there are no heat pumps installed, this value is zero.
- Global energy balance: The results below are obtained using the expressions (10.3), (10.4), obtained from Fig. 10.28, generating in Fig. 10.42.

  Obviously heat losses are greater in extreme winter than other periods of the year, which does not necessarily mean that energy efficiency is going to be lower during this period.

  It is interesting to see the effect of consumption profile change from a working day to a weekend day in the midseason segment (at 120 h). There is a drop in the weekend consumption, leading to a DHN water flow reduction and therefore a sudden rise of the heat losses.

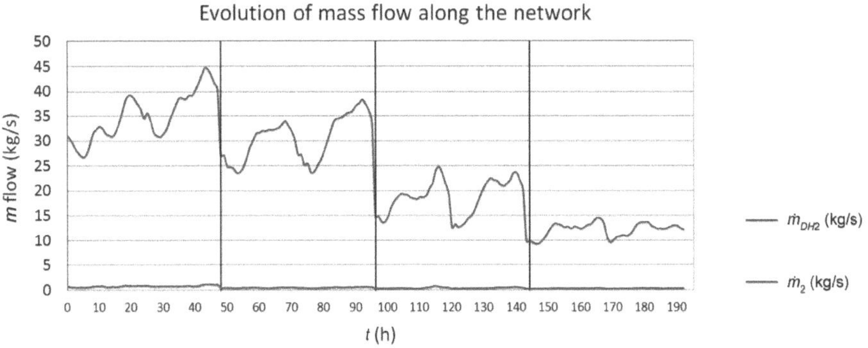

Fig. 10.41 Evolution of mass flow throughout the network.

Table 10.5 Building numbers and their corresponding names

Building number	Building name
2	GN2
8	GN1
14	GK
20	GN
26	GJ
32	GI
38	GM1
44	GM
50	GL
56	GM2
62	GG
68	GE
74	GC
80	GH
86	GD[a]
92	GF[b]
98	GB
104	GA

[a]Buildings GD1 and GD2.
[b]Building GF1.

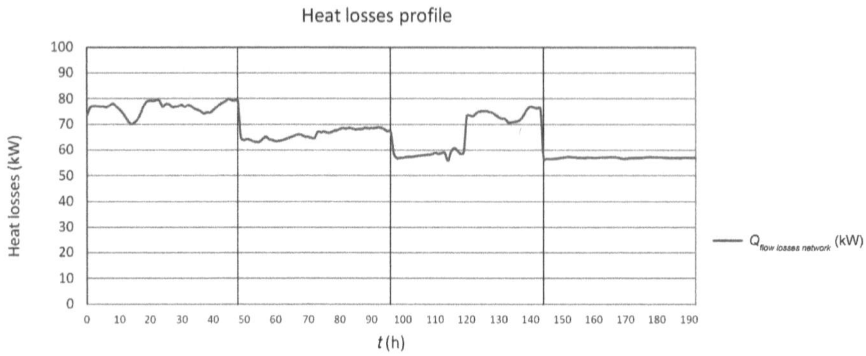

Fig. 10.42 Heat losses profile. Subsystem: district heating network.

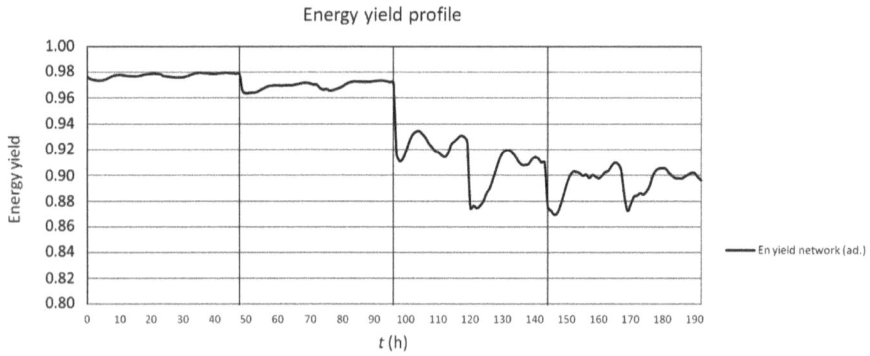

Fig. 10.43 Energy yield profile. Subsystem: district heating network.

Due to high absolute consumption in winter, the efficiency is higher than in low consumption summer season despite higher heat losses (Fig. 10.43).

- Global exergy balance (Fig. 10.44): Fig. 10.45 compares the exergy losses of the substation (ex flow losses substation) and the buildings (ex flow losses substation), that is to say, for those thermodynamic subsystems in which heat exchangers are involved. Higher exergy losses are produced during winter and extreme winter, matching with higher temperatures along the system.

  However, in the case of the network (ex flow losses substation) the situation is just the opposite. As we move from extreme winter to summer, exergy losses are decreasing. This is because in this thermodynamic subsystem, the exergy contained is only associated with the heat flow. Consequently, the destruction of exergy is intimately associated with the heat losses of the circulating water in the pipeline. This is the reason why both variables in Figs. 10.43 and 10.45 describe similar patterns.

  As indicated above, in the presence of heat exchangers the temperature differences generate the exergy losses. The greater the temperature difference on both sides of the

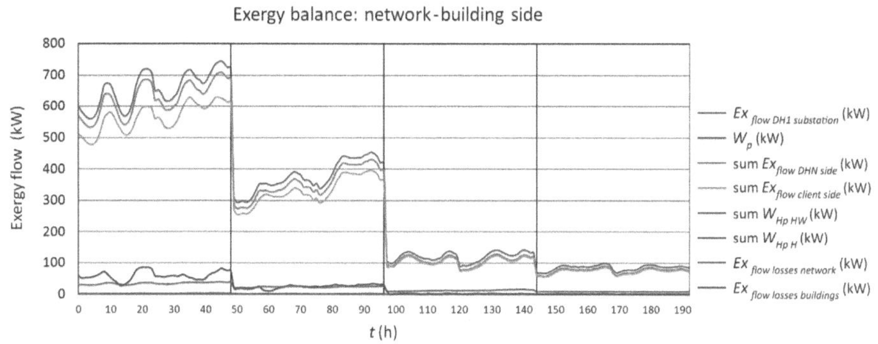

Fig. 10.44 Exergy balance for subsystems: DHN and buildings.

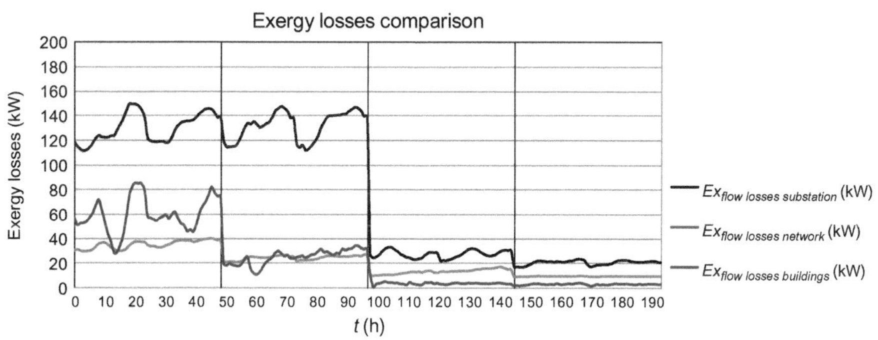

Fig. 10.45 Evolution of the exergy losses.

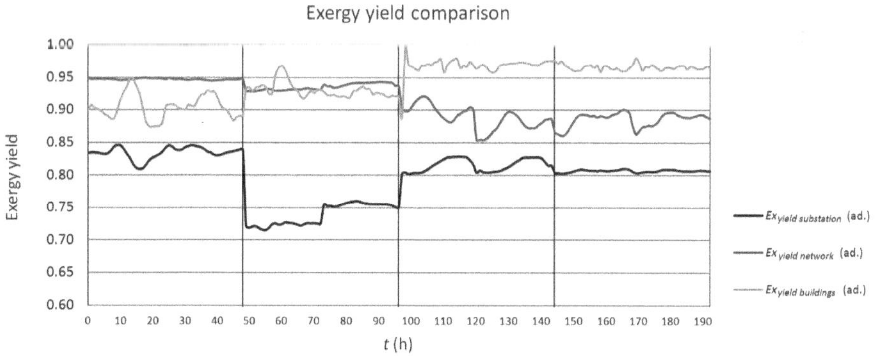

Fig. 10.46 Evolution of the exergy yield.

heat exchanger is the greater the exergy destruction and the worse the performance, as it can be seen in Figs. 10.45 and 10.46.

Then, continuing with the thermodynamic system of the buildings, a detailed exergy analysis is made for both services, domestic hot water and heating (Figs. 10.47 and 10.48).

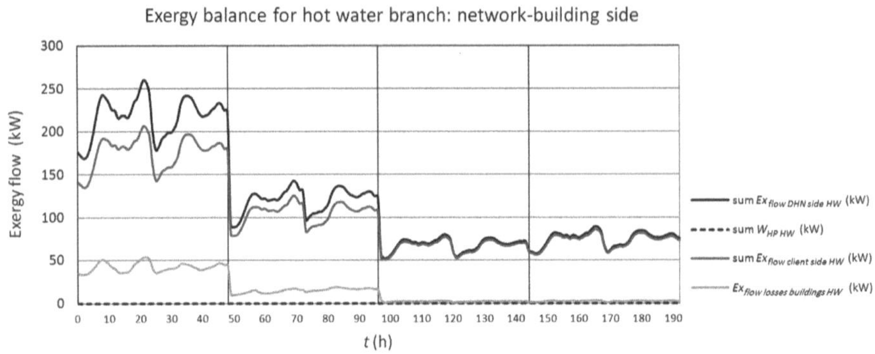

Fig. 10.47 Exergy balance for buildings' subsystem. Domestic hot water branch.

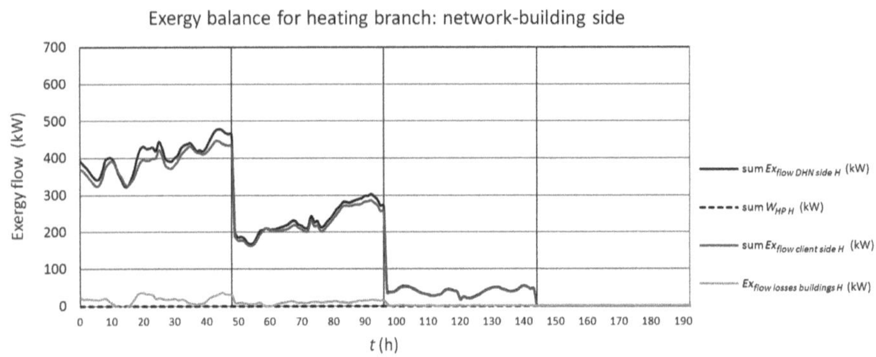

Fig. 10.48 Exergy balance for buildings' subsystem. Heating branch.

The main difference between both branches is the temperature differences between both heat exchangers. For example, in the DHW branch, the heat exchanger receives water at almost 90°C on the network side in winter or extreme winter; meanwhile, on the client side the temperature is fixed, equal to 60°C, in order to avoid the formation of Legionella. Thus, the temperature difference is 30 K. In the heating branch, the heat exchanger also receives water at almost 90°C, but on the client side temperature is much higher, approximately 80–83°C. In this case the temperature difference is only 10 K.

Using the same argument as before, the higher the temperature difference, the greater the destruction of exergy, because the high quality of the energy carried by a high temperature stream is being wasted by transferring the heat inefficiently from high to lower temperatures, even if heat losses are zero.

In other words, the more efficient the energy transfer in the heat exchangers, the less exergy will be destroyed and the greater the exergy yield.

Consequently, from the exergetic point of view, the heating branch is more efficient than the DHW branch, as shown in Fig. 10.49.

- Comparative results for all scenarios: This section presents a comparative overview describing how each variable evolves for each calculated hypothesis, giving an idea about all pros and cons for the various improvements and changes, which can be applied in this district heating network, described in Section 10.3 (refurbishment, new regulation of the DHN temperature, introduction of heat pumps, etc.).

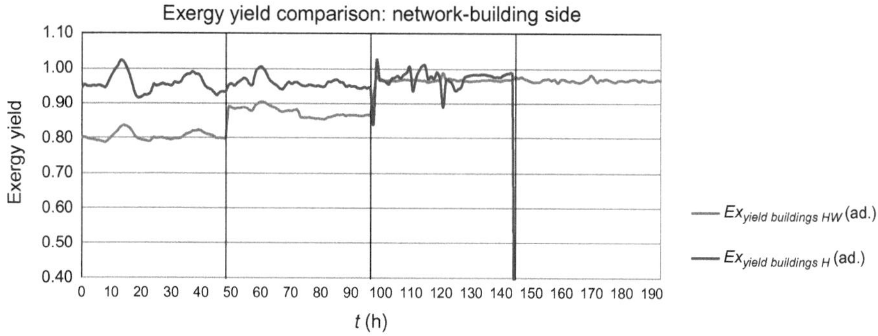

Fig. 10.49 Exergy yield for both branches.

Instead of hourly resolved profiles, for each scenario a single value has been calculated corresponding to the minimum external temperature of the day.

- Operation variables: Changing from SCENARIO 1 to SCENARIO 2, the temperature of the network (T_{DH2}) is reduced to a fixed value of 65°C, which means a reduction of 26% and its consequent decrease of $T_{r\ DH2}$ by 37% regard to SCENARIO 1 (Fig. 10.51). On the other hand, radiators are replaced by radiant floors. These two facts give rise to the following consequences:
  1. In the domestic hot water branch, $\Delta T_{DHN\ side\ HW\ i}$ decreases, leading to an increase of $\dot{m}_{HW\ i}$ to supply the same heat flow to this consumption point.
  2. In the heating branch the opposite effect occurs; $\Delta T_{DHN\ side\ Hi}$ increases, allowing a reduction of $\sum_i^n \dot{m}_{H\ i}$.

  Both effects counteract, resulting in a slight increase of the total mass flow ($\sum_i^n \dot{m}_i$) of 1.14% (Fig. 10.52). This is translated into a small pumping pressure increase of 1% (Fig. 10.50).

  In SCENARIO 3 T_{DH2} is reduced again from 65 to 40°C as a fixed temperature for all the year. This fact obliges to install a heat pump before the hot water heat exchanger, raising $T_{p\ in\ i}$ from 40 to 63°C, approximately. In this way, a Low Temperature District Heating Network is achieved, reaching a 60% decrease of T_{DH2} regarding to the actual situation (Fig. 10.51). The flow of DHN water arriving to each building ($\dot{m}_i$) rises reduction of $\Delta T_{DHN\ side\ H\ i}$ respect to SCENARIO 1 (Fig. 10.52), influencing the pumping pressure, namely increasing it up to 5.25% (Fig. 10.50).

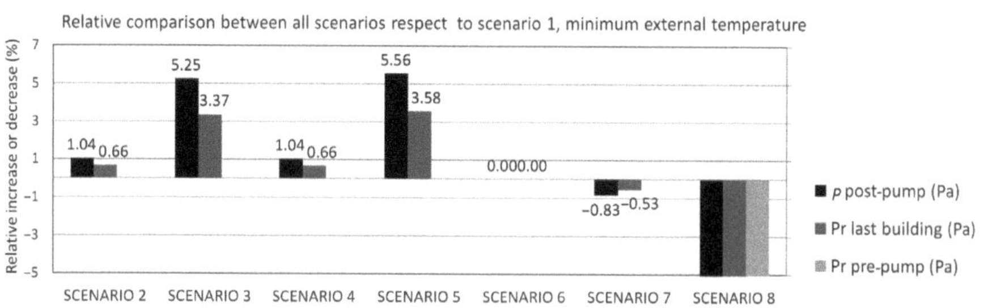

Fig. 10.50 Relative pressures comparison between each scenario and the reference SCENARIO 1.

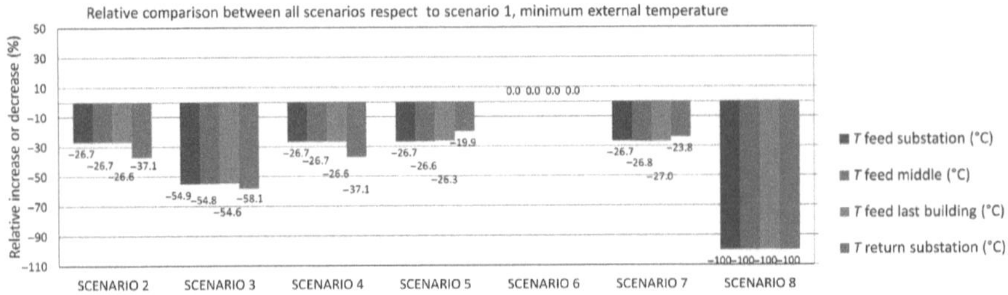

Fig. 10.51 Relative temperatures comparison between each scenario and SCENARIO 1.

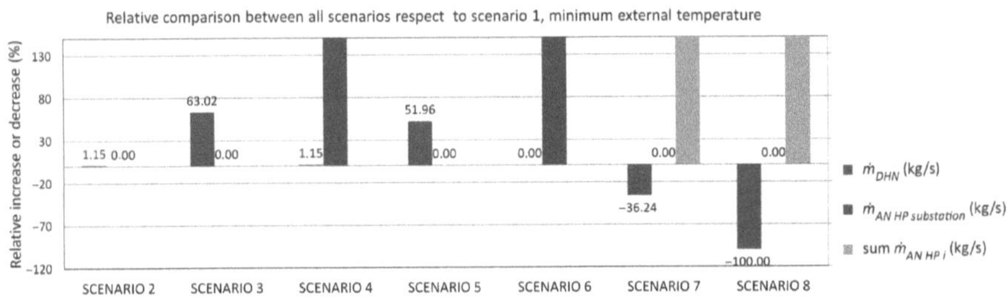

Fig. 10.52 Relative mass flows comparison between each scenario and SCENARIO 1.

SCENARIO 4 has the same characteristic as SCENARIO 2. The only difference is that the necessary heat is not provided from the primary DH network, but obtained from a heat pump. This new circumstance requires a new variable, the mass flow of anergy network's water ($\dot{m}_{AN\ SUBSTATION}$) (Fig. 10.52).

In SCENARIO 5, although radiators are used in the buildings, T_{DH2} is reduced to 65°C, like in the other cases (Fig. 10.51), hence, heat pumps are necessary again, in this case for the heating branch, in order to supply the radiator's temperatures. As a consequence in the hot water branch $\Delta T_{DHN\ side\ HW\ i}$ decreases, generating a rise of $\dot{m}_{HW}$ to an increase of $\dot{m}_i$ (Fig. 10.52), increasing the pressure in its turn (Fig. 10.50).

SCENARIO 7 is a special situation, in which only the DHW branch uses the district heating network as a heat source, while the heating branch for each building will follow a de-centralized strategy based on reversible heat pumps combined with floor heating or cooling. As shown in Fig. 10.51, the return temperature decreases less than the rest of scenarios. This is because $T_{p\ out\ i}$ comes only from DHW branch temperature $T_{p\ out\ HW\ i}$. Since the heating branch disappears for each building ($\dot{m}_{H\,i} = 0$), $\dot{m}_i$ will be reduced (Fig. 10.52), reducing pumping power (see *P* post-pump in Fig. 10.50).

Finally, SCENARIO 8 is the most extreme case, in which the district heating network disappears and only an anergy network is available for the heat pumps installed in the buildings. It only considers the use of decentralized individual systems for both heating and hot water. In the absence of DHN, the values of pressure and mass flow related to the network are zero, and in the three graphs (Figs. 10.50–10.52), values *P* postpump T_{DH2} and $\dot{m}_i$ are reduced by 100% relative to SCENARIO 1. The only existing magnitude is the sum of all anergy network water streams used for the installed heat pumps ($\sum_i^n \dot{m}_{AN\ HP\ i}$) (Fig. 10.52, SCENARIO 8).

– Global energy balance: In case of SCENARIOS 3 and 5 no anergy is extracted from the environment outside of the borders of the thermodynamic system, because the stream circulating through heat pump's evaporator comes directly from the heat exchanger, thus:

$$\dot{B}_{AN\,HW\,i} = 0$$
$$\dot{B}_{AN\,H\,i} = 0$$

Resulting in the energy balance:

$$\sum_{i}^{n} \dot{Q}_{DHN\,side\,i} + W_{HP\,HW\,i} + W_{HP\,H\,i} = \sum_{i}^{n} \dot{Q}_{client\,side\,HW\,i} + \dot{Q}_{client\,side\,H\,i} \tag{10.22}$$

In case of SCENARIOS 7 and 8 there is extraction of anergy from the environment, although these values have not been calculated by the simulation, the extended energy balance is therefore fulfilled:

$$\sum_{i}^{n} \dot{Q}_{DHN\,side\,i} + \sum_{i}^{n} \dot{B}_{AN\,HW\,i} + \sum_{i}^{n} \dot{W}_{HP\,HW\,i} + \sum_{i}^{n} \dot{B}_{AN\,H\,i} + \sum_{i}^{n} \dot{W}_{HP\,H\,i}$$
$$= \sum_{i}^{n} \dot{Q}_{client\,side\,HW\,i} - \sum_{i}^{n} \dot{Q}_{client\,side\,H\,i} \tag{10.23}$$

In Fig. 10.53 it is interesting to note how the hydraulic pumping power grows, due to the rise of $\dot{m}_i$ and P postpump for SCENARIOS 2, 3, 4, and 5 (Figs. 10.50 and 10.52).

Fig. 10.53 shows the impact of the heat pumps both for substations (SCENARIOS 4 and 6) and individual buildings (SCENARIOS 3, 5, 7, and 8).

The units are W/m^2, showing the density of energy or exergy flow for the total heating surface.

The following explanation revolves around the results concerning the energy losses in the network ($\dot{Q}_{losses\,NETWORK}$) (see Eq. 10.3) as well as the exergetic performance of the mentioned thermodynamic subsystem ($\eta_{ENERGY\,NETWORK}$) (see Eq. 10.4).

As previously discussed, with the exception of SCENARIO 6, which practically equals SCENARIO 1, the rest of the hypotheses are based temperature of the network (T_{DH2}), which reduces the energy losses along the pipes.

The graphs show that SCENARIO 3 is the most favorable if the idea is to reduce energy losses in the network (Figs. 10.54 and 10.55), creating a gain of 1.33% for the energetic yield regarding to the current situation (Fig. 10.56).

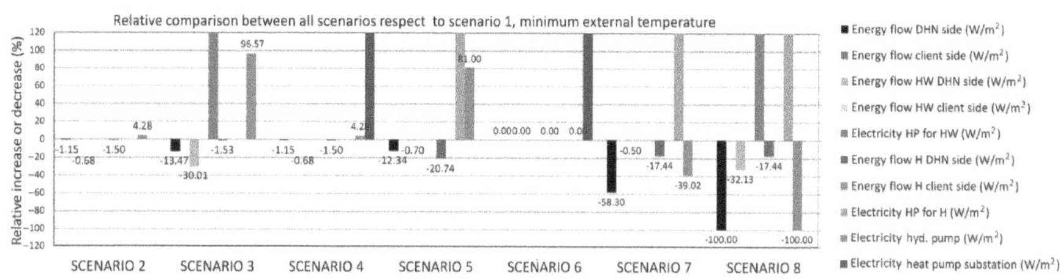

Fig. 10.53 Relative energy balance comparison between each scenario and SCENARIO 1.

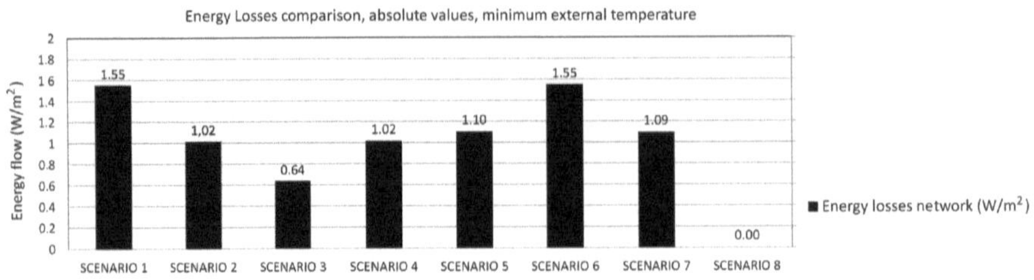

Fig. 10.54 Energy losses comparison between scenarios.

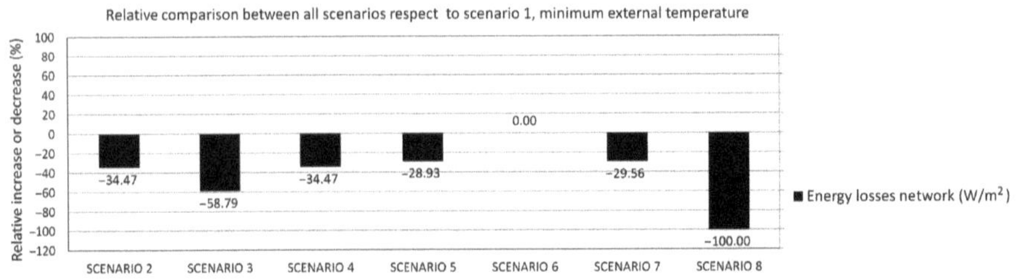

Fig. 10.55 Relative energy losses comparison between each scenario and SCENARIO 1.

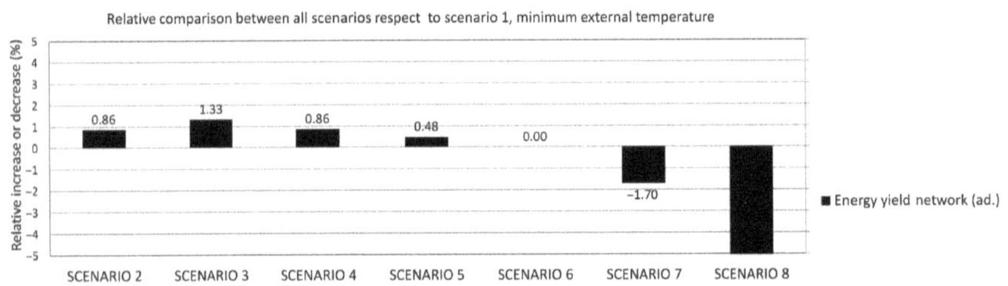

Fig. 10.56 Relative energy yield comparison between each scenario and SCENARIO 1.

– Global exergy balance: In Section 10.5.4 it was explained that exergy is the maximum useful work that can be obtained from a given thermodynamic system, which is not in equilibrium with the environment. The farther this system is from the environmental conditions, the more useful work can be extracted from it, and therefore the greater the exergy content will be. Therefore, it is a thermodynamic magnitude, which gives an idea of the quality of the energy, and not the quantity (Fig. 10.59).

Figs. 10.57 and 10.58 show exergy and electric power for each scenario, with exergy differences between $\sum_i^n \dot{E}x_{DHN\ side\ HW\ i} - \sum_i^n \dot{E}x_{client\ side\ HW\ i}$ and $\sum_i^n \dot{E}x_{DHN\ side\ H\ i} - \sum_i^n \dot{E}x_{client\ side\ H\ i}$. The greater the difference between each couple, the greater the exergy losses, and the exergy efficiency will be smaller.

For example in SCENARIO 3, the transformation to LTDHN generates a 32% reduction of $\dot{E}x_{losses\ BUILDINGS\ H}$ (Fig. 10.60). This will be the optimal situation of the heating system among all the available options. The $\dot{E}x_{losses\ NETWORK}$ also reduce prominently by 79%. However, Fig. 10.57 shows the SCENARIO electrical heat pump power

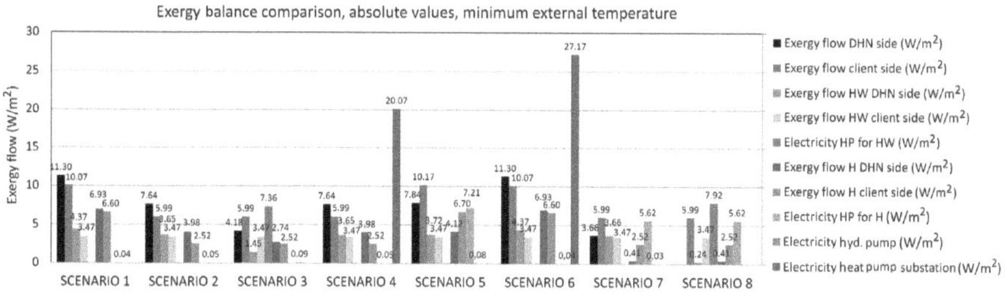

Fig. 10.57 Exergy balance comparison between scenarios.

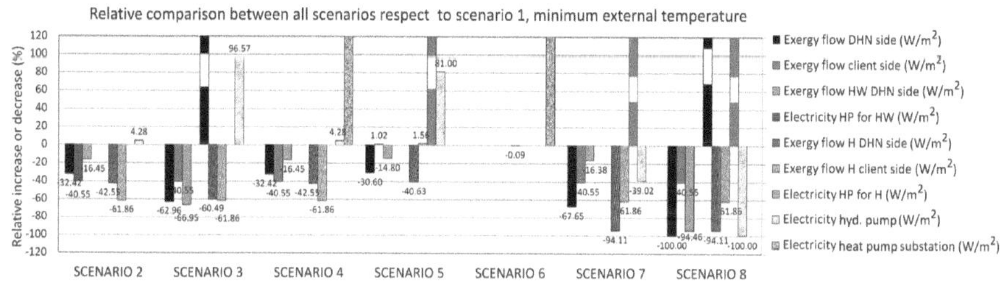

Fig. 10.58 Relative exergy balance comparison between each scenario and SCENARIO 1.

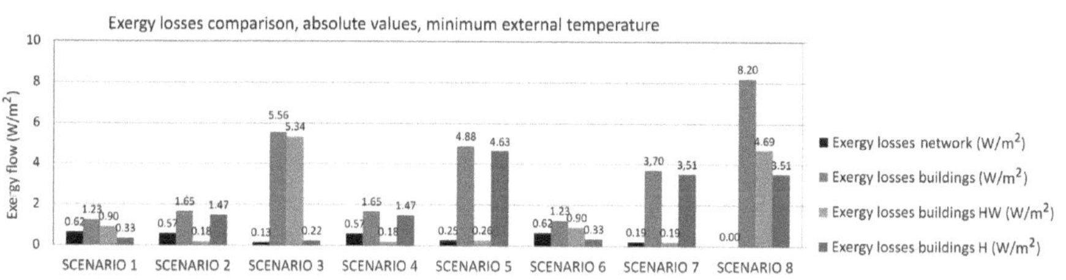

Fig. 10.59 Exergy losses comparison between scenarios.

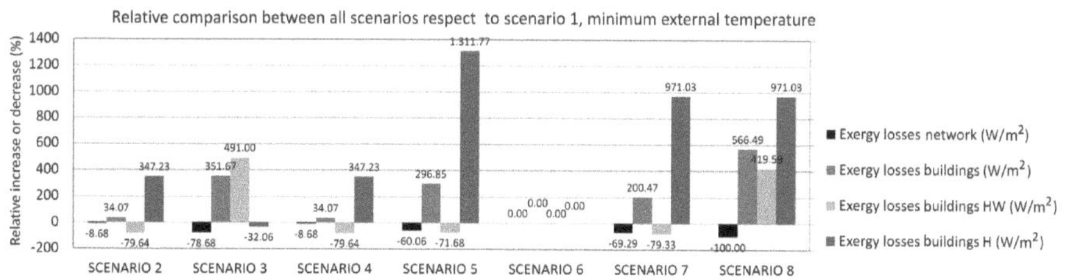

Fig. 10.60 Relative exergy losses comparison between each scenario and SCENARIO 1.

($\sum_i^n \dot{W}_{HP\ HW\ i}$) that is necessary to heat the domestic water. This circumstance makes $\dot{E}x_{losses\ BUILDINGS}$ 350% higher than those at SCENARIO 1 (Fig. 10.60).

A similar trend is observed in SCENARIO 5. As shown in Fig. 10.60, the heat exchanger for hot water decreases the exergy losses more than a 71%, the network

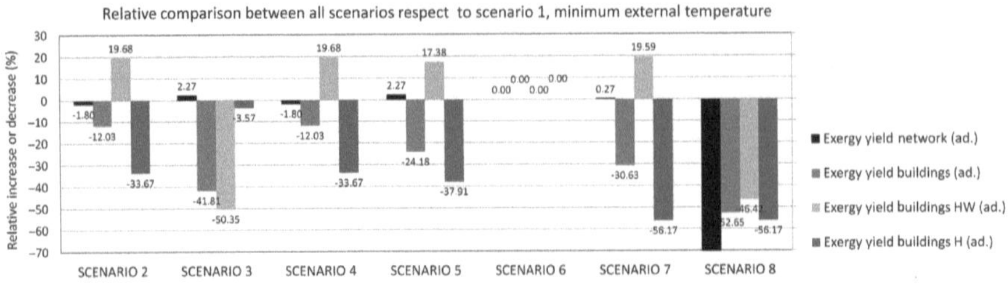

Fig. 10.61 Relative exergy yield comparison between each scenario and SCENARIO 1.

exergy losses is also reduced by 60%, but the effect of high power consumption increases the exergy losses in the heating branch (1312% respect to actual situation).

Returning to Fig. 10.57, SCENARIO 7 shows a drastic reduction of $\sum_i^n \dot{E}x_{DHN\ side\ i}$ because the DH network is only partially used for hot water production; instead, the heating branch needs a reversible heat pump, which consumes electricity $\left(\sum_i^n \dot{W}_{HP\ H\ i}\right)$ leading to a building exergy efficiency $\eta_{EX\ BUILDINGS}$ of -31% regarding the actual situation (Fig. 10.61, SCENARIO 7).

SCENARIO 8 is similar to SCENARIO 7 but the supply of exergy by the thermal network completely disappears (Fig. 10.57), hence $\sum_i^n \dot{E}x_{DHN\ side\ i} = 0$. This means that the buildings are totally dependent on electricity and anergy to obtain the heat, so the total electricity consumption will be even higher. Again, this circumstance reduces $\eta_{EX\ BUILDINGS}$—by 53 (Fig. 10.61, SCENARIO 8).

10.4.1 Conclusions

The work methodology followed has been useful to organize the large amount of initial data and information, classifying the necessary parameters, clearly defining the different thermodynamic systems, raising and feeding the physical and mathematical models, and finally proceeding to run the simulation. The simulator worked correctly and without failures for the eight studied scenarios, being able to resolve with acceptable accuracy all pressure, mass, energy, and exergy balances. The obtained results are realistic and coherent.

Adding heat pumps to the energy system allows the introduction of electricity into the heating sector, diversifying to other possible (renewable) energy sources.

Exergy balances are useful to inform about the power consumption and all possible inefficiencies produced by the heat pumps.

As a final result it can be affirmed that of all simulated and studied hypotheses SCENARIO 3 with a low temperature district heating network, floor heating in the buildings and decentral heat pumps for DHW shows the highest energy savings at the network area and the highest exergetic efficiency. However, as DHW requires a heat pump with significant electricity requirements, the overall exergetic efficiency does not improve, unless renewables are used for the electricity generation. This scenario shows the possibility of adapting the current Viennese district heating network to a Low Temperature District Heating Network (LTDHN) with 40°C supply temperature, advancing in this way toward the fourth generation of district heating network.

Both SCENARIOS 4 and 6 are using central heat pumps in the substation without any primary network connection. These kind of options are useful especially for urban areas that are located far from the principal thermal network. These scenarios would also be useful in cases such as the residential area of "Les Cherpines," explained in Chapter 2. In this example, the heat pump of the substation can use water coming from industries or wastewater as an anergy source. Those industrial effluents are still thermally utilizable, having temperatures around 25–30°C. Thereby, they act as an energetic link between industrial and urban areas. SCENARIO 6 is even more feasible since it does not require any refurbishment of the buildings or installation works, allowing radiators to remain. In principle this option will be the cheapest in terms of fixed and installation costs.

SCENARIOS 7 and 8 involve a gradual disappearance of the district heating network. In particular, SCENARIO 7 is designed for a case in which decentralized heating systems are chosen, but at the same time a connection to the district heating network must be available.

By contrast, SCENARIO 8 has no district heating network anymore, only a hydraulic network remains at moderate or ambient temperature, providing anergy to the individual heat pumps installed in the buildings. The consequence would be a total reconversion of the district heating network, moving into an anergy network, advancing toward the fifth generation of district heating networks.

This work informs urban planners, technicians, and decision makers on thermal networks in combination with various urban energy systems.

Annex

10.5 Energy performance of decentralized solar thermal feed-in to a Vienna Case Study District Heating Network by Pilar Monsalvete Álvarez de Uribarri

In the previous sections, the DHS of Vienna and, specifically, the branch named Girasolstrasse were presented. This section aims to analyze the effect of installing distributed solar thermal energy systems at the consumer side and thus turn them into prosumers. The solar thermal potential and whether it is possible to feed in all the surplus solar energy into the net are analyzed. Also a comparison of the solar thermal system with a system based on PV and heat pumps is discussed.

10.5.1 Background

Research over the last years addressed the potential of decentralized energy systems in DHN as well as some of the possible technical problems that this connection could provoke. Excess energy has first been discussed for the industrial sector in terms of the thermal energy contribution to DHN (Broberg et al., 2012; Persson et al., 2014). More recently the concept has been extended to the residential sector. An interesting overview of the state of the art by Brand et al. includes distributed solar energy in DHN in combination with seasonal storage or using heat pumps (Brand et al., 2014). The influence of excess heat production with seasonal storage has also been analyzed (Nielsen and Möller, 2012). Others investigate the benefits of local thermal storage for direct utilization of heat from solar collectors in a dwelling connected to a DHN, but with no injection into the net (Heikkinen et al., 2014); or connected and using the net as a virtual storage system (Di Pietra et al., 2015). The impacts of feed-in to the net have also been studied, in particular how the temperatures, pressure, or flow rates and velocities can be affected by the introduction of small-scale prosumers into the DHN (Brand et al., 2014). These past studies tend to treat the network as an ideal storage facility without constraints on capacity or dynamics, so that prosumers can release excess solar energy into the network and use it later when it is needed (Di Pietra et al., 2015; Vesaoja et al., 2014; Heikkinen et al., 2014; Nielsen and Möller, 2012). This idealization may lead to optimistic results especially for the warm seasons. Thus, to develop this case study, the possible thermal saturation of the net has been taken into account by simulating it dynamically together with the prosumers. The simulation software INSEL8.2 (Luther and Schumacher-Gröhn, 1991) has been used to model the network thermodynamics (Hassine and Eicker, 2014) as well as the thermal behavior of the prosumers. The weather and the heat demands are considered as inputs to the model.

10.5.2 Case study and model description

10.5.2.1 Geographical area and heat demand

As previously presented in this chapter, the network under study belongs to the District Heating Network of the city of Vienna. It is a secondary branch connected to the main network through two heat exchangers in the supply substation (GUFO). More than 25 buildings are connected to the Girasolstrasse branch through 19 substations (Fig. 10.62).

All buildings connected to one substation have been aggregated and treated as a single entity, that is, their demands have been added together, and also the solar areas and the storage volumes so they are seen from the primary side as just one consumer, not the separated buildings.

The database for the simulation was very complete, providing the main parameters needed to define the network: length and diameters of pipes; together with 4 years with 15 min resolution of temperatures and flows at the substations, the power supplied by the GUFO substation and external temperature. Data of the year 2015 have been selected to develop the case study because it was the most complete. Fig. 10.63 shows the monthly demand of all substations along the year 2015. It can be seen how the demand strongly decreases during summer season according to the decrease of heating demand.

This information was combined with solar radiation data provided by the Austrian Central Institution for Meteorology and Geodynamics (ZAMG by its German acronym). Fig. 10.64 shows the monthly solar radiation on the city of Vienna during the year 2015.

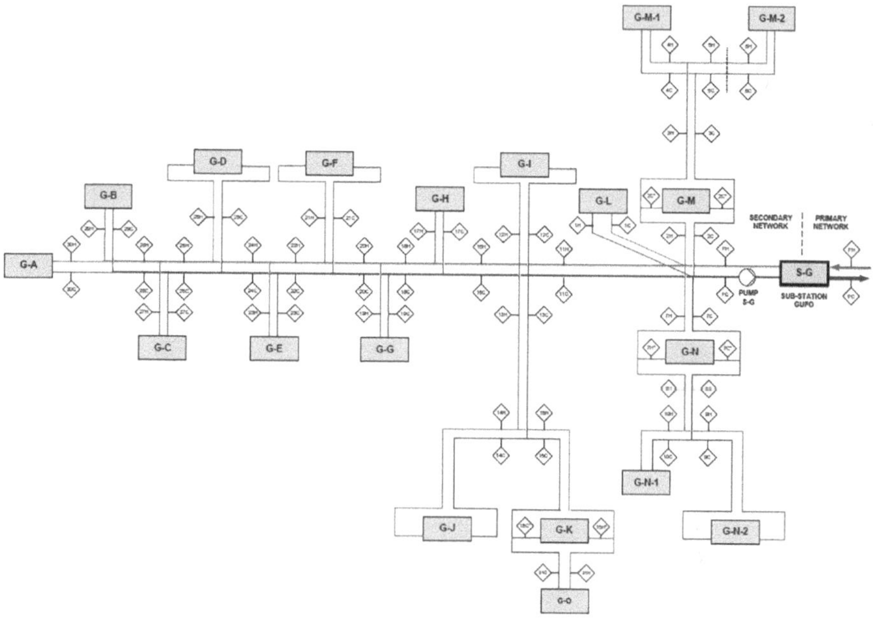

Fig. 10.62 Schema of the Girasolstrasse branch. The schema includes the GUFO substation on the right that connects the branch with the main network and the 19 substations where one or more buildings are connected. The connection of the substations to the branch is done in parallel.
(By M. Potente.)

The foot print areas of the buildings range from around 200 m^2 to not much more than 2000 m^2. This range together with Figs. 10.63 and 10.64 give a first impression of how difficult it may be to cover the complete demand of the network only with solar energy. Nevertheless, there is still the possibility of being an autonomous micro-grid during summer. Mostly the idea is to not modify anything but the prosumers, meaning that the network branch stays as it is and the buildings are neither refurbished nor have any change in the systems. Knowing that the buildings are not new, the heating systems in all buildings are assumed to be based on radiators. Therefore, the new solar systems installed must produce water of at least 65°C and the capacity of the prosumers to feed the net depends also on the diameters of the network pipes.

Two months, January in winter and July in summer, have been selected to do the study. These two months have the lowest and highest ratio of solar radiation to heat demand, respectively.

10.5.2.2 The solar system models

The hydraulics of the solar system model is presented in Fig. 10.65. The solar thermal system is mainly composed of solar panels connected to a storage tank through a heat exchanger. On the solar panels side, a standard water-glycol mixture is used

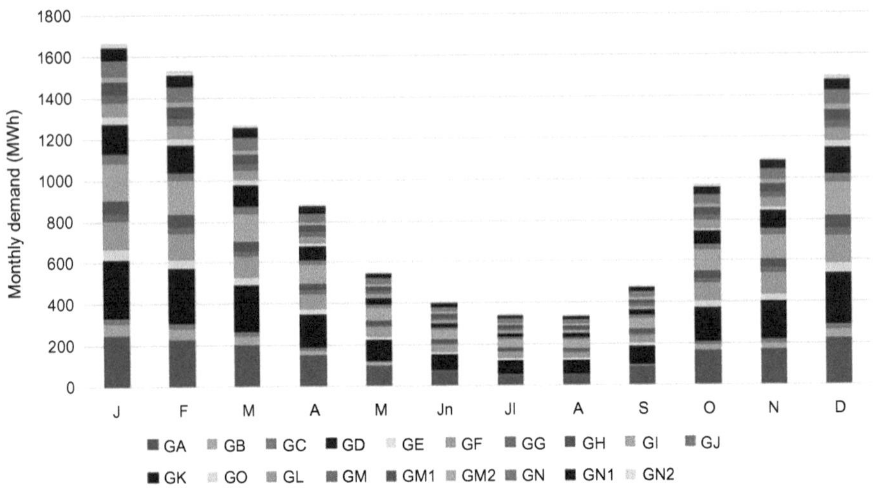

Fig. 10.63 Monthly demand of the substations of Girasolstrasse. Every substation is connected to one or more than big buildings. Their demands are aggregated per substation.

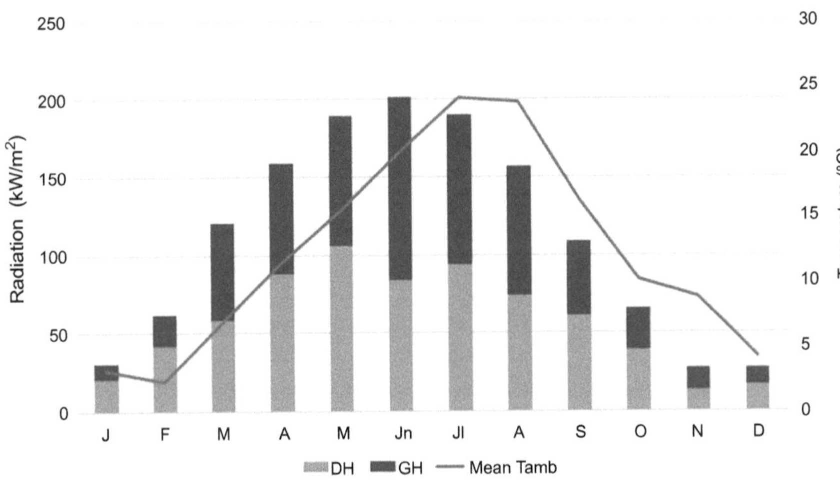

Fig. 10.64 Monthly meteorological data of Vienna in 2015.
(From ZAMG.)

to prevent freezing in winter. In the rest of the system, pure water is used. The system is hydraulically connected through a second heat exchanger to the DHN. The central component of the system is the storage tank: the solar input goes directly there as well as the input from the network. Both mass flows converge thanks to a mixing valve (V1), allowing the activation of pumps B2 and B3 in parallel. At the same time, the demand is always covered from the tank. This configuration is widely used in

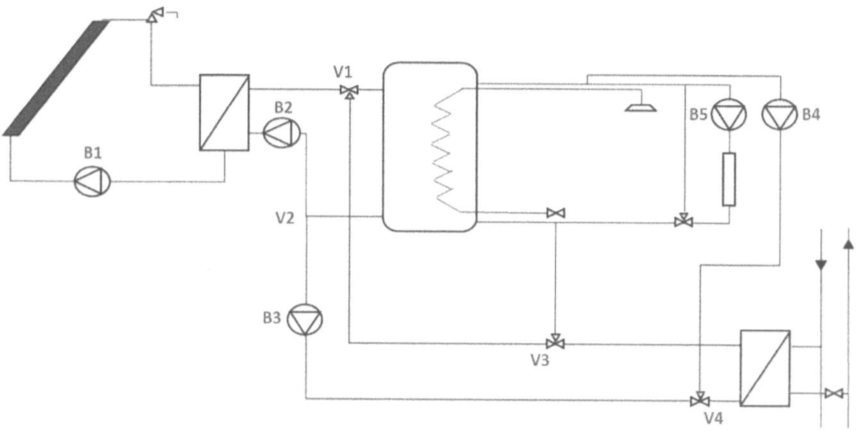

Fig. 10.65 The schema of the solar thermal system. The solar thermal system, mainly composed by solar panels connected to a storage tank through a heat exchanger, is in turn connected to the DHN.

the sector to smooth the peaks in demand. The possible working modes are the following:

1. solar heating mode: pumps B1 and B2 on;
2. network heating mode: pump B3 on;
3. solar surplus to net: pump B4 on; and
4. covering demand mode: pump 5 on.

Valves V3 and V4 separate modes 2 and 3, so they cannot occur at once. There is no other hydraulic restriction, the activation-deactivation of other modes depend on the control system, which is based on tank temperatures and solar radiation. The control strategies are explained below.

The same components of the system were selected for all substations. Tables 10.6 and 10.7 present the simulation parameters of the solar thermal collectors and tanks. A very efficient evacuated tube collector from those found on the market has been chosen for the study.

Table 10.6 Solar thermal collector parameters

Fluid	–	Water-glycol mixture
Orientation	–	South
Inclination of solar collector	Degree	30.0
a_0 at normal incidence	–	0.642
Linear heat loss coefficient	W/m^2 K	0.89
Quadratic heat loss coefficient	W/m^2 K^2	0.001
IAM	–	0.9

Table 10.7 Storage tank parameters

Fluid	–	Water
Overall heat loss coefficient	W/m^2 K	2.247
Ambient temperature around the tank	°C	8

For the heat exchangers, two counterflow HXs with an overall heat transfer coefficient of 3500 W/m^2K for the one on the solar side and 3000 W/m^2K for the one connecting with the DHN have been chosen. As the network model considers not only the thermal but also its hydraulic behavior, the pressure losses at the primary side of the corresponding heat exchanger had to be defined. They were set to 200 kPa only on the primary side.

Finally, a control system for pumps and valves with the following strategy has been included:

- Whenever the radiation is high enough (threshold 150 W/m^2), the solar collector pumps start working to feed the tank.
- The solar field has a maximum working temperature of 99°C due to pressure restrictions. Whenever this temperature is reached, a security valve opens loosing the energy.
- If the tank is full (temperature at the top of the tank greater than 75°C), its energy is supplied to the net.
- If the tank is empty and the solar input is insufficient (temperature on top of the tank lower than 60°C), the deficit is supplied from the net.
- The pumps on the demand side work whenever there is demand.

10.5.2.3 System dimensioning

The system dimensioning has been carried out starting from the solar collectors maximum size and calculating the rest of the components according to this. The maximum size, as was presented in Section 10.5.2.1, will probably not be enough for covering the complete demand, so there is no problem with oversizing the system.

The maximum collector area per building has been calculated based on the work developed by Romero Rodríguez et al. (2017) where a systematic process to evaluate the usable roof area for PV and solar thermal depending on the available geometrical data of the building is defined. Due to data privacy reasons, the actual buildings connected to every substation were not known, only some characteristics of them. Table 10.8 presents the known parameters per building, always aggregated by substation.

The footprint has been estimated as follows:

- For those buildings with available information, the footprint has been calculated as the total heated surface divided by the number of floors.
- For those buildings, where the number of floors was unknown, four floors were assumed and the footprint has been calculated in the same way as the previous case.
- Finally, when the heated surface area was also unknown, then:

Table 10.8 Substation parameters

	Floors	Heating surface (m^2)	Annual demand (MWh)	Estimated footprint (m^2)
GA	4	8320	1753	2080
GB	2	704	284	352
GC	6	1640	179	273
GD	6	19,212	1930	3202
GE	9	2770	324	308
GF	5	11,650	943	2330
GG	8	2408	244	301
GH	6	4036	429	673
GI	?	10,835	1194	2709
GJ	?	2748	292	687
GK	8	4227	838	528
GO	?	?	169	321
GL	?	4662	446	1166
GM	?	2491	278	623
GM1	?	3351	406	838
GM2	5	2569	190	514
GN	33	5377	562	1792
GN1	5	3255	418	651
GN2	3	1766	163	589

1. The total specific demand of every other building was calculated as

$$q_h = \frac{Q_h}{A_h}$$

2. The mean specific demand of the group $\overline{q}_h$ was calculated.
3. Finally the heated surface of the unknown building was estimated as

$$A_{fp} = \frac{Q_h}{\overline{q}_h \cdot n_{floors}}$$

The roof was considered to be flat and with the same area as the footprint in the absence of further information. Than, according to Romero Rodríguez et al. (2017), the reduction factors of the roof area used in the case study are follows:

- *Construction restrictions*: $C_{CON} = 0.8$ for flat roofs
- *Protected buildings*: It was assumed that the district has no protected buildings, $C_{PROT} = 1$

- *Shadow effects*: $C_{SH} = 0.7$ for flat roofs
- *Orientation losses*: All collectors are assumed to be oriented south and tilted 30°C. Then, the software calculates the solar radiation for that orientation without applying any other reduction factor: $C_{AZ} = 1$ and $C_{SL} = 1$
- *Separation of panels*: $C_{GCR} = 0.46$
- *Service area*: This factor is already included in C_{GCR}
- *Ratio of thermal to photovoltaic panels*: The study considers only thermal panels, so $C_{ST} = 1$
- *Ratio of PV panels*: For the same reason $C_{PV} = 0$

In all cases a solar volume flow rate of 54 L/(h m^2) was considered.

For the dimensioning of the tanks, the conclusions extracted from Monsalvete Álvarez de Uribarri et al. (2017) were followed, and two cases have been considered with a ratio of tank volume to solar area of 0.05 m^3/m^2 and 0.2 m^3/m^2, respectively (the typical ratio in systems to cover only DHW is from 0.05 to 0.08 m^3/m^2).

Table 10.9 summarizes the sizes of the systems connected to each substation.

Table 10.9 Solar systems of each substation

	Solar collector area (m^2)	**Storage volume case 1 (m^3)**	**Storage volume case 2 (m^3)**
GA	536	26.8	107.2
GB	91	4.5	18.2
GC	71	3.5	14.2
GD	825	41.2	165.0
GE	80	4.0	16.0
GF	600	30.0	120.0
GG	78	3.9	15.6
GH	173	8.7	34.6
GI	698	34.9	139.6
GJ	177	8.8	35.4
GK	136	6.8	27.2
GO	83	4.1	16.7
GL	300	15.0	60.0
GM	160	8.0	32.0
GM1	216	10.8	43.2
GM2	132	6.6	26.4
GN	462	23.1	92.4
GN1	168	8.4	33.6
GN2	152	7.6	30.4

10.5.3 Results and discussion

Table 10.10 summarizes the results of cases 1 and 2 for both months, January and July.

The solar energy used refers to the total amount of solar energy collected during the month and used either to cover demand or to feed the net. And the solar energy lost accounts for the losses due to exceeding the maximum temperature in the collector field. The high amount of these losses is caused by the high temperatures in the tank required by the heating systems in the buildings. This leads, in turn, to high temperatures in net, 65°C in summer and 70–85°C during winter. That leaves only a small margin for the solar collectors to operate.

Heat imported refers to the total energy extracted from the net and the heat exported that excess of energy released to the net.

The losses to the ground are presented in total amount of energy and in percentage of the increase with respect to the initial value (original case without consumers).

Finally, the energy consumed by the pumps has been calculated assuming that the pumps operate with a constant efficiency always, as the dynamics of the prosumer are not simulated. A correlation of the power of the pumps as a function of the mass flow has been extracted from a water pump catalog.

$$E_p[Wh] = \sum_i P_i \cdot h_i \tag{10.24}$$

where h_i are the hours of operation of each pump and P_i its power calculated with the correlation as in Eq. (10.25).

$$P_i\,[\mathrm{W}] = 7.897 + 35.104 \cdot \dot{m}_i \tag{10.25}$$

Table 10.10 Summary results solar thermal

	January		July	
	Case 1	Case 2	Case 1	Case 2
Total collector area (m^2)	5138	5138	5138	5138
Total tank volume (m^3)	1027.6	256.9	1027.6	256.9
Total demand (MWh)	1665.1	1665.1	345.8	345.8
Solar used (MWh)	68.71	62.65	403.21	396.25
Solar lost (MWh)	86.77	90.22	106.72	112.18
Heat imported (MWh)	1564.46	1596.36	26.04	18.24
Heat exported (MWh)	0.0	0.68	74.83	64.92
Losses to the ground (MWh)	65.86	65.74	51.24	53.83
Increase of losses to the ground (%)	2.12	1.93	40.18	47.29
Pumps consumption (MWh)	0.3	0.3	2.7	3.3

Effects of the storage volume on the solar use. Increasing the storage volume has two opposite effects. On the one hand, the storage capacity ensures a bigger solar energy collection; but on the other hand, it implies bigger thermal losses. These two effects may lead, as extracted form the results, to a low improvement or even worsening system efficiency.

Restrictions to a bigger solar energy use:

- Low solar area: One of the main points, as predicted in the introduction, is the low available solar area, which is a common issue in cities. This cannot even be compensated with the high efficiency of the evacuated tube collectors.
- High temperatures of the net: The set temperature of the net is very high for an efficient solar energy use, mainly in winter when the temperatures of the net range between 70 and 85°C, while in summer they stay constant at 65°C. For example, a reduction of 10°C of the net temperature in winter could increase the solar use during the month of January by 77 and 45 MWh for cases 1 and 2, respectively, that means more than double the energy for case 1 and an increase of almost 73% for case 2. Nevertheless, this reduction could only be done after a refurbishment of the buildings envelope and a modification of the heating systems inside them.
- Restrictions to high temperatures in the net: This case study does not contemplate an upper threshold on net temperatures, so the maximum temperature is defined by the solar field (99°C). But it is usual that the net does not allow temperatures above a limit to ensure the proper functioning of downstream substations. A limitation on the solar injection temperatures could highly decrease the harnessing of solar energy.
- Increasing losses: Finally, the increase of the temperatures of the net due to the solar injection increases in turn the thermal losses to the ground (between 40% and 47% in July in this study).

10.5.4 Comparison with a system based on PV and heat pumps

In this section, a new option is considered instead of the solar thermal system and the results are compared in terms of primary energy.

The system considered is a decentralized water-water heat pump installation. The condensers of the heat pumps are connected to the DHN, whose temperature are lowered to 40°C. The heat pumps are driven by electricity generation of photovoltaic panels when possible. There is no storage for the PV but if there is electrical surplus, this can be fed back into the grid.

These systems have two main advantages. First, the COP of the water-water heat pumps is very high, so the electrical consumption compared to air-water heat pumps is much lower. Second, the losses in the network are reduced with the temperature. Nevertheless, there is an extra electrical consumption from the heat pumps that may not be completely covered with PV.

The calculation of this system has been done in a simplified way. Both the PV panels and the heat pump have been considered to have constant efficiency:

$$\eta_{PV} = 0.15$$

$$COP_{HP} = 4.5$$

Considering that the temperatures of condenser and evaporator of the HP remain almost constant along the year, the assumption is suitable.

The electricity produced by the PV panels in every substation, P_{PVi} has been calculated as:

$$P_{PVi}(t)[W] = A_i \cdot \eta_{PV} \cdot I_{gn}(t) \tag{10.26}$$

where A_i are the suitable PV areas at each substation i and I_{gn} is the global radiation on the PV panels that has been simulated for every hour. The same suitable areas have been assumed for PV as for thermal collectors (see Table 10.9). Then, every hour the needs of electricity have been compared with the PV production and two results per substation have been obtained:

$$\text{If } \frac{\dot{Q}_i(t)}{COP_{HP}} > P_{PVi}(t),\ P_{Gi}(t)[W] = \frac{\dot{Q}_i(t)}{COP_{HP}} - P_{PVi}(t) \tag{10.27}$$

$$\text{Else, } P_{PV\_Gi}[W] = P_{PVi}(t) - \frac{\dot{Q}_i(t)}{COP_{HP}} \tag{10.28}$$

where $\dot{Q}_i$ is the thermal demand for heating and DHW, P_{Gi} is the electricity needed for driving the HP taken from the net, and P_{PV_Gi} is the extra electricity generated by the panels and sent back to the grid at each substation.

Furthermore, the savings on the thermal side have been calculated as:

$$\dot{Q}_{save}(t)\ [\mathrm{W}] = \dot{Q}_i(t) \cdot \frac{1}{COP_{HP}} + \dot{Q}_{lTs} - \dot{Q}_{l40} \tag{10.29}$$

With $\dot{Q}_{lTs}$ being the losses on the DHN at the original supply temperature and $\dot{Q}_{l40}$ the losses at 40°C. Both losses have been dynamically simulated for the complete network.

Finally, in order to be able to compare electricity with heat, the final energy has been converted to primary energy using the following Primary Energy Factors (Leoncini, 2013):

$$PEF_{elec} = 1.91$$
$$PEF_{heat} = 1.01$$

For the comparison between the two systems, only the best solar thermal case (Case 1) was chosen.

Table 10.11 summarizes the results from System 1, the combination PV plus Heat Pump (PV + HP), and System 2, Case 1 of the solar thermal system (ST), for the comparison. The results are presented in Primary Energy by applying the corresponding Primary Energy Factors. Negative values refer to energy saved and positive values to extra energy demanded.

The thermal savings due to the system refers, in case of System 1, to the savings in thermal energy due to the high COP of the heat pumps. In case of System 2, the

Table 10.11 Comparison solar thermal vs. PV plus HP

	January		July	
	System 1	System 2	System 1	System 2
	PV + HP	ST case 1	PV + HP	ST case 1
Demand covered by solar (%)	12.12	1.49	100	90.28
Thermal savings due to system (MWh)	−409.47	−68.02	−102.11	−467.99
Thermal savings for lowering Tsup (MWh)	−35.39	–	−24.25	–
Extra electricity consumption (MWh)	621.11	0.56	−137.96	5.15

thermal savings are the solar energy used and exported minus the extra losses to the ground. System 1 has an extra amount of thermal energy demand saved due to the reduction of the network temperature to 45°C. Finally, the extra electricity consumption of System 1 is due to the electrical consumption of the heat pumps minus the electricity generated by the PV panels. When positive (January), it means that the system has consumed more electricity from the grid than generated. When negative (July), the opposite is true, that is, the PV panels have generated more electricity and this extra amount is sold to the grid. In case of System 2, the extra electricity consumption is due to the pumps activity.

Figs. 10.66 and 10.67 show how the savings in thermal energy in System 1 (PV + HP) come at the expense of high electrical energy consumption. Due to the limited solar

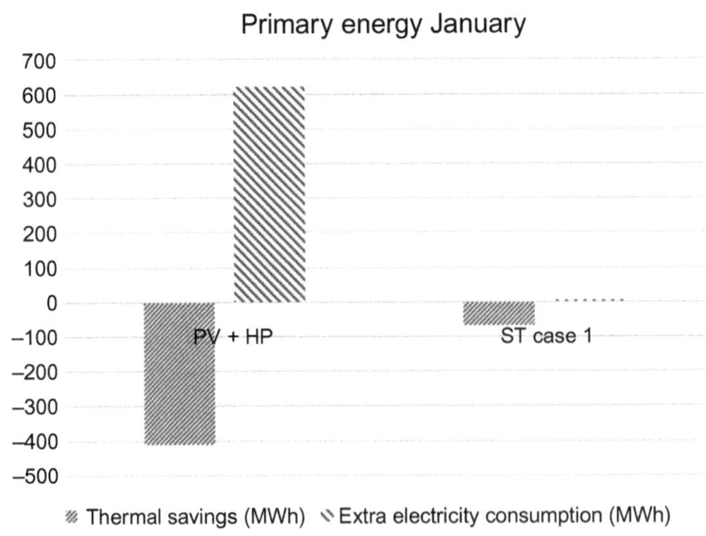

Fig. 10.66 Comparative results in primary energy in January.

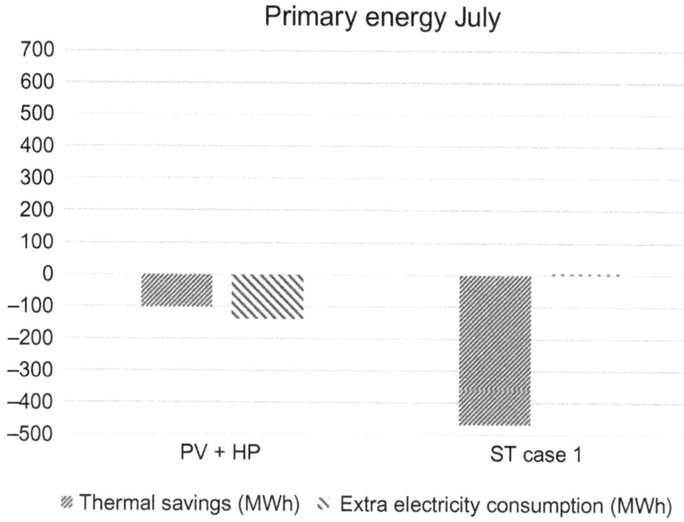

Fig. 10.67 Comparative results in primary energy in July.

area, a big portion of this consumption comes from the grid. On the other hand, during summer all demand can be covered with solar energy and some extra electricity can be sold to the grid.

The energy mix in the grid plays a big role in the selection of the best option, as it gives the clue to decide whether the thermal savings compensate the energy consumed from the grid or not. A better mix (with more renewable) can decrease significantly the Primary Energy Factor and therefore, the "Extra electricity consumption" bar.

Fig. 10.68 resumes the total primary energy consumption for both months and both systems. It can be clearly seen how depending on the month of the year with changing relation between heat demand and solar supply, System 1 consumes extra energy in winter compared with the original case or can save just a small amount of thermal energy in summer. At the same time, System 1 exports electricity to the grid in summer. System 2 can save primary energy for heating and DHW in both months and that apart from pumping electricity has nearly no impact on the electrical grid.

The selected months, January and July, are those that have a high and low relation of demand to solar supply. It can be stated that the thermal savings along the year for System 2 will be always negative (i.e., always some energy saved), but not very high in absolute terms. As the savings of these 2 months amount to almost 26% of the total demand, it can be assumed that the yearly percentage will be around that value.

In terms of total primary energy savings, System 2 (Solar Thermal) performs better than the PV + HP system. Nevertheless, none of them is an optimum solution: System 1 still needs a high amount of electricity consumed from the grid, and the high working temperatures of System 2 do not provide the best conditions for an optimal solar thermal system operation.

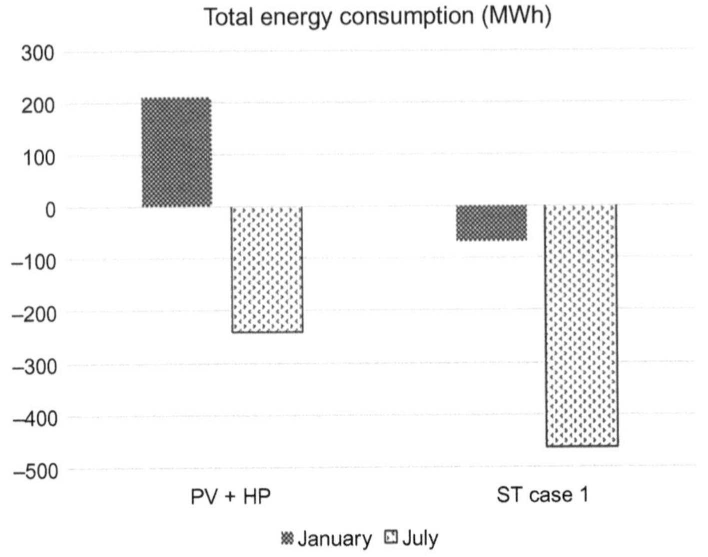

Fig. 10.68 Resume in primary energy.

10.6 Nondeterministic calibration of district scale building energy models by Giovanni Tardioli

The section is divided into three main parts. Section 10.6.1 describes the district case study. Section 10.6.2 describes the steps involved in the creation of an automated urban scale building energy modeling approach. Section 10.6.3 describes the methodology adopted for the development of the energy modeling calibration framework, with particular attention to the processes required for the creation of emulators, metamodels of the dynamic simulation engine.

10.6.1 Case study

The city of Geneva (Switzerland) is considered as the case study for the development and testing of the methods presented in this chapter. In particular, the district of Meyrin will be used as reference district for the following analysis. The district comprises a total of about 3370 buildings as classified in the city cadastral database. Fig. 10.69 shows the location of the district within the city and Fig. 10.70 provides a visualization of the buildings within the district.

The online repository (SITG, n.d.) provides building information organized by the identification code "EGID" and organized in different layers. Information of the building stock is gathered from various datasets and additional resources. Fig. 10.71 describes the typologies of buildings within the district and their period of construction.

Most of the buildings are residential constructions (for the majority multifamily buildings). Other typologies cover a large portion of the total number of buildings

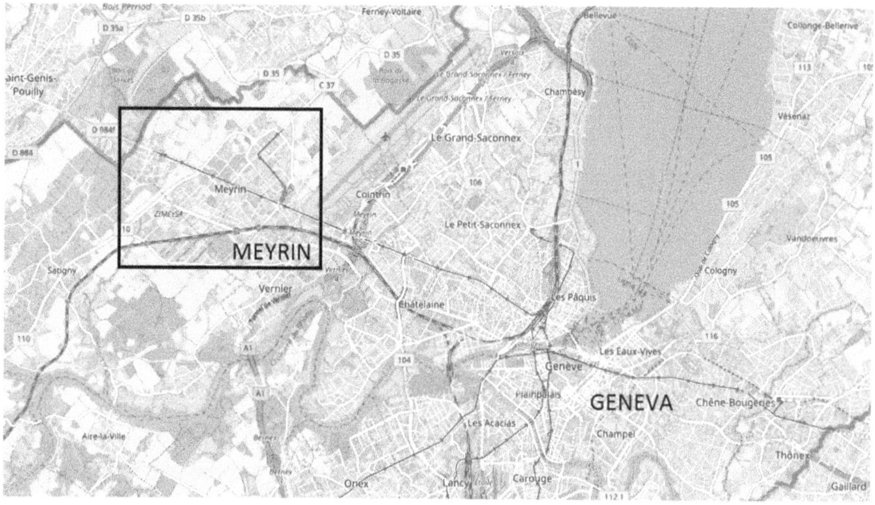

Fig. 10.69 Location of the district.
(Map data: ©OpenStreetMap contributors.)

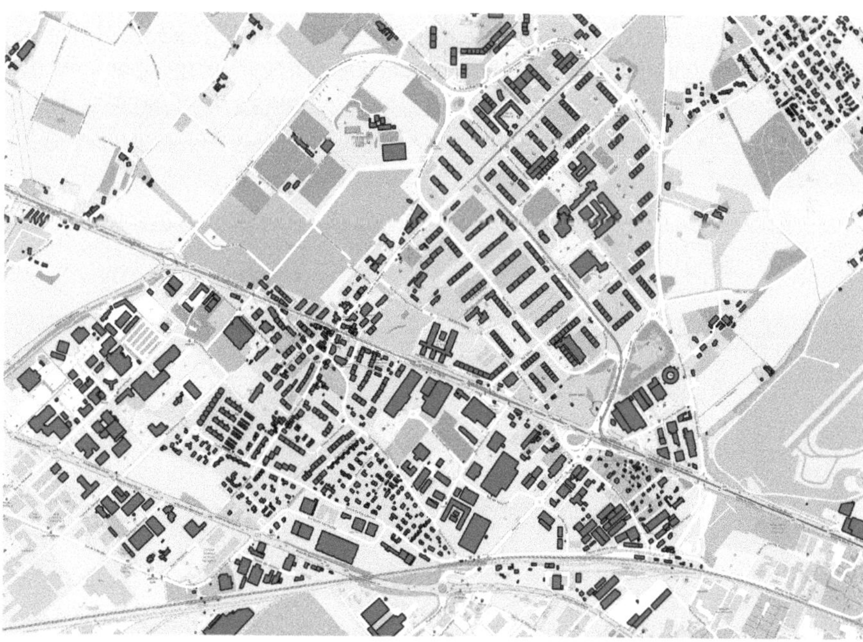

Fig. 10.70 GIS visualization of the buildings in the district.
(Map data: ©OpenStreetMap contributors.)

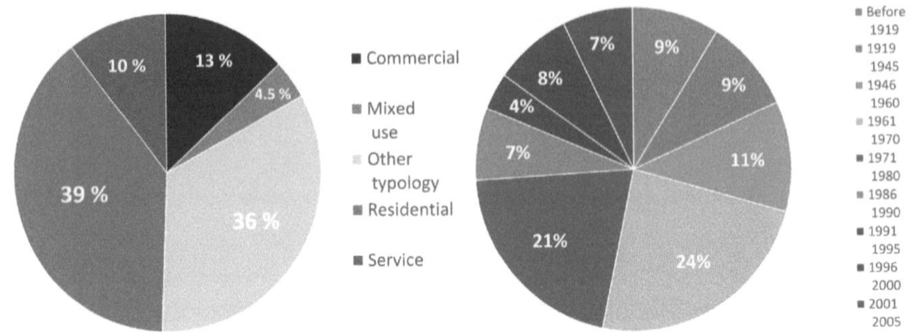

Fig. 10.71 Typology and period of constructions of buildings within the district of Meyrin.

and it includes a variety of small constructions (e.g., verandas, garages, spaces <20m^2). In addition, commercial, service, and mixed-use buildings cover a smaller part of the total number of buildings. Regarding the period of construction, most of the buildings were constructed during the years 1960–80. A quite large portion of the building stock was built in the period before 1919–60. In terms of building energy use, the dataset provides an indication of building consumption as a sum of heating and DHW requirements for each building considered in the analysis.

After data preprocessing, a normalization of the total consumption of the dataset by the number of buildings in the district of Meyrin and for the dataset of a large portion of city of Geneva provides the distribution reported in Figs. 10.72 and 10.73. The two normalized distributions in terms of total consumption and energy index overlap, thereby the district is quite representative of the overall energy distribution of the city case study.

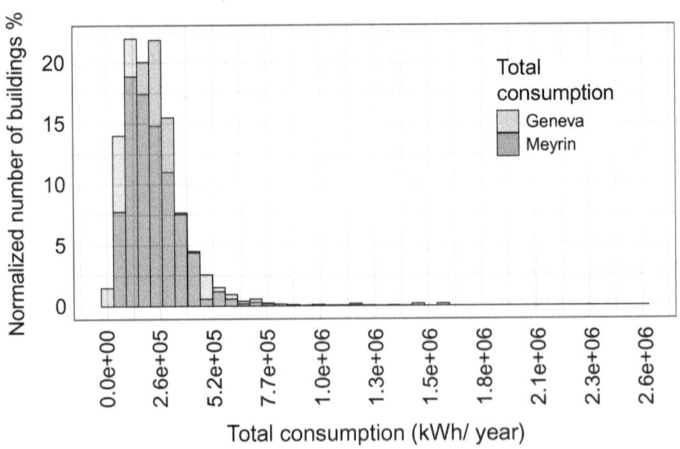

Fig. 10.72 Normalized distribution of the total consumption for the district and a large portion of the buildings in Geneva (Heating-DHW).

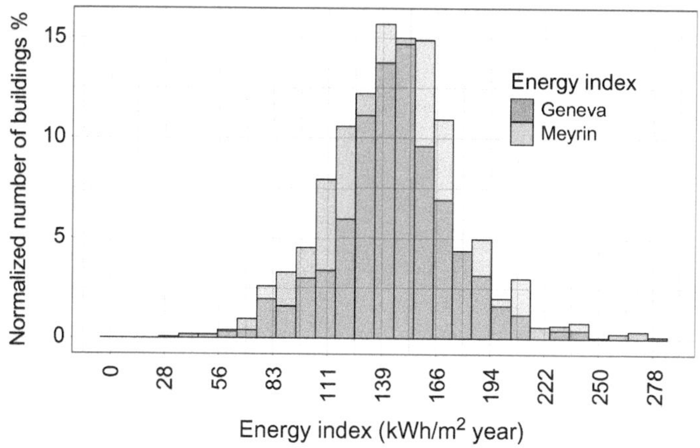

Fig. 10.73 Normalized distribution of the energy index for the district and a large portion of the buildings in Geneva (Heating-DHW).

10.6.2 AUSBEMA

An Automatic Urban Scale Building Energy Modeling Approach (AUSBEMA) has been developed starting from geometric data of the city case study. A 3D model of the city defined in the CityGML standard has been exploited for this purpose. The final objective of this step is to use the information relative to the geometry of the buildings and to connect it with other possible input data from other databases of the city. This enables the creation of input files for a dynamic building energy simulation software (EnergyPlus), which will be used to run dynamic simulation of the buildings in the district case study. The automatic approach has been created using a python script able to modify the input file template of the dynamic building simulation software. The following section explains the main steps for the creation of the automatic modeling approach (Fig. 10.74).

10.6.2.1 Parsing CityGML information

The 3D model of the city stores geometry information in a CityGML standard with an LOD2 level of details. This means that the model is able to represent complicate roof shapes and therefore to provide a more reliable indication of the heating and cooling volumes of the buildings with respect to simple footprint extrusion methods. The large CityGML file of the city is partitioned in different subfiles. Each subfile contains the 3D model information of an area of the city (called "COMMUNE"). The dataset "Scane.Commune" (SITG, n.d.) contains geo-referenced data for the majority of the buildings in the city. The location of a building, in terms of area membership, can be extracted using a building ID, named "EGID," and merging the information with the 3D model. This allows to parse the right portion of the file containing information for the selected urban area. Building data are stored as polygon information in

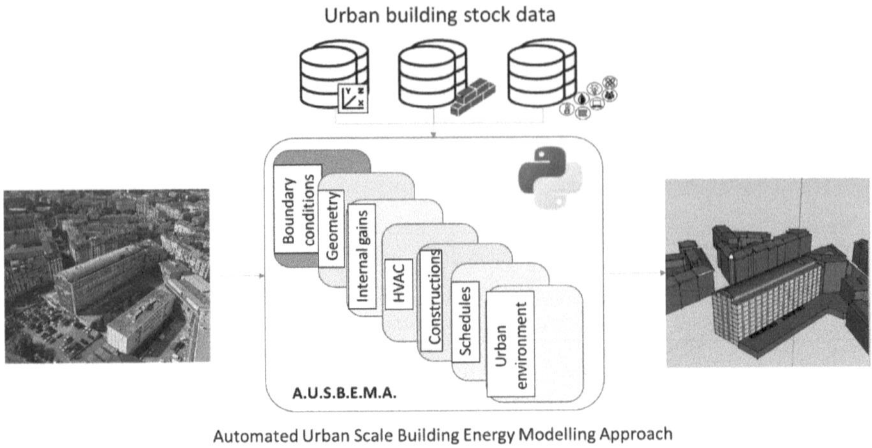

Fig. 10.74 Automated Urban Scale Building Energy Modeling Approach.

the CityGML description file; building geometry is described by roof, ground, and floor surfaces. Each polygon is characterized by various vertices. Therefore, surface information requires processing before it can be used as input file of the dynamic simulation software.

10.6.2.2 Surface processing

Building surfaces are classified in roof, ground, and exterior walls information. Each building is described as single geometric volumes and does not contain building floors geometric information. For this reason, some initial geometric calculations are required to define the (approximated) height of each floor and to determine its relative volume. Different building datasets (such as the one named "CAD.BATIMENT" (SITG, n.d.)) provide information of number of floors of each building and can be used to evaluate the relative height of each floor. Such data are then used to create partitions of the initial building volume and to enrich the CityGML surface information with ceiling and floor geometric polygons and vertices data. Nevertheless, for some buildings, the CityGML file contains information of unheated volumes, such as rooftop technical areas. For this reason some initial calculations are required to estimate the height of such spaces and curtail them from the overall building height in order to better estimate the height of each subvolume. The LOD2 CityGML file does not contain information of building openings such as doors or windows. However, windows may have a high impact on the overall heating and cooling requirement of a building. For this reason, an estimation of windows area is performed considering a windows to wall ratio. The ratio allows the evaluation of the overall windows area of the building. The automatic script allows to define sides and shape of the windows. With this information windows polygons are generated and imported directly in the polygons description file. On site surveys, digital panoramic views or surveys on representative buildings (such as the one identified in Chapter 2 by clustering techniques) can enable a better estimation of the windows to wall ratio.

10.6.2.3 Zoning and boundary conditions

Once polygons and vertices are defined, the simulation input file requires the definition of all the boundary conditions and the association of each surface to each thermal zone. First, all the surfaces are classified in glazing, floor or ceiling, roof, ground and external walls. Then zoning and boundary condition definition are performed in an automatic fashion by scripting. External walls, roof, and ground floors require definition of different type of boundary conditions. This can be performed easily in the scripting environment considering the position of each surface. Nevertheless, for the internal partitions such as ceiling-to-floor adjacencies and building-to-building adjacencies more sophisticated calculations are required to estimate relative orientations and contact areas between different surfaces. Building adjacencies are evaluated considering the relative position of the buildings and the surrounding area of the target building. The script allows to define automatically the boundary conditions. An example of building adjacency and complex boundary condition definition is shown in a SketchUp visualization of the building energy model in Fig. 10.75.

Thermal zones are defined considering the derived information of each floor, and each surface is linked to the relative thermal zone. Further work and a higher level of information are required to define internal partitions and to include a better division of each thermal zone. A possible future development could be the definition of core and perimeter zones following the standard ASHRAE 90.1 annex G (ASHRAE, 2010). In addition, information regarding the internal partitions of the building is required to generate accurate thermal zone taking into account unheated surfaces and volumes.

10.6.2.4 Urban environment characterization

Urban features such as adjacencies and shading scenarios play an important role in the estimation of building energy consumption. Together with many other modeling parameters, they are relevant in defining the heterogeneity of the building stock.

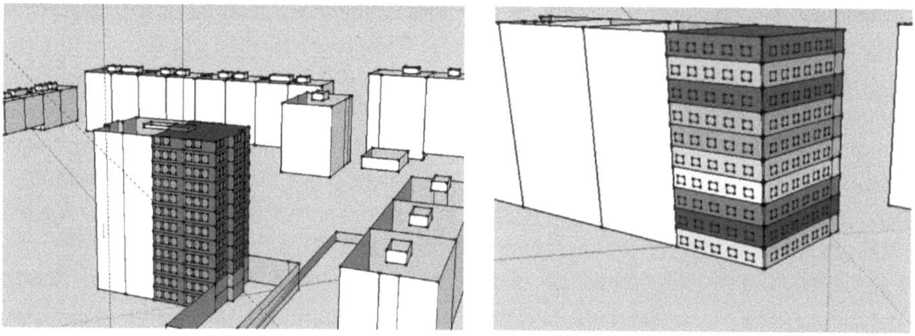

Fig. 10.75 AUSBEMA: surface processing and zoning.

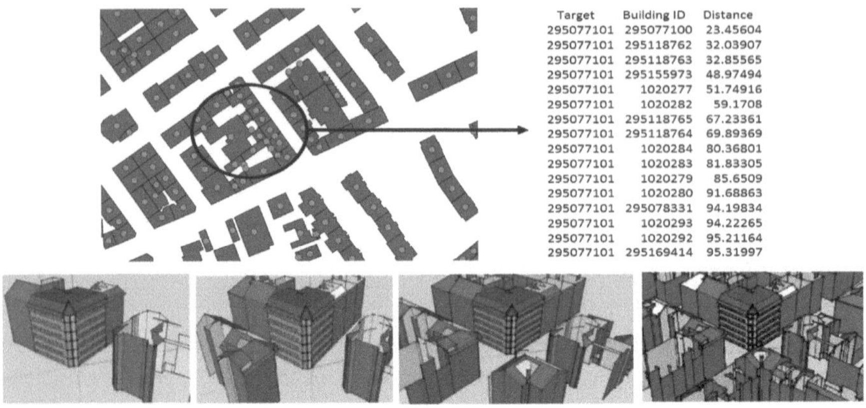

Fig. 10.76 AUSBEMA: shading objects.

For this reason, it is important to take into account building relative positioning in the modeling approach. This can be performed considering relative building distances in the definition of the simulation input file. The inclusion of surrounding buildings as shading objects of a selected target is performed in an automatic way. The estimation of the number of shading objects is defined considering first the geo-referenced position of each building, and then evaluating their distance from the target of the simulation. This allows the modeler to select the number of objects close to the target building as shading objects in the simulation input file. Future work may include the use of view factors, which can enable the estimation of the relative position of each building with respect to the sun daily pattern and, therefore, to consider only surfaces and buildings, which effectively cast a shadow on the target, reducing the size of the input file and consequently simulation and rendering time (Fig. 10.76).

10.6.2.5 Modeling input characterization

The definition of the geometry is only one of the many steps required to create the model. A number of inputs must be provided to the dynamic engine to perform the simulation. This task requires the establishment of links between information relative to the building stock and the target building. This is performed creating automatic selection mechanisms able to filter information in databases based on the building target characteristics. For the case study under consideration, the majority of the data have been gathered considering values available in the literature and from normative values. In some cases data have been enriched considering expert knowledge or engineering assumptions based on building typologies and periods of construction. A close collaboration with the city council energy department (DALE) allowed the definition of some of the modeling parameters such as constructions and windows U-values. Two main categorical variables have been used to sample data from building repositories and to enrich the simulation input file: building period of construction and building typology. In this way, libraries of input data have been generated for building construction, equipment use, lighting, ventilation, occupancy, heating systems use, DHW use, and building set point. In a similar way, internal gains definition has been performed defining a value in W/m^2 for lighting and equipment and average space allocation for a person in the building.

10.6.2.6 HVAC and DHW uses

An automatic approach is considered for the definition of HVAC and DHW systems. For each building a centralized supply system with definition of heating and DHW loops is considered. The demand side is modeled by definition of zone equipment lists and connections. The method employed is based on the use of HVAC templates extracted from the current EnergyPlus distribution. Scaling is performed defining connections and demand side equipment loops for each thermal zone and developing a script environment, which enables automation.

The definition of the DHW loop is characterized by three major parts: (i) the plant DHW loop, (ii) the water heater side, and (iii) the DHW side. The plant loop defines the characteristics of the system. The water heater side defines the supply side and its characteristics. The demand side defines the final uses and nodes connection must be defined for each end-use such as showers, sinks, etc. A scripting approach is used to define automatically the structure of the DHW system and the connections with the demand side with estimation of the number of end-users in each thermal zone. A general structure is defined for the supply and loop sections and a variable structure is used for the demand side. This allows to define the equipments for each thermal zone. Each of these parts requires the definition and interconnection of different objects to work together both for the supply (i.e., pipes, controls, boilers, tank, and pumps) and the demand connections (i.e., connectors, splitters, mixers, etc.). Tag names related to each thermal zone are used to define different objects in the input file. This specific notation allows the automatic interconnection of the parts of the systems. Schedules and equipment description are added to the equipment for each thermal zone.

An automatic definition of the HVAC system of a building is a very challenging task since it requires the definition of all the interconnections of the supply, demand, and loops, which characterize the systems. To partially overcome these difficulties a simpler method is used in the current version of the automated modeling approach. HVAC systems are defined starting from templates definition and enriching the input file with information of the demand side for each thermal zone of the building. The script can generate automatically the demand side part of the heating systems, while maintaining the general structure used in the Template definition. This allows to scale the system to the required number of zones. However, only a limited number of predefined systems are available. Future work may include the definition of other template systems or the study of a more complex automatic HVAC definition (Bres et al., 2017).

The weather file is selected based on the location of the target building. Various weather stations are available in the city case study and this allowed to select the closest dataset using geo-referenced data for the building target. The period of simulation have been set to yearly with an hour time resolution.

The automatic model has been used to generate building energy models for the entire district case study of Meyrin in Geneva.

10.6.3 Urban scale model calibration

Traditional engineering bottom-up modeling approaches rely on deterministic methods to model building energy consumption. This means that for a set of input values, defined for each simulation, the same output is generated by the simulation framework. The quality of the simulation inputs directly affects the reliability of the results. For this reason, deterministic simulation methods require an accurate definition of the modeling inputs.

Deterministic approaches should be adopted when uncertainties of the input modeling data are negligible. This could be achieved performing accurate surveys, gathering metered data, and exploiting data analysis techniques. In the absence of accurate data, other sources of information such as normative values, guideline recommendations, or educated guesses supported by expert knowledge are generally adopted to characterize building simulations. Nevertheless, the aforementioned source of modeling information may introduce large errors and uncertainties compromising the simulation results. In this context, calibration approaches, performed by fine tuning input parameters, attempt to reduce the gap between simulation and actual building consumption. Most of the time calibration methods are applied as result-driven or evidence-based methods. In these applications, the uncertain parameter (or set of parameters) strongly related with the target variable is manually tuned until there is a convergence of the results to a defined value or range of acceptance. Most of the times, calibration is a time-consuming process. It involves expert knowledge and a deep understanding of the energy model under analysis (Coakley et al., 2014).

While bottom-up engineering simulation software, such as EnergyPlus, IES-VE, INSEL, eQUEST, and ESP-r, could work with varying inputs or parameter distributions, most user only perform deterministic simulations.

It is challenging to extend single building calibration methods at a large scale such as at district or city level. Discrepancies between simulated and actual building consumption are likely to be large for large-scale building energy models such as at district and city scale for disaggregated case studies. This is due, among other reasons, to different levels of quality and quantity of data, which is possible to collect from the urban environment. Data at large scale may not be accessible due to privacy issues, economic interests, or simply because they are scarce and, therefore, missing. Other sources of uncertainties are related to the complexity of representing the building stock from a physical point of view due to its heterogeneity, to the high computational requirements for a simulated representation of a large number of buildings, to the complexity in representing interconnected urban physical phenomena and to the difficulties in providing scalable and replicable solutions. To overcome these major issues substantial research effort is required as well as disclosure of building data from utilities, energy departments, and building owners. On the other hand, the definition of novel cross-disciplinary modeling frameworks able to include the increasingly large

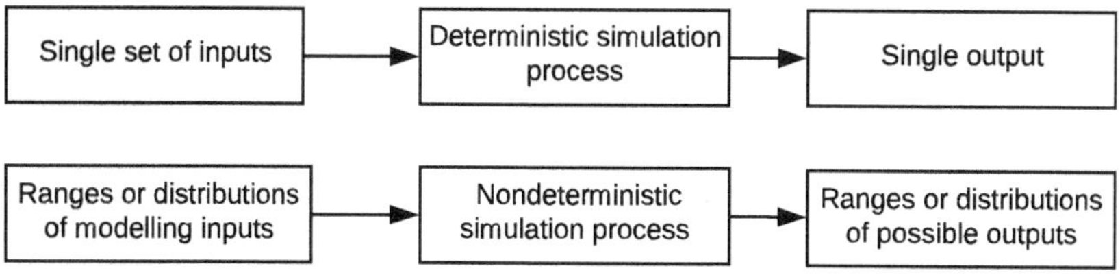

Fig. 10.77 Deterministic and nondeterministic simulation approaches.

amount of data available from buildings is paramount, as well as the definition of approaches able to handle uncertainties of the modeling process.

In this context, bottom-up stochastic engineering approaches may play a key role. Such methods are able to include uncertainties related to the case study, to the modeling approach adopted, and to the metered data. Stochastic building energy modeling overcomes the main limitation of deterministic approaches considering ranges of values for the definition of input parameters of a building energy model. This means that inputs are sampled from a distribution rather than considering a single value. Consequently, the output will be a range of values. Metered data may be included in the stochastic framework as source of information to derive the most likely distributions of the input parameters. Fig. 10.77 provides a simple explanation of the aforementioned concepts.

An example of a technique able to handle uncertainties of the modeling parameters is the Bayesian method. Bayesian inference is a statistical method, which uses the Bayes theorem to obtain a distribution of modeling parameters given observed data. The method is able to infer the most plausible ranges of values of different modeling parameters. The uncertain parameters are estimated in order to match simulations to metered data. Therefore, such a method evaluates the so-called "posterior distributions" of modeling parameters after estimating their probability given their prior distribution and their likelihood compared to metered data. Such an approach is able to directly incorporate expert knowledge of the energy modelers in the calibration process by defining prior belief in terms of distributions which the input parameters may have.

10.6.3.1 The Bayesian method

Bayesian calibration has been used during the past years to calibrate models in a variety of different research fields and to take intrinsically into account uncertainties related to modeling techniques (Kumar et al., 2012; Yamamoto et al., 2012; Van Oijen et al., 2013; Mebane et al., 2013; Hawkins-Daarud et al., 2013). Bayesian approaches are statistical methods, which can be used to derive probability distributions of sets of variables (Bishop, 2006). Bayesian calibration relies on the Bayes theorem reported in Eq. (10.30):

$$p(\theta|y) \propto p(\theta)p(y|\theta) \quad (10.30)$$

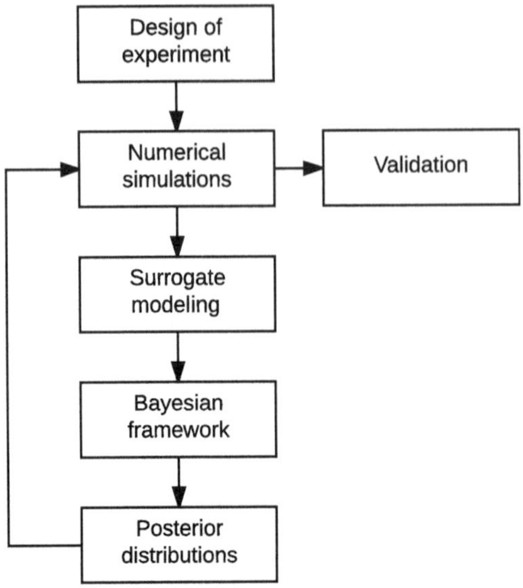

Fig. 10.78 Bayesian workflow.

where $p(\theta|y)$ is the posterior distribution of the modeling parameter, $p(\theta)$ is the prior distribution of the parameter, and $p(y|\theta)$ is the likelihood function, which provides information about how close model outputs are to metered data.

Generally, the Bayesian approach is implemented within a workflow, which allows to test the accuracy of the calibration method for a simulation approach. A general framework for the implementation of the Bayesian approach is described in Fig. 10.78.

A design of experiment is required to define parameters and their value ranges considered in the analysis (Bastos and O'Hagan, 2009). In this phase sensitivity analysis is employed to determine important parameters affecting the target variable of the modeling process (Saltelli et al., 2008). Numerical simulations in the form of parametric simulations are performed to explore sets of input parameters and collect outputs. Generally a limited number of parameter variations are defined as combination to generate simulations and produce model outputs.

In the case of time-consuming simulations, emulators can be used to reduce the amount of time required to simulate many variations of a baseline model. In this context, regression approaches (such as Gaussian process, linear or nonlinear techniques such as neural networks) can be used to replicate the behavior of the dynamic simulation software and, thus, shorten the time required to generate a large number of simulations. This is achieved by a training process of the regression method using the simulation outcomes. Therefore, in these applications emulators map the dynamic behavior of the time-consuming simulation process.

Following this, the Bayes rule is applied on the results of the modeling technique. In most of the cases it is not possible to evaluate analytically the posterior distribution of modeling parameters. For this reason, a sampling approach is used to generate the posterior distribution of each parameter. There are several methods to sample from the posterior distributions. Markov Chain Monte Carlo (MCMC) methods are one of the most popular. With respect to the sampling methods, random walk algorithms such as the Gibbs method and the Motropolis-Hastings algorithm are generally considered.

In the last step, validation is performed. Once the posterior distribution of the modeling parameters is available, parametric simulation is repeated for the set of parameters considered, this time sampling from the identified posterior distributions. Finally, results are compared to metered data to understand if the modeling workflow is able to generate reliable results.

10.6.3.2 Literature review

There are various examples in the literature of the use of Bayesian calibration for the building energy sector. The majority of the studies focus on single building level applications for model calibration and to handle uncertainty. In the last few years, there has been a clear intent of the research community in extending Bayesian modeling approaches at large scale such as at district or urban level.

A comprehensive review of Bayesian methods for the building energy sector can be found in Lim (2017). In summary, at single building level, applications focus on: calibration of unknown parameters, retrofit analysis, comparison with traditional calibration methods, study of simple emulators, influence of uncertainties in the input data, and metamodeling comparison.

In terms of scale of applications literature studies are quite diversified: Bayesian calibration has been used for single component level to entire cities (Chong et al., 2017). In terms of emulators, Gaussian processes are among the most adopted techniques (Heo et al., 2013, 2015; Kang and Krarti, 2016). Nevertheless, there are several studies, which try to employ more lightweight data-driven models to reduce the computational burdens of Gaussian process emulators (Zhao et al., 2016; Li et al., 2016). This has been attempted considering linear regression methods. However, there is a lack in applying advanced machine learning techniques such as neural networks and support vector machines especially at large scale where computational time is important. In terms of resolution, Bayesian calibration studies focus at yearly or monthly temporal granularity (Tian et al., 2016).

10.6.3.3 Methodology

The current research work aims to apply Bayesian calibration methods on building clusters of a large urban case study. In particular, the objective is to create a modeling simulation framework for district scale building energy models with the use of machine learning techniques acting as emulators of a dynamic simulation software. The automated urban scale building energy modeling approach presented in Section 10.6.2 will be used to generate hundreds of building energy models for a district case study. Particular attention will be taken in assessing the quality of the emulators using different accuracy measures. Fig. 10.79 presents the main steps of the research methodology.

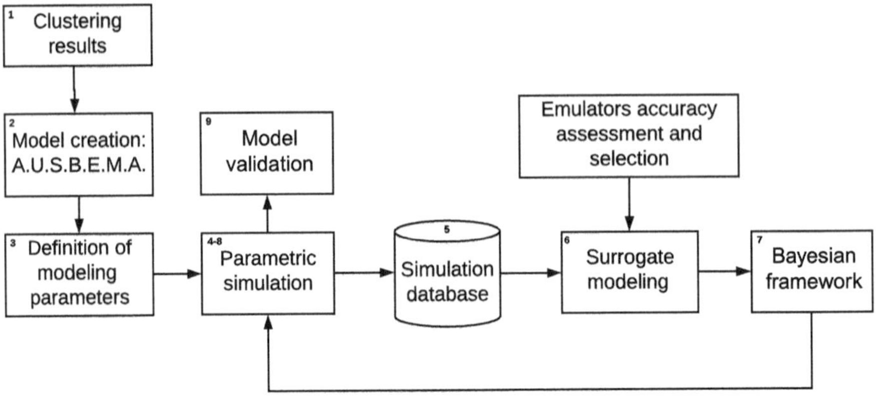

Fig. 10.79 Methodology overview.

10.6.3.3.1 Clustering results

Initially, results derived from the application of clustering algorithms on the city datasets are used to classify buildings in large clusters. The methodology and the results relative to building clustering are presented in Chapter 3.

10.6.3.3.2 AUSBEMA

Building classification information is forwarded to the large-scale modeling approach AUSBEMA for automatic creation of template input files for all the buildings in different clusters. The modeling approach AUSBEMA has been presented in Section 10.6.2.

10.6.3.3.3 Definition of modeling parameters

The identification of important parameters related to building energy consumption is performed considering previous works in the literature. Nevertheless, sensitivity analysis could be employed in this step. Once identified, input parameters are initially defined using normative values, previous studies found in the literature and data directly gathered from the building stock. Possible initial ranges of values for the modeling parameters are defined in this step. These values are generated considering expert knowledge and possible values for the typology of building under analysis. The selected parameters for this study are: internal gains related to people (m^2/person), lighting (W/m^2), and equipment (W/m^2); infiltration rates (ACH); heating set point (°C); DHW uses (m^3/s); and DHW use efficiency (%). Constant distributions are initially considered to define the initial range of variation of the selected uncertain inputs.

10.6.3.3.4 Parametric simulation

Parametric simulations are then employed to generate large sets of simulation results, which are collected in a database. In this research study the interface JEplus (Zhang and Korolija, 2010) is used for the parametric analysis. In addition, a scripting approach using Python is developed to create input file for the parametric simulator. The parametric tool generates all the possible combinations of the modeling parameters. Eq. (10.31) provides a mathematical description of the process:

$$Ns = N_b \cdot \prod_{k=1}^{K} n_k \tag{10.31}$$

where Ns is the number of simulations, K is the total number of variables, n is the number of values the variable k can have, and N_b is the number of buildings considered in the analysis. The number of simulations generated from the parametric analysis grows quickly considering various parameters, many variations of the parameters or a large number of buildings. For this reason, a Latin Hypercube Sampling (LHS) (Iman, 2008) approach is useful to limit the total number of outputs but at the same time to correctly explore the entire range of variability of the modeling parameters. This is particularly important to ensure a correct training of the emulators. Table 10.12 provides information on the selected parameters, their range of variation, and incremental value.

10.6.3.3.5 Dynamic simulation and simulation database

Dynamic simulations are performed using the input files modified by the parametric simulator. The parametric tool calls the simulation engine allowing tasks parallelization and optimizing the computational time. The results of thousands of simulations are stored in a simulation database.

Table 10.12 Definition of experiments: parameters, distributions, range of variations, and increments

Name	Units	Distribution	Range	Increment	Sampling
Lighting intensity	W/m^2	Uniform	∈IR: (1–10)	1	LHS
Occupancy density	m^2/person	Uniform	∈IR: (1–50)	10	LHS
Equipment intensity	W/m^2	Uniform	∈IR: (1–40)	10	LHS
Infiltrations	ACH	Uniform	∈IR: (0.1–1.5)	0.1	LHS
Heating set point temp.	C	Uniform	∈IR: (17–23)	1	LHS
DHW use	W/m^2	Uniform	∈IR: (0.5×10^{-6} to 1.5×10^{-5})	0.1×10^{-6}	LHS
DHW syst. efficiency	%	Uniform	∈IR: (0.7–0.95)	0.05	LHS

10.6.3.3.6 Emulators

Following this, emulators are trained on the simulation outcomes to map the dynamic models. It is important to create predictors with high accuracy in order to replicate the behavior of the dynamic simulation software in the best way possible. The quality of the emulators is very important and can bias the final results of the Bayesian framework. For this reason, it is crucial to employ emulators able to correctly map the dynamic behavior of the simulation engine. With reference to the literature review analysis presented in Section 10.6.3.2, the majority of the studies to date employ Gaussian process emulators or linear models. The first have been found suitable to produce high prediction accuracy, the latter have been adopted due to their "lightweight" properties and, therefore, their suitability to be included in high demanding computational frameworks. Nevertheless, Gaussian processes are highly time demanding, especially for large datasets and at high dimensionality, and linear models may be characterized by low prediction accuracy.

In this study one emulator for each cluster is developed as prediction method. Whole building features (i.e., building heating surface and building volume) are considered in the training process as explanatory predictors of the total building consumption.

Considering the aforementioned reasons, different families of emulators have been considered in this study. In particular neural network (NN), support vector machines (SVM), generalized linear models (GLM), and random forests (RNF) have been used as emulators. In the following section results of the prediction accuracy achieved by each algorithm in each cluster are presented.

The overall dataset of simulations is split in two chunks: 70% for model training and 30% for model testing. Tenfold cross-validation is employed for automatic tuning of the algorithmic parameters. Final prediction accuracy is evaluated on the testing sample considering three different quality measures: root mean square error (RMSE), mean absolute percentage error (MAPE), and coefficient of determination (R^2):

$$MAPE = \frac{1}{N}\sum_{i=1}^{N}\left|\frac{y_i - p_i}{y_i}\right| \cdot 100 \tag{10.32}$$

where MAPE is the mean absolute error, N is the total number of predictions, y is the actual value of measurement i, and p is its predicted value;

$$RMSE = \sqrt{\frac{1}{N}\sum_{i=1}^{N}(y_i - p_i)^2} \tag{10.33}$$

where RMSE is the root mean squared error, N is the total number of predictions, y is the actual value of measurement i, and p is its predicted value;

$$R^2 \equiv 1 - \frac{SS_{res}}{SS_{tot}} = 1 - \frac{\sum_{i=1}^{N}(y_i - p_i)^2}{\sum_{i=1}^{N}(y_i - \bar{y})^2} \tag{10.34}$$

where SS_{res} is the residual sum of squares and SS_{tot} is the total sum of squares. N is the total number of predictions, y is the actual value of measurement i, and p is its predicted value. $\bar{y}$ is the mean value. The training process has been repeated across all the clusters and the accuracy of the emulators has been evaluated.

10.6.3.3.7 The Bayesian framework

Emulators are directly used in the Bayesian framework to further explore the parameter distribution and relationship with the target variable. The Bayesian framework uses metered data to infer the most likely distribution of the modeling parameters.

Bayesian calibration allows for a comprehensive evaluation of measurement errors, modeling input uncertainties and model discrepancies. In this study a variation of the traditional Kennedy and O'Hagan (2001) framework is used. The main change to the traditional use of the framework is to consider different families of emulators rather than a Gaussian process as suggested by Li et al. (2016). This requires the introduction of a residual error due to the inaccuracy of the regression approach.

Measurements of a physical process can be analytically described as Higdon et al. (2004):

$$y(x_i) = \zeta(x_i) + \epsilon_m(x_i)\; i = 1, \ldots, n \tag{10.35}$$

where $y(x_i)$ indicates the value of the observation, x_i indicates the set of independent variables considered in the analysis, $\zeta(x_i)$ the actual physical system, and $\epsilon_m(x_i)$ the measurement error. Eq. (10.35) can be written as:

$$y(x_i) = \eta(x_i, \theta) + \delta(x_i) + \epsilon_m(x_i)\; i = 1, \ldots, n \tag{10.36}$$

where the physical system has been substituted by the outcome of a simulation method (dynamic or not) $\eta(x_i, \theta)$ for variables x_i and simulation parameters θ. $\delta(x_i)$ indicates the model inaccuracies or discrepancies between the actual physical process and the model. Model discrepancies are assumed to be dependent only to model inputs. Random observation error is considered to follow a Gaussian distribution $\epsilon_m(x_i) \sim N(0, \sigma_m^2)$. In case that the dynamic simulation process is time consuming and a large number of simulations are required in the analysis, an alternative way is to use regression-based models, called in this application emulators, or meta-model, to estimate the output of the dynamic simulation approach. Considering emulators, Eq. (10.36) is expanded in:

$$y(x_i) = f(x_i, \theta) + \delta(x_i) + \epsilon_m(x_i)\; i = 1, \ldots, n \tag{10.37}$$

Traditionally, emulators and model discrepancies are modeled with a Gaussian process defined as follows:

$$f(x_i, \theta) = N(\mu, \Sigma),\; \delta(x_i) = N(\mu_\delta, \Sigma_\delta) \tag{10.38}$$

The Gaussian process is used as regression algorithm in place of the time-consuming simulation model. This family of predictive methods offers the advantage of providing a perfect fit on the training dataset. For this reason there are not residuals when quality is assessed on the training dataset. In turn, no regression error is added to the general framework. However, training of GP emulators is computationally and time demanding. For this reason, when the scale of simulation is large in terms of dataset, in terms of number of buildings, or in terms of simulations to perform, other "lightweight" emulators may be introduced. Nonetheless, regression emulators different from GP generally do not offer a perfect fit on the training dataset, therefore a residual error is added to the general equation describing the framework defined as follows:

$$y(x_i) = f(x_i, \theta) + \epsilon_r(x_i) + \delta(x_i) + \epsilon_m(x_i)\ i = 1, \ldots, n \tag{10.39}$$

where $f(x_i, \theta)$ is a regression function of variables x_i and modeling parameters θ. $\epsilon_r(x_i)$ is the regression error introduced to take into account the limits of the regression approach. The regression error is considered to follow a Gaussian distribution. This requires verification through inspection of the residuals: $\epsilon_r(x_i) \sim N(0, \sigma_r^2)$ where σ_r^2 indicates the variance of the residuals of the predictive approach. The model discrepancy $\delta(x_i)$ is modeled as an unbiased Gaussian distribution $\mu_\delta = 0$ and covariance matrix $\Sigma_\delta = C(x, x)$ evaluated using a squared exponential kernel:

$$c_{ij} = \exp\left(-\frac{1}{\lambda}\sum_s \omega_s (x_{is} - x_{js})^2\right) \tag{10.40}$$

The matrix and its elements are defined once the hyperparameters ($\phi = (\lambda, \omega)$), λ (length parameter), and ω (precision) are evaluated for all the set of independent variables.

The observations are assumed to form a joint Gaussian distribution. Considering a matrix notation, it follows:

$$y \sim N(\mu, \Sigma),\ \mu = f(x, \theta),\ \Sigma = \Sigma_\delta + (\sigma_m^2 + \sigma_l^2)I \tag{10.41}$$

where μ is the mean value of the regression function and Σ is the total covariance matrix.

With the adoption of this model, it is possible to apply the Bayes rule, which states that the posterior distribution for design parameters θ and hyperparameters ϕ, conditioned on the data y is defined as:

$$P(\theta, \phi | y) \simeq p(\theta, \phi) \cdot L(y | \theta, \phi) \tag{10.42}$$

The prior probabilities p are defined with distribution of the modeling parameters and hyperparameters. The likelihood function is defined as:

$$L(y|\theta, \phi) = |\Sigma|^{1/2} \exp\left(-\frac{1}{2}\left[(y - \mu)^T \Sigma^{-1} (y - \mu)\right]\right) \tag{10.43}$$

An analytic solution of the posterior distribution is possible only in trivial cases, in most of the analysis a sampling approach is required to obtain the final values of the distributions. The posterior distribution then will indicate the most likely distributions of the modeling parameters.

10.6.3.3.8 Markov Chain Monte Carlo sampling

It is impractical to derive an analytical expression of the posterior distribution of the modeling parameters especially in case of high dimensionality of the modeling input and of the hyperparameters (Gilks et al., 1995). Nevertheless, it is possible to create a group of "realizations," which form an estimate of the analytic distributions of the parameters. This is achieved constructing a chain of samples using an MCMC technique whose convergence distribution is the posterior distributions of the parameters. In this study, the Metropolis-Hasting algorithm is used as random walk technique to derive the posterior distributions (Metropolis et al., 1953). The algorithm generates a random walk of the parameter space. At each iteration, the algorithm computes the probability of a sample to be accepted as new candidate in the random walk. This is done by evaluating the ratio of a distribution proportional to the target probability distribution of the previous step and the suggested new point (Chib and Greenberg, 2007). The ratio of the probabilities of the two points is compared with a random number generated from a uniform distribution ($r_n \in (0, 1)$). The new point is accepted (acceptance criterion) only if the ratio of the probabilities is greater than the random number (r_n). The algorithm requires an initial phase to start generating samples from the actual posterior distributions. For this reason a number of samples should be initially discarded because not representative of the distribution ("burn in samples").

10.6.3.3.9 Parametric simulation

Once the posterior distribution of the parameters has been identified, it is used in the parametric simulation framework to validate the model by comparing the final results of the modeling approach with the metered data. A scripting approach has been developed to sample a finite number of point from the posterior distribution of the parameters and to generate the parametric input file.

10.6.3.3.10 Validation

For each building, the outcomes of the parametric simulation are stored in a database. All the simulation outcomes for a particular building are averaged. The final value is stored for comparison. Outcomes are visualized and compared to the metered data in order to assess the reliability of the model.

10.6.3.4 Results

The methods described in the previous sections are applied on the district case study of Meyrin described in Section 10.6.1. In particular, the residential buildings are selected to test the steps of the methodology being one of the largest typology within the district boundaries.

10.6.3.4.1 Quality of the emulators

Quality of the emulators is evaluated on the test dataset as explained in Section 10.6.3.3.6 using different accuracy measures for each building cluster.

Fig. 10.80 shows the results of the emulators on the testing dataset. Results are grouped for each building cluster.

In terms of overall accuracy for the building clusters considered in the analysis the emulators perform quite well. Nevertheless, it is possible to identify strong differences in the accuracy of different types of emulators. In particular, the linear models are the one which in general perform worst in the majority of the case studies. Neural networks perform quite well with small MAPE and RMSE. On average the models show good coefficient of determination. For the majority of the emulators MAPE is between 5% and 10% and coefficient of determination on average equal to 0.98. The RMSE is on average in the range 10,000–20,000 kWh.

A further step is to visualize the result of the prediction on a graph to compare simulation outputs and predictions to appreciate the accuracy of the emulators over the entire spectrum of results of the simulations.

Fig. 10.81 shows the results of the four emulators for a specific cluster providing a comparison between the simulated and predicted values. From the graph it is possible to appreciate the quality at different levels of consumption. It is evident that the four methods perform quite well. Nevertheless, the generalized linear model has bad

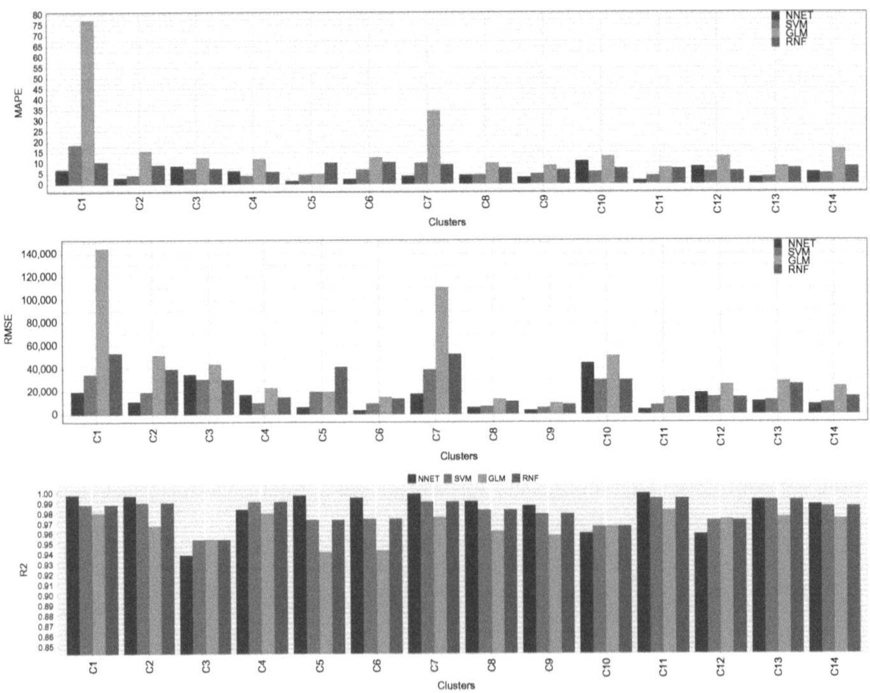

Fig. 10.80 RMSE, MAPE, and R^2 of various emulators in each building cluster.

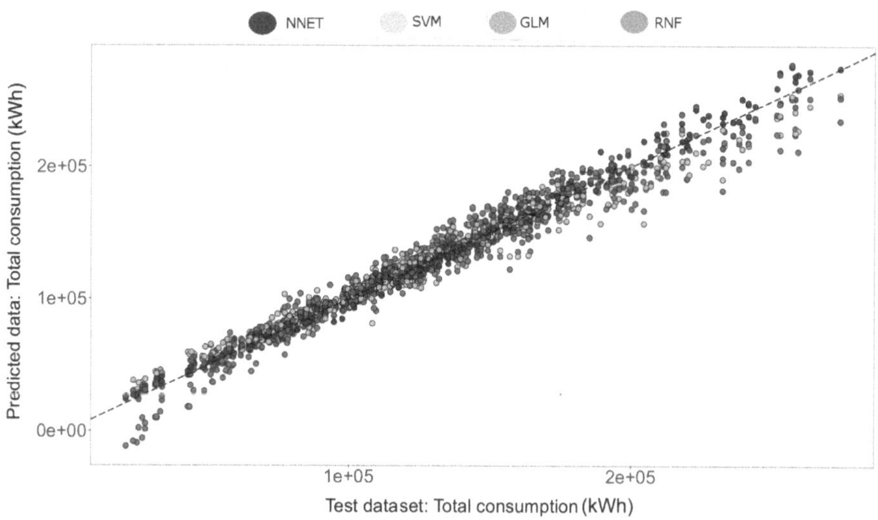

Fig. 10.81 Simulated versus predicted consumption.

performances for small and high consumption values, underestimating and over-estimating the actual consumption. The other three methods perform well with neural networks, showing the best performances over the entire range of consumption.

10.6.3.4.2 Selection of the best emulators

The analysis of the results of the accuracy measures allows to select the best performing models in each cluster. Results of this operation are shown in Table 10.13.

From Table 10.13 it is evident that NNET and SVM are the best performing methods. Results range between 1.5% and 6.8% in terms of MAPE, between 3368

Table 10.13 Selected emulators for each cluster

Cluster	**1**	**2**	**3**	**4**	**5**	**6**	**7**
Emulator	NNET	NNET	SVM	SVM	NNET	NNET	NNET
MAPE	6.8	3	7.2	4	1.5	2.4	3.8
RMSE	19,907	10,854	30,701	9570	5899	3368	17,509
R^2	0.997	0.996	0.953	0.99	0.997	0.995	0.999
Cluster	**8**	**9**	**10**	**11**	**12**	**13**	**14**
Emulator	NNET	NNET	SVM	NNET	SVM	NNET	SVM
MAPE	4.2	3.2	5.8	1.6	5.9	3.1	4.9
RMSE	5662	3485	29,704	3964	14,975	10,789	9978
R^2	0.991	0.987	0.966	0.999	0.972	0.993	0.986

and 30,701 kWh in terms of RMSE, and between 0.95 and 0.999 in terms of R^2. Interestingly, a closer look at the results shows that in general for high values of the coefficient of determination there are not always high values of the other accuracy measures, for this reason it is advised to not use the coefficient of determination as only quality index to assess the quality of the emulators.

10.6.3.4.3 Posterior distribution

The selected emulators are included in the Bayesian framework. Their use allows to further explore the overall range of consumption of the buildings in each cluster. For this reason, it is important to be particularly careful in the initial definition of the range of variability of the consumption defined by the parametric simulation. The emulators show the evaluated quality only in the defined range of variability of the consumption, but there is not evidence that they can provide accurate prediction outside of these ranges.

The 15,000 MCMC simulations are run for each building cluster to generate samples from the posterior distributions of the modeling parameters considering a 25% of initial samples as "burn-in" and therefore discarded. In terms of calculation performances the nonlinear emulators performed quite well although the neural networks required an higher computational time at prediction phase than the support vector machines. The final result of the application of the Bayes rule is a set of posterior distributions of the modeling parameters characterizing each group of buildings of each cluster. An example of posterior distribution for the investigated modeling parameters is shown in Fig. 10.82.

Overall, the calibration approach is able to create well-defined distributions of the parameters in different clusters providing useful information to the energy modeler.

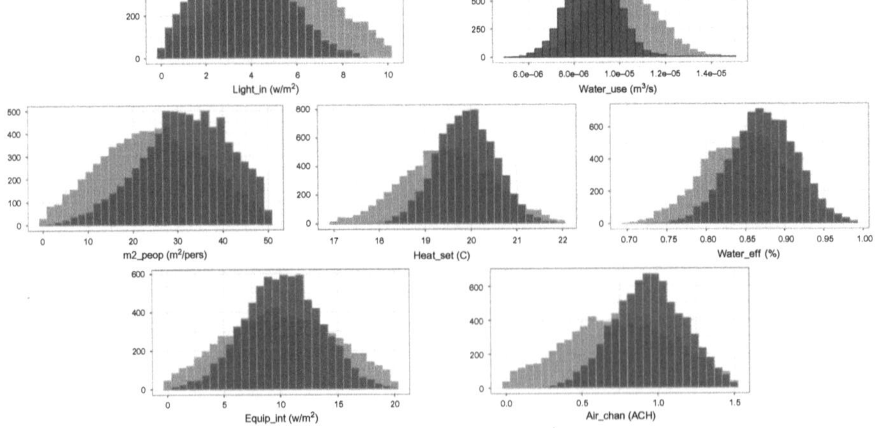

Fig. 10.82 Posterior distribution of modeling parameters.

10.6.3.4.4 Validation

Result validation is performed considering the posterior distributions of each cluster. This involves repeating the parametric simulation with the values of the modeling parameters sampled from the posterior distributions. This step is performed using a scripting approach able to provide input to the parametric simulator in terms of parameters combinations, randomly sampling from the posterior distribution of the modeling parameters. A number of simulations are set for each building and results are averaged over the total number of simulation results. The results are visualized with comparison with the metered data of the district case study.

In terms of accuracy, with reference to Fig. 10.83 it is evident that the calibrated large-scale modeling approach is able to correctly reproduce the distribution of the district and the clusters of buildings under analysis. In terms of total consumption, the model is able to cover the overall range of distributions from the smallest energy users to outlier buildings with large overall consumption.

Considering the energy index it is possible to appreciate similar results. The Bayesian calibration approach is able to tune the modeling parameters and match quite correctly the distribution of the energy intensities of the district (Fig. 10.84).

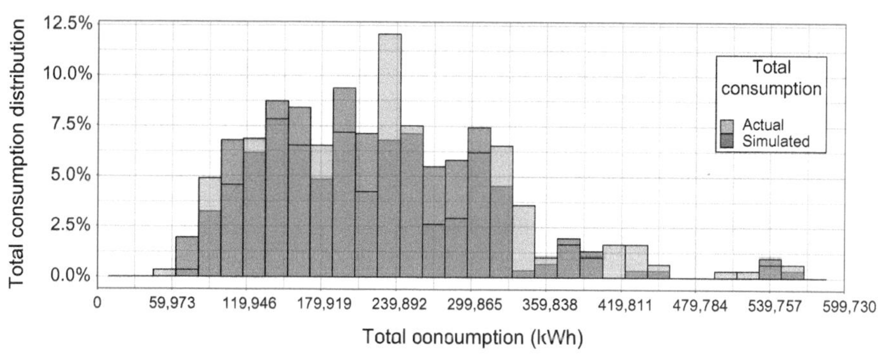

Fig. 10.83 Comparison simulated and metered total consumption.

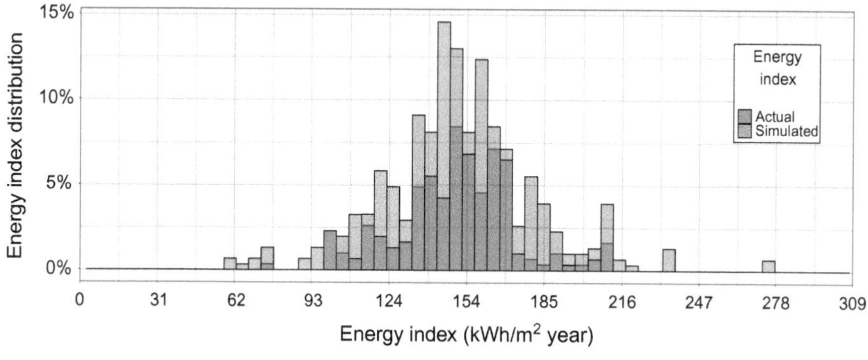

Fig. 10.84 Comparison simulated and metered energy indices.

10.6.3.5 Conclusions

This section has shown the potential of merging a large-scale modeling approach, results of clustering algorithms applied on urban datasets, a nondeterministic calibration method, and the use of advanced data-driven predictive techniques. The framework has been created to address large-scale challenges, and therefore to provide a useful and at the same time accessible decision support tool for energy planners and urban decision makers. The inclusion of building classification driven by clustering approaches, a dynamic simulation engine, and nonlinear predictive algorithms constitutes a source of innovation for the traditional Bayesian calibration framework. The development of an automatic modeling approach (AUSBEMA) allows to generate complex scenarios and to take into account the heterogeneity of the urban environment reducing sources of uncertainties of prototypical "stand-alone" models.

The application of Bayesian methods at large scale is a time-consuming process, which requires thousands of simulations. Emulators can shorten time requirements when included in the framework. In this study nonlinear models such as neural networks and support vector machines have shown a higher level of accuracy compared to traditional linear algorithms with acceptable computational time requirements. With reference to the existing literature the current research work has shown the importance of accuracy assessment on the testing dataset. In this context it is fundamental to deliver emulators with high accuracy, estimated with different quality measures and visualization tools.

Model calibration has shown the possibility to calibrate clusters of buildings considering common modeling parameters. The validation phase shows high quality of the results with the possibility of matching the distributions of the district in terms of total consumption and energy intensity.

Nevertheless, the current approach has some limitations. In this context, it is important to continue the investigation of automatic approaches for model creation including complex HVAC scenarios, control systems, as well as adopting advanced solutions for zoning and more realistic solutions to take into account stochastic occupant behavior. The quality of the modeling solution is directly affected by the available data, hence disclosure of large building stock datasets at different levels of granularity is a key to generate reliable energy models. Although the model shows good performance for model calibration at annual level of resolution, it is important to consider alternative approaches and analysis at a finer time resolution. In this context the development of automatism in model calibration at finer temporal resolution may be a key to the problem. For this reason, results shown in Chapter 3 for the creation of emulators of dynamic building simulation engines may be relevant.

In the current application, once calibrated the energy models could be used to assess the effect of different Energy Conservation Measures at district scale thereby providing decision support for energy planning. The disaggregated nature of the modeling approach allows to scale the results at different level of spatial granularity starting from block of buildings to clusters, to entire districts, and possibly to the entire city. In conclusion, this section showed the importance of integrating data-driven modeling techniques in traditional physics-based simulation workflows for large-scale applications.

In this context, the use of predictive modeling approaches can enhance the quality of the results of traditional simulation methods, without compromising their flexibility for scenario evaluation and innovative technologies assessment.

References

ASHRAE, 2010. Energy standard for buildings except low-rise residential buildings. ASHRAE Standard 90.1-2010.

Bastos, L.S., O'Hagan, A., 2009. Diagnostics for Gaussian process emulators. Technometrics 51 (4), 425–438.

Bishop, C.M., 2006. Pattern Recognition and Machine Learning. vol. 4, ISBN: 9780387310732, p. 738. https://doi.org/10.1117/1.2819119. Available from: http://www.library.wisc.edu/selectedtocs/bg0137.pdf.

Brand, L., Calvén, A., Englund, J., Landersjö, H., Lauenburg, P., 2014. Smart district heating networks—a simulation study of prosumers' impact on technical parameters in distribution networks. Appl. Energy 129, 39–48.

Bres, A., Judex, F., Suter, G., de Wilde, P., 2017. A Method for Automated Generation of HVAC Distribution Subsystems for Building Performance Simulation. IBPSA, San Francisco, CA.

Broberg, S., Backlund, S., Karlsson, M., Thollander, P., 2012. Industrial excess heat deliveries to Swedish district heating networks: drop it like it's hot. Energy Policy 51, 332–339.

Chib, S., Greenberg, E., 2007. Understanding the metropolis-Hastings algorithm. Am. Stat. 49 (4), 327–335.

Chong, A., Lam, K.P., Pozzi, M., Yang, J., 2017. Bayesian calibration of building energy models with large datasets. Energy Build. 154, 343–355.

Coakley, D., Raftery, P., Keane, M., 2014. A review of methods to match building energy simulation models to measured data. Renew. Sust. Energ. Rev. 37, 123–141.

Di Pietra, B., Zanghirella, F., Puglisi, G., 2015. An evaluation of distributed solar thermal "net metering" in small-scale district heating systems. Energy Procedia 78, 1859–1864.

FERNWÄRME WIEN GmbH, 2013. Technical Documentation TR-TAB. https://www.wienenergie.at/media/files/2015/technische%20richtlinie%20tr-tab%20technische%20auslegungsbedingungen_140557.pdf, abgerufen (accessed on August 4, 2018).

Gilks, W.R., Richardson, S., Spiegelhalter, D., 1995. Markov Chain Monte Carlo in Practice. CRC Press, Boca Raton, FL.

Hassine, I.B., Eicker, U., 2014. Control aspects of decentralized solar thermal integration into district heating networks. Energy Procedia 48, 1055–1064.

Hawkins-Daarud, A., Prudhomme, S., van der Zee, K.G., Oden, J.T., 2013. Bayesian calibration, validation, and uncertainty quantification of diffuse interface models of tumor growth. J. Math. Biol. 67 (6–7), 1457–1485.

Heikkinen, J., Rämä, M., Klobut, K., Laitinen, A., 2014. Solar thermal integration into a district heated small house. In: Proceedings of the 14th International Symposium on District Heating and Cooling, Stockholm, Sweden, pp. 530–533.

Heo, Y., Augenbroe, G., Choudhary, R., 2013. Quantitative risk management for energy retrofit projects. J. Build. Perform. Simul. 6 (4), 257–268.

Heo, Y., Graziano, D.J., Guzowski, L., Muehleisen, R.T., 2015. Evaluation of calibration efficacy under different levels of uncertainty. J. Build. Perform. Simul. 8 (3), 135–144.

Higdon, D., Kennedy, M., Cavendish, J.C., Cafeo, J.A., Ryne, R.D., 2004. Combining field data and computer simulations for calibration and prediction. SIAM J. Sci. Comput. 26 (2), 448–466.

Höckner, E., 2014. Wien Energie—district heating. Presentation for the 2nd Training Module, CI-NERGY European Project, October, Vienna.

Iman, R.L., 2008. Latin Hypercube Sampling. In: Melnick, E.L., Everitt, B.S. (Eds.), Encyclopedia of Quantitative Risk Analysis and Assessment. https://doi.org/10.1002/9780470061596.risk0299.

Kang, Y., Krarti, M., 2016. Bayesian-emulator based parameter identification for calibrating energy models for existing buildings. Build. Simul. 1996-3599. 9 (4), 411–428. https://doi.org/10.1007/s12273-016-0291-6.

Kennedy, M.C., O'Hagan, A., 2001. Bayesian calibration of computer models. J. R. Stat. Soc. Ser. B (Stat Methodol.) 63 (3), 425–464. https://doi.org/10.1111/1467-9868.00294.

Kumar, N.C., Subramaniyan, A.K., Wang, L., 2012. Improving high-dimensional physics models through Bayesian calibration with uncertain data. In: Proceedings of ASME Turbo Expo.

Leoncini, L., 2013. The primary energy factors play a central role in European 2020 targets achievement. In: International Conference Portugal SB13-contribution of sustainable building to meet EU 20-20-20 targets, pp. 113–120.

Li, Q., Augenbroe, G., Brown, J., 2016. Assessment of linear emulators in lightweight Bayesian calibration of dynamic building energy models for parameter estimation and performance prediction. Energy Build. 03787788. 124, 194–202. https://doi.org/10.1016/j.enbuild.2016.04.025.

Lim, H., 2017. Review on stochastic modeling methods for building stock energy prediction. Build. Simul. 10, 607–624. https://doi.org/10.1007/s12273-017-0383-y.

Luther, J., Schumacher-Gröhn, J., 1991. INSEL—a simulation system for renewable electrical energy supply systems. Tenth EC Photovoltaic Solar Energy Conference, pp. 457–460.

Mebane, D.S., Bhat, K.S., Kress, J.D., Fauth, D.J., Gray, M.L., Lee, A., Miller, D.C., 2013. Bayesian calibration of thermodynamic models for the uptake of CO_2 in supported amine sorbents using ab initio priors. Phys. Chem. Chem. Phys. 15 (12), 4355–4366.

Metropolis, N., Rosenbluth, A.W., Rosenbluth, M.N., Teller, A.H., Teller, E., 1953. Equation of state calculations by fast computing machines. J. Chem. Phys. 21 (6), 1087–1092.

Monsalvete Álvarez de Uribarri, P., Eicker, U., Robinson, D., 2017. Energy performance of decentralized solar thermal feed-in to district heating networks. Energy Procedia 116, 285–296.

Nielsen, S., Möller, B., 2012. Excess heat production of future net zero energy buildings within district heating areas in Denmark. Energy 48 (1), 23–31.

ÖNORM B 5019, 2011. Hygienerelevante Planung, Ausführung, Betrieb, Überwachung und Sanierung von zentralen Trinkwasser-Erwärmungsanlagen.

ÖNORM EN 806, 2011. Technische Regeln für Trinkwasser Installationen.

Persson, U., Möller, B., Werner, S., 2014. Heat Roadmap Europe: Identifying strategic heat synergy regions. Energy Policy 74, 663–681.

Recknagel, H., Sprenger, E., Schramek, E.R., 2006. Taschenbuch für Heizung und Klimatechnik 2005/2006: Einschließlich Warmwassererzeugung und Kältetechnik. Oldenbourg. ISBN-10: 3486265601.

Romero Rodríguez, L., Duminil, E., Sánchez Ramos, J., Eicker, U., 2017. Assessment of the photovoltaic potential at urban level based on 3D city models: a case study and new methodological approach. Sol. Energy 146, 264–275.

Saltelli, A., Ratto, M., Andres, T., Campolongo, F., Cariboni, J., Gatelli, D., Saisana, M., Tarantola, S., 2008. Global Sensitivity Analysis: The Primer. John Wiley & Sons, New York, NY.
Schmidt, E., 1982. Properties of Water and Steam in SI Units. Springer-Verlag, Berlin.
SITG: Le système d'information du territoire à genève, http://ge.ch/sitg/, n.d., Accessed 1 March 2017.
Tian, W., Yang, S., Li, Z., Wei, S., Pan, W., Liu, Y., 2016. Identifying informative energy data in Bayesian calibration of building energy models. Energy Build. 03787788. 119, 363–376. https://doi.org/10.1016/j.enbuild.2016.03.042.
Van Oijen, M., Reyer, C., Bohn, F., Cameron, D., Deckmyn, G., Flechsig, M., Härkönen, S., Hartig, F., Huth, A., Kiviste, A., et al., 2013. Bayesian calibration, comparison and averaging of six forest models, using data from Scots pine stands across Europe. For. Ecol. Manag. 289, 255–268.
Vesaoja, E., Nikula, H., Sierla, S., Karhela, T., Flikkema, P.G., Yang, C.-W., 2014. Hybrid modeling and co-simulation of district heating systems with distributed energy resources. In: 2014 Workshop on Modeling and Simulation of Cyber-Physical Energy Systems (MSCPES), pp. 1–6.
Yamamoto, S., Miyazaki, M., Iwano, T., Kitazawa, S., 2012. Bayesian calibration of simultaneity in audiovisual temporal order judgments. PLoS ONE 7 (7), e40379.
Zhang, Y., Korolija, I., 2010. Performing complex parametric simulations with jEPlus. 9th SET Conference Proceedings, Shanghai, China.
Zhao, F., Lee, S.H., Augenbroe, G., 2016. Reconstructing building stock to replicate energy consumption data. Energy Build. 03787788. 117, 301–312. https://doi.org/10.1016/j.enbuild.2015.10.001. http://www.sciencedirect.com.ucd.idm.oclc.org/science/article/pii/S0378778815303157?via%3Dihub.

Further reading

Akkaya, B., Romanchenko, D., 2013. Modeling and Analysis of a District Heating System Containing Thermal Storage. Case Study of the District Heating System of Boras. Chalmers University of Technology, Gothenburg, pp. 9–51.
Baker, D.K., Sherif, S.A., 1997. Heat transfer optimization of a district heating system using search methods. Int. J. Energy Res. 21, 233–252.
Çomakli, K., Yuksel, B., Çomakli, O., 2004. Evaluation of energy and exergy losses in district heating network. Appl. Therm. Eng. 24, 1009–1017.
FERNWÄRME WIEN GmbH, 2009. Technical documentation TR-SZT. https://www.wienenergie.at/media/files/2015/technische%20richtlinie%20tr-szt%20schemen,%20zeichnungen,%20tabellen_140556.pdf, abgerufen.
Gudmundsson, O., Brand, M., Thorsen, J.E., 2014. Ultra-low temperature district heating and micro heat pump application. In: The 14th International Symposium on District Heating and Cooling, Danfoss A/S, District Energy Division, Stockholm.
Hoffmann, K., Pearson, D.F., 2011. Ammonia heat pumps for district heating in Norway—a case study. In: To be presented before the Institute of Refrigeration at London Chamber of Commerce and Industry, London.
Kofinger, M., Basciotti, D., Schmidt, R.R., Meissner, E., Doczekal, C., Giovannini, A., 2016. Low temperature district heating in Austria: energetic, ecologic and economic comparison of four case studies. Energy 110, 95–104.

Kotas, T.J., 1995. The Exergy Method of Thermal Plant Analysis. Krieger Publishing Company, Malabar. pp. 113–137, 197–204.

Lauka, D., Julija Gusca, J., Blumberga, D., 2015. Heat pumps integration trends in district heating networks of the Baltic states. In: 5th International Conference on Sustainable Energy Information Technology.

Lindenberger, D., Bruckner, T., Groscurth, H.-M., Kummel, R., 2000. Optimization of solar district heating systems: seasonal stage, heat pumps and cogeneration. Energy 25, 591–608.

Lund, H., Werner, S., Wiltshire, R., Svendsen, S., Thorsen, J.E., 2014. 4th generation district heating (4GDH) integrating smart thermal grids into future sustainable energy systems. Energy 68, 1–11.

Marx, R., Bauer, D., Drueck, H., 2014. Energy efficiency integration of heat pumps into solar district heating system with seasonal thermal energy storage. Energy Procedia 57, 2706, 2715.

Ommen, T., Markussen, W.B., Elmegaard, B., 2014. Heat pumps in combined heat and power systems. Energy 76, 989–1000.

Rämä, M., Heikkinen, J., Klobut, K., Laitinen, A., 2014. Network simulation of low heat demand residential area. In: Proceedings of The 14th International Symposium on District Heating and Cooling, Stockholm, p. 4.

Schmidt, D., 2009. Low exergy systems for high performance buildings and communities. Energy Build. 41, 331–336.

The Danish Energy Agency, 2010. Project EUDP 2010-II. Guidelines for Low Temperature District Heating. .

Zvingilaite, E., Ommen, T., Elmegaard, B., Franck, M.L., 2012. Low temperature district heating consumer unit with micro heat pump for domestic hot water preparation. In: Proceedings of the 13th International Symposium on district heating and cooling, Copenhagen.

Application of the planning support system URB^{io}

11

Nils Schüler, Sébastien Cajot
Industrial Process and Energy Systems Engineering, École Polytechnique Fédérale de Lausanne, Lausanne, Switzerland

11.1 Description of the case studies in Geneva

The planning support system URB^{io}, presented in Chapter 2, is demonstrated via the application to two case studies. Both case studies are located in the canton of Geneva, Switzerland, one of the two municipal partners of the CI-NERGY project. Thus many of the input parameters have common values. First, a brief description of each case study is provided, including the project context, specific targets, and constraints.

11.1.1 Greenfield planning project: "Les Cherpines"

The case study, based on which the planning support system URB^{io} was initially developed (see Chapter 2), is an ongoing greenfield development project aimed at transforming a rural zone into a neighborhood named "Les Cherpines." This zone is located in the south-eastern part of the canton of Geneva, Switzerland, about 5 km away from the city center. The climate of Geneva is temperate with cool winters and warm summers. The development area comprises 58 ha and shall host about 3000 dwellings and 2500 jobs by 2030. A neighborhood master plan has been issued in 2013 and defines several goals for the new development (Office de l'urbanisme, 2013). To cope with the increasing demand for residential and office space in the canton (de l'urbanisme, 2011), it shall achieve a high floor area ratio (FAR). Parcels located close to the major street bordering the neighborhood on its south-eastern side shall be preferably used for commercial and office purposes to increase their visibility.

The new development shall meet "eco-district" standards by covering at least 75% of the energy demand from renewable sources and constructing buildings according to the requirements of the Swiss Energy Standard MINERGIE-P for thermal insulation. An industrial zone is located toward the southeast of the area with already installed pipes for a heating network. Consequently an attractive option could be to employ the incidental waste heat for satisfying the new neighborhood's energy demands. The waste heat is taken as being constantly available over the year with a power of 10.75 MW at a temperature of 20°C. Further envisaged energy sources are geothermal heat and electricity from PV. As the new development is located within the city

Urban Energy Systems for Low-Carbon Cities. https://doi.org/10.1016/B978-0-12-811553-4.00011-1

boundaries, some infrastructure is already existing, for example, electric lines are passing by all blocks. Consequently no installation costs for an electric grid are considered.

The spatial allocation of buildings is performed based on a preliminary site layout, which is defined by the master plan (Office de l'urbanisme, 2013). It indicates the foreseen locations of major streets and some specific functionalities as an industrial zone in the southwestern part and a sports ground in the northern part of the area. In order to limit the initial scope of this study, the aforementioned elements were taken as fixed so that the remaining elements (i.e., buildings for residential and commercial purposes and for offices) are to be determined by the optimization. Based on the layout of the master plan, a map is generated (see Fig. 11.1) showing the various elements considered by the optimization model.

The areas between the streets, which are already sketched in the preliminary layout (Fig. 11.1), are defined as blocks and all blocks not containing any predefined building functions were meshed with a regular grid of evenly sized parcels. As a first approximation these parcels are taken to be quadratic. The parcels' alignment deviates about 32.1 degrees from global north toward west.

Based on Table 11.1 the footprint of mixed-use buildings and single family houses (SFH) is assumed to be 250 and 125 m^2, respectively. Afterward the parcel size was set to 1000 m^2 where each parcel is foreseen to host two buildings of the same type. These both decisions were taken out of several considerations: First, this is about the smallest size of existing adjacent parcels, which lie in the northeast of the new development (Fig. 11.1), and allows thus a continuous urban form. Second, both decisions reflect the targeted trade-off between spatial detail and computational tractability since one building per parcel would mean doubling the number of parcels. Third, this parcel size allows to extend the model by the definition of typical office or commercial buildings, which still would fit into a parcel each. And fourth, the resulting building area ratio of one parcel is for mixed-use buildings 0.5 and for SFH 0.25, which is in line with common values for central and rural developments, respectively (Ruzicka-Rossier, 2005). The chosen parcel size resulted in a separation of the case study area into 233 parcels.

11.1.2 Brownfield planning project: "Les Palettes"

Together with the project partners of the canton of Geneva a second planning project was identified, which served to adapt and extend the planning support system URB^{io} to the context of brownfield planning. This project is named "Les Palettes." Although this is the name of a certain neighborhood, the project comprises as well the two adjacent neighborhoods "Les Semailles" in the north and "Le Bachet" in the east.

The project site lies about 1.5 km east of "Les Cherpines" and covers an area of approximately 50 ha housing about 10,000 inhabitants (Fig. 11.2).

Again, most of the data is available at the SITG (1957): The planning project includes 428 buildings, which are mainly used for residential or partly residential purposes. Furthermore, there are a number of buildings for educational purposes like schools and kindergartens. Fig. 11.3 shows the distribution of building types.

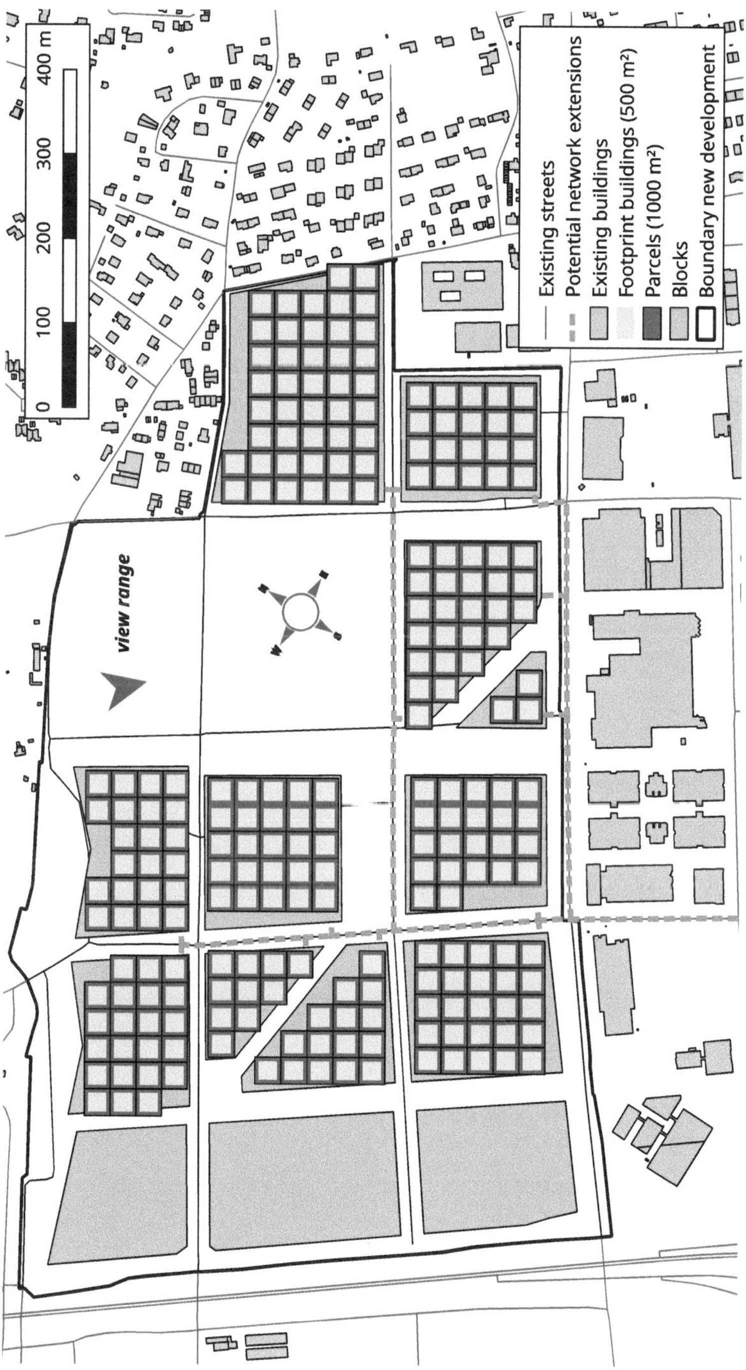

Fig. 11.1 Map of the case study "Les Cherpines."

Table 11.1 Average values for form-related building parameters of the building stock of Geneva, for which the full set of listed parameters is available

Building function	Buildings	Floor height (h_{fl}) (m)	Floors (n_{fl})	Footprint (A_{fp}) (m^2)
SFH	23,954	4.1	2.29	113.8
MFH	8153	3.8	5.32	249.7
Office	1182	4.8	4.35	655.2
Commercial (excl. malls)	191	5.0	2.57	817.8

Fig. 11.2 Map of the case study "Les Palettes."

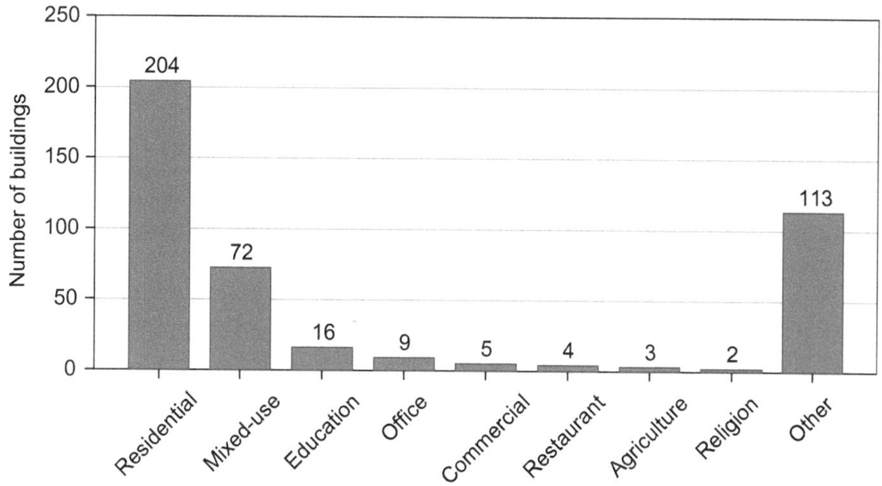

Fig. 11.3 Distribution of building types for the 428 buildings in the case study "Les Palettes."

The category "other" comprises mainly garages and unclassified buildings of less than 20 m^2, which are excluded from the energy analyses for their nonexisting or undefinable energy demands. The number of floors is available for 271 of the remaining 315 buildings.

Measurements of annual heat demand for both room heating and domestic hot water preparation are available for 151 buildings. The building parameters required as input for the multiple linear regression model presented by Schüler et al. (2015) are available for 265 of the 271 buildings with floor information. Concerning the existing energy conversion systems, only information about installed boilers is available: Of the 123 boilers in the planning perimeter, about 50.4% are running with oil, 48.1% are fueled by natural gas, and 1.5% by wood. It is possible that some of these boilers each supply several buildings. However, this information is not readily available. Electric and gas networks are already in place so that no according investment costs have to be taken into account. Furthermore, the installation of a new heating network is foreseen.

The planning zone is subject to several master plans from the cantonal level down to the neighborhood level (Cajot et al., 2017a). The communal master plan of the according community "Lancy" states overarching objectives (de Lancy, 2013): Existing neighborhoods should be densified, while preserving their social and functional variety. Densification is especially foreseen in form of the construction of new floors on top of existing buildings. The district should be further transformed to meet ecological standards, for example, by thermal refurbishment of buildings or replacement of existing energy conversion systems. Related to these energetic aspects is the cantonal goal to reduce the number of oil boilers. In addition to the above, urban heritage conservation should be respected. The communal master plan also lists a number of noteworthy objects, mainly buildings, which reinforce the identity of the neighborhood, and whose visibility should be preserved or improved.

11.2 Common input data for both case studies

11.2.1 Urban form and function

Statistical analyses of the building stock of Geneva were carried out using data provided by SITG (1957), to identify representative values for parcel and building parameters. These values are especially relevant for the greenfield planning case in order to start from realistic assumptions where otherwise data are missing. The results are summarized in Table 11.1.

The permissible building height in the canton of Geneva is 21 m (Le grand conseil de la république et canton de Genève, 1988), which can be increased to 27 m to allow for an extra floor, if this floor is used for residential purposes.

11.2.2 Social

Table 11.2 lists assumed surface values related to residential use. An average gross floor area (GFA) per job of $A_{jobs} = 30\ m^2$ was taken from Rey and Lufkin (2013).

11.2.2.1 Landmark view factor

A mountain range named "Jura" lies in the north-west of Geneva. Its highest peak is "Cret de la Neige" at 1720 m and coordinates 2,484,462.5, 1,125,493.8 in the Swiss coordinate system CH1903+/LV95 (EPSG:2056). The coordinates are used to calculate the distance to each building. The landmark's radius is taken as 9 km. It is further assumed that a floor is considered to have a view on the landmark if at least the uppermost 25% of it are visible. Under the assumptions that one wants to see at least the uppermost 25% of the landmark.

Table 11.2 Dwelling surface, occupancy rate, and surface ratio for two building types (GFA, gross floor area; ERA, energetic reference area)

Building type (bt)	**GFA/dwelling** (A_{dw})	**Inhabitants/dwelling** ($n_{inhab/dw}$)	**ERA/GFA** (sr_{ERA})
MFH/mixed-use	100[a] m^2	2.57[b]	0.84[d]
SFH	213[a] m^2	2.85[c]	0.73[d]

[a]Average values for the neighboring municipalities Confignon and Plan-les-Ouates (SITG, 1957).
[b]Values for the neighboring municipalities Confignon and Plan-les-Ouates (OCSTAT, 2009a).
[c]Values for the neighboring municipalities Confignon and Plan-les-Ouates (OCSTAT, 2009b).
[d]Schneider et al. (2016)

11.2.3 Energy

11.2.3.1 Energy demand

The energy demand for four different categories of end use is considered: room heating, room cooling, hot water, and electricity for appliances, lighting and ventilation. Most of these demands are adapted from national norms specifying the demand of the different categories in dependence on building and/or occupancy type, building state (i.e., existing, refurbished, or new) and the according energy standard. Since some of these norms are not free of charge, Table 11.3 indicates only the tables and pages of those norms containing the values used in this work.

Since the cantonal energy law in Geneva strongly restricts the cooling in buildings of specific occupancy types (l'énergie (LEn), 1986), the room cooling demands of residential and educational floor areas are set to zero. Measurements of the annual heat demand for room heating and domestic hot water preparation are used for existing, unrefurbished buildings where available. Otherwise and if enough building parameters are available, the annual heat demand is estimated with the multiple linear regression model presented by Schüler et al. (2015). Only if both are not the case, a building is not considered. Design heating demands are estimated using the energy signature model (Girardin, 2012) with the design temperature specified in Table 11.5.

Supply temperatures of 12°C for room cooling and 55°C for domestic hot water preparation and room heating (Girardin, 2012, p. 78) were assumed to estimate coefficients of performance of chillers and heat pumps, respectively.

Table 11.3 References and refurbishment factors for the estimation of energy demands

	Unrefurb.	MoPEC 2014	MINERGIE-P
Annual demand			
Room heating	[a]	[b]	$= 0.7 \cdot$ MoPEC[c]
Room cooling	[d]	[e]	[e]
Hot water	[a]	[e]	[e]
Electricity	[d]	[e]	[e]
Design demand			
Room heating	[f]	[g]	[g]
Room cooling	[h]	[i]	[i]
Refurb. factor	–	1.5[b]	9/7[c]

[a]Schüler et al. (2015).
[b]EnDK (2014, p.22).
[c]Minergie (2017, p.11).
[d]Schweizerischer Ingenieur- und Architektenverein (2015, Table16): values "existing."
[e]Schweizerischer Ingenieur- und Architektenverein (2015, Table16): values "standard."
[f]Girardin (2012).
[g]EnDK (2014, p.22) where available, otherwise from (Schweizerischer Ingenieur- und Architektenverein, 2015, Table11): values "standard."
[h]Schweizerischer Ingenieur- und Architektenverein (2015, Table13).
[i]EnDK (2014, p.27).

11.2.3.1.1 Refurbishment

The requirements for thermal building refurbishments are usually less strict than the requirements for new buildings. The used Swiss norms account for this by a factor with which the limit for annual room heating demand of new buildings is to be multiplied (see Table 11.3). These refurbishment factors are used to estimate not only the heating demands of refurbished buildings, but also their cooling demands, although the latter are not specified by those norms.

11.2.3.2 Energy supply

For this study, only centralized CHP technologies are foreseen, while chillers and PV panels are installed only at the buildings. Boilers and heat pumps can be installed both centrally or decentrally. Two different annual efficiencies are used for heat pumps to account for higher efficiencies of larger, centralized technologies. Table 11.4 lists the different efficiencies for all considered energy conversion technologies.

The numerical assumptions regarding the estimation of PV potential include the panels' conversion efficiency (see Table 11.7), the available roof surface area for PV panels, the panels' tilt angle, and the skydome. The roof surface can be partly shaded by roof superstructures such as chimneys or machine rooms for elevator systems. Furthermore the roof surface available for PV panels might be reduced by windows. After reduction of these factors the remaining surface is still not equal to the collector surface, since, in the case of flat roofs, the panels should be mounted with a certain tilt angle to maximize the incident annual irradiation on the panels, depending on the geographical location. Consequently, it is assumed here that the PV panel area is 50% of the footprint area. In addition, a panel tilt angle of 30 degrees is taken (Desthieux et al., 2014, p. 19) that maximizes the annual solar irradiation exploitation for the latitude of Geneva.

In principle a skydome containing the 145 sky patches of the Tregenza sky could be used to assess the solar potential. The piecewise linear formulation, however, requires a binary variable for each sky patch. Thus the Mixed Integer Linear Programing

Table 11.4 Efficiencies of energy conversion technologies

	Boiler			CHP[b]		Heat pump		Chiller[c]	PV[e]
Location (η_{th})	**Gas**[a] (η_{th})	**oil**[a] (η_{th})	**Wood**[a] (η_{th})	(η_{th})	(η_{el})	**Water-water**[c] (η_{COP})	**Air-water**[d] (η_{COP})	(η_{COP})	(η_{el})
Decentral	0.9	0.85	0.85	–	–	0.5	0.34	0.40	0.16
Central	0.95	0.873	0.864	0.46	0.44	0.6	0.38	–	–

[a]Moret (2017).
[b]Voll (2014).
[c]Henchoz (2016).
[d]Girardin (2012).
[e]Desthieux et al. (2014).

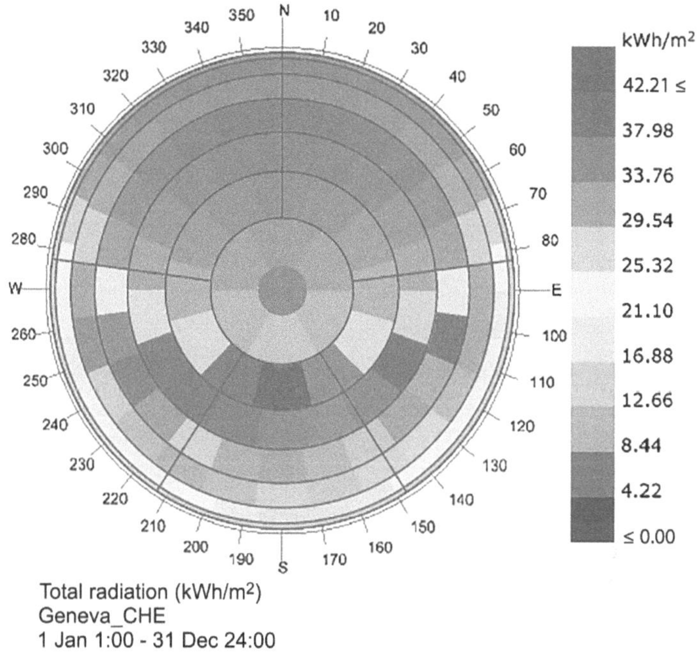

Fig. 11.4 Top view of a spheric skydome indicating the total annual irradiation on a surface tilted 30 degrees toward south for the location of Geneva: Aggregation into larger patches indicated by *gray lines* (cumulative sky created with Ladybug; Roudsari et al., 2013).

problem would become already for small numbers of parcels quite large. This can be counteracted by disaggregating the sky vault into a smaller number of patches (Fig. 11.4).

11.2.3.3 Energy networks

Losses in the electric grid are assumed as 5% of the supplied electricity (Best et al., 2015, p. 164) while losses in the gas grid are neglected (Keirstead et al., 2012, Table 2). Losses in the heating network are estimated as 4.3% of supplied heat per kilometer of network distance (Keirstead et al., 2012, Table 2). Its nominal supply temperature is set to 60°C at the heating center.

11.2.4 Environment

Table 11.5 lists the meteorological parameters that influence the design of the energy conversion technologies. The EnergyPlus weatherfile available for Geneva is used to generate the cumulative sky (EnergyPlus, 2017). Furthermore a constant soil temperature of 10°C is assumed (Weber and Shah, 2011).

Table 11.5 Meteorologic parameters

	Heating	Cooling
Design temperature[a] (°C)	−7	30
Average temperature[b] (°C)	7.3	22.6
Threshold temperature[c] (°C)	16	18
Annual duration[b] (h)	6386	1755

[a]Henchoz (2016).
[b]Reference year 2016 (Federal Office of Meteorology and Climatology, 2016).
[c]Girardin (2012).

For the calculation of the RES shares on parcel and neighborhood scale, an RES share of 47.1% for the Swiss supplier electricity mix is taken, considering only electricity from verifiable sources and further considering waste incineration as renewable (Frischknecht et al., 2012).

11.2.5 Costs

Assumed interest rates are: 1.7% for owner-occupants (Nationalbank, 2017, reference year 2016) and 6% for a local energy provider (OFEV, 2016).

11.2.5.1 Refurbishment

Costs for refurbishments are estimated based on the values of Meier (2015). These costs depend on the building type, the energy standard, and the refurbishment measure are listed in Table 11.6. The specified costs for multifamily houses (MFH) are assumed for mixed-use and school buildings.

Table 11.6 Costs for different refurbishment measures and energy standards (in CHF)

	MoPEC 2014	MINERGIE-P
SFH		
Planning	8000	18,000
Roof	80	180
Basement ceiling	80	150
Façades	300	400
Windows	800	1000
MFH		
Planning	40,000	90,000
Roof	95	120
Basement ceiling	85	100
Façades	215	250
Windows	720	800

11.2.5.2 Energy supply

An economic lifetime of 15 years is taken for all energy conversion technologies (Henchoz, 2016, p. 82). For the economic analysis cost functions of several authors were compared. Nonlinear cost functions were linearized with up to five segments where the choice of number and range of segments were made in order to achieve a good approximation of the original function. The resulting ranges, intercepts, and slopes of these segments are listed in Table 11.7. A conversion factor of 1.1 CHF/EUR is used for the costs of Voll (2014) and Henchoz (2016).

The same cost function is used for air-water heat pumps and water-water heat pumps, neglecting the additional costs of air-water heat pumps due to larger heat exchangers. The cost function for GSHP includes the costs for the geothermal probe. The cost function of Voll (2014) for CHP technologies is extrapolated below 0.5 and above 3.2 MW. In order to convert the reference quantity of the cost function for heat exchangers given by Henchoz (2016) from area to power, his assumed heat transfer coefficient of 2.5 kW/m^2 K and an estimated logarithmic mean temperature difference of 5 K was used. For PV panels a fixed-cost coefficient of 4000 CHF/kW$_{peak}$ is taken (Desthieux et al., 2014, p. 33).

11.2.5.3 Energy networks

For the calculation of operation costs, an electricity price of 0.2076 CHF/kWh (Eidgenössische Elektrizitätskommission, 2016) is assumed for all actors. Further prices are listed in Table 11.8. The heat price at which the local supplier buys is the price for waste heat of the industrial zone and is assumed as 0.1 CHF/kWh

For the calculation of investment costs an economic lifetime of 40 years is taken. Consequently repeated investments in conversion technologies with lifetimes shorter than 40 years (see Section 11.2.5.2) are taken into account for the calculation of, for example, the annual rate of return (Henchoz, 2016, p. 82). If a new gas grid has to be installed, the related investment costs are estimated with a factor of 375 CHF/m based on (Keirstead et al., 2012, Table 2). To estimate the investment costs for a new heating network the cost function of (Henchoz, 2016, Table 1.12) is used with a length and diameter specific cost factor of 5670 CHF/m^2 and a length-specific cost factor of 613 CHF/m. An average diameter of the network pipes of 0.15 m is assumed based on existing networks in comparable neighborhoods. The resulting length-specific costs are in the range of the costs given by (Keirstead et al., 2012, Table 2).

11.3 Interactive optimization workflow

The interactive optimization process adapted within URB^{io} is theoretically laid out in Section 2.1. This section provides a short summary of the workflow from a user perspective. Concrete examples will then be given in the consecutive sections.

As the process is iterative, all solutions are not immediately available in the parallel coordinates. The user can therefore specify where the next solutions should be generated, reflecting their preferences and expectations. This steering of the optimization

Table 11.7 Linearized cost functions of energy conversion technologies

Seg.	Param.	Unit	Gas boiler[a]	Oil boiler[c]	Wood boiler[d]	CHP[a]	HP[b]	GSHP[b]	CH[a]	HTS[c]
1	Q_{min}	kW_{th}	0	0	0	–	0	0	0	0
	$c_{inv,\ fix}$	CHF	6694	6020	0	–	18,501	38,628	1181	658.7
	$c_{inv,\ lin}$	CHW/kW	220.2	242.5	494	–	2823.2	12,085	274.7	73.5
2	Q_{min}	kW_{th}	50	50	100	100	50	50	–	50
	$c_{inv,\ fix}$	CHF	11,150	10,587	0	50,045	31,994	77,473	–	1338
	$c_{inv,\ lin}$	CHW/kW	126.3	146.5	123	403.8	1701.3	8885.1	–	58.9
3	Q_{min}	kW_{th}	100	100	–	500	100	100	–	100
	$c_{inv,\ fix}$	CHF	23,271	24,065	–	103,775	72,066	232,124	–	16,355
	$c_{inv,\ lin}$	CHW/kW	46.5	58.9	–	294.6	671.1	5102.6	–	31.7
4	Q_{min}	kW_{th}	1000	1000	–	1000	1000	1000	–	1000
	$c_{inv,\ fix}$	CHF	53,235	47,270	–	166,093	224,108	78,141	–	122,825
	$c_{inv,\ lin}$	CHW/kW	17.4	32.8	–	234.9	204.6	3079.2	–	20.6
5	Q_{min}	kW_{th}	5000	–	–	2000	–	5000	–	50,000
	$c_{inv,\ fix}$	CHF	88,650	–	–	372,960	–	1,678,988	–	315,784
	$c_{inv,\ lin}$	CHW/kW	10.0	–	–	153.4	–	2332.0	–	16.8
	Q_{max}	kW_{th}	10,000	2000	500	10,000	10,000	10,000	100	125,000

[a] Voll (2014).
[b] Bochatay et al. (2005).
[c] Henchoz (2016).
[d] (Moret, 2017, Section A.3), the reference size for decentralized wood boilers was increased from 10 to 100 CHW/kW.

Table 11.8 Energy prices in CHF/kWh

Resource	Nat. ⇔ Loc. supplier	Loc. supplier → End user	End user → Loc. supplier
Electricity	0.1056[a]	0.2076[b]	0.0944[b]
Gas	0.0602[c]	0.0844[d]	–
Heat	–	0.15	–
Oil	0.08655[e]	0.08655[e]	–
Wood chips	0.065[f]	0.065[f]	–

[a]Moret et al. (2016, Supplementary Material, Table 4).
[b]Eidgenössische Elektrizitätskommission (2016).
[c]Moret et al. (2016).
[d]Eidgenössisches Departement für Wirtschaft, Bildung und Forschung (2016): consumer category IV—multifamily housing, average demand 100,000 kWh/a, heating and DHW.
[e]Moret (2017, TableD3), local suppliers are assumed to only buy from national suppliers and not sell.
[f]ForêtSuisse (2017), local suppliers are assumed to only buy from national suppliers and not sell.

can be done through several actions described hereafter. Each of these actions is performed directly within the parallel coordinates interface, by brushing (i.e., dragging the pointer over) the axes on screen in the areas of interest.

- Objectives are what the solver shall maximize or minimize. Therefore, a preferred direction is defined for each criterion, for example, costs are to be minimized, while the share of renewable energy should be maximized.
- A criterion can constrain the solution space with upper or lower bounds. These are defined by the user by brushing an axis below or above the desired threshold.
- When a criterion is marked as "range," the corresponding constraint will be systematically varied within that range according to the requested number of solutions. The numeric values of these constraints are automatically determined by either a systematic or a quasirandom sampling method (Copado-Méndez et al., 2016; Cajot et al., 2018). Known as the ϵ-constraint method, this approach allows to generate solutions, which simultaneously optimize multiple criteria (Branke, 2008).

Attributes refer to the criteria which do not play an active role in the optimization process. As such, these are values which are calculated after the optimization finished, based on the resulting values of the decision variables.

11.4 Demonstration of the developed planning support system URB[io]

11.4.1 Greenfield planning

This section describes the application of URB[io] to the presented greenfield planning project, thus providing an example of the actions and thought process when iterating through the spiral in Fig. 11.5. Les Cherpines is one of the large ongoing urban development projects in Geneva, aiming to accommodate 3000 dwellings and 2500 jobs in a

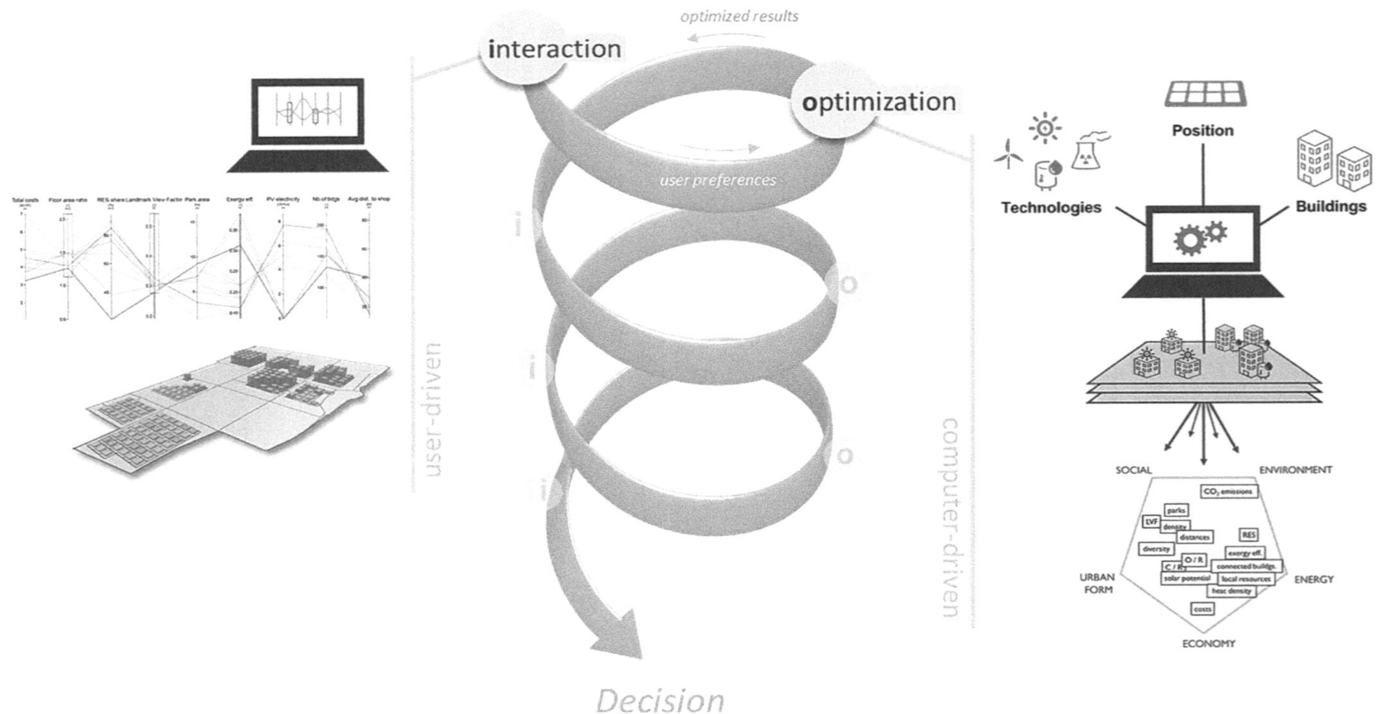

Fig. 11.5 URB^{io}: an interactive optimization framework supporting urban planning. The workflow is an iteration between a user-driven phase, where results are explored and preferences input, and a computer-driven phase, where an optimization model generates urban configurations (Cajot et al., 2017b).

mixed-use eco-district. Stemming from political goals specified in the cantonal and neighborhood master plans, several criteria were implemented, nine of which are depicted in the axes on Fig. 11.6. The first axis represents the total costs (i.e., investment and operation) of the entire energy system required for the development. The FAR plotted on the second axis is a substantive indicator for building density. The third axis (RES) represents an environmental indicator in form of the part of energy demand satisfied via renewable resources. The fourth axis shows the electricity produced by all photovoltaic (PV) panels installed on the roofs of buildings. The fifth indicates the share of connected buildings to a heating network, the sixth reflects the amount of natural gas required normalized by GFA, the seventh represents how much of the area is covered by parks, the eighth indicates the average walking distance from residential buildings to the nearest tram stop, and the ninth indicates the landmark view factor (LVF), or the share of building floors with direct view on predefined landmarks.

The process begins with the visualization of 25 precalculated solutions, chosen loosely to cover a wide sample of the solution space. They were established by setting two 5-fold ϵ-constraints on, respectively, the FAR (between 0.5 and 2.5) and RES share (between 10% and 100%), while minimizing total costs for each configuration of geo-referenced energy technologies and building types.

In the first iteration, the user might begin the exploration of these results by considering the FAR axis, density being a central urban planning parameter (Fig. 11.6). First, a negative correlation between costs per built square meter and FAR is noticeable by the crossing lines. For this reason, and possibly guided by other political or contextual reasons, they can limit the search to solutions with densities of at least 1.2. To force the generation of multiple additional solutions also for higher densities, a 10-fold ϵ-constraint is set between 1.2 and 2.5. Aiming to improve further criteria, the user might realize that parks are still underrepresented in the neighborhood. They can request additional parks in the next solutions by setting the park area share as the new objective to maximize.

In the second iteration, new solutions maximizing the possible park area are provided (Fig. 11.7). However, these configurations also present lower shares of renewable energy. This can partly be explained by the fact that less roof tops are available for PV panels, which contribution must be replaced by, for example, gas boilers or imports from the partly fossil-based electricity grid. Solutions which improve on RES shares can be requested by setting a 10-fold ϵ-constraint on RES share, while keeping the previous objective on park area (Fig. 11.8).

In a third iteration, the new ϵ-constraint allowed to identify solutions, which improve both urban and energy goals (Fig. 11.8). Although this may come at the expense of other important criteria (such as density or costs) which should be further investigated, some tradeoffs have been made clear and quantified. The iterations continue until the user identifies a satisfying and efficient solution in regard to the criteria of interest.

11.4.2 Brownfield planning

While many cities are still growing in terms of occupied surface and thus need to be extend on to green fields, a more frequent planning problem at least in the European

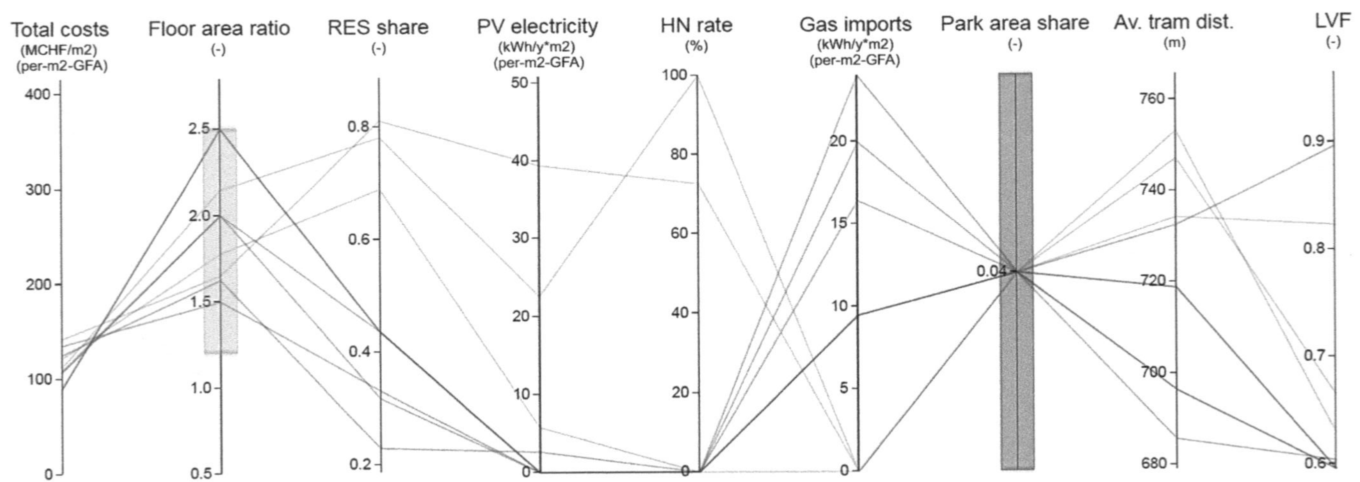

Fig. 11.6 In the first iteration, an objective (*park area share*) and an ϵ-constraint (*floor area ratio*) are set to generate 10 solutions. *Light lines* indicate high-RES shares, *dark lines* low-RES. *GSF*, gross floor area; *RES*, renewable energy source; *PV*, photovoltaic; *HN*, heating network; *LVF*, landmark view factor (Cajot et al., 2017b).

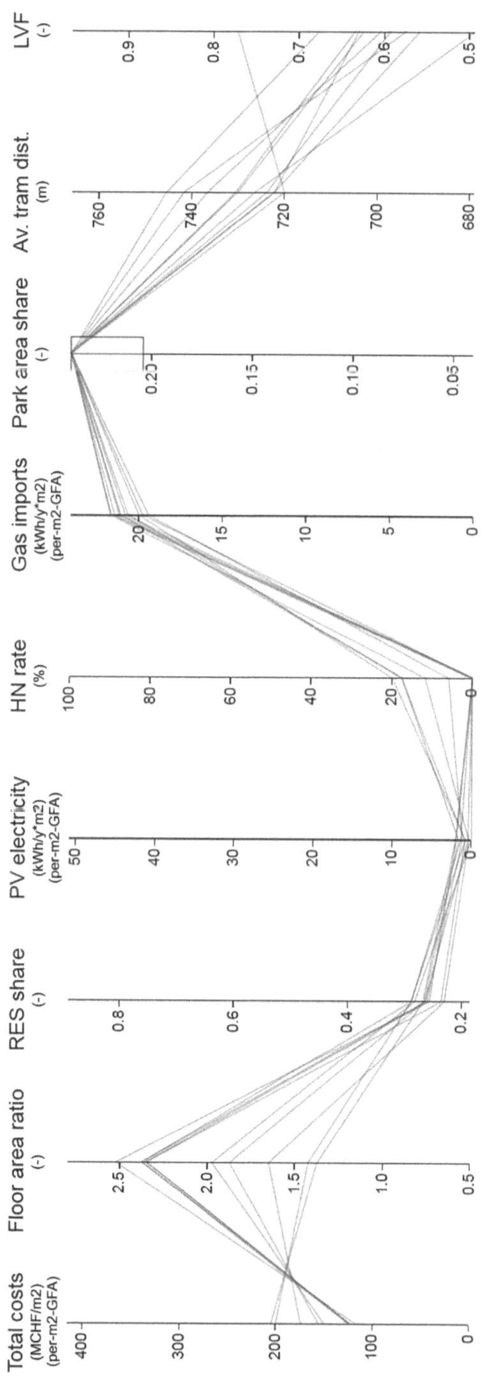

Fig. 11.7 In the second iteration, new urban configurations have been calculated, which now include a maximum of park areas with respect to other constraints. However, they do not yet contain high shares of RES (Cajot et al., 2017b).

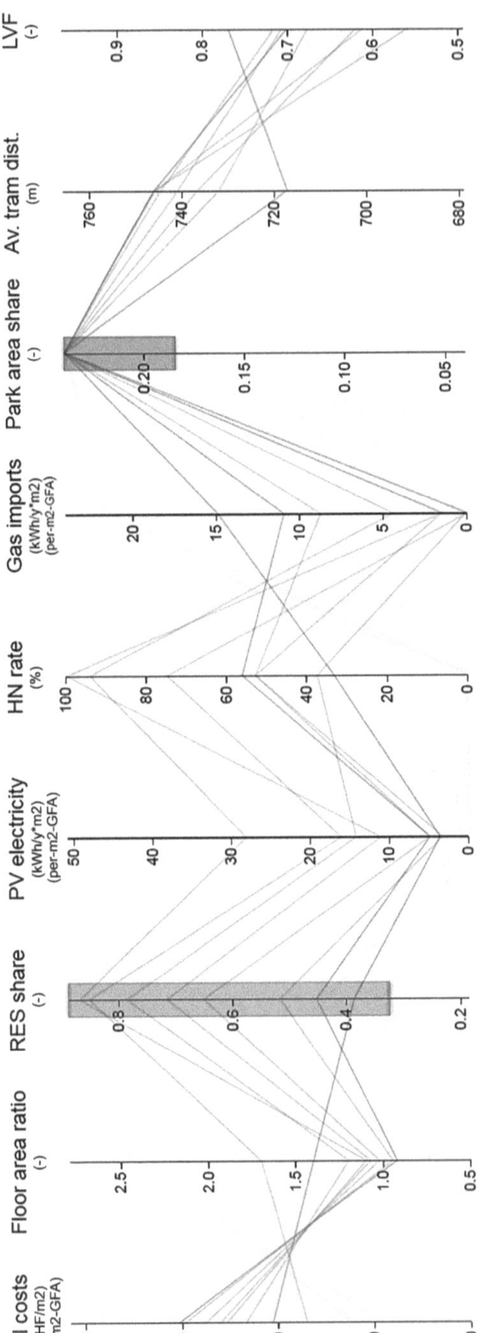

Fig. 11.8 In the third iteration, a 10-fold E-constraint is included on RES to search for improved solutions in this criteria. These are found at the expense of other criteria (e.g., FAR or costs). Note: The boxes were added again after calculation of solutions for illustration purpose (Cajot et al., 2017b).

context is the redevelopment of existing districts to, for example, house more people or keep up with increasing social and environmental standards. This section demonstrates the application of URB^{io} to an according planning project. Note that the models presented in Chapter 2 are not containing all equations required to solve the brownfield planning tasks presented in the following. The complete model can be found in the dissertation of the first author of that chapter.

11.4.2.1 Densification potential

Starting from the current state of existing buildings, the maximum density is determined by choosing the according axis in URB^{io} as objective to maximize. Since no new constructions are foreseen, the density can only be increased by building additional floors on top of existing buildings. However, these elevations are limited by overall and building-type-specific maximum height constraints and if buildings are listed. Thus the identified potential for densification is about 8% (Fig. 11.9).

11.4.2.2 Current energy system

In order to evaluate future improvements to the energy system, it is required to establish reference values, namely (a) the current demand by building and end use; (b) the type and size of existing conversion systems; (c) the according energy carrier (net) imports; and (d) the corresponding GHG emissions. While the norms (Section 11.2.3.1), the available measured annual demands (Section 11.1.2) and the statistical demand model (Schüler et al., 2015) allow to estimate the demands of almost all considered buildings, information about existing energy conversion systems is available for only about 40% of the buildings (Section 11.1.2).

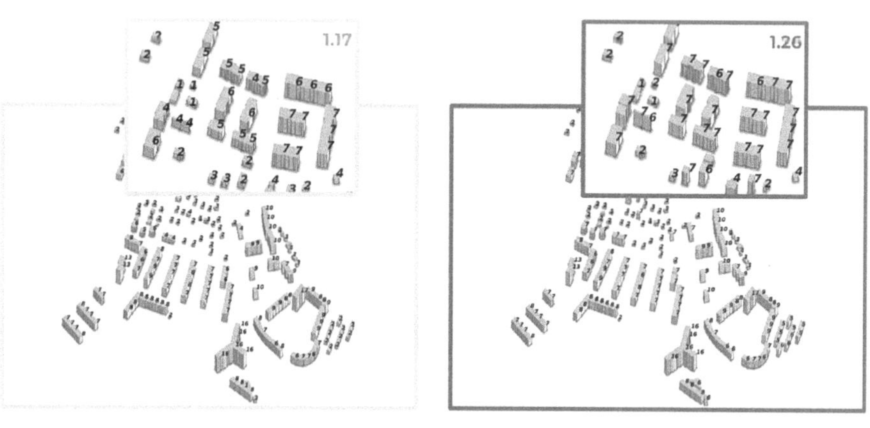

Fig. 11.9 Number of floors of buildings and the according FAR: maximizing the latter leads from the current state (*left*) to the maximum density without constructing new buildings and respecting permissible building heights (*right*). The enlarged maps illustrate buildings which would have to be extended.

Therefore, additional assumptions must be made. Here, it is assumed that the shares of known existing energy conversion systems reflect the current state in the entire planning zone and that these unknown energy conversion systems are distributed in order to minimize overall investment costs. Consequently, the option of a connection to a district heating network is excluded, since this network is not yet in place. The minimization of the investment costs for energy systems leads to the type and size of systems shown in Fig. 11.10.

Fig. 11.11 (light line) and Table 11.9 show the values of key criteria for the reference scenario. It is pointed out that the investment costs are not equal zero, as energy conversion systems have to be "installed" where information on existing systems is not available.

Fig. 11.10 Annual energy supply by conversion system estimated by applying the shares of known energy conversion systems to the entire planning perimeter and assuming that the cheapest system in terms of investment costs is in place.

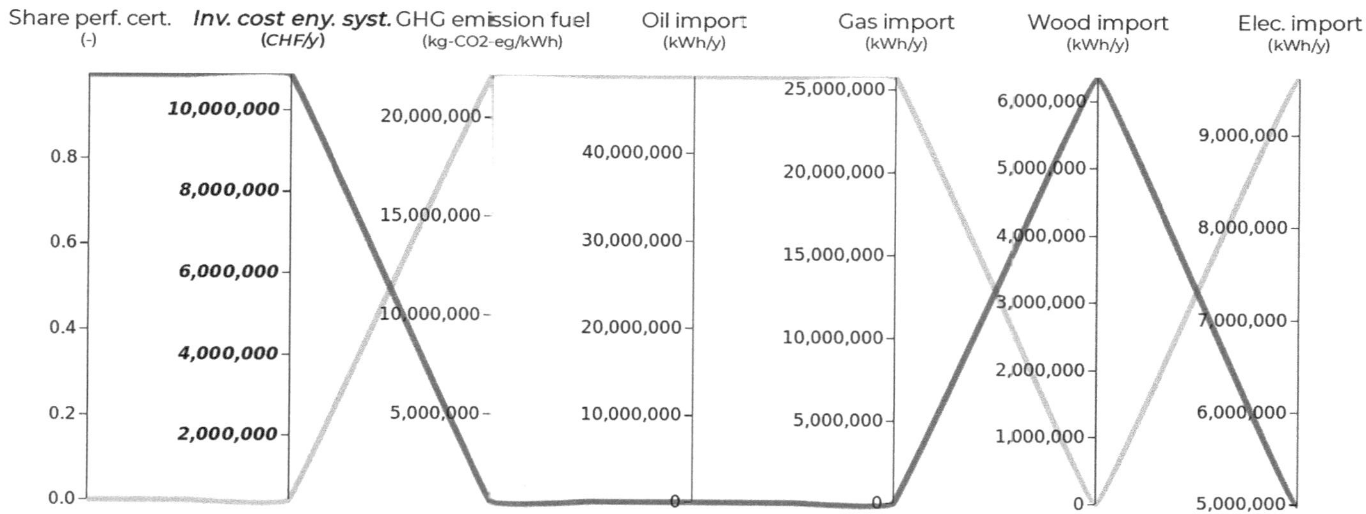

Fig. 11.11 Key criteria for the reference scenario (*light*) and the scenario with minimum GHG emissions (*dark*).

Table 11.9 Numerical results for the reference scenario and the scenario with minimum GHG emissions

Criteria	Unit	Reference	Min. emissions
Investment costs	MCHF/year	0.393	10.89
GHG emissions	t-CO_2-eq/kWh	21,680	538.34
RES share	%	7	89
Oil import	GWh/year	45.19	0
Natural gas import	GWh/year	28.42	0
Wood import	GWh/year	0.873	6.35
Electricity import	GWh/year	9.62	5.00
Room heating demand	GWh/year	64.7	5.48
Room cooling demand	MWh/year	332.9	1020
Electricity demand	GWh/year	9.13	9.13
Refurbishment rate	%	0	99.6

11.4.2.3 Reduction of GHG emissions

Proceeding from the reference scenario, the maximum achievable reduction of GHG emissions is determined by minimizing the latter. The results are depicted in Table 11.9 and Fig. 11.11 (*red line*). To achieve such low emissions it is necessary to renovate all buildings that are not listed, according to the strictest considered energy standard (Fig. 11.12). In addition, all buildings are equipped with PV panels and connected to the district heating network (Fig. 11.13). It can be seen that the annual energy output by conversion systems installed in SFHs in the center and the south-east, respectively, of the planning perimeter, is dominated by electricity from PV panels. The reason therefore is that smaller buildings have a larger roof surface in relation to their overall energy demands. This electricity is either consumed within those buildings or exported to the local grid.

11.4.2.4 Cost-effective increase of share of RES

While it is indeed highly desirable to minimize GHG emissions drastically, it is still important to consider the implied costs. Thus, in a further iteration scenarios are explored that are both environmentally and economically sustainable. To achieve this, the total energy system costs (i.e., investment and operation) are minimized while the minimum permissible share of energy from RES is systematically varied (Fig. 11.14). This approach allows to generate pareto-optimal solutions regarding the two considered objectives.

First, it can be noted that reducing costs is indeed in conflict with increasing energy from RES. This is indicated by the colored lines, which are ordered synchronously on the axes of "RES share" and "total costs." Besides, an increase in the share of RES results in a progressive decrease of natural gas imports and an increase of wood and electricity imports. The latter is used to drive heat pumps since, if buildings

Fig. 11.12 Selected energy standard to minimize GHG emissions: All buildings that are not listed are refurbished according to the standard MINERGIE-P.

are neither enlarged (i.e., a constant density) nor refurbished, the electricity demand for end uses remains constant. Since here the total costs were considered as objective, they are indeed much lower than those for a single-objective minimization of GHG emissions.

Looking at the refurbishment rate ("share performance certificates") reveals that a thermal refurbishment of building envelopes does not constitute the most cost-effective way to reduce emissions or increase the share of RES. However, currently only complete refurbishments of buildings are considered. A future improvement of the model and results would be to differentiate between the various measures (roof, facades, glazing, etc.), which could lead to the identification of the most cost and energy efficient measures.

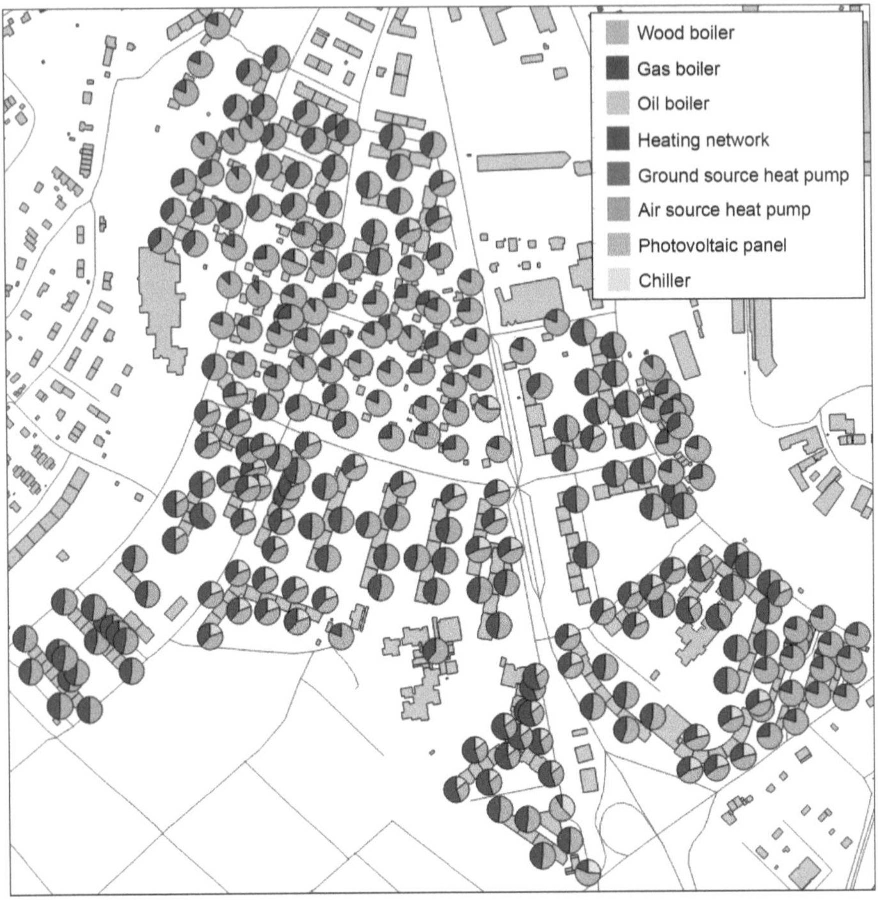

Fig. 11.13 Annual energy supply of installed conversion systems that minimize GHG emissions: All buildings are mainly supplied by the heating network and PV panels.

11.4.2.5 Refurbishment

While the above results indicate that complete refurbishment is not necessarily the most competitive measure, higher refurbishment rates might nevertheless be desirable for (a) ecological reasons, as reduction of energy demand should be preferred to increase of system efficiency, or (b) political reasons, like the availability of dedicated cantonal or national funds. In this direction, the task would rather be to determine the ideal allocation of financial resources to achieve the highest energy savings. This question can be answered with URB^{io}, by minimizing refurbishment costs while varying the maximum permissible heating demand (Fig. 11.15).

A closer look at the results shows, in dependence on the refurbishment rate, that in general large buildings are more attractive than small ones (Fig. 11.16) and which single buildings are the most effective to be refurbished (Fig. 11.17).

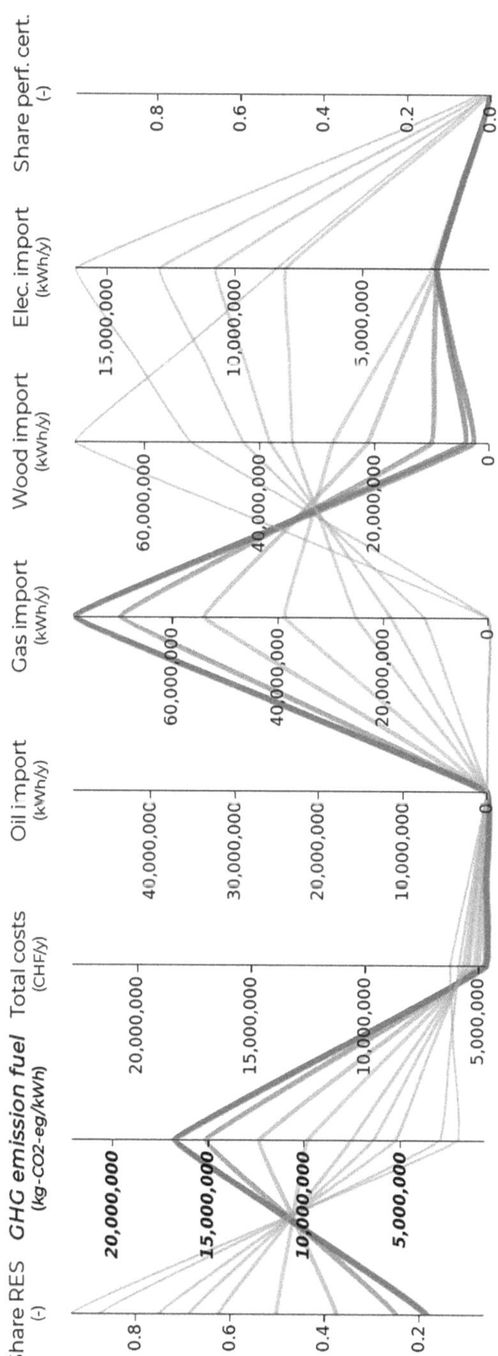

Fig. 11.14 Parallel coordinate plot of scenarios generated by a minimization of total costs while systematically varying the share of energy from renewable sources (RES) (*line color and thickness* based on second axis).

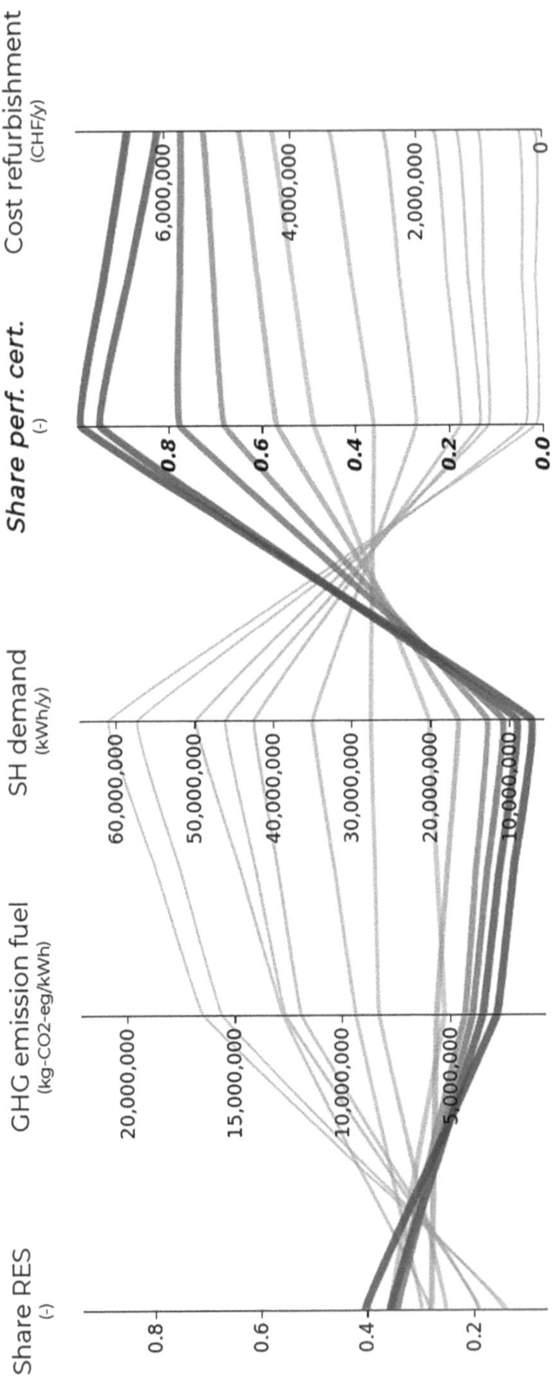

Fig. 11.15 Parallel coordinate plot of scenarios generated by a minimization of refurbishment costs while systematically varying the permissible heat demand (*line color and thickness* based on fourth axis). *SH*, space heating; *Share perf. cert*, share of buildings with a performance certificate (i.e., refurbished).

Fig. 11.16 Selected energy standards for an optimal allocation of financial resources for a higher refurbishment rate (78%): Large buildings are more attractive to refurbish than small ones.

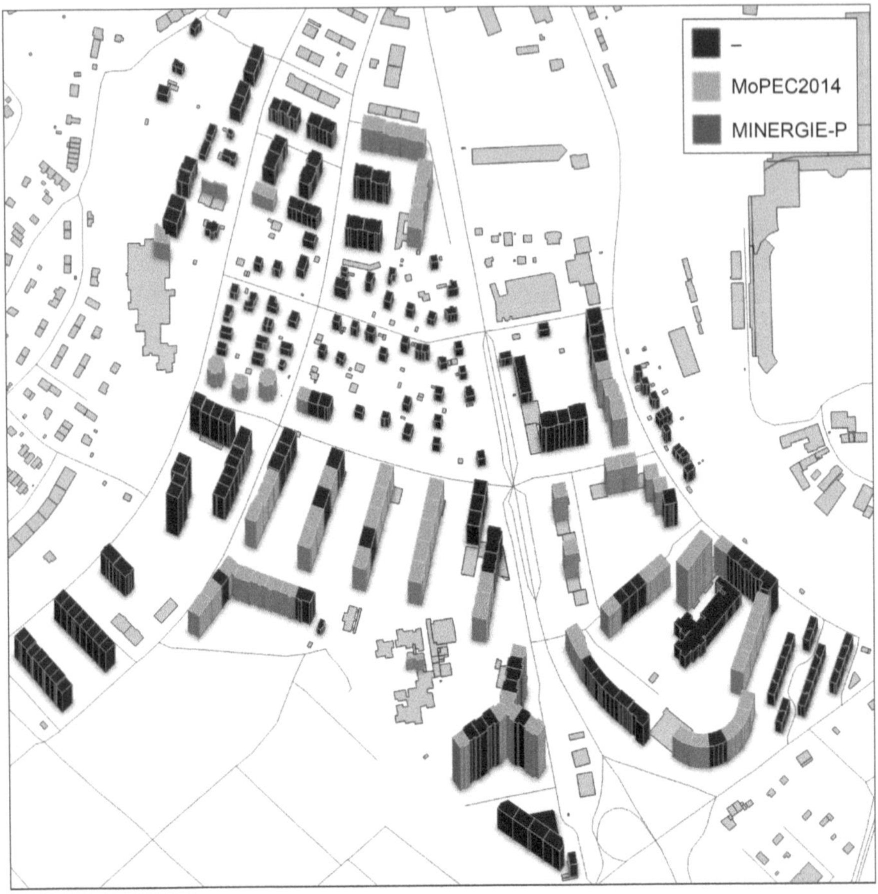

Fig. 11.17 Selected energy standards for an optimal allocation of financial resources for a lower refurbishment rate (27%): Identification of the most attractive buildings to refurbish.

Nomenclature

For a nomenclature and explanation of notation see Chapter 2.

Acknowledgments

The authors would like to thank the urban planning and energy offices from the canton of Geneva for their constructive inputs in the elaboration of this work.

References

Best, R.E., Flager, F., Lepech, M.D., 2015. Modeling and optimization of building mix and energy supply technology for urban districts. Appl. Energy 159, 161–177.

Bochatay, D., Blanc, I., Jolliet, O., Maréchal, F., Manasse-Ratsimandresy, T., Beck, R., 2005. Projet PACOGEN, Evaluation Économique et Environnementale de Systèmes Énergétiques á Usage Résidentiel. http://infoscience.epfl.ch/record/59329.

Branke, J., 2008. Multiobjective Optimization: Interactive and Evolutionary Approaches. Lecture Notes in Computer Science, State-of-the-Art Survey, 5252S. pringer. ISBN 978-3-540-88907-6.

Cajot, S., Peter, M., Bahu, J.M., Guignet, F., Koch, A., Maréchal, F., 2017. Obstacles in energy planning at the urban scale. Sustain. Cities Soc. 30, 223–236.

Cajot, S., Schüler, N., Peter, M., Koch, A., Maréchal, F., 2017. Interactive optimization for the planning of urban systems. In: Scartezzini, J.-L. (Ed.), Energy Procedia, vol. 122. Elsevier, Amsterdam, pp. 445–450.

Cajot, S., Schüler, N., Peter, M., Koch, A., Maréchal, F., 2018. Interactive optimization with parallel coordinates: exploring multidimensional spaces for decision support. Front. ICT. (submitted).

Copado-Méndez, P.J., Pozo, C., Guillén-Gosálbez, G., Jiménez, L., 2016. Enhancing the ϵ-constraint method through the use of objective reduction and random sequences: application to environmental problems. Comput. Chem. Eng. 87, 36–48.

de l'urbanisme, O., Septembre 2011. Plan Directeur Cantonal—Etude des Besoins et de l'Offre en Logements. In: Département des constructions et des technologies de l'information, Genèvep. 33. http://ge.ch/amenagement/media/amenagement/files/fichiers/images/documents/etude_des_besoins_et_de_l_offre_en_logements_light.pdf.

de Lancy, V., 2013. Plan directeur communal de Lancy. Genève, Switzerland, Available from: http://www.lancy.ch/habitants/territoire-domaine-public/plan-directeur-communal-0.

Desthieux, G., Gallinelli, P., Camponovo, R., 2014. Analyse Du Potentiel de Production Énergétique Par Les Panneaux Solaires Thermique et PV. Cadastre solaire du Canton de Genève, 2,HEPIA, Geneva. Rapport final 2.

Eidgenössische Elektrizitätskommission, 2018. Strompreis-Webseite der ElCom: Vergleichen Sie Ihren Strompreis. Available from: https://www.strompreis.elcom.admin.ch/Map/ShowSwissMap.aspx.

Eidgenössisches Departement für Wirtschaft, Bildung und Forschung, 2018. Eidg. Volkswirtschaftsdepartement EVD, Preisüberwachung. Available from: http://gaspreise.preisueberwacher.ch/web/index.asp?z=9.

EnDK, 2014. Mustervorschriften der kantone im energiebereich (Muken). Available from: https://www.endk.ch/de/ablage/grundhaltung-der-endk/muken2014-d20150109-2.pdf.

EnergyPlus, 2018. Weather Data. Available from: https://energyplus.net/weather.

Federal Office of Meteorology and Climatology, 2018. MeteoSwiss IDAWEB: Welcome at IDAWEB! Available from: https://gate.meteoswiss.ch/idaweb/system/welcome.do.

ForêtSuisse, 2017. Marché du bois. Available from: http://www.energie-bois.ch/fileadmin/user_resources/01_Holzenergie/Energieholz_Richtpreise/1712_La_Foret_prix_bois-energie.pdf.

Frischknecht, R., Itten, R., Flury, K., 2012. Treibhausgas-Emissionen Schweizer Strommixe, v1.4. 426 Uster. Bundesamt für Umwelt BAFU, Switzerland. Available from: http://www.bafu.admin.ch/klima/09608/index.html?lang=de&download=NHzLpZeg7t,lnp6I0NTU04212Z6ln1acy4Zn4Z2qZpnO2Yuq2Z6gpJCGfIF5fWym162epYbg2c_JjKbNoKSn6A-.

Girardin, L., 2012. A GIS-Based Methodology for the Evaluation of Integrated Energy Systems in Urban Area.

Henchoz, S., 2016. Potential of Refrigerant Based District Heating and Cooling Networks. Ph.D. thesis.

Keirstead, J., Samsatli, N., Shah, N., Weber, C., 2012. The impact of CHP (combined heat and power) planning restrictions on the efficiency of urban energy systems. Energy 41 (1), 93–103.

Le grand conseil de la république et canton de Genève, 1988. Loi sur les constructions et les installations diverses (lci). Available from: http://www.ge.ch/legislation/rsg/f/s/rsg_L5_05.html.

l'énergie (LEn), L.S., 1986. Collaborator With Geneva, Available from: http://www.geneve.ch/legislation/rsg/f/s/rsg_l2_30.html.

Meier, R., 2015. Wirtschaftlichkeit energetischer Sanierungsmassnahmen - Modellrechnungen für Ein- und Mehrfamilienhäuser. Energie-cluster.ch, Bern, Switzerland, p. 121. Available from: https://www.energie-cluster.ch/admin/data/files/file/file/1291/20160429-bfe-wirtschaftlichkeit-energetischer-sanierungsmassnahmen_.pdf?lm=1464330144.

Minergie, 2017. Produktreglement zu den Gebäudestandards MINERGIE/MINERGIE-P/MINERGIE-A—Version 2017.3. Minergie Schweiz. Available from: https://www.minergie.ch/media/171215_produktreglement_minergie_p_a_v2017.3_de.pdf.

Moret, S., 2017. Strategic Energy Planning Under Uncertainty. Available from: https://infoscience.epfl.ch/record/231814?ln=en. Ph.D. thesis.

Moret, S., Peduzzi, E., Gerber, L., Maréchal, F., 2016. Integration of deep geothermal energy and woody biomass conversion pathways in urban systems. Energy Convers. Manag. 129, 305–318.

Nationalbank, S., 2017. Zinssätze, renditen und devisenmarkt. Available from: https://data.snb.ch/de/topics/ziredev?_escaped_fragment_=/doc/explanations_ziredev#!/cube/zikreddet?fromDate=2016-01&toDate=2016-12&dimSel=D0(K,OBB,MBB,FH,FI).

OCSTAT, 2009. Population, logements, nombre de pièces et taux d'occupation dans les bâtiments á plusieurs logements, par commune, en 2009. Office cantonal de la statistique Genève. Available from: http://www.ge.ch/statistique/tel/domaines/09/09_03/T_09_03_1_08.xls.

OCSTAT, 2009. Population, logements, nombre de pièces et taux d'occupation pour les maisons individuelles et les maisons á deux logements, par commune, en 2009. Office cantonal de la statistique Genève. Available from: http://www.ge.ch/statistique/tel/domaines/09/09_03/T_09_03_1_11.xls.

OFEV, 2016. Référentiel de 6 % pour les réseaux de chauffage á distance. Newsletter sur la compensation des émissions de CO2 en Suisse. Available from: https://www.bafu.admin.ch/bafu/fr/home/themes/climat/newsletter/newsletters–compensation-des-emissions-de-co2-/7e-newsletter-sur-la-compensation-des-emissions-de-co2-en-suisse.html.

Office de l'urbanisme, 2013. Plan directeur de quartier - Les Cherpines. État de Genève and Commune de Confignon and Commune de Plan-les-Ouates, Genève. 29897. Available from: http://ge.ch/amenagement/pdq-cherpines.

Rey, E., Lufkin, S., 2013. Durabilité á l'échelle du quartier. In: Architecture Albums, pp. 161–178.

Roudsari, M.S., Pak, M., Smith, A., 2013. Ladybug: a parametric environmental plugin for grasshopper to help designers create an environmentally-conscious design. In: Proceedings of the 13th International IBPSA Conference Held in Lyon, France, August. http://www.ibpsa.org/proceedings/BS2013/p_2499.pdf.

Ruzicka-Rossier, M., 2005. Développement territorial et urbanisme i.

Schneider, S., Khoury, J., Lachal, B.M., Hollmuller, P., 2016. Geo-dependent heat demand model of the Swiss building stock. In: Habert, G., Schlueter, A. (Eds.), In: Expanding Boundaries Systems Thinking in the Built Environmentvdf Hochschulverlag AG, pp. 286–290. http://archive-ouverte.unige.ch/unige:84240.

Schüler, N., Mastrucci, A., Bertrand, A., Page, J., Maréchal, F., 2015. Heat demand estimation for different building types at regional scale considering building parameters and urban topography. In: Energy Procedia, vol. 78. Elsevier, Amsterdam, pp. 3403–3409. http://linkinghub.elsevier.com/retrieve/pii/S187661021502490X.

Schweizerischer Ingenieur- und Architektenverein, 2015. Données d'utilisation des locaux pour l'énergie et les installations du bâtiment.

SITG, 1957. Le territoire genevois à la carte. L 1 35. http://ge.ch/sitg/.

Voll, P., 2014. Automated Optimization-Based Synthesis of Distributed Energy Supply Systems, first ed. Aachener Beiträge zur Technischen Thermodynamik1Wissenschaftsverl. Mainz. ISBN 978-3-86130-474-6.

Weber, C., Shah, N., 2011. Optimisation based design of a district energy system for an eco-town in the United Kingdom. Energy 36 (2), 1292–1308.

Conclusions and outlook

Prof. Ursula Eicker, CI-NERGY Coordinator, University of Applied Sciences, Stuttgart

During four years an international team of PhD and PostDoc researchers worked on methods to plan, model, optimize and operate energy efficient and renewable urban districts. As a result this book was produced, which summarizes the main methods and lessons learned from the work.

Cities are the main players of the future and crucial to prevent climate change and develop alternative renewable solutions to today's fossil fuel world. The complex interactions between the building sector consumption, renewable and decentral generation, storage, and demand side management require sophisticated tools that were developed and applied to the two major European cities Geneva and Vienna by the PhD network. Further work is still needed to include other major energy users in cities such as the transport sector, the water infrastructure, and the industry sector. Also a much closer connection to the urban planning community has to be sought so that the tools and methods enter the daily practice. Last but not least, the citizens themselves should have their say in how they envisage life in future cities. This multi- and trans-disciplinary research area will keep us busy for years to come, and we hope to meet and discuss our ideas within an international community of sustainable city developers. The book concludes with remarks and outlooks from some of the participating fellows and their supervisors.

Sebastién Cajot and Markus Peters, EIFER, Karlsruhe

The task of transitioning urban energy systems to efficient, low-carbon ones is undeniably challenging. The recognition of this challenge as a multifaceted one entails and encourages a multifaceted solution. Hardin (1968) made the following remark in his seminal article: "An implicit and almost universal assumption of discussions published in professional and semipopular scientific journals is that the problem under discussion has a technical solution." Multiple facets of the solutions proposed in this book are, indeed, only partly technical. They are crucial; however, they are not sufficient. Advanced digital models of cities, buildings, and energy infrastructure help us better understand how they work and enable us to propose consequent technical improvements. However, if the bridges between researchers and practitioners and between urban planners and technical experts are not consolidated, the problems will persist. The interactive tool presented in Chapter 2 is just one example illustrating the necessary synergy between the technical solution (the underlying models) and its human counterpart (the expertise and local knowledge of the practitioner). The broad and persistent network established within CI-NERGY is another example. Bridges were built between countries, universities, and disciplines and, most importantly, between research and practice. The regular feedback from the partner cities allowed to paint a realistic picture of practical problems toward which to direct our efforts and was one of the keys to the project's success. To keep information and ideas flowing across these bridges was and is, in fact, the real challenge.

Challenges and outcomes include the development of a methodology to support urban energy planning for distributed urban resources, using applied, graduated modules on different temporal planning instances and variable, appropriate spatial scales. The aim was to analyze, define, and overlap suitable urban (energy) planning instances and to integrate to existing levels and procedures of urban planning.

Integrating an advanced urban energy planning process is a major issue for upcoming urban design challenges in the transformation of cities and leads to new requirements for interdisciplinary competences. Establishing a decision support methodology for integrated urban energy planning implies the integration of energy needs in existing urban planning approaches and the handling of conflicting demands from various actors and at various scales.

Intensive local exchange and interaction with the case-study cities were crucial to identifying overlaps in the work of urban planners and urban energy planning, demonstrating the needs for shifting administration structures; making new planning methods and collaborations; integrating modeling results (from simulation and optimization) into the relevant stages of the planning process; and accounting for local structure, timing, scale, and planning cultures.

The new decision support approach uses analysis and optimization techniques to assist in the selection and the characterization of the most promising options (ecological, social, and economic) and the definition of potential markets for energy service companies and industry. The developed model based the approach for urban design, supporting local urban planning, energetic goals, and constraints. Resulting modeling outputs are supporting the urban design and planning process and can initially determine the specific role and influence of the urban form. Thus, the application of the solution areas and spatial spaces can be tested in early stages of the urban planning and design.

Pilar Monsalvete Álvarez de Uribarri, HFT, Stuttgart

The urban developer community is getting interested in the use of modeling tools to understand and predict energy flows in cities and to design urban district with a more efficient energy management. This interest has led to much work done on the urban modeling tool field in the last years. But two crucial problems still have to be solved to make these tools more manageable and useful: how to provide the data efficiently and how to avoid a too high specialization of the tools. The main problem of having specialized tools is that they are not flexible enough to simulate the complexity of the building sector and its energy supply systems.

Both problems are closely related, as the level of detail of the available data strongly influences the modeling technique and the software structure. Therefore, the specialization can be due to the availability of data. The second reason for specialization is, of course, the target and application field of the tools.

These two points have been addressed within the CI-NERGY project. On the one hand, the data are normally manually set by the user, sometimes with the help of other tools, for example, based on CAD or GIS. Even then, the provision of data is really tedious and time-consuming. Standards help to automatize the data handling, thus avoiding possible human errors and highly reducing the time required.

On the other hand, the set of software tools developed within CI-NERGY cover a big range of spatial and temporal resolution, from a broad overview of the city quarter used for optimization to a close look on very specific problems, like demand side management or the design of district heating systems. Furthermore, a big effort was done on developing modular tools. The idea behind the modularity is the division of a complex system into simpler and more manageable modules. The modules can then be independently combined and used in different systems or for different purposes. Furthermore, they can be separately improved, updated, or modified and allow the developer to easily add new capabilities to the software.

The target of having a set of tools that cover the broad domain to be analyzed for an energy-efficient development of a city is still far away from been fulfilled, but the work done within CI-NERGY, partially covered in this book, made a step forward to achieve it.

Solène Goy, UCD, Dublin

Urban building energy modeling has been used to understand the building stock dynamics and current demand, to test the impact of new policies, and to quantify the potential of novel technical solutions. Modeling the building stock is complex as it requires interdisciplinary expertise and is influenced by the data availability. Work within the CI-NERGY project has shown that it can be addressed from various perspectives, using different methods over various spatial scales.

To address the global energy challenges, the role of buildings and their occupants is changing. In that context, advanced control techniques can be used to provide a more efficient, flexible, and sustainable energy usage. More specifically, model predictive control (MPC) has the potential to address several concerns from multiple stakeholders and from building occupants to network operators. However, the potential flexibility and energy/cost savings are case-study-dependent, thus the importance of preassessment tools. The analysis of these potential gains over multiple building characteristics gives an insight into the influencing design and operational parameters and can be used to design the buildings of the future. Nonetheless, the deployment of such novel control technique will require the acceptance and inclusion of all stakeholders to ensure a successful implementation.

Dr. Donal Finn, UCD, Dublin

Urban energy modeling, when undertaken at scale, presents significant opportunities for a wide variety of stakeholders. Nevertheless, significant open challenges continue to exist and need to be addressed. One of key challenges is the issue of identification of representative buildings from urban-scale energy datasets. Identification of representative buildings is critical as they act as the basis for subsequent scaling and analysis. Such representative buildings and associated data act as input to urban energy models that are subsequently used to generate representative energy models. These models are the basis by which energy-consumption patterns are estimated, thereby influencing potential policy decisions on future energy conservation measures and demand side management.

Traditionally, representative buildings have been identified using statistical approaches or expert knowledge. However, such methods do not allow the assessment of multiple features in parallel or may skew the selection of representative buildings toward specific features. Data mining techniques such as clustering allow for the analysis of different building features in parallel taking into account building heterogeneity in urban contexts. When combined with traditional classification techniques, predictive clustering algorithms have the potential to overcome the challenge of data scarcity in urban environment, thereby facilitating the classification of large building stocks with improved accuracy.

The implementation of energy conservation and demand management measures will benefit from the use of advanced control techniques. The research in the CI-NERGY project showed that the use of model predictive control, when considered at scale, provided the opportunity for flexibility provision in urban energy building stock. Nevertheless, there continue to remain significant challenges to the widespread adoption of such technologies, including the need to deploy MPC systems at scale in urban environments and obtain industry adoption of these emerging technologies.

Prof. Karl Ponweiser, Technical University of Vienna

The change of district heating networks of the third generation to the networks of the fourth generation with lower temperatures in supply and return lines is in progress. Concepts of networks with producer and consumer (prosumer networks) are upcoming. Besides the simulation of the thermal behavior of the network, the simulation of the hydraulic behavior will become crucial, since the feed and subtraction of heat-transfer fluid with pumps into and out of a liquid-based intermeshed network can cause high demands on the capability of the control systems.

Therefore, the modeling and simulation of the networks as part of district heating systems are essential in urban energy research.

For the future, many things point to an increase in the use of energy networks, coupled to seasonal thermal storage units. These systems are working with heat pumps between the net and the clients, which are used for heating or cooling just if they are needed to overcome disadvantageous temperature differences. These systems can provide heating and moderate cooling for office buildings and for residential buildings. Thermal storages between the heat pumps and the prosumers might be necessary for peak shaving. To cover the differences in demands for heating and cooling, low-temperature heat, for example, waste heat from industrial processes, can be used. Waste heat from data centers, for example, can be used as well with the advantage of free cooling up to a certain temperature level.

Concerning the system design, the development may go to computer-aided automatic design of whole HVAC systems for buildings and to the automatic design of networks for development areas.

Mario Potente Prieto, Technical University of Vienna

If any fundamental conclusion can be reached, it is that there is no fixed or single established vision in the urban energy planning field. In this way, researchers are mistaken to believe that it is possible to create a unique and global methodology, like a "user's guide" to carry out urban energy planning tasks automatically for any case.

Throughout the four years in the Marie Curie PhD program, all people involved in this project have verified the great diversity of strategic, economic, logistical, political, and even cultural factors to be considered, in a way that each country and each city have their own preferences and methods in both the energy field and other issues.

However, the topic offers excellent opportunities for researchers, engineers, scientists, and technicians. The challenge consists of supporting urban planners' work throughout all their stages: from data collection, through map development or network simulation and calculation, to creating useful tools that must be increasingly linked between each other and interactive for their users. Thus, the urban energy planning process will merge with other areas, such as services, urban mobility, or environmental and sustainability issues. They will not be separate fields anymore and rather become central issues inside the strategy for each city, resulting in a positive impact from the economic and environmental point of view.

Prof. Volker Coors, HFT, Stuttgart

From the geoinformatics point of view, one fundamental research topic in CI-NERGY was the impact of uncertainty of geospatial data, for example, on the result of heating demand simulation based on 3-D city models. Based on the definition of an ontology for urban energy system simulation, a data model has been designed and implemented in an open-source data management system. It significantly influenced the CityGML energy application domain extension development on international level. This data model plays a fundamental role in the design of a modular distributed urban energy simulation platform based on a lightweight service-orientated architecture. A first prototype of such a system has been implemented and evaluated in the project. Questions such as the impact of a multiresolution model of a building geometry with its uncertainties due to automatic data capturing and generalization in different scales have been investigated using this simulation platform.

Based on these results and insights, the data model and web services shall be integrated in the upcoming concept of an urban data platform as the backbone of smart cities in future. One of the key research questions in this context will be how to semi-automatically populate the data model from unstructured data flows in order to make use of the vast amount of geospatial and nongeospatial data in urban energy simulation. Design of information flows and a good balance of human interaction and deep machine learning are core elements to tackle this challenge.

Dr. Giorgio Agugiaro, Austrian Institute of Technology, Vienna

In terms of 3-D city modeling, the work in the CI-NERGY program has contributed to the following:

- Generate a first version of semantic 3-D city models of Vienna (district of Meidling) and of the whole Canton of Geneva, allowing to test on different cities—if modeled according to the same standard—the same tools, thus contributing to replicability and validation of the results.
- Analyze and understand the available data sources—mostly coming from the public administration—in hindsight of their harmonization and integration toward a CityGML-compliant city model. Particular attention was paid to those datasets that are of particular relevance in energy-related applications. Such datasets cover not only the building sector but also supply networks and centralized and distributed sources of renewable energy.
- Identify issues and criticalities for data integration when it comes to semantic and structural properties of the input data. Solutions and workarounds to solve the abovementioned problems have been defined and implemented.

In terms of data modeling for cities, the strongly interdisciplinary nature of project CI-NERGY has allowed to profit from the joint expertise of energy, urban planning, and IT specialists, leading to an intense cooperation and scientific exchange of knowhow.

In particular, when it comes to CityGML, said cooperation has substantially contributed to the further development and testing of the energy application domain extension (ADE), which extends the CityGML data model with specific entities and properties for energy-related applications.

The energy ADE data model not only has been enhanced from the conceptual point of view but also has been implemented as a database solution in order to simplify further tests and augment the number of potential users (and therefore of constructive feedback), eventually speeding up the improvement of energy ADE.

In general, the project CI-NERGY has contributed to raise the awareness about the strong synergies and potentials that exist between heterogeneous disciplines when it comes to the integrated and sustainable planning of resilient cities.

Reference

Hardin, 1968. G. Hardin, The tragedy of the commons, Science 162 (1968) 1243–1248.

Index

Note: Page numbers followed by *f* indicate figures and *t* indicate tables.

A

Activity modeling, 107
ADEs. *See* Application domain extensions (ADEs)
Advanced metering systems (AMS), 142
Agent learning for DR optimization, 153–155
Agent representation, software orchestration with, 152–153
Agents, type of, 152–153
Agglomerative method, 85–86
Aggregation method, 122
Airborne laser scanners (ALS), 269
Algorithm testing, 84
Alternative-focused thinking, 39
AMS. *See* Advanced metering systems (AMS)
Analysis paralysis, 36–37
Appliance modeling, conclusion on, 116
Application domain extensions (ADEs), 243, 247, 452
Aquifers, 192, 193*f*
Archetypes, 199, 199*f*
  identification of, 83
Artificial neural network (ANN) method, 123–124
AUSBEMA. *See* Automatic Urban Scale Building Energy Modeling Approach (AUSBEMA)
Automatic Urban Scale Building Energy Modeling Approach (AUSBEMA), 391, 400
  HVAC and DHW uses, 395–396
  modeling input characterization, 394–395
  parsing CityGML information, 391–392
  surface processing, 392–393, 393*f*
  urban environment characterization, 393–394
  zoning and boundary conditions, 393

B

Battery discharge operation, 156–158, 162
Bayesian calibration, 403
  approach, 409
  framework, 410
  methods, 399
Bayesian framework, 403–405
Bayesian methods, 397–399, 410
Bayesian workflow, 398, 398*f*
Bayes rule, 398–399, 408
Belief-Desire-Intention (BDI) framework, 149
Bentley® sisHYD, 206
Bezug, 326
Biomass energy resources, 308–309
Bottom-up engineering simulation software, 396
Bottom-up physics based building models, 116–117
Bottom-up stochastic engineering approaches, 397
Brayton cycle, 185–187
  open gas turbine unit based on, 187, 187*f*
Brownfield planning project, 416–419, 429–441
Brownfield projects, 55
"Brute force" approach, 116–117
"Building archetype,", 116–117
Building case study, 126*f*
Building energy demand modeling
  large scale approach, 79–80
  occupant behavior modeling, 80
  techniques for, 79
  WB model equations, 80
Building energy modeling, 81, 125–127
Building Information Model (BIM), 242, 242*f*
Building load management with MPC, 167–173
  advantages and challenges, 171–172
  application, example of, 172–173

Building load management with MPC *(Continued)*
formulation, 168–170
thermal load control, 167–168
Building model development automation, 104
Building performance simulation (BPS), 257–258, 284
case study designs, 266–268
problem statement and objectives, 261
qualitative measures, 262–264, 278–280
quantitative measures, 264–266, 280–284
uncertainty assessment approaches in, 259–261
Building Physics module, Energy ADE, 248
Building refurbishments, 422
Building sector, 79
Buildings' subsystem, exergy balance for, 366, 368*f*
Building stock of Geneva, 416, 418*t*
Building thermal load control, 167–168
Building *vs*. grid programs, 141

C

Calibration procedure, 107
Candidate techniques, 108
Cantonal directive plan, 311
Cantonal energy law in Geneva, 421
Cantonal Office of Energy (OCEN), 310–311
Carnot cycle, 189
Cascading failures, utility infrastructures, 224
Case studies cities
Geneva, 314–315, 314*f*, 415–419
Vienna, 328–329, 329–330*f*
Catalogue Service for the Web, 292
Central energy systems, 182–183, 182*f*
Central heat pump
with refurbishment, 352*f*
without refurbishment, 353*f*
Centrality analysis, graph-theoretical algorithms, 225
Centralized *vs*. decentralized programs, 140–141
CGE. *See* General energy conception (CGE)
CHP. *See* Combined heat and power (CHP)
CI-NERGY project, 314–315, 326–327, 447–450, 452
CI-NERGY Q20 Project, 236
City Geographic Markup Language (CityGML), 242–243, 242*f*, 289–290, 294, 297
characteristics of, 245–247
energy application, 451–452
standards of, 244
3D City Database extensions, 253
City information model, 289–290
City infrastructure networks modeling
infrastructure dependency multigraph, 222
interdependencies types, 220–221
maintenance, 224–225
multimodal utility network, 222, 225–230
optimization control, 223
planning future network expansions, 224
utility networks, 220, 221–222*f*, 222
vulnerability, 223–224
CityObject, 294
Classical MC technique, 265–266
Classic thermal power plants, 184–185, 185*f*
Clustering methods, 90*f*
conclusions on, 91–92
for representative building identification in urban contexts, 81–92
CO_2-equivalent emission per person in Vienna, 319, 322*t*
Cogeneration, 183–184
advantages of, 183
disadvantages of, 183
elements of, 184
types of, 184–185, 185–186*f*
Cogeneration engine, 189, 189*f*
Coherence, 58–59
Coherent modeling of semantics, CityGML, 246
Combined heat and power (CHP)
cycle, 188–189, 188*f*
plants, 336
Common-cause failures, utility infrastructures, 224
Community detection, graph-theoretical algorithms, 226
Constraints, 170
Construction and Materials module, Energy ADE, 248
Consumption profiles classification, 83
Continuous
cyclical process, 25
and interconnected planning cycles, 26*f*

Conventional control techniques, 167–168
Core module
  Energy ADE, 248
  Utility Networks ADE, 250
Cost-effective increase of RES share, 436–437, 439*f*
Cost function for GSHP, 425
Courant-Friedrichs-Lewy (CFL) condition, 205
Current energy system, 433–435, 434–435*f*, 436*t*

D

Data analytics methodologies, 228–229
Data-driven modeling approach, 122–131
Data-driven stochastic models, 148
Data gathering, 83
Data mining techniques, 450
Data quality (DQ), 233–236, 235*f*
Dataset analysis, 87
Decentralized DR schemes, 141
Decentralized energy systems, 182–183, 182*f*
Decentralized solar thermal feed-in, energy performance of, 375–387
Decentralized *vs.* centralized programs, 140–141
Decision support, 37–43
  approach, 448
  concepts, 37–39
  modeling, simulation and optimization, 39–42
  for urban energy system planning, 37–43
Demand response of electrical appliances, using multiagent simulation
  literature review
    DR and multiagent simulation, 145–146
    DR simulation software tools, application into, 146–147
    proposed modeling framework, 147–148
  methodology, 154*f*
    agent learning for DR optimization, 153–155
    electricity demand models, 151–152
    mechanisms, 155–159
    software orchestration with agent representation, 152–153
  No-MASS background, 148–151
    for DR, 150–151
    flowchart, 150*f*
    workflow, 149–150
  self-consumption, of community of buildings
    implementation, 160–161
    reward functions, 161–162
    scenario description, 159–160
  software requirements, formulation of, 144–145
Demand Side Management (DSM) strategies, 105
Densification potential, 433, 433*f*
Descriptive statistics, 226–227
Design temperatures from radiators, 348
Detailed microsimulation approaches, 106
DHW. *See* Domestic hot water (DHW)
Dialogue mapping, 35–36
Digital revolution, 31
Dispatchable *vs.* nondispatchable programs, 139
Disposal reliable models, 116
Distributed Energy Resources, 151
Distributed urban energy infrastructure, 291–292
  business level, 291
  computational level, 291
  decision level, 291
  GIS, 292, 292*f*
  open interface, 291 292, 291*f*
District heating network (DHN), 203–204, 343, 450
  advantages of, 196
  consumers types, 195
  disadvantages of, 196
  dynamic approach, 204–205
  fifth generation, 203, 204*f*
  first generation, 196
  fourth generation, 198–203, 198*f*, 201–202*f*
  gas turbines and use for, 185–188, 187*f*
  second generation, 196–197, 197*f*
  steady-state approach, 204
  thermal power plant operation for, 336–339
  third generation, 197–198
  typologies and reference system, 341–350
  in Western Europe, 196
District heating systems, 340
District scale building energy models, 388–411

Diversity, 59
Divisive approach, 86
Domestic air conditioning-heating appliances, 190
Domestic hot water (DHW), 374, 395–396
  branch, return temperatures from, 348
  services, 341
DR optimization, agent learning for, 153–155
DR program classification, 138–141
  centralized *vs.* decentralized programs, 140–141
  dispatchable *vs.* nondispatchable programs, 139
  grid *vs.* building programs, 141
  optimized *vs.* nonoptimized programs, 139–140
DR simulation software tools, application into, 146–147
DR software requirements, formulation of, 144–145
Dry deposits, 192
DSM/DR deployment, 141–143
Dynamic approach, 204–205
Dynamic simulation and simulation database, 401

E

Eco-district standards, 415–416
Efficiency, defined as, 207
EGID, 86–87
Electrical appliance modeling, 105–116
Electrical power system, 137
Electric battery storage system, 152
Electricity consumption by end-user type, 306–307, 306*f*
Electricity demand models, 151–152
Emulators, 402–403
  quality of, 406–407
  selection of, 407–408
End-users, 141, 295–297
Energy ADE, 243, 247–248
  characteristics of, 248–249
  3D City Database extensions, 253
Energy cascade, 199, 200*f*
Energy conversion
  and district heating systems in Vienna, 333–341
  technologies, 422, 422*t*
    linearized cost functions of, 425, 426*t*
Energy demand, 64, 421–422, 421*t*
Energy directive plan (PDE), 310–311
Energy efficiency, 138
Energy/exergy performance evaluation, 206–214, 209*f*, 211*f*, 213*f*
Energy flowchart of Vienna, 319, 320*f*
Energy flow paths, 93*f*
Energy losses comparison, 372*f*
Energy networks, 423, 425
  block scale, 67–68
  building scale, 67
  neighborhood scale, 68, 71
Energy performance, 260
  of decentralized solar thermal feed-in, 375–387
Energy planning, 21*t*, 31–32
EnergyPlus engine, 119, 260–261
EnergyPlus weatherfile, 423
Energy prices in CHF/kWh, 425, 427*t*
Energy storage tanks, 341
Energy supply, 422–423, 425
Energy system in building blocks, 350–353
Energy system module, Energy ADE, 249
Engineering/bottom-up physics based building models, 116–117
Enhanced geothermal systems (EGS). *See* Dry deposits
ESRI Geodatabase-based dataset, 327
ESRI shapefile file format, 230–231
EU Energy Performance of Buildings Directive (EPBD), 322
European CORDIS database, 20–21
European Institute for Energy Research, 20–21
Exchange technology, 67
Exergy balance
  applied to heat exchanger, 358, 358*f*
  comparison, 372, 373*f*
  for subsystems, 366, 367*f*
Exergy losses
  comparison, 372, 373*f*
  evolution of, 366, 367*f*
Exergy yield, evolution of, 366, 367*f*
Exploratory data analysis methods, 233
Extract, transform, and load (ETL) tool, 251

F

Fifth generation district heating networks, 203, 204*f*
Final energy consumption per person in Vienna, 319, 321–323*t*
Finite-Difference Approximation Method, 96
First generation district heating networks, 196
First-order RC model, 99*f*
Flat roof-type case, 278*t*, 279
Flexible consumers, 137
Floor area ratio (FAR), 59
Fossil energy sources, 319
Fossil fuels, 338
Fourth generation district heating networks, 198–203, 198*f*, 201–202*f*
Fractional energy states, Monte Carlo simulation of, 110–112
Functional characteristics module, Utility Networks ADE, 250
Functional Mockup Interface standard, 148

G

Gabled roof-type case, 278*t*, 279
Gas turbines, 185–188, 187*f*
Gaussian process emulators, 399
General energy conception (CGE), 310
Generic objects and attributes concept, 247
Geneva
  case studies, 314–315, 314*f*
    Brownfield planning project, 416–419
    costs, 424–425
    energy, 421–423
    environment, 421–422
    Greenfield planning project, 415–416
    social, 420, 420*t*
    urban form and function, 420
  city goals, actors and tools, 310–311
  climate of, 304–305, 305*f*
  construction and urban planning in, 311, 312*t*
  data, 311–314
  energy demand, resources and infrastructure, 305–310, 306*f*
  general description, 303–305, 304*f*
  urban and energy planning documents in, 311, 313*t*
Geneva canton GIS data, 232, 232*t*
Geographic information system (GIS), 292, 292*f*
Geographic interdependencies, 221
Geography Markup Language 3 (GML3), 244
Geometry plugin, 294
Geothermal energy, 191–194, 193*f*
Geothermal potential of Geneva, 309–310
Geysers, 192
GFA. *See* Gross floor area (GFA)
GHG emissions, reduction of, 436, 437–438*f*
GIS. *See* Geographic information system (GIS)
Global energy balance, 354, 364, 366, 371
Global energy consumption, 319
Global exergy balance, 358, 372
Graph model, 231–232, 232*t*
Green building XML (gbXML), 242–243
Greenfield planning, 427–429, 428*f*, 430–432*f*
  project, 415–416
Greenfield projects, 25
Grid-oriented analysis, 141
Grid *vs.* building programs, 141
Gross floor area (GFA), 57
Gross inland energy consumption in Vienna, 319, 321*t*

H

Heat exchangers, function of, 344
Heat generation
  breakdown of, 338, 339*t*
  and installed capacities, breakdown of, 338, 339–340*f*
Heating and cooling, urban energy systems for
  central and decentralized, 182–183, 182*f*
  cogeneration, 183–185
  combined heat and power cycle, 188–189, 188*f*
  gas turbines, 185–188, 187*f*
  geothermal energy, 191–194, 193*f*
  heat pumps, 189–191, 190*f*, 192*f*
  internal combustion/cogeneration engine, 189, 189*f*
  solar thermal energy, 194–195, 195*f*
Heating consumption index, 313
Heating curve, 346–347, 347*f*, 360
Heat losses, evolution of, 364, 366*f*

Heat pumps, 189–191, 190*f*, 192*f*, 203
comparison with system based on PV and, 384–387
Hierarchical clustering, 85–86
Household electricity survey data set, 112
Household member agents, 149
HVAC, 395–396
Hydraulic balance, 349*f*, 350
Hydraulic network, 344
Hydraulic pumps, 338
Hydraulic system of the secondary network, 348, 349*f*

I

Import/Export tool, 252
Individual heat pumps for heating without refurbishment, 353*f*
Industrial heat integration, fourth generation DHN, 199, 200*f*
Industry Foundation Classes (IFC), 242
Information and Communication Technology (ICT), 289, 292
Infrastructure dependency multigraph, 222
Inhomogeneous Markov chains, 107
Input workflow step, 294
INSEL8.2 software, 375–376
INSEL template, 294–295, 295*f*
INSPIRE, 242, 242*f*
INSPIRE Generic Network Model, 249
INSPIRE Utility Networks model, 249
Integrated semantic 3-D virtual city models, 241
Interaction model, 21
Interactive optimization
motivation for, 51–55
workflow, 425–427
Intercept theorem, 63
Internal combustion engine, 189, 189*f*

K

Kagran thermal plant, 333–334
KliP balance calculation method, 319
K-means clustering, 85

L

Landmark view factor (LVF), 61–64, 420, 427–429
LargeApplianceLearning agent, 156
Large-scale building energy modeling, 81
Large-scale demand profiling of buildings, data-driven modeling approach for, 122–131
Large-scale physics-based BEMs, 82
Latin Hypercube Sampling (LHS), 400–401
Law-driven approaches, 40–41
Levels of detail (LoDs), CityGML, 245–246, 246*f*
LHS. *See* Latin Hypercube Sampling (LHS)
Linear mathematical function, 346
Linear Time Invariant (LTI) state-space model, 101–102
Load management in buildings, 137–144
definitions, 138
DR program classification, 138–141
centralized *vs.* decentralized programs, 140–141
dispatchable *vs.* nondispatchable programs, 139
grid *vs.* building programs, 141
optimized *vs.* nonoptimized programs, 139–140
DSM/DR deployment, 141–143
summary, 143–144
Load shifting optimization, 155–156
Local sensitivity analysis (LSA), 259
LoD0 model, 245–246
LoD1 model, 245–246
LoD2 model, 245
LoD3 model, 245–246
LoD4 model, 246
Low-load appliances, 108
Low Temperature District Heating Network (LTDHN), 201, 202–203*f*, 374
Low voltage network model, 152
LTDHN. *See* Low Temperature District Heating Network (LTDHN)
LVF. *See* Landmark view factor (LVF)

M

Manhattan distance, 60, 60*f*
Markov Chain Monte Carlo (MCMC)
methods, 398–399
sampling, 405
MATLAB, 206

MC-based evaluation, of multidimensional integral, 265
MCMC. *See* Markov Chain Monte Carlo (MCMC)
Mehrzweckkarte (MZK), 326
Metropolis-Hasting algorithm, 405
Meyrin, 315, 316*f*
MILP. *See* Mixed-Integer Linear Programming (MILP)
Mixed-Integer Linear Programming (MILP), 52
Modeling and data analysis algorithms, 233
Modeling buildings, 79
Modeling of energy use, techniques for, 79
Modeling parameters, definition of, 400
Model integration, 233, 234*f*
Model Predictive Control (MPC), 144, 449
  building load management with, 167–173
    advantages and challenges, 171–172
    application, example of, 172–173
    formulation, 168–170
    thermal load control, 167–168
Monte Carlo simulation of fractional energy states, 110–112
Morris design, 262–264
MPC. *See* Model predictive control (MPC)
Multiagent simulation (MAS) approach, 144
Multi-Agent Stochastic Simulation framework, 144
Multiagent systems, 145–146
Multicriteria decision analysis (MCDA) methods, 35–36
Multifamily houses (MFH), 424
Multimodal networks, 218–219
Multimodal utility network, 222
  graph-theoretical algorithms, 225–226
  statistical algorithms, 226–230, 229*f*
Multiobjective optimization, 51
Multi-parametric programming for urban planning, 71–73
Multiple electricity tariff schemes, 172
Multiple linear regression model, 419
Multivariate linear regression model, 260
Municipal waste incineration plant (MWIP), 337*f*, 338
MWIP. *See* Municipal waste incineration plant (MWIP)

N

Network components module, Utility Networks ADE, 250
Network data and models, 67–68
  competitivity and complementarity of, 236
  condition monitoring and forecasting services, 236–237
  correlation and causalities, 236
  data quality assessment and improvement, 233–236, 235*f*
  graph model, 231–232, 232*t*
  model integration, 233, 234*f*
  numerical data, 232–233
  spatial model, 230–231, 231*f*
Network properties module, Utility Networks ADE, 251
Network's pressure, 363, 364*f*
Network's temperatures, 363, 364*f*
Network vulnerability analysis, 223–224
Nodal method approach, 94–95
The node method, 205
No-MASS. *See* Nottingham Multi-Agent Stochastic Simulation (No-MASS)
Nondispatchable *vs.* dispatchable programs, 139
Nonlinear cost functions, 425
Nonoptimized *vs.* optimized programs, 139–140
Nottingham Multi-Agent Stochastic Simulation (No-MASS), 148–151
  for DR, 150–151
  flowchart, 150*f*
  workflow, 149–150
Novel control technique, 449
Numerical data, 232–233

O

OASIS, 244
Occupant behavior, 80, 83
Occupant Behavior module, Energy ADE, 248–249
OCEN. *See* Cantonal Office of Energy (OCEN)
OGD. *See* Open government data (OGD)
One-At-Time (OAT) design, 262
One-dimensional heat-transfer model, 94–95

Open data-modeling standards, city-wide energy simulations, 241–244
CityGML, 244–247
Energy ADE, 247–249
existing tools, 251–253
Utility Networks ADE, 249–251
Open geospatial consortium (OGC), 244, 292
Open government data (OGD), 326
Open interface, 291–292, 291*f*
Operation variables, 363
Optimization-based thought process, 54–55
Optimization methods, 39, 41–42, 140–141
vs. simulation, 40*t*
Optimization model, 55–71
economic aspects, 69–71
building scale, 70–71
neighborhood scale, 71
energy aspects
block scale, 67–68
building scale, 65–67
floor scale, 64
neighborhood scale, 68
environmental aspects, 69
social aspects
block scale, 58–59
neighborhood scale, 59–64
subfloor scale, 57–58
urban form and function, 56–57
building scale, 56–57
floor scale, 56
Optimized *vs.* nonoptimized programs, 139–140
ORC. *See* Organic Rankine cycles (ORC)
Organic Rankine cycles (ORC), 309–310
Osmose Platform, 236

P

PAM. *See* Partitioning around medoids (PAM)
Parallel coordinates, 52–53, 425–427, 439–440*f*
Parametric simulation, 127, 128*t*, 400–401
Pareto-optimal solutions, 51–53
Partitioning around medoids (PAM), 86
Path analysis, graph-theoretical algorithms, 226
PDE. *See* Energy directive plan (PDE)
Peak load district heating plants (PL DH), 337*f*, 338
"Performance bound" (PB), 170
PhD network CINERGY, 19–20
Photovoltaic (PV) model, 65–67, 71, 151
Physical dynamic building models, 81–98
Physical interdependencies, 221
Physical models of buildings, 116–117
Physics plugin, 294
PipelineML, 249
Planners, 19
Planning and implementation process (PIP) diagrams, 21
Planning-design gap, 42–43
Planning processes in reviewed documents, 24*t*
Planning support system URB, 415–416
Brownfield planning, 429–441
Greenfield planning, 427–429, 428*f*, 430–432*f*
Plate heat exchangers, 341–342
Point-based vector datasets, 327
Point, spatial model, 230
Polygon-based vector map, 327
Polygon, spatial model, 231
PolyLine, spatial model, 231
Posterior distribution, 397, 408, 408*f*
Predictive modeling accuracy, 91*t*
Predictive modeling approaches, 410–411
Predictive modeling framework, 83
Predictive modeling performance, 128–129
Preprocessing workflow step, 294
Price-incentive-based DR control strategies, 159–160
Primary energy consumption per person in Vienna, 319, 322*t*
Primary Energy Factor, 385, 387
Proposed approach, advantages and limitations of, 131
Proposed modeling framework, 147–148
Prosumers, 137
Proximity parameter on block scale, 60–61
Psychological convergence, 54
PV and heat pumps, comparison with system based on, 384–387
Python, 400–401

Q

QGIS, 231
Q-learning algorithm, 146, 153–155
  for battery discharge, 156–157
Qualitative measures, BPS, 262–264
  case study designs, 266–268, 278–280
Quantitative measures, BPS, 264–266
  case study designs, 267–268, 267*f*, 280–284, 281–283*f*

R

Random sample consensus (RANSAC), 271
Rankine cycle, 184
  steam power plants with classic, 184, 186*f*
RC model complexity, 103–104
Ready-made package for RC model parameter estimation, 102–103
Recknagel-Schramek-Sprenger method, 344–345
Reduced-order RC models, 98–105
Reference buildings, 119
  model based on, 116–122
Reference energy surface (SRE), 307, 307*t*
Refurbishments, 422, 438–441, 440–442*f*
  building, 422
  costs for, 424, 424*t*
Regression analysis, 233
Regression-based models, 403
Relative energy balance comparison, 371, 373*f*
Relative energy losses comparison, 372, 372–373*f*
Relative energy yield comparison, 372, 372*f*, 374*f*
Renewable energy sources, share of, 69
Representative building identification in urban contexts, 81–92
RES share, cost-effective increase of, 436–437, 439*f*
Return temperatures
  from DHW branch, 348
  from radiators, 348
Reversible heat pumps
  configuration, 203–204*f*
  for heating/cooling, 354*f*
Reward function, 155
  appliance reschedule, 161
  for battery discharge, 157–158
  for load shifting, 156
Root mean square error (RMSE), 272–273

S

Sampling methods, 53
Scopus database and scanning, 20–21
Second generation district heating networks, 196–197, 197*f*
Self-consumption maximization, 158–159
Self-consumption, of community of buildings
  implementation, 160–161
  reward functions, 161–162
  scenario description, 159–160
Semantic 3-D city models, 241–242
Sequential model, 21
Sequential operation, model integration, 233
Service consumer, 290
Service-oriented architecture (SOA), 289–290, 290*f*
  distributed urban energy infrastructure, 291–292, 291–292*f*
  input workflow step, 294, 297, 298*f*
  INSEL template, 294–295, 295*f*
  preprocessing workflow step, 294, 297, 298*f*
  runtime data flow, 295, 296*f*
  simulation workflow step, 294, 297, 298*f*
  urban scale data process, 293
  workflow, 293
Service provider, 290
Service registry, 290
Simple Object Access Protocol (SOAP), 290
SimStadt, 268
SiMulated Adaptable Smart Home (SMASH), 147
Simulated audio-visual appliances, 114–115
Simulation *vs.* optimization, 40*t*
Simulation workflow step, 294
Simulink®, 206
Simultaneous operation, model integration, 233
SITG platform, 311–312
Small appliances, modeling of
  Monte Carlo methods, 110–112
  survival analysis models, 109–110
  tasks, 108–109

Small-scale systems, 152
Smart appliance, 159, 162–164
Smart City Wien framework, 329
strategy, 324–325
Smart City Wien Initiative, 324
Smart Grid, 137
SMASH. *See* SiMulated Adaptable Smart Home (SMASH)
SOA. *See* Service-oriented architecture (SOA)
Sobol sequence, 53
Software architectures, 141
Software orchestration, with agent representation, 152–153
Solar energy, 309
Solar system models, 377–380
Solar thermal energy, 194–195, 195*f*
Solar thermal system, 377–379, 379*f*
Spatially enhanced relational database management system (SRDBMS), 252
Spatial model, 230–231, 231*f*
Squared exponential kernel, 404
STANET®, 206
Steady-state approach, 204
Steam power plants
with classic Rankine cycle, 184, 186*f*
for district heating, 186*f*
features of, 184
Stochastic building energy modeling, 397
Stochastic formulations, 101–102
Stochastic methods, 108
Stokes' model, 106
Storage tanks, 348
Subgraph isomorphism, graph-theoretical algorithms, 226
Supporting classes module, Energy ADE, 249
Surface shares, 59–60
Survival analysis models for appliance modeling, 109–110
Swiss Energy Standard MINERGIE-P, 415–416
Symmetry of ignorance, 43–44
Synthetic Environment Data Representation and Interchange Specification (SEDRIS), 249
System dimensioning, 380–382
System model thermal networks, 214–218, 216–217*f*

T

Tackling wicked problems, 34*t*
Tame problems, 32
Temporal scale, 102
Tenfold cross-validation, 402–403
Thermal building refurbishments, 422
Thermal loads, 167–168
Thermal power plants, 338
operation for district heating, 336–339
Thermodynamic systems, 358*f*
Thermodynamic variables, 354, 357*f*
Thermo-hydraulic network models
commercial network modeling tools, 206
district heating network models, 203–205
Third generation district heating networks, 197–198
3D building, 297
3D building geometry, 293
3D City Database (3DCityDB), 251–252
extensions, 253
3D CityDB-Web-Map-Client, 252
3D city modeling, 289–290, 452
3D objects, CityGML, 246–247
Time-consuming process of modeling appliances, 116
Time-consuming simulations, 398
Time series analysis, 229
Time-use-survey (TUS) datasets, 106–107
Top-down model-driven approach, 270, 270*f*
Top-down techniques, 79–80
Total effect index, 266
Total mass flow, 364, 365*f*
Traditional engineering bottom-up modeling, 396
Transmitted diffuse solar radiation, 97*f*
Transmitted direct solar radiation, 96*f*
Transportation network, 71

U

Uncertainty analysis framework
application of, 268–277
uncertainty characterization, 268–277
Uncertainty analysis study, 259
Uncertainty assessment approaches, in BPS, 259–261
Uncertainty characterization, building geometry, 268–277

indicators and structure assignment, 274–277
ISPRS benchmark test datasets for 3D building reconstruction, 272*t*
uncertainty issues, 270–273
Unified Modeling Language (UML) diagram, 244, 245*f*
Universal Discovery, Description and Integration (UDDI), 290
Universal topographical information model, 247
Urban building energy modeling, 449
Urban developer community, 448–449
Urban Development Plan STEP2025, 325–326
Urban Energy Efficiency Program, 324
Urban energy management, 290
Urban energy planning process, 448
Urban Information Modeling (UIM), 243
Urban planning processes
mapping of, 27
multi-parametric programming for, 71–73
phases, 43*f*
review of, 20–28
Urban scale, 102
Urban scale data processing, 293
Urban scale model calibration
Bayesian method, 397–399
literature review, 399
methodology, 399–405
results, 405–409
URB^{io} workflow, 53–54*f*
Utility networks, 220, 221–222*f*, 222
graph-theoretical model of, 224
models, 219–220
Utility Networks ADE, 243, 249–250
characteristics of, 250–251
3D City Database extensions, 253

V

Value-focused thinking, 39
Variance-based global sensitivity analysis (GSA), 259–260
Vector-based maps, 327
Vienna
case studies, 328–329, 329–330*f*
city goals, actors and tools, 324–326
data, 326–328
energy conversion and district heating systems in, 333–341
energy demand, resources and infrastructure, 318–324
general description, 316–318, 317*f*
Viennese district heating network, 340
Viennese energy system, 333
Visualization tools, 88–90

W

Wastewater treatment plant in Vienna, 323
WB model equations, 80
W3C Consortium, 244
Web Coverage Service, 292
Web 3-D Consortium, 244
Web Feature Service (WFS), 292
Web Map Service, 292
Web Processing Service (WPS), 292
Web Service Description Language (WSDL), 290
Web Services, 290. *See also* Service-oriented architecture (SOA)
based distributed GIS infrastructure, 292, 292*f*
Weibull distributions, 109, 110*f*, 274, 275*t*
Wicked problems, 32, 33*t*
managing, 33–37
Wiener Null, 326
Wiener Wärme Kataster (WWK) data, 328

FSC
www.fsc.org
MIX
Papier | Fördert gute Waldnutzung
FSC® C083411

Kolibri 360 GmbH
Ferdinand-Jühlke-Straße 7
99095 Erfurt, Deutschland
produktsicherheit@kolibri360.de